Fundamentals of Engineering Thermodynamics

SI VERSION

Academic Reviewers

McGraw-Hill and the authors wish to express their sincere thanks to the following people, who have reviewed all or various portions of the manuscript.

William Bathie	*Iowa State University*
James Beck	*University of Central Florida*
Adrian Bejan	*Duke University*
Michael A. Boles	*North Carolina State University*
Van P. Carey	*University of California, Berkeley*
Bart Conta	*Cornell University*
Thomas L. Eddy	*Georgia Institute of Technology*
Said Elghobashi	*University of California, Irvine*
Nanak Grewal	*University of North Dakota*
John J. Henry	*Pennsylvania State University*
Sanford A. Klein	*University of Wisconsin*
Shankar Lal	*University of Southern California, Los Angeles*
David E. Lamkin	*Northern Arizona University*
Thomas J. Love	*University of Oklahoma, Norman*
Alister K. Macpherson	*Lehigh University*
Stuart T. McComas	*University of Notre Dame*
Michael Muller	*Rutgers University*
James W. Murdock	*Drexel University*
William Murphy	*Texas A&M University*
Charles Proctor	*University of Florida, Gainesville*
Charles R. St. Clair, Jr.	*Michigan State University*
Mohammad E. Taslim	*Northeastern University*
Eric F. Thacher	*Clarkson University*
Timothy R. Troutt	*Washington State University*

Fundamentals of Engineering Thermodynamics

SI VERSION

John R. Howell

Department of Mechanical Engineering
University of Texas at Austin

Richard O. Buckius

Department of Mechanical and Industrial Engineering
University of Illinois at Urbana-Champaign

McGraw-Hill Book Company
New York St. Louis San Francisco Auckland Bogotá
Hamburg Johannesburg London Madrid Mexico Milan
Montreal New Delhi Panama Paris São Paulo Singapore
Sydney Tokyo Toronto

Fundamentals of Engineering Thermodynamics, SI Version

1 2 3 4 5 6 7 8 9 0   H A L H A L   8 9 8 7 6

ISBN 0-07-079662-9

This book was set in Times Roman by Progressive Typographers, Inc.
The editors were Anne Duffy and J. W. Maisel; the design was done by Caliber Design
Planning; the production supervisors were Marietta Breitwieser and Phil Galea. The
drawings were done by Fine Line Illustrations, Inc.
Arcata Graphics/Halliday was printer and binder.

Library of Congress Cataloging-in-Publication Data

Howell, John R.
  Fundamentals of engineering thermodynamics.

  Includes bibliographies and index.
  1. Thermodynamics.   I. Buckius, Richard O.
II. Title.
TJ265.H726 1987b   621.402′1   86-7446
ISBN 0-07-079662-9 (set)
P/N 030615-X (text)
P/N 831516-6 (disk)

About the Authors

John R. Howell received his B.S. (1958), M.S. (1960), and Ph.D. (1962) in chemical engineering from Case Institute of Technology (now Case Western Reserve University) in Cleveland, Ohio. He worked at the NASA Lewis Research Center in Cleveland on fundamental heat-transfer research from 1961 until 1968, when he joined the Department of Mechanical Engineering at the University of Houston. He joined the Department of Mechanical Engineering at the University of Texas at Austin in 1978, where he is presently E.C.H. Bantel Professor of Professional Practice.

Dr. Howell has published widely in the areas of heat transfer and solar energy, including more than 100 technical papers and reports as well as text and reference books. He has twice been voted the outstanding Mechanical Engineering advisor by the undergraduate students at the University of Texas, and was also given the Outstanding Service Award by the Graduate Engineering Council. He is a Fellow of the ASME.

Richard O. Buckius received his B.S. (1972), M.S. (1973), and Ph.D. (1975) in mechanical engineering from the University of California, Berkeley. He then joined the Department of Mechanical and Industrial Engineering at the University of Illinois at Urbana-Champaign as an assistant professor. He was promoted to professor in 1984 and he is presently associate head of the department.

Dr. Buckius has published numerous technical articles in the areas of heat transfer and combustion. He has received various teaching awards, including the Campus Award for Excellence in Undergraduate Teaching from the University of Illinois and the Western Electric Fund Award from the American Society for Engineering Education.

Contents

CHAPTER 3 Properties Of Common Substances   64

CHAPTER 4 First Law of Thermodynamics   110

Contents

CHAPTER 5 Entropy and the Second Law of Thermodynamics   164

CHAPTER 6 Thermodynamic Cycles and Common Energy Systems 244

CHAPTER 7 Analysis Using the Second Law of Thermodynamics 318

CHAPTER 8 General Property Relations and Equations of State 348

Preface

This text provides an introduction to engineering thermodynamics from the so-called classical approach. The organization follows a logical sequence which is considerably different from the historical development of thermodynamics. However, we proceed in a manner that allows the student to build an understanding of the fundamentals and applications, proceeding from simple but useful relations and applications for simple substances to the necessarily more complex relations for mixtures and materials with chemical reactions.

We have included many fully worked examples to illustrate the application of the theory presented in the text, and we have found these very helpful to the student. In these examples, we have followed the problem-solving methodology presented in Chaps. 1, 4, and 6. This methodology emphasizes the careful structuring of the problem, the use of property diagrams to visualize the processes involved, and the use of state tables for defining the process end states. These allow the student to see exactly what information is given and what must be generated through use of thermodynamic relations. Such an approach can help the student to see to the heart of the problems posed and to develop orderly solution procedures.

Because of the availability of computerized property data for microcomputers, a wide range of problems can be treated by students that were not feasible for an earlier generation. We have provided a mix of problems; most can be solved by "hand" calculation, and others require so much interpolation from tabular data that only the use of computerized tables makes complete solution feasible. These problems have been flagged in the problems sections so that the instructor will not inadvertently assign them. We believe that the "hand" problems, making use of tabular data where necessary, are required to help the student gain the necessary insight into thermodynamics. The problems requiring computerized data generally show the behavior of a particular control mass or device under parametric variations of conditions; such problems also can provide great insight but have often been ignored in introductory texts because of the time formerly involved in solution. However, care must be taken to help the student develop the ability to critically examine what the computer is providing, so that erroneous computer results will not be blindly accepted and used.

The computerized tables available with this text cover the range of properties encompassed by the problems and are discussed in some detail in Appendix G.

At the beginning of each chapter are photographs, cutaways, and diagrams of the equipment analyzed in the text. These are included because many students entering the engineering curriculum have not seen such equipment and have little concept of its scale and complexity. These pictures give some idea of what lies within the blocks shown on the cycle diagrams in the body of the text.

Two versions of this text are available — this one, which is solely in SI and one that contains both SI and U.S. conventional (USCS) systems of units. Appendix E contains property data in USCS units for use with extra problems that might be assigned by the instructor.

Although the presentation in the first 11 chapters is generally from the classical viewpoint (with some microscopic interpretations where they appear helpful), Chap. 12 deals with the statistical interpretation of thermodynamics. It is organized so that a more detailed statistical viewpoint can be introduced along with the classical material if the instructor desires. Alternatively, Chap. 12 can be used to review the classical relations from another viewpoint following completion of the first 11 chapters. In any case, the statistical treatment is presented as an aid to understanding how properties can be computed from a fundamental understanding of structure; the interpretation of entropy in terms of uncertainty; the ideas of the increase of the entropy of the universe from the microscopic viewpoint; and a microscopic interpretation of the first and second laws of thermodynamics. No attempt is made to provide a complete treatment of statistical thermodynamics.

Finally, we have observed that thermodynamics is often the first course in which mathematical concepts from courses in partial differential equations are applied to engineering problems. We have tried to aid in the transition from abstract concepts to concrete applications with a section in Chap. 1 on the mathematics needed in this course. Some instructors may wish to omit this section, using it for reference as required.

We are indebted to our colleagues at the University of Texas at Austin and the University of Illinois at Urbana-Champaign. Their comments, criticisms, and suggestions have helped us improve this final product. We are also indebted to the academic reviewers listed on page ii. We acknowledge the efforts of Kumbae Lee and Larry Lister for their detailed review of the problems and text. We are also grateful for the excellent typing of the entire manuscript and revisions by Angela Ehrsam.

We now understand the reason that authors invariably thank their families for their encouragement and support. Their contribution is a very real one. We are very grateful to Susan and Kathy, who have endured many periods of doubt and second thoughts about this project with great grace and understanding. And we are very grateful to our children, Reid, Keli, and David, who have passed to adulthood, and Sarah and Emily, who have passed through their preschool years, during this writing.

John R. Howell
Richard O. Buckius

Nomenclature

a	activity, specific Helmholtz function, acceleration
A	Helmholtz function, area
AFR	air-fuel ratio
c_P	specific heat at constant pressure
c_P^*	temperature mean specific heat at constant pressure
c_v	specific heat at constant volume
COP	coefficient of performance
d	distance
e	specific energy
E	energy, Young's modulus
$\mathbf{E}$	electric potential
f	fugacity
f^*	ideal solution fugacity
$\mathbf{F}$	vector force
F	force
$\mathbf{F}$	generalized force
FAR	fuel-air ratio
g	acceleration due to gravity, specific Gibbs function, degeneracy
g_c	constant that relates force, mass, length, and time in USCS units
G	Gibbs function
h	specific enthalpy, Planck's constant
H	enthalpy
i	electric current specific irreversibility
I	irreversibility
k	specific heat ratio c_P/c_v, rate constant, Boltzmann constant
k_s	spring constant
K	equilibrium constant
KE	kinetic energy
L	length
m	mass
$\dot{m}$	mass rate of flow
M	molecular weight
n	number of moles, polytropic exponent
N	number
N_a	Avogadro's number
N_p	number of particles

P	pressure
P_i	partial pressure of component i
PE	potential energy
q	heat transfer per unit mass
Q	heat transfer
$\mathbf{Q}$	charge
$\dot{Q}$	rate of heat transfer
$\underline{R}$	particular gas constant
$\overline{R}$	universal gas constant
$\mathbf{s}$	displacement
s	specific entropy
S	entropy
S_{gen}	entropy generation
t	time
T	temperature
u	specific internal energy
U	internal energy
v	specific volume
V	volume
$\mathbf{V}$	velocity
V	speed
w	work per unit mass
W	work
$\dot{W}$	rate of work, or power
W_{rev}	reversible work between two states
x	quality, mass fraction
$\mathbf{X}$	generalized displacement
y	mole fraction
z	compressibility factor
Z	elevation, partition function

Greek Letters

α	residual volume, extent of reaction
β	coefficient of thermal expansion
ε	strain, particle energy
η	efficiency
θ	angle between surface normal and direction of vector
κ	isothermal compressibility
μ	chemical potential, Joule-Thomson coefficient, degree of saturation
υ	stoichiometric coefficient
ρ	density
σ	stress, surface tension
τ	sheer stress
ϕ	nonflow availability per unit mass, relative humidity

Φ	nonflow availability
ψ	flow availability per unit mass, wave function
Ψ	flow availability
ω	humidity ratio, acentric factor

Subscripts

a	actual, air
$comb$	combustion
c	components, compressor
C	carnot
cr	critical point
CM	control mass
CV	control volume
d	dew point
f	formation, fuel
g	property of saturated vapor
H	high-temperature source
in	state of a substance entering a control volume
i	component
irr	irreversible
j	phase
L	low-temperature source
l	property of saturated liquid
lg	difference in property for saturated vapor and saturated liquid
out	state of a substance leaving a control volume
p	product
ph	phases
pr	properties
r	reduced property, reactant
rev	reversible
s	isentropic process
v	vapor
0	property of the surroundings, zero pressure

Superscripts

	bar over symbol denotes property on a molal basis, partial molal property
^{o}	property at standard reference state
$\cdot$	dot over symbol denotes rate

Fundamentals of Engineering Thermodynamics

SI VERSION

1

Introduction

Philosophy is written in this one grand book — I mean the Universe — which stands continually open to our gaze, but it cannot be understood unless one first learns to comprehend the language and interpret the characters in which it is written. It is written in the language of mathematics . . . without which it is humanly impossible to understand a single word of it; without these, one is wandering around in a dark labyrinth.
Galileo

The turbine shaft and bucket (blade) assembly from a utility steam turbine during maintenance. *(General Electric.)*

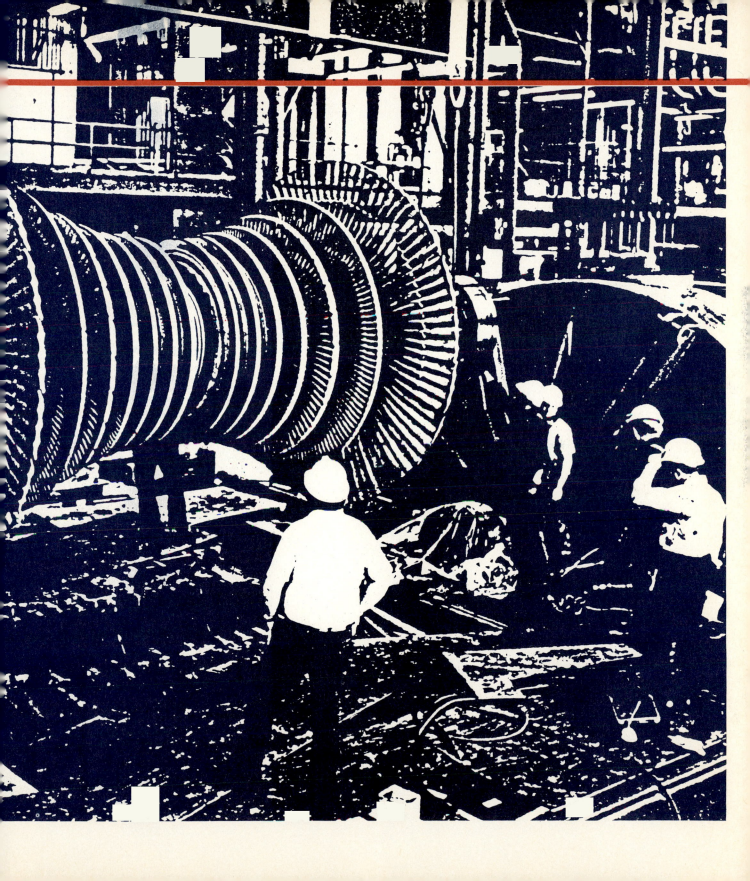

1.1 Energy and Society

Thermodynamics is defined as the study of energy, its forms and transformations, and the interactions of energy with matter. Before we begin this study, it is useful to think about the place and usefulness of this discipline, not only within the curriculum of the engineer and scientist but also within the framework of society itself.

1.1.1 Value of Energy

The availability of energy and people's ability to harness that energy in useful ways have transformed our society. A few hundred years ago, the greatest fraction of the population struggled to subsist by producing food for local consumption. Now, in many countries a small fraction of the total work force produces abundant food for the entire population, and much of the population is thus freed for other pursuits. We are able to travel great distances in short times by using a choice of conveyances (including trips to earth orbit as well as our nearest natural satellite for some); we can communicate instantaneously with persons anywhere on earth; and we control large amounts of energy at our personal whim in the form of automobiles, electric tools and appliances, and comfort conditioning in our dwellings.

How did these changes come about? They have resulted from a combination of inventiveness and ingenuity, coupled with a painstaking construction of theory by some of the great scientists and engineers throughout the years. The story of this development of the underlying science and engineering that so affect us now is one of interest and inspiration, but is too lengthy to reproduce here. A short history of the development of classical thermodynamics is included in App. A.

As a result of the development of the science and application of thermodynamics, our ability to obtain energy, transform it, and apply it to society's needs has brought about the change from agrarian to modern society. From our definition of thermodynamics, then, clearly the study of this discipline not only is useful to engineers in their professional lives but also has played, and will continue to play, a vital role in how society evolves.

1.1.2 Need to Understand Energy and Its Forms

Because of its generality, thermodynamics is the underlying science that forms the framework for the study of most other engineering subjects. The most obvious are *heat transfer,* the study of how energy is transferred from a material or location at a certain temperature to another material or location at a different temperature; *fluid mechanics,* which deals with the motion of fluids under externally applied forces and the transformation of energy between mechanical and thermal forms during this motion; many parts of *materials science,* such as those that consider the relative amounts of various structural forms of materials present in solids and how these relative amounts change under different conditions; and, in a sense, all subjects that treat energy in any of its forms.

Another way of seeing the pervasiveness of thermodynamics in studies of interest to engineers is to examine the many and diverse areas of application. These include power plants (fossil fuel, nuclear fission, nuclear fusion, solar, geothermal, etc.); engines (steam, gasoline, diesel, stationary and propulsion gas turbines, rockets, etc.); air-conditioning and refrigeration systems of all sorts; furnaces, heaters, and chemical process equipment; the design of electronic equipment (e.g., to avoid overheating and failure of individual components, circuit boards, and larger assemblies and to understand the chemistry of semiconductor behavior); design of mechanical equipment (e.g., in lubrication of bearings to predict the overheating and subsequent failure due to excessive applied loads and in brake design to predict lining wear rates due to frictional heating and erosion); and in manufacturing processes (where, for example, the wear of tool bits is often due to frictional heating of the cutting edge). Indeed, it is fairly easy to make a case that thermodynamics in its broadest sense is the underlying science in most fields of engineering. Even the fields of pure mechanics use energy conservation relations which are subsets of more general thermodynamic principles.

1.2 Energy Balance Approach — Applications in Engineering

The principle "energy can be neither created nor destroyed" is one of the conservation relations that is explored in detail in later chapters. It is one of the building blocks which, when carefully expanded and explored, forms the basis for a good deal of our study of thermodynamics. The conservation-of-energy principle can be made true in any situation by simply changing or redefining what we mean by energy, so that it is indeed conserved in all situations. This was, in fact, what happened in the historical development of thermodynamics.

The basic conservation principle has two important suppositions. The first is that energy is something that is "contained." A certain defined system "has" energy. The second supposition is that there is a well-specified system where the energy is contained. To apply the energy conservation principle, the user must define the space or material of interest that "has" the energy.

The accounting system that we set up to allow for the transformation of energy should be a useful one for the types of problems that we plan to solve. The various energy transfer mechanisms that have been found to be most useful for our purposes are explored in Sec. 1.3 as well as in greater detail in Chap. 2.

   Without worrying at this point about how we classify energy, we can immediately invoke the conservation-of-energy principle to solve some problems in thermodynamics. For example, let us consider a power plant used for the production of electricity.

Example Problem 1.1

A power plant uses 1 unit of fuel energy to produce 0.4 energy units of electricity. What is the net energy transferred to the environment in the conversion of fuel to electricity?

System Diagram

Solution

   For this type of problem, the energy inside the power plant boundary remains constant according to the conservation-of-energy principle (since it cannot be created or destroyed). We could write in this case that

$$\left(\begin{array}{c} \text{Energy} \\ \text{transferred} \\ \text{in} \end{array} \right) + \left(\begin{array}{c} \text{Energy} \\ \text{transferred} \\ \text{out} \end{array} \right) = 0$$

where *in* and *out* refer to the direction in which the energy is crossing the plant boundary. Now we can expand these terms to include each of the energy transfer arrows on the diagram, or

$$\left(\begin{array}{c}\text{Energy} \\ \text{transferred} \\ \text{in}\end{array}\right)_{\text{fuel}} + \left(\begin{array}{c}\text{Energy} \\ \text{transferred} \\ \text{out}\end{array}\right)_{\text{electricity}}$$

$$+ \left(\begin{array}{c}\text{Net energy} \\ \text{transferred} \\ \text{out}\end{array}\right)_{\substack{\text{cooling} \\ \text{water}}} + \left(\begin{array}{c}\text{Energy} \\ \text{transferred} \\ \text{out}\end{array}\right)_{\substack{\text{stack} \\ \text{gas}}} = 0$$

or

$$1 \text{ unit} - 0.4 \text{ unit} + \left(\begin{array}{c}\text{Net energy} \\ \text{transferred} \\ \text{out}\end{array}\right)_{\substack{\text{cooling} \\ \text{water}}} + \left(\begin{array}{c}\text{Energy} \\ \text{transferred} \\ \text{out}\end{array}\right)_{\substack{\text{stack} \\ \text{gas}}} = 0$$

or, finally,

$$\left(\begin{array}{c}\text{Energy} \\ \text{transferred} \\ \text{out}\end{array}\right)_{\text{environment}} = \left(\begin{array}{c}\text{Net energy} \\ \text{transferred} \\ \text{out}\end{array}\right)_{\substack{\text{cooling} \\ \text{water}}} + \left(\begin{array}{c}\text{Energy} \\ \text{transferred} \\ \text{out}\end{array}\right)_{\substack{\text{stack} \\ \text{gas}}}$$

$$= -0.6 \text{ unit}$$

Comments

This very simple example illustrates a number of points about the energy conservation principle. First, when we apply the principle, we must carefully define the location to which we are going to apply it, in this case the power plant. Second, we must define a convention for how we assign the sign of the energy transfers. Here, we simply chose energy transfers into the plant as carrying a positive sign, so that energy transferred out is negative. Finally, we have tacitly kept the units (dimensions) of each quantity in the energy balance consistent.

The power of a simple energy balance is obvious from Example Problem 1.1; however, in more practical problems, the energy transfers in the various terms of the energy balance have different forms. For example, in the problem above, the energy transferred in might be chemical energy in a fossil fuel such as coal, oil, or natural gas; it might be the binding energy of the nucleus of atoms for a nuclear plant; it might be the energy transfer from the sun in a solar power plant or from energy stored in the earth in a geothermal plant. The electric energy is in the form of an electric current carried by the transmission lines leaving the plant. The energy transfer to the cooling water is usually in the form of thermal energy added to cooling water or the atmosphere, which then leaves the plant. Finally, the energy from the stack is transported in the hot flowing gas from the plant into the atmosphere. Thus, we need to classify the energy transfer across the plant boundary so that we can properly set up our energy accounting system.

One additional point is that we assumed (without stating the assumption) that the energy entering the plant was exactly balanced at each instant by the energy leaving. However, this is not always the case. Consider a new system boundary for a coal-fired plant that includes a coal stockpile, where coal is often delivered to the plant for later use at periods of high electrical demand. In such a case, our energy conservation equation must be extended to account for an energy storage term. Alternatively, the boundary could be redrawn to include only the powerhouse itself, excluding the coal pile, so that the original conservation equation could still be used. Thus, how we draw the boundary around our energy system defines the form of the equation to be used.

Let us now look at a useful way of classifying energy transfers for engineering problems.

1.3 Work and Heat Transfer

The conservation of energy is tied to a defined system. As indicated in Example Problem 1.1, the plant boundary was the considered system, and there were energy transfers into and out of the plant. The electricity that is transported across the plant boundary can be viewed as a form of *work.* The transfer to the cooling water is through an energy transfer mechanism termed *heat transfer.*

Chapter 2 addresses these two energy transfers, *work and heat transfer,* in detail, but an important distinction between the two is made here. Work is viewed as an organized transfer mechanism. This transfer can be used to raise weights, move diaphragms, turn shafts, etc., which is often the desired product of a thermodynamic system. An energy transport in the form of a heat transfer is viewed as a disorganized transfer mechanism. Heat transfer cannot be directly employed to raise a weight, turn a shaft, etc. This distinction is important for classifying the energy terms that make up the conservation-of-energy principle.

A last point concerning these energy transfers is that they are *not* "contained" by a space or material. They should be thought of as transfers and are, therefore, necessarily coupled with system boundaries where these transfers take place. This point is developed in much more detail in Chap. 2.

1.4 Macroscopic versus Microscopic Viewpoint

A microscopic viewpoint is used to attempt to understand a process or system by considering the particle nature of matter. This viewpoint might focus on molecules, atoms, or even an electron and nucleus. A complete description would require an enormous effort with suitable approximations. A macroscopic consideration addresses the appropriate observable averages of the microscopic phenomena. For example, the microscopic momentum transfer between gas molecules and a surface is observed on the macroscopic level as the gas pressure on the surface. Clearly, the macroscopic viewpoint is of direct consequence to the engineer.

Classical thermodynamics is a macroscopic science. The fundamental statements, or laws, concern the macroscopic properties of matter. Any atomic or microscopic concept must be exhibited in the macroscopic behavior of the system. This does not imply that a microscopic viewpoint is inappropriate for thermodynamics. A clear understanding of macroscopic phenomena is often possible through microscopic concepts. Yet the overriding goal of engineering thermodynamics is to address macroscopic properties.

This text stresses the fundamental concepts from a macroscopic viewpoint. References are made to microscopic behavior where it is helpful to clearly present the material.

1.5   Problem Solving

One of the main objectives of this text is to present a logical methodology to solve engineering problems. The subject of thermodynamics is composed of a few basic principles which can be applied to many different problems. Some problems are quite complex; yet, through a careful and logical approach, the solutions are generally straightforward.

We present an approach to solving problems which can be as important as the solution obtained. The student must learn the basic principles as well as the method to apply the principles and must not view the solution to problems as a routine substitution into an appropriate equation. Further, the student should seek the generality in the problem, whether it is an engineering problem being solved or an example in the text. Numerous examples are presented throughout the text to demonstrate these concepts and approaches.

There are many ways to subdivide the general approach to solving problems. The specific categories are not as important as the inclusion of all the basic steps. Grouping these basic steps into three categories yields the following elements:

1. *Problem statement*
   Carefully evaluate the information presented. What are the unknowns? What is given or known? Determine the principal parts of the problems and those parts which are secondary. An essential element is a figure of the physical system with the considered boundaries and diagrams of the states (precisely defined later) indicating known and unknown information. Carrying out the details on a problem that is not clearly understood might yield a correct answer, but will not help

in learning the fundamental principles or applying the approach to new problems.

Large complex systems involving many complicated subcomponents require a systematic methodology to obtain the desired result. The given information for the component processes should be carefully presented so that the known and unknown information is clearly defined. The individual processes are then considered separately, and the entire system's behavior is built from the subcomponents.

2. *Analysis*

The plan of attack to obtain the unknown must be formulated and carried out. The plan is composed of a mix of physical laws or principles, material properties, and assumptions. The specific proportion depends on the problem and its complexity. This planning stage generally turns out to be an iterative procedure, particularly at the early stages of this course. Try to relate the given information to the basic principles or to a previously considered problem. Restating the problem differently might yield a possible direction for solution. Once the plan is formulated, the solution must be carried out correctly. Make sure that *each* step is correct. It is frustrating to abandon a correct approach because of an incorrect step. Recheck each step to ensure that it is necessary.

Quite often it is useful to carry a solution as far as possible in algebraic form, because many quantities may cancel or simplify. Early substitution of numerical values offers extra chances for numerical errors.

3. *Review*

This essential step is often neglected. It is important from the viewpoint of both obtaining the correct result and solving the problem. First, try an alternate approach to the problem to check the results. Second, does the result make sense physically? Does the result have the correct dependencies? This type of thinking will be valuable in our approach to new problems. Last, try to generalize the analysis and consolidate the knowledge. What are the key elements of the problem?

An instrument used in the first chapters of this book is a tabular presentation of the states and processes. The approach is not essential, but it has been found very useful by the first-time student. This table forces you to carefully understand the problem statement. When this table is combined with a diagram of the states and processes, a direction for the solution of the problem should be evident.

1.6   Units

Engineers and scientists need to communicate with their peers not only through carefully defined words but also through numeric descriptions of the magnitude of certain quantities. The magnitude of a quantity such as

volume depends on the system of units used to make such a description; i.e., we can describe volume in terms of cubic centimeters, cubic feet, gallons, barrels, etc. So we must define our quantities carefully, but we must be equally careful to use a set of measurement units that is universally understood and accepted. Two unit systems are in general use among scientists and engineers: the U.S. Conventional System (USCS), or English system of units (sometimes called the customary engineering system), and the SI (Système International d'Unités), or international system. The latter system is used almost universally outside the United States. Despite efforts to make SI accepted worldwide, in the United States much of engineering practice is still carried out according to the USCS.

It is generally agreed that SI has certain advantages over the USCS, notably that fewer conversion factors must be memorized and that the decimal base of the system leads to simple choices of the scale of units needed to describe a given quantity.

Any system of units can be subdivided into *base units* and *derived units.* The base units are prescribed, and then the derived units are obtained from the prescribed set. Table 1.1 shows the base units and some of the derived units for both USCS and SI. Conversion factors between the SI and USCS values are given in App. B. The key points to note are the two dashes in this table. The USCS specifies force as a base unit while SI treats force as a derived unit. In the USCS, the original definition of certain units leads to the necessity of carrying a conversion factor in many equations.

This is made clear by an examination of Newton's second law for a constant mass subjected to a single force F and accelerated at a in the direction of F. In USCS, force and mass are unfortunately both given in units called pounds. To denote the fundamental difference between these quantities, the unit of force is always called the *pound force* (lbf), and the unit of mass is called the *pound mass* (lbm). To avoid confusion, both words should always be used in discussing these quantities, i.e., *pound force* or *pound mass,* never simply *pound.* Newton's second law in USCS would now be written as

$$F \text{ lbf} = m \text{ lbm} \times a \text{ ft/s}^2 \tag{1.1}$$

However, one difficulty remains. The pound force is defined as a *base* unit in the USCS. One pound force is defined as the force that accelerates a mass of 1 lbm at a rate of 32.1740 ft/s². If these values are substituted directly into Eq. (1.1), the result is

$$1 \text{ lbf} = 1 \text{ lbm} \times 32.1740 \text{ ft/s}^2$$

which, because both mass and force are derived units, is inconsistent in both defined units and magnitude. Thus, Eq. (1.1) must be modified by including the appropriate conversion factor, often explicitly denoted by the symbol g_c, so that Eq. (1.1) becomes

TABLE 1.1

Quantity	USCS	SI
	Base Units	
Length	foot, ft	meter, m
Mass	pound mass, lbm	kilogram, kg
Time	second, s	second, s
Temperature	degree fahrenheit, °F	kelvin, K
Force	pound force, lbf	——
	Derived Units	
Force	——	newton, N
Pressure	atmosphere, 1 atm $= 14.696$ lbf/in^2	pascal, Pa
Energy	Btu $= 778.16$ lbf $\cdot$ ft	joule, J
Power	Btu/s	watt, W
Specific heat	Btu/(lbm $\cdot$ °F)	J/(kg $\cdot$ K)
Area	ft^2	m^2
Volume	ft^3	m^3
Density	lbm/ft^3	kg/m^3
Velocity	ft/s	m/s

	SI Multipliers	
	Number	**Prefix**
	10^{12}	tera, T
	10^{9}	giga, G
	10^{6}	mega, M
	10^{3}	kilo, k
	10^{-2}	centi, c
	10^{-3}	milli, m
	10^{-6}	micro, μ
	10^{-9}	nano, n
	10^{-12}	pico, p

$$F = \frac{ma}{g_c} \tag{1.2}$$

where g_c has the value 32.1740 ft $\cdot$ lbm/(lbf $\cdot$ s^2). In SI, this conversion problem is avoided, because the unit of force, the newton, is a *derived* unit, defined as the force that will accelerate 1 kg of mass at a rate of 1 m/s^2. Substituting these values into Eq. (1.2) will show that the value of g_c is then simply 1 kg $\cdot$ m/(N $\cdot$ s^2), and g_c is not considered in equations that use exclusively SI units.

Example Problem 1.2

A 1-kg mass is attached to a spring scale on the earth's surface. The scale and the mass are then moved to the surface of the moon by an enterprising astronaut. The gravitational acceleration on the surface of the moon is one-sixth that on the surface of the earth. What will the spring scale read in each case?

Solution

The mass of an object is unaffected by the local gravitational field. In both cases, the force exerted by the mass on the spring scale equals the force exerted on the mass by gravity (the weight), or

$$F = ma = mg$$

On the earth's surface, given $g = 9.8$ m/s^2, the scale will then read

$$F = (1 \text{ kg})(9.8 \text{ m/s}^2) = 9.8 \text{ kg} \cdot \text{m/s}^2 = 9.8 \text{ N}$$

and on the moon

$$F = (1 \text{ kg})[(\tfrac{1}{6})(9.8 \text{ m/s}^2)] = 1.63 \text{ N}$$

Comments

Note that the mass is not changed in either case and that the weight indicated by the spring scale F depends on the gravitational field in which the scale is used. Now think about using a chemical beam balance to determine the mass and weight of the object used in the above experiment.

1.7 Mathematical Preliminaries*

In this section we review the mathematics needed for the study of thermodynamics. The mathematics required for thermodynamics is minimal, yet it is important to have a clear understanding of the necessary elements. This makes it possible to distinguish between the physical laws of thermodynamics and the mathematical manipulations. So some of the difficulties that arise in the understanding of thermodynamics can be attacked at the source — by a clear division between mathematical manipulation and physical principle. This section should be reconsidered throughout the text as needed.

In this section we review the necessary mathematics rather than offer a rigorous development of the subject. Excellent references are available for further details [1,2,3].

1.7.1 Function Representation

Consider the function

$$f(x, y, z) = c = \text{constant} \tag{1.3}$$

---

* This section may be referred to as needed in subsequent chapters.

which represents a surface in the three-dimensional space of coordinates x, y, z. An example of such a function is shown in Fig. 1.1. Any two of the three variables uniquely specify the value of the third. Thus, an alternate representation is

$$z = z(x, y) \tag{1.4}$$

where the value of z is uniquely specified by the values of x and y. A general surface is mathematically specified in terms of n variables. The physical representation of the general surface with n variables is difficult to show pictorially if n is greater than 3.

The intersection of the surface with a plane that is parallel to two of the coordinates forms a line. For example, in Fig. 1.1, an xz plane can intersect the surface at y coordinates of y_1, y_2, y_3, etc. The lines of intersection are shown as $y = $ constant lines. These lines can be projected onto

Figure 1.1 Surface representing $f(x,y,z) = c.$

the plane of the xz coordinates to form a two-dimensional representation of the surface. Figure 1.1 shows the xz and xy representations and a similar representation on the yz surface.

A general coordinate system is composed of *independent* variables since each can be varied independently without varying the others. The coordinates x, y, and z are independent variables for a general coordinate system. The concept of a *dependent* variable arises when a function like Eq. (1.3) is introduced. As indicated above, two variables uniquely specify the third for this surface, so one is dependent on the two other independent variables. Standard convention indicates the independent variables in parenthetical functional notation; so Eq. (1.4), for example, represents the dependent variable z as a function of the independent variables x and y. The denotation of independent and dependent variables is arbitrary. Thus, alternative forms are

$$x = x(y, z) \tag{1.5}$$

and

$$y = y(x, z) \tag{1.6}$$

Equation (1.3) is called an *implicit* representation of the function or surface. The dependent and independent variables are not directly indicated and can be varied from consideration to consideration. Equations (1.4) through (1.6) are termed *explicit* representations.

1.7.2 Partial Derivatives

A *partial derivative* represents the rate of change of the dependent variable with respect to a single independent variable, with all other independent variables held constant. This is mathematically denoted as

$$\left(\frac{\partial z}{\partial x}\right)_y \equiv \lim_{\Delta x \to 0} \frac{z(x + \Delta x, y) - z(x, y)}{\Delta x} \tag{1.7}$$

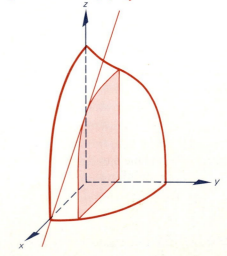

Figure 1.2 Geometric representation of a partial derivative $(\partial z / \partial x)_y$.

for the surface prescribed in Eq. (1.4). The derivative is taken with respect to the independent variable x, with y held constant. This is geometrically represented by the slope of a curve obtained by passing a plane parallel to the x and z coordinates (at constant y) through the surface. This is shown in Fig. 1.2. For the three-dimensional example in Eq. (1.4), there are two independent variables, so that another possible partial derivative is

$$\left(\frac{\partial z}{\partial y}\right)_x = \lim_{\Delta y \to 0} \frac{z(x, y + \Delta y) - z(x, y)}{\Delta y} \tag{1.8}$$

The partial derivatives generally depend on the particular values of the independent variables; i.e., the slope of the line on the surface depends on the specific y plane considered and the specific x point on the line. Thus, partial derivatives of the partial derivatives are possible. The second derivatives are denoted as

$$\left[\frac{\partial}{\partial x}\left(\frac{\partial z}{\partial x}\right)_y\right]_y = \frac{\partial^2 z}{\partial x^2} \tag{1.9}$$

$$\left[\frac{\partial}{\partial y}\left(\frac{\partial z}{\partial x}\right)_y\right]_x = \frac{\partial^2 z}{\partial y\,\partial x} \tag{1.10}$$

For single-valued, continuous functions (the usual case in thermodynamics), the result is independent of the order of differentiation, so

$$\frac{\partial^2 z}{\partial y\,\partial x} = \frac{\partial^2 z}{\partial x\,\partial y} \tag{1.11}$$

Example Problem 1.3

The ideal gas relation is $P = mRT/V$, where P is pressure, V is volume, m is mass, R is the gas constant, and T is temperature. Verify Eq. (1.11) with m and R constant.

Solution

Let $P = P(V, T)$. The first derivatives for the two independent variables V and T are

$$\left(\frac{\partial P}{\partial V}\right)_T = -\frac{mRT}{V^2}$$

$$\left(\frac{\partial P}{\partial T}\right)_V = \frac{mR}{V}$$

The second derivatives are

$$\left[\frac{\partial}{\partial T}\left(\frac{\partial P}{\partial V}\right)_T\right]_V = -\frac{mR}{V^2}$$

$$\left[\frac{\partial}{\partial V}\left(\frac{\partial P}{\partial T}\right)_V\right]_T = -\frac{mR}{V^2}$$

This satisfies the equality in Eq. (1.11).

Total Differential

The partial derivatives represent the slopes of lines tangent to the surface for a plane parallel to the coordinate axes. The change in z corresponding to an infinitesimal change in x for constant y is

$$dz = \left(\frac{\partial z}{\partial x}\right)_y dx \tag{1.12}$$

If the change in z is sought for simultaneous changes in x and y, then

$$\Delta z = z(x + \Delta x, y + \Delta y) - z(x, y) \tag{1.13}$$

Adding and subtracting $z(x, y + \Delta y)$ yield

$$\Delta z = \frac{z(x + \Delta x, y + \Delta y) - z(x, y + \Delta y)}{\Delta x} \Delta x$$

$$+ \frac{z(x, y + \Delta y) - z(x, y)}{\Delta y} \Delta y \tag{1.14}$$

and considering infinitesimal changes (taking the limits as $\Delta x \to 0$ and $\Delta y \to 0$) yields

$$dz = \left(\frac{\partial z}{\partial x}\right)_y dx + \left(\frac{\partial z}{\partial y}\right)_x dy \tag{1.15}$$

This is termed the *total differential* of z. The partial derivative and total differential expressions can be extended to more independent variables.

Relations between Partial Derivatives

As in the case of ordinary derivatives, the following product identities are valid:

$$\left(\frac{\partial z}{\partial x}\right)_y \left(\frac{\partial x}{\partial y}\right)_z \left(\frac{\partial y}{\partial z}\right)_x = -1 \tag{1.16}$$

and

$$\left(\frac{\partial y}{\partial x}\right)_z \left(\frac{\partial x}{\partial y}\right)_z = 1 \qquad \text{so} \qquad \left(\frac{\partial y}{\partial x}\right)_z = \frac{1}{(\partial x/\partial y)_z} \tag{1.17}$$

Another important relation among partial derivatives is needed to change from one set of independent variables to another. If $z(x, y)$ is desired in the form $z(x, w)$, then y must be expressed as $y = y(x, w)$. The total differential of $z(x, y)$ is

$$dz = \left(\frac{\partial z}{\partial x}\right)_y dx + \left(\frac{\partial z}{\partial y}\right)_x dy \tag{1.18}$$

With $y = y(x, w)$,

$$dy = \left(\frac{\partial y}{\partial x}\right)_w dx + \left(\frac{\partial y}{\partial w}\right)_x dw \tag{1.19}$$

Substituting Eq. (1.19) into (1.18) yields

$$dz = \left[\left(\frac{\partial z}{\partial x}\right)_y + \left(\frac{\partial z}{\partial y}\right)_x \left(\frac{\partial y}{\partial x}\right)_w\right] dx + \left(\frac{\partial z}{\partial y}\right)_x \left(\frac{\partial y}{\partial w}\right)_x dw \tag{1.20}$$

This last equation is exactly the form desired, $z(x, w)$. The total differential of $z(x, w)$ is

$$dz = \left(\frac{\partial z}{\partial x}\right)_w dx + \left(\frac{\partial z}{\partial w}\right)_x dw \tag{1.21}$$

Comparison of Eqs. (1.20) and (1.21) shows that

$$\left(\frac{\partial z}{\partial w}\right)_x = \left(\frac{\partial z}{\partial y}\right)_x \left(\frac{\partial y}{\partial w}\right)_x \tag{1.22a}$$

or

$$\left(\frac{\partial z}{\partial y}\right)_x \left(\frac{\partial y}{\partial w}\right)_x \left(\frac{\partial w}{\partial z}\right)_x = 1 \tag{1.22b}$$

and

$$\left(\frac{\partial z}{\partial x}\right)_w = \left(\frac{\partial z}{\partial x}\right)_y + \left(\frac{\partial z}{\partial y}\right)_x \left(\frac{\partial y}{\partial x}\right)_w \tag{1.23}$$

This last expression is called the *substitution rule,* and it is used to change independent variables.

1.7.3   Integration

Two types of integration commonly arise in thermodynamics. The first type is ordinary integration, which parallels the ordinary derivative in differentiation. The second type is the line integral, which parallels the partial derivative in differentiation. This parallelism is not complete but does serve as a distinction for the two types of integration.

The ordinary integral of the continuous function $f(x)$ between two limits x_1 and x_2 is denoted as

$$\int_{x_1}^{x_2} f(x)\,dx \tag{1.24}$$

This is geometrically displayed as the area under the curve $f(x)$, as indicated in Fig. 1.3. The ordinary integral is thought of as an infinite sum of infinitesimal strips forming the area indicated. The fundamental theorem of calculus connects differentiation and integration as

$$\int_{x_1}^{x_2} f(x)\,dx = g(x_2) - g(x_1) \tag{1.25}$$

where

$$\frac{dg(x)}{dx} = f(x) \tag{1.26}$$

and $g(x)$ is often termed an *antiderivative.* An alternate representation for ordinary integration is given by combining Eqs. (1.25) and (1.26):

$$\int_{x_1}^{x_2} dg(x) = g(x_2) - g(x_1) \tag{1.27}$$

The extension of ordinary integration to more than one variable represents the summation of a differential on a specified curve. This is written as

Figure 1.3   Ordinary integral.

$f(x)$

$$\text{Area} = \int_{x_1}^{x_2} f(x)\,dx$$

$x_1 \qquad x_2$

$$\int_{x_1,y_1}^{x_2,y_2} dg(x, y) \tag{1.28}$$

The line integral does not have the area representation, as is the case for ordinary integration. Line integration is most easily presented by an example.

Example Problem 1.4

The work required to move a particle through a path from position 1 to 2 in a force field $\mathbf{F}$ is

$$_1W_2 = \int_1^2 \mathbf{F} \cdot d\mathbf{s}$$

where $d\mathbf{s}$ is the differential length and the boldface center dot denotes the dot product. For the general path and force indicated below, the components are given as $\mathbf{F} = F_x\hat{\mathbf{i}} + F_y\hat{\mathbf{j}}$ and $d\mathbf{s} = dx\,\hat{\mathbf{i}} + dy\,\hat{\mathbf{j}}$. Thus, the expression for the work is

$$_1W_2 = \int_1^2 (F_x\,dx + F_y\,dy)$$

Evaluate the work done for the two paths from 1 to 2 indicated below if $F_x = y$ N and $F_y = c = 1$ N.

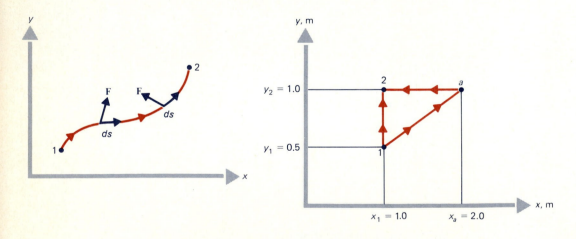

Solution
The required line integral for the *direct* path is

$$_1W_2 = \int_1^2 (y\,dx + c\,dy)$$

Along this path $x = \text{constant} = x_1$, so $dx = 0$ and

$$_1W_2 = \int_1^2 y\,dx^{\,0} + c\int_1^2 dy = c(y_2 - y_1)$$
$$= (1)(1 - 0.5)\ \text{N} \cdot \text{m} = 0.5\ \text{N} \cdot \text{m} = 0.5\ \text{J}$$

The required line integral for the *indirect* path through position a is

$$_1W_2 = \int_1^a (y\,dx + c\,dy) + \int_a^2 (y\,dx + c\,dy)$$

Each integral is considered separately, and each has particular characteristics. The first integral on the right-hand side is along a line given by

$$\frac{y - y_1}{x - x_1} = \frac{y_2 - y_1}{x_a - x_1} = \text{slope} \equiv m$$

or

$$y = y_1 + \left(\frac{y_2 - y_1}{x_a - x_1}\right)(x - x_1)$$
$$= y_1 + m(x - x_1)$$

Thus, this first integral is

$$\int_1^a (y\,dx + c\,dy) = \int_1^a [y_1 + m(x - x_1)]\,dx + \int_1^2 c\,dy$$
$$= y_1(x_a - x_1) + \frac{m}{2}(x_a - x_1)^2 + c(y_2 - y_1)$$

Substituting m yields

$$\int_1^a (y\,dx + c\,dy) = y_1(x_a - x_1) + \tfrac{1}{2}(y_2 - y_1)(x_a - x_1) + c(y_2 - y_1)$$

The second integral is along a path parallel to the x axis, so $y = \text{constant} = y_2$. Thus,

$$\int_a^2 (y\,dx + c\,dy) = \int_a^2 y\,dx + c\int_a^2 dy^{\,0}$$
$$= y_2(x_2 - x_a) = y_2(x_1 - x_a)$$

Therefore, the work done for the indirect path is

$$_1W_2 = y_1(x_a - x_1) + \tfrac{1}{2}(y_2 - y_1)(x_a - x_1) + c(y_2 - y_1) + y_2(x_1 - x_a)$$
$$= (y_1 - y_2)(x_a - x_1) + \tfrac{1}{2}(y_2 - y_1)(x_a - x_1) + c(y_2 - y_1)$$
$$= (-0.5)(1) + (0.5)(0.5)(1) + (1)(0.5) = 0.25\ \text{J}$$

Comments

The line integration is written for each variable, and the rules of ordinary integration are used. The work done is different along each path. Work generally depends on the path considered, and special notation will be used to denote such functions in subsequent chapters.

Line integration depends on the direction of integration, so that for a given path

$$\int_{x_1, y_1}^{x_2, y_2} dg(x, y) = -\int_{x_2, y_2}^{x_1, y_1} dg(x, y) \tag{1.29}$$

1.7.4 Exact and Inexact Differentials

Exact and inexact differentials play important roles in thermodynamics. Knowledge of whether a differential is exact indicates certain facts about the quantity. Consider a general differential expression given by

$$dz(x, y) = g(x, y) \, dx + h(x, y) \, dy \tag{1.30}$$

This is an *exact differential* if there exists a function $z(x, y)$ such that $dz(x, y)$ is a total differential [see Eq. (1.15)]. This is an *inexact differential* if there exists no function of x and y yielding Eq. (1.30). Listed here are three conditions which are valid for exact differentials:

$$\left[\frac{\partial g(x, y)}{\partial y}\right]_x = \left[\frac{\partial h(x, y)}{\partial x}\right]_y \tag{1.31a}$$

$$\int_{x_1, y_1}^{x_2, y_2} dz(x, y) = \text{function of endpoints only and independent of path} \tag{1.31b}$$

$$\oint_c dz(x, y) = 0 \qquad \text{The integral around every closed path } C \text{ is zero} \tag{1.31c}$$

Equation (1.31a) is useful to test for exactness. Complete proofs of these conditions are not presented, but illustrative examples that demonstrate these conditions follow.

Inexact differentials are denoted by δz to clearly distinguish these path-dependent functions from exact differentials throughout this text.

Example Problem 1.5

Determine whether the following differentials are exact or inexact, using the condition given in Eq. (1.31a):

(a) $y \, dx + x \, dy$

(b) $y \, dx - x \, dy$

Solution

Comparing these differentials with Eq. (1.30) and performing the derivatives indicated in Eq. (1.31a) yield

(a) $g(x, y) = y$ and $\left[\dfrac{\partial g(x, y)}{\partial y}\right]_x = 1$

 $h(x, y) = x$ and $\left[\dfrac{\partial h(x, y)}{\partial x}\right]_y = 1$

Then Eq. (1.31a) indicates that this differential is *exact*.

(b) $g(x, y) = y$ and $\left[\dfrac{\partial g(x, y)}{\partial y}\right]_x = 1$

 $h(x, y) = -x$ and $\left[\dfrac{\partial h(x, y)}{\partial x}\right]_y = -1$

so that this differential is *inexact*.

Comment

Is the differential in Example Problem 1.4 exact or inexact? (It is inexact.)

Example Problem 1.6

For the two paths between 1 and 2 shown, determine which of the differentials in Example Problem 1.5 satisfy the condition given in Eq. (1.31b).

Solution

(a) The integral along path 1-*a*-2 is

$$_1I_2 = \int_1^2 (y \, dx + x \, dy)$$

For path 1-*a*-2,

$$_1I_2 = \int_1^a (y \, dx + x \, dy) + \int_a^2 (y \, dx + x \, dy)$$

where y is constant at y_1 from 1 to a and x is constant at x_2 from a to 2. Thus,

$$_1I_2 = y_1 \int_1^a dx + x_2 \int_a^2 dy$$

$$= y_1(x_2 - x_1) + x_2(y_2 - y_1) = x_2 y_2 - x_1 y_1$$

For path 1-*b*-2,

$$_1I_2 = \int_1^b (y \, dx + x \, dy) + \int_b^2 (y \, dx + x \, dy)$$

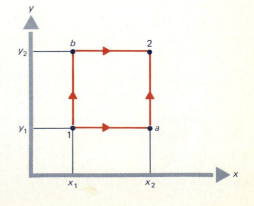

where $x = x_1$ from 1 to b and $y = y_2$ from b to 2. Thus

$$_1I_2 = x_1 \int_1^b dy + y_2 \int_b^2 dx$$
$$= x_1(y_2 - y_1) + y_2(x_2 - x_1)$$
$$= x_2y_2 - x_1y_1$$

The integrals are independent of path since this differential is *exact*.

(b) The same process is repeated for this differential, so some of the steps can be eliminated. For path 1-*a*-2,

$$_1I_2 = y_1 \int_1^a dx - x_2 \int_a^2 dy$$
$$= y_1(x_2 - x_1) - x_2(y_2 - y_1)$$
$$= 2x_2y_1 - x_1y_1 - x_2y_2$$

For path 1-*b*-2,

$$_1I_2 = -x_1 \int_1^b dy + y_2 \int_b^2 dx$$
$$= -x_1(y_2 - y_1) + y_2(x_2 - x_1)$$
$$= -2x_1y_2 + x_1y_1 + x_2y_2$$

This integral depends on the path since the differential is *inexact*.

Example Problem 1.7

Consider the differentials in Example Problem 1.5 for the closed path 1-*a*-2-*b*-1 shown in Example Problem 1.6. Determine which differentials satisfy the condition given in Eq. (1.31*c*).

Solution

The cyclic integral can be decomposed into the four separate sides, and much of the detail in Example Problem 1.6 can be used. A key point is the negative sign arising from the direction of integration, as indicated in Eq. (1.29).

(a) The integral is

$$\oint (x\,dy + y\,dx) = \left[\int_1^a (x\,dy + y\,dx) + \int_a^2 (x\,dy + y\,dx) \right]$$
$$+ \left[\int_2^b (x\,dy + y\,dx) + \int_b^1 (x\,dy + y\,dx) \right]$$

Using the details in Example Problem 1.6 yields

$$\oint (x\,dy + y\,dx) = (x_2y_2 - x_1y_1) - (x_2y_2 - x_1y_1) = 0$$

This is the third condition for an *exact* differential.

(b) The integral is

$$\oint (y\,dx - x\,dy) = \left[\int_1^a (y\,dx - x\,dy) + \int_a^2 (y\,dx - x\,dy) \right]$$
$$+ \left[\int_2^b (y\,dx - x\,dy) + \int_b^1 (y\,dx - x\,dy) \right]$$

Using the details in Example Problem 1.6 yields

$$\oint (y\,dx - x\,dy) = (2x_2 y_1 - x_1 y_1 - x_2 y_2) - (-2x_1 y_2 + x_1 y_1 + x_2 y_2)$$
$$= 2(x_2 y_1 + x_1 y_2) - 2x_1 y_1 - 2x_2 y_2$$
$$\neq 0$$

This is indicated by the third condition for an *inexact* differential.

Comments

These three example problems are applications of the conditions for exact and inexact differentials. Reconsider the differential in Example Problem 1.4. This differential is inexact, as can be found from Eq. (1.31).

The following chapters discuss thermodynamic properties and energy transfer processes. We see that the differentials of thermodynamic properties are exact and that the differentials of energy transport processes are inexact.

1.8 Scope of Text

Thermodynamics deals with energy transformations. The presentation of the concepts requires an initial understanding of the basic definitions of systems, processes, and states. To understand the energy transformations of a system, energy transfer mechanisms in the form of work and heat transfer must be defined. Then the fundamental principles of thermodynamics can be presented. The portion of thermodynamics that includes the basic definitions, concepts, and laws forms the core of the subject.

Another important aspect of any introductory thermodynamics course is the set of required thermodynamic properties. This aspect includes the graphical, tabular, and equation forms of the quantities that describe the states of a substance. The use of thermodynamic properties is essential to solving problems, yet the detailed properties are not part of the foundations of the laws of thermodynamics.

The third portion of the study of thermodynamics is the application of the core concepts to particular problems of interest to the engineer.

These three elements are the basis of any presentation of engineering thermodynamics. Clearly the core element including the basic concepts must be presented first. A logical presentation might then discuss properties followed by applications. This text has a slightly different, yet equally logical organization. The core concepts, properties, and applications are presented simultaneously, so a working knowledge of the properties is obtained as the basic concepts are introduced. Practical applications of the fundamental principles can be carried out as the laws are presented.

Problems

1.1 One system of units takes the base units of length, time, and force as foot, second, and pound force, respectively. The set of derived units includes mass, and the derived mass unit is called a *slug*. With Newton's second law $F = ma$, determine the relation between the slug and the length, time, and force units above.

1.2 Determine the SI values for the following quantities: 6 ft, 200 lbm, 70°F, 25 lbf/ft², and 1 atm.

1.3 Determine the SI values for the following energy-related quantities: 1 Btu/(ft² · h), 13,000 Btu/lbm, and 50,000 Btu/h.

1.4 A function is represented as

$$P = C\frac{T}{V}$$

where P is pressure, T is temperature, V is volume, and C is a constant.
(a) Is this an implicit or explicit representation?
(b) What are the dependent and independent variables?
(c) Graph the function.

1.5 What are the mathematical representations in terms of partial derivatives for the following infinitesimal changes? Consider a gas enclosed in a tank. All processes take place at constant pressure.
(a) The temperature of the gas is altered by changing the gas mass. The tank is at a constant volume during this process.
(b) The temperature of the gas is altered by changing the volume of the gas. The tank mass is maintained at a constant value during this process.
Write the total differential of the dependent variable in terms of the partial derivatives.

1.6 The implicit representation is given for a function as

$$f(U, S, V) = 0$$

where U is internal energy, S is entropy, and V is volume.

(a) Write the functional form, indicating that U is the dependent variable and S and V are the independent variables.

(b) Write the expression for the total differential of U, dU, in terms of partial derivatives and differentials.

(c) Sketch the figure that represents the two partial derivatives in (b).

1.7 Sketch the following partial derivatives for an arbitrary surface: $\left(\dfrac{\partial z}{\partial x}\right)_y$, $\left(\dfrac{\partial T}{\partial P}\right)_V$, and $\left(\dfrac{\partial U}{\partial S}\right)_V$.

1.8 What is the total differential of the function $z = z(y_1, y_2, y_3, \ldots, y_n)$ where there are n independent variables? Write the total differential dz in terms of the appropriate partial derivatives.

1.9 Develop the relationship given in Eq. (1.16) as

$$\left(\frac{\partial z}{\partial x}\right)_y \left(\frac{\partial x}{\partial y}\right)_z \left(\frac{\partial y}{\partial z}\right)_x = -1$$

Start by considering the total differential of $z = z(x, y)$ for the special case of a constant z ($dz = 0$).

1.10 Evaluate the following ordinary integrals, and represent the values as areas on the appropriate diagrams:

(a) $\displaystyle\int_{x=1}^{x=2} x \, dx$

(b) $5 \displaystyle\int_{x=1}^{x=2} dx$

(c) $\displaystyle\int_{x=0}^{x=3} (3 - x) \, dx$

1.11 Evaluate the line integrals from A to C for the paths indicated in Fig. P1.11. (Note that the ideal gas relation, $PV = mRT$, is used in this problem with m and R as constants.)

(a) $\displaystyle\int_{A-B-C} P \, dV$

(b) $\displaystyle\int_{A-C} P \, dV \quad$ where $P = \dfrac{mRT}{V} = \dfrac{\text{constant}}{V}$

(c) $\displaystyle\int_{A-B-C} \left(\frac{P}{mR} \, dV + \frac{V}{mR} \, dP \right)$

(d) $\displaystyle\int_{A-C} \left(\frac{P}{mR} \, dV + \frac{V}{mR} \, dP \right)$

Figure P1.11

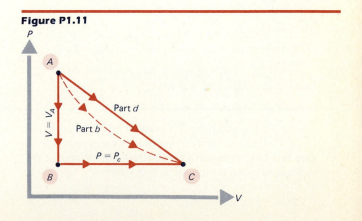

1.12 Evaluate the line integrals from A to C for the specific paths indicated in Fig. P1.12 (take m and R as constants).

(a) $\displaystyle\int_{A-B-C}\left(\frac{mR}{P}\,dT-\frac{mRT}{P^2}\,dP\right)$

(b) $\displaystyle\int_{A-B'-C}\left(\frac{mR}{P}\,dT-\frac{mRT}{P^2}\,dP\right)$

1.13 Evaluate the line integrals from A to B for the specific paths indicated in Fig. P1.13.

(a) $\displaystyle\int_{A-B} P\,dV$ along the path given by

$$P=\frac{\text{constant}}{V}=\frac{C}{V}$$

(b) $\displaystyle\int_{A-B} P\,dV$ along a straight path from A to B.

1.14 Which of the following functions $z=z(x,y)$ are exact and which are inexact? Prove your answers.

(a) $dz = x\,dx + x\,dy$

(b) $dz = y\,dx$

(c) $dz = \sin y\,dx - \cos x\,dy$

(d) $dz = xy^2\,dx + x^2y\,dy$

1.15 The differential of the function $z(x,y)$ is given by

$dz = xy^3\,dx + \tfrac{3}{2}(xy)^2\,dy$

(a) Is dz an exact differential? Prove your answer.
(b) What is the value of $z_2 - z_1$ between the points ($x_1 = 1$, $y_1 = 0$) and ($x_2 = 3$, $y_2 = 5$) along the path shown in Fig. P1.15?

1.16 What relations are required between partial derivatives for the following differentials to be exact?

(a) $-P\,dV - S\,dT$

(b) $V\,dP - S\,dT$

1.17 The total differential of a property called *enthalpy H* is an exact differential, and is given by

$dH = T\,dS + V\,dP$

where T is temperature, S is another property called *entropy*, V is volume, and P is pressure. Represent temperature and volume as partial derivatives, and determine the relation between temperature and volume.

Figure P1.12

Figure P1.13

Figure P1.15

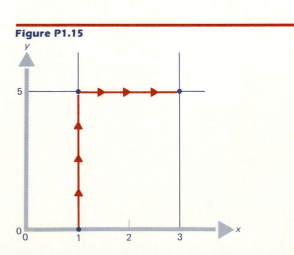

1.18 It is given that dU is an exact differential and that

$$dU = T\, dS - P\, dV$$

where U is internal energy, S is entropy, and T, P, and V are temperature, pressure, and volume, respectively.
(*a*) What are the dependent and independent variables?
(*b*) Express T and P as partial derivatives.
(*c*) What is the relation between T and P?

1.19 It is given that

$$dz = -2\, dx + \tfrac{3}{2}x^2 y^2\, dy$$

and

$$dw = xy^3\, dx - y\, dy$$

(*a*) Are dz and dw exact differentials?
(*b*) Is $dz + dw$ an exact differential?

1.20 Determine whether any values of $A(y)$ exist such that $d\phi$ and $d\mu$ as given here are exact differentials. Then find any values of $A(y)$ that will make $dz = d\mu + d\phi$ an exact differential.

$$d\mu = A(y)y\, dx + dy$$

$$d\phi = A(y)\, dx + x\, dy$$

References

1. C. R. Wylie, Jr., *Advanced Engineering Mathematics,* McGraw-Hill, New York, 1966.
2. P. H. Badger, *Equilibrium Thermodynamics,* Allyn and Bacon, Boston, 1967.
3. S. M. Blinder, "Mathematical Methods in Elementary Thermodynamics," *Journal of Chemical Education,* vol. 43, no. 2, 1966, pp. 85–92.

2

Energy and Energy Transfer

Normally screws are so cheap and small and simple you think of them as unimportant. But now, as your Quality awareness becomes stronger, you realize that this one, individual, particular screw is neither cheap nor small nor unimportant. Right now this screw is worth exactly the selling price of the whole motorcycle, because the motorcycle is actually valueless until you get the screw out. With this reevaluation of the screw comes a willingness to expand your knowledge of it.

Robert M. Pirsig, *Zen and the Art of Motorcycle Maintenance*, William Morrow and Co., Inc., New York, 1974, used with permission.

Cutaway of a hot-water to air heat exchanger for a commercial-scale heating system. *(The Trane Company.)*

2.1   Introduction

All specialized fields, such as law and economics, use terms with quite specific meanings, and these meanings may not be the same as those used in everyday conversation. To be sure of accurate communication in the classroom as well as with other engineers and scientists, the definitions used in thermodynamics must be held in common among us, and they must be as complete and accurate as possible.

The purpose of this chapter is to define the basic terms of thermodynamics. In later chapters, additional terms and concepts arc introduced and defined; however, the terms defined here are sufficient to begin the study of thermodynamics.

2.2   Concepts and Definitions

To begin our study of thermodynamics, we must have in mind some definable quantity of matter which we wish to consider. Then the behavior of energy as it interacts with this matter can be examined.

2.2.1   System and Surroundings

We might want to determine, for example, the temperature of shaving cream as it leaves a pressurized container or to find the work output of a steam turbine in a power plant. What "definable quantity of matter" shall we examine in each of these cases?

Any such quantity of matter is obviously a subclass of all matter. Suppose we begin by defining the *universe* as the totality of matter that exists.

Now, to get down to the part of the universe that we want to study for a particular reason, we define a *system* as that part of the universe set apart for examination and analysis. In most interesting problems that we examine in thermodynamics, the system under study interacts with the rest of the universe through an exchange of energy and/or mass. Yet the great bulk of the universe can be considered to be unaffected when, for example, we allow shaving cream to expand from its container. It is useful, then, to look at a further subset of the universe, called the *surroundings,* that is made up of that portion of the universe which strongly interacts with the system under study. These definitions are shown schematically in Fig. 2.1.

In summary, then, we perform a thermodynamic study of a system

Figure 2.1   The system, the surroundings, and the universe.

The universe

The system

The surroundings

that interacts with its surroundings; the system and its surroundings form a part of the universe.

It is critical in all thermodynamic studies to carefully define the particular system under examination. It sounds as if this task would be obvious and easy, and in many cases it is. In other cases, judicious choice of the system to be studied can greatly reduce the problem-solving effort. For example, in the shaving cream example, what system should be studied? Is it the container? The container plus its contents? The contents only? If the last, how do we treat the fact that the separation between the system and its surroundings is moving in space as the contents expand out of the can? All these choices have benefits and deficiencies that will be addressed soon; for now, though, clearly we need some further ways to describe the system. In particular, how can the separation between the system and the surroundings be described?

Generally, we define the *system boundary* as the surface that separates the system from its surroundings. Thus the shaving cream as a system is separated from the air (the surroundings) as the cream leaves the container by an imaginary surface, or boundary. The boundary expands with the shaving cream, so the system always contains all the initial mass of the system (the shaving cream).

That approach seems straightforward enough. However, consider the other example mentioned earlier, the power plant turbine. What do we use as the system and its boundary? If we follow the shaving cream example, we might define the steam flowing through the turbine as the system. The system boundary, however, would then extend throughout all components of the power plant that contain the steam (and the water from which it is generated) before and after the steam passes through the turbine. Such a system definition is not useful for many problems. To study hardware characteristics, it is often more useful to look at a boundary that is roughly coincident with the hardware being studied (the turbine casing, for example). In that case, however, mass is crossing the boundary!

Our definition of a system does not preclude mass crossing the system boundary. We must simply realize that the system under study has different mass within its boundary at each instant of time; alternatively, we can imagine the study to be of a series of constant-volume systems, each with a different inventory of mass.

We have seen that special circumstances will determine how to treat the system boundary. In the first case (shaving cream), no mass crosses the system boundary, and the boundary is allowed to expand in space so that the system mass is always contained within the boundary and so it is constant. In the second case, it is more convenient to fix the boundary in space and to allow the inventory of mass within the system to be different at each time. In such a case, the *amount* of mass within the system may or may not be constant, but it will be *different* mass at each time. We will see that there is no fundamental difference in the basic principles as we analyze these different cases. The former type of system (no mass crossing

Control volume

Crossing
boundary

Inside
boundary

Crossing
boundary

Figure 2.2   Control volume showing the system boundary composed of inside and crossing boundaries.

the boundary) is sometimes referred to as a *closed system,* or *control mass* (CM), while a system with mass crossing the boundary is called an *open system,* or *control volume* (CV).

In a system that has mass crossing its boundary, usually a part of the boundary allows no mass to cross, while another part of the boundary allows mass flow. Such a boundary configuration is typical of engineering hardware such as a turbine or compressor. In such equipment, the inside of the casing of the equipment is impervious to mass flow (the *inside boundary*) while the inlet and exit ports to the equipment encompass all the mass flow streams that enter or leave the system (the *crossing boundary*). A complete volume that is fixed in space and has a crossing boundary is called a control volume. Figure 2.2 shows the control volume and its boundaries for an idealized pump.

2.2.2   System Description

Now that we have defined the nomenclature that describes the matter under study, we need some means of describing the behavior of the system. In other words, how do we tell whether changes have occurred in the system? What will allow us to quantitatively describe these changes?

Let us examine for a moment a particular system that has no mass crossing its boundary and has as well no energy crossing its boundary. Such a system is called *isolated.* Assume further that the system is uniform. From experience, we would expect such a system, after sufficient time had passed, to be unchanging. But how do we define *unchanging?* Our experience indicates that certain measurable quantities such as pressure, temperature, and volume will be constant for an isolated system. If we measure these or certain other quantities, a particular *set* of measured values will result. Let us agree that this set of values defines the *state* of the system. If we allow mass or energy to cross the system boundary, some of the measured values will change, and the system will be in a new state.

What quantities define the state of a system, and how many such quantities are necessary and sufficient to define the state? The answer is given later, but here we note some attributes of the quantities describing the state of a system. These attributes attempt to describe the *condition* of the system.

Let us define a *property* as any quantity that describes the state of a system and, conversely, as any quantity whose value depends solely on the state of the system under study. This is a quantity that is "possessed" by a system or that the system "has." A number of familiar quantities meet this definition. For example, the volume of the shaving cream before and after expansion from the container depends on only the states at the beginning and end of the expansion process; therefore, volume is a property. Pressure and temperature also meet the requirements of being properties. Other less familiar properties will be defined as they are found useful.

Properties have other characteristics that result from this definition. Since properties fix the state of a system, they are independent of how the system reaches a given state. Thus, all properties have the mathematical characteristics of exact differential quantities; i.e., the change in their values between two system states is independent of how the system state was changed. The mathematical properties presented in Eq. (1.31) are valid for properties. These characteristics are used in later chapters.

Properties are conveniently divided into two categories. The first contains those properties with values that depend on the amount of mass contained in the system in a given state. From the list of properties mentioned so far, volume falls in this category. If the mass of the system is doubled while temperature and pressure are constant and uniform within the system, then the volume will also double. Properties that depend on the system mass, or the "extent" of the system, are called *extensive* properties. Properties that are independent of the mass contained in the system boundaries are called *intensive* properties.

Both intensive and extensive properties are always denoted by capital letters. Thus P will stand for pressure (intensive), T for temperature (intensive), and V for volume (extensive). Extensive properties of a given system in a given state, however, can be denoted in an alternate intensive form. The volume of the shaving cream, for example, can be assigned the symbol V. However, for a system in a given state with properties that are uniform throughout, the volume is directly proportional to the system mass m. Therefore, the quantity V/m is also a property (since the ratio depends solely on the system state). The ratio V/m is much more convenient to tabulate than V itself, since one parameter, m, is eliminated from the tables. This is true of all extensive thermodynamic properties. Whenever any extensive property X is used or tabulated in the form X/m, the ratio is assigned the corresponding lowercase letter x, and the property is called a *specific* property. Thus, $v \equiv V/m$ is the *specific volume,* which is the inverse of the density $v = 1/\rho$. Specific extensive properties are intensive properties, since their value no longer depends on the mass of the system.

Example Problem 2.1

The mass of a system is a property which must be specified when the state of the system is being described. Because it is possible to divide certain equations by the system mass, thus converting the extensive properties in those equations to intensive specific properties, the system mass can sometimes be handled in such a way that it need not be specifically given or found. Demonstrate this for the ideal gas relation.

Solution

   An ideal gas in a closed container is to be treated as a system. The gas has its properties related by

$$PV = mRT$$

where the symbols are as used earlier. Dividing the equation through by m results in

$$Pv = RT$$

Thus, the mass has disappeared as an explicit system property, and the resulting ideal gas equation provides a relation among the temperature, pressure, and specific volume of the system. Specifying any two of the system properties P, v, and T apparently fixes the other, without regard to the mass of the system.

2.2.3   Equilibrium States and Quasi-Equilibrium Processes

The concept of equilibrium is important to thermodynamics and is closely tied to the definitions of properties and states. For a system, the properties describing the state of an equilibrium system will be constant if the system is not allowed to interact with the surroundings or if the system is allowed to interact completely with unchanging surroundings. Such a state is termed an *equilibrium* state, and the properties are equilibrium properties. When a system is in equilibrium with its surroundings, it will not change unless the surroundings change.

   Specific types of equilibrium refer to individual properties. When a single property is unchanging within the system, the equilibrium is specific to that property. Typical examples include *thermal equilibrium* (constant T), *mechanical equilibrium* (constant P), and so on. When all possible properties are unchanging, the system is in *thermodynamic equilibrium*.

Much of the study of classical thermodynamics deals with equilibrium states. The properties that we use (and develop in Sec. 2.3 and Chap. 3) must be equilibrium properties. In fact, we refer to properties of particular materials with the understanding that there is a system that contains the material and that the system is in equilibrium. We then represent the locus of states for these materials as surfaces in a space with coordinates denoted by the properties.

Various systems that will be addressed may not meet all the conditions for equilibrium when the entire system is viewed. Such a system can be subdivided to yield local, smaller systems that can be treated as being in equilibrium. This will be important when we analyze more complex problems, particularly open systems or control volumes where changes occur throughout the volume.

The equilibrium state is described by equilibrium properties and is viewed as a surface with coordinates of properties. A particular state is then a point on this surface (see Fig. 1.1). If the system is altered so that the state of the system moves along the surface from one equilibrium position to another equilibrium position, the process is termed a *quasi-equilibrium process*. Every position on this surface is in equilibrium, and so every step of the process is in equilibrium. Given the concept of equilibrium defined by unchanging properties in relation to the surroundings, the system must be allowed to reach equilibrium at every step or state. This is achieved by visualizing the process as occurring at an infinitely slow rate, so that there is only a slight difference in properties between the system and surroundings, and equilibrium is achieved at each state along the process. This is an idealization, but it is very useful in many problems.

2.3   Some Common Properties

We have discussed one property of a system that is quite familiar, the volume. We now discuss specific volume and some other familiar properties in more detail. All the properties discussed in this chapter have a misleading attribute—they are all quantities with which we have extensive experience before beginning the study of thermodynamics. In addition, the properties of volume, temperature, and pressure are all measurable quantities.† The student should not infer that properties are all directly measurable in the sense that P, V, and T are. Indeed, many of the properties useful in thermodynamics are *not* directly measurable. However, every property that is introduced in classical thermodynamics does share the attributes noted in our general definition.

---

† The concept of "measurable" is not as straightforward as implied here. Temperature, for example, is really not measured directly, but is found by measuring the length of a mercury column or some other quantity proportional to temperature.

2.3.1 **Pressure** P

Because many systems studied in thermodynamics involve gases or vapors, pressure is a useful property for describing the state of the system. *Pressure* is defined as the normal force exerted per unit area on a real or a fictitious surface within the system.

In classical thermodynamics, we do not consider effects that occur on a microscopic scale, so that we can treat only pressures that exist on areas which are large relative to intermolecular spacings. Thus the fluid is viewed as continuous, and this is the so-called continuum approximation. This approximation must be questioned in vacuum systems where molecular spacings become large.

With the restriction that the area over which the force is applied must not become smaller than some minimum value a (because of the continuum approximation), the mathematical definition of the local pressure is

$$P \equiv \lim_{\Delta A \to a} \frac{\Delta F}{\Delta A} \tag{2.1}$$

For engineering work, pressures are often measured with respect to atmospheric pressure rather than with respect to an absolute vacuum. The former, or gauge pressure, is related to the atmospheric pressure by

$$P_{\text{gauge}} = P_{\text{abs}} - P_{\text{atm}} \tag{2.2}$$

Unless specifically noted otherwise, all pressures used in the text and tables of this book are absolute pressures.

Some pressure-measuring devices measure the height of a fluid column to evaluate the pressure at a particular point. The relation for the pressure change within a fluid is obtained from a force balance on a fluid element. Figure 2.3a presents a portion of fluid within the fluid located at an arbitrary position y above a reference plane. The pressure forces act on the top and bottom surfaces, and these forces balance the gravitational force on the fluid within the element. The force balance yields

$$P_y A - P_{y+dy} A - mg = 0 \tag{2.3}$$

The mass is expressed in terms of the fluid density as $m = \rho A\, dy$, where ρ is the fluid density. The pressure at P_{y+dy} is expressed in a Taylor series about the point y, so

$$P_y A - \left(P_y + \frac{dP_y}{dy}\, dy + \cdots\right) A - \rho A\, dy\, g = 0 \tag{2.4}$$

Canceling the common terms yields

$$-\left(\frac{dP_y}{dy}\, dy + \cdots\right) - \rho\, dy\, g = 0 \tag{2.5}$$

Dividing by dy, noting that the higher-order terms have dy to power 1 or greater, and taking the limit as dy goes to a point ($dy \to 0$) yield

Figure 2.3 **Pressure measurement. (a) Force balance in a fluid element.** $dP_y/dy = -\rho g$ **(b) Pressure difference between two elevations.** $P_2 - P_1 = -\rho g(y_2 - y_1)$

(a)

(b)

$$\frac{dP_y}{dy} = -\rho g \tag{2.6}$$

This relates the change in pressure with elevation to the density of the fluid element and the gravitational acceleration. Integration between the elevations y_1 and y_2 for a constant-density fluid yields

$$P_2 - P_1 = -\int_{y_1}^{y_2} \rho g \, dy = -\rho g (y_2 - y_1) \tag{2.7}$$

This expression gives the *difference* in pressure at two elevations in terms of density of the fluid and difference in elevation and is diagramed in Fig. 2.3b.

In SI, force is given in newtons and area in square meters. The derived unit for pressure is the pascal (Pa), where 1 Pa is defined as 1 N/m^2. This turns out to be a very small unit for engineering purposes, so most pressures are tabulated in terms of kilopascals (1 kPa $= 1 \times 10^3$ Pa) or megapascals (1 MPa $= 1 \times 10^6$ Pa). One standard atmosphere (1 atm) is equal to 101.325 kPa. Another common unit for pressure is the bar, which is 1×10^5 Pa. This is not strictly an SI unit, but it is used in some applications. In the USCS, pressures are given in pounds force per square foot or pounds force per square inch (psi).

2.3.2 Specific Volume v

Specific volume must also be defined in classical thermodynamics in light of the continuum restriction. Thus, the mathematical definition of *specific volume* is

$$v = \lim_{\Delta m \to \mu} \frac{\Delta V}{\Delta m} \tag{2.8}$$

where μ is a minimum amount of mass that is large compared with the mass of the individual molecules of which it is composed. Again, such a restriction causes few difficulties unless gases at high vacuum conditions or very small volume systems are examined. The specific volume is the inverse of density, or $v = 1/\rho$.

2.3.3 Temperature T

Temperature may be the property which we think is most familiar, yet it is among the most difficult properties to define exactly. Our senses are not very reliable in determining temperature. For example, we might go swimming on what we would describe as a hot day. Upon emerging from the water, we find the air suddenly feels quite chilly, and we prefer to stay in the water that felt cold when we first entered. We doubt that either the air or water temperature has actually changed, even though our senses tell us that they seem to be different.

We might also observe that when we take a cold bottle of milk from the refrigerator and set it on the table next to a hot cup of coffee, both liquids will approach room temperature if we wait long enough. The coffee and milk are then said to be in *thermal equilibrium* with the room, and our senses tell us that thermal equilibrium is attained when all the materials are at the same temperature. We have observed here a general law based on this and other experiences: *Two bodies which are each in thermal equilibrium with a third body are in thermal equilibrium with each other*. This observation, because it is the basis for the measurement of temperature and precedes the first and second laws of thermodynamics, is often called the *zeroth law.*

The zeroth law now assures us that if we place a thermometer or other temperature sensor in thermal equilibrium with body (or system) A, and we similarly place the sensor in thermal equilibrium with system B, then systems A and B are at the same temperature if the sensor reads the same temperature in both cases.

We now need to define an adequate scale of temperature, so that all engineers can present their measurements on a common basis. For most such measurements, it is convenient to define a scale that is a linear function of some measurable quantity (such as the length of a mercury column), at least over some range of temperatures between fixed points. There are good reasons in various theoretical studies to use other "temperature" scales, such as one that is the inverse of the common scales.

The two scales most useful in thermodynamics are the so-called absolute scales. The absolute scale for SI is the *Kelvin scale,* named after William Thomson (1824–1907), who became Lord Kelvin. This scale is a 1-point scale based on the second law of thermodynamics. This is presented in Chap. 5. The single point is the triple point of water, where ice, liquid water, and water vapor coexist in a closed system in the absence of air. The Kelvin scale replaced the original scale which was based on a linear function between two selected points. Note that the kelvin unit does not use the degree symbol, but simply the symbol K. Temperatures on this scale are referred to, for example, as "36 kelvins," not "36 degrees kelvin." The other absolute scale is called the *Rankine scale* after W. J. M. Rankine (1820–1872). It is related to the Kelvin scale by

$$T, \, °R = 1.8T, \, K \tag{2.9}$$

where a degree Rankine is denoted by °R.

Two other commonly used scales are the Fahrenheit scale, after Gabriel D. Fahrenheit (1686–1736), and the Celsius scale, after the Swedish astronomer Anders Celsius (1701–1744).

The Fahrenheit scale is also linear, and it was originally based on two defined points: 32° as the temperature at which a system of air–saturated water and ice coexist, and 212° as the temperature of a system containing water and steam at a pressure of 1 atm. The symbol for the Fahrenheit degree is °F. This original definition has been replaced by the following relations to the Kelvin scale:

Figure 2.4 Comparison of temperature scales.

$$T, \degree C = T, K - 273.15 \qquad (2.10a)$$

$$T, \degree F = 1.8T, \degree C + 32 \qquad (2.10b)$$

where a degree on the Celsius scale is denoted by $\degree C$.

The Celsius scale was originally defined by a single fixed point and a defined size for the degree. The fixed point was the triple point of water, and it was defined to have a temperature of $0.01\degree C$. The size of the degree chosen for the Celsius scale comes from the Kelvin scale and yields the temperature of the steam point to be $100.00\degree C$ at 1 atm. This original definition was replaced by the relation with the Kelvin scale given in Eq. (2.10a).

Figure 2.4 compares the various scales. Note that the size of 1 kelvin is the same as 1 degree on the Celsius scale; similarly, the size of the degrees on the Rankine and Fahrenheit scales are equal.

2.4 Energy

Now that we have defined a system and can describe the state of that system by some set of its properties, we can discuss the quantity called *energy*. In particular, we discuss the various classifications of the forms of energy.

First, let us look again at the special case of an isolated system. Remember that such a system by definition has a fixed state, and no mass or energy is crossing its boundary. What forms of energy does such a system have? From studies of mechanics in introductory physics courses or later in engineering mechanics courses, we know that the mass of the system carries kinetic energy that depends on the overall system speed $\mathbf{V}_{ref}$ relative to its surroundings (which we take as the reference frame) of $m\mathbf{V}_{ref}^2/2$. In addition, the position of the system relative to a reference plane in the surroundings imparts the system with potential energy that depends on the local gravitational acceleration g as well as the height of the system above the reference plane Z. The potential energy has magnitude mgZ. The potential energy is a property of the system, so the gravitational field is viewed as part of the system. This is convenient since many problems in engineering are considered within the gravitational field.

Are other forms of energy possessed by the isolated system? Consider an isolated system composed of a compressed gas at high pressure. Certainly if we remove the restriction of the system being isolated, we could conceive of a number of ways of using this high pressure to operate some form of engine to perform useful work. This stored energy could be tapped without altering the system's velocity, position, or gravitational field and, therefore, does not fall in the categories of kinetic or potential energy. Nevertheless, it is apparently an energy that is possessed by the isolated system, and it seems to depend on the state of the system. We call this stored energy *internal* energy, and we give it the symbol U. *Internal energy incorporates the microscopic forms which result from the molecular motions.* We examine the attributes of U in later chapters.

The total energy contained within the system boundary can be given the symbol E and is made up of the kinetic energy (KE), the potential energy (PE), and the internal energy U:

$$E = \text{KE} + \text{PE} + U = \tfrac{1}{2}m\mathbf{V}_{ref}^2 + mgZ + U \qquad (2.11)^\dagger$$

The intensive form for the total energy, or the specific total energy, is

$$e = \frac{E}{m} = \frac{\text{KE}}{m} + \frac{\text{PE}}{m} + \frac{U}{m} = \tfrac{1}{2}\mathbf{V}_{ref}^2 + gZ + u \qquad (2.12)$$

where u is the specific internal energy.

2.5 Energy Transfer

To change the value of the system energy E, we must examine a system that is not isolated from its surroundings. How can energy be changed for the system?

Note that energy crossing the system boundary could be categorized

---

† In USCS units, the factor g_c must be included as

$$E = \frac{mg\mathbf{V}_{ref}^2}{2g_c} + \frac{mgZ}{g_c} + U$$

in many ways. The categories historically chosen are used because they help in problem solving, and allow the engineer or scientist to easily tie the results of calculations to the performance of equipment such as turbines, compressors, and heat engines. Thus we use these categories here.

2.5.1   Work

It is useful to define *work* as an energy transfer across the system boundary in an organized form such that its sole use could be the raising of a weight. Such a definition encompasses the definitions used in mechanics, such as a force acting through a distance. In mathematical terms, the amount of work δW done in moving through a differential distance $d\mathbf{s}$ is

$$\delta W = \mathbf{F} \cdot d\mathbf{s} \tag{2.13}$$

where the boldface center dot denotes the dot product. The work done over a finite path between points s_1 and s_2 is

$$_1W_2 = \int_1^2 \mathbf{F} \cdot d\mathbf{s} \tag{2.14}$$

where $\mathbf{F}$ is the external force in the direction of $\mathbf{s}$ by the surroundings on the system. Additionally, however, the thermodynamic definition of work includes other phenomena; electricity flowing across the system boundary could be used to turn an electric motor and thus lift a weight. Electric energy thus is classified as work when it crosses the system boundary.

Note that unless the path is specified from the initial to the final state of the system, we are not able to calculate the work done. In other words, the work done in going between the initial and final states can have *any* value, depending on the path chosen. This is not surprising in the light of mechanics. Consider the frictional work done in pushing a sandpaper block from point A to point B; certainly the amount of work will depend on whether the path taken is direct or roundabout. Work is, therefore, called a *path function* in thermodynamics and an *inexact differential* quantity in mathematics (as discussed in Sec. 1.7.4). It is therefore denoted by δW. The value of work depends on not only the initial and final states of the system but also the path followed. Work is obviously not a property, since it cannot be specified from knowing only the state of the system.

Also note the sign convention for work relative to the system boundary. The sign convention adopted in this text is that work done *on* the system *by* the surroundings is positive, while work done *by* the system *on* the surroundings is negative. Thus, any energy crossing the system boundary into the system in the form of work has a positive sign. We follow this convention for all forms of energy that enter the system.

Also, the process used to compute the value of work is based on the force by the surroundings and is equal to the force within the system for a quasi-equilibrium process. Recall that a quasi-equilibrium process is an idealized process which occurs very slowly. If the particular application is the motion of a piston in a cylinder, then the quasi-equilibrium process is motion due to a force which is counterbalanced at every step of the process.

Finally the units for work must be specified. Work is an energy transfer, so the units are the same as for energy. In SI, this is a joule (J); in USCS, it is a pound force-foot (lbf · ft) or the British thermal unit (Btu).

P dV **Work**

Many problems of interest in thermodynamics are carried out in terms of P, v, and T rather than in terms of force and distance. So it is convenient to cast our mathematical relations for work in terms of variables P, v, and T. This is conveniently done by multiplying the numerator and denominator of the work equation by the cross-sectional area A_x normal to which the force is applied, to obtain

$$_1W_2 = + \int_1^2 \mathbf{F} \cdot d\mathbf{s} = + \int_1^2 \left[\frac{\mathbf{F}}{A_x} \cdot (d\mathbf{s}\, A_x) \right] \qquad (2.15)$$

Figure 2.5   Pressure vs. specific volume.

With the definition of pressure as the normal force per unit area and with $d\mathbf{s}$ denoted by dx which is in the direction of $\mathbf{F}$, the dot product in the integrand becomes $- P\, dV$ (since volume decreases as work is done *on* the system, so that the negative sign is required by our convention), and

$$_1W_2 = - \int_1^2 P\, dV \qquad (2.16)$$

If Eq. (2.16) is divided by mass, then the work per unit mass $_1w_2$ is

$$_1w_2 = - \int_1^2 P\, dv \qquad (2.17)$$

This pressure is the normal force of the system on the container, so an expansion is negative work by the system.

Work is expressed in terms of the system pressure times the negative of the change in volume, integrated over the P-v curve between the initial and final states of the system. The work done in going between the initial volume v_1 and the final volume v_2 is a line integral that has been reduced to an ordinary integral and is visualized by using a diagram of pressure versus specific volume (or volume). Such a plot is shown in Fig. 2.5. The negative of the area under the curve of pressure vs. specific volume is, according to Eq. (2.17), simply the work done per unit mass in going between the initial and final specific volume.

Example Problem 2.2

A gas has an initial volume of 0.3 m³ and expands to a final volume of 0.6 m³. The initial pressure of the gas is 0.1 MPa. Find the work done if the gas is held at a constant temperature of 300 K and the pressure between the initial and the final states of the system is (*a*) constant, (*b*) inversely proportional to volume, (*c*) given by the ideal gas relation.

System Diagram

Control mass

Solution

The state information is tabulated as follows:

State	P, MPa	V, m³	Process
1	0.1	0.3	
2a	—	0.6	$P = \text{constant}$
2b	—	0.6	$P = c/V$
2c	—	0.6	$P = mRT/V$

The processes from state 1 to the final state are denoted at the final state.

In each case, the work is given by Eq. (2.16). For (*a*), the pressure is constant at P_1, and we can simply write

$$_1W_2 = -\int_1^2 P\,dV = -P_1(V_2 - V_1)$$

$$= -0.1\ \text{MPa} \times (0.6\ \text{m}^3 - 0.3\ \text{m}^3)$$

$$= -0.03\ \text{MJ} = -30\ \text{kJ}$$

For (b), the pressure varies so that P cannot be removed from the integral, and we must write that $P = c/V$, where c is a proportionality constant. The work is then

$$_1W_2 = -\int_1^2 P\,dV = -\int_1^2 \frac{c}{V}\,dV = -c\ln\frac{V_2}{V_1} = -c\ln 2$$

Because both the pressure and volume are given at state 1, the value of c must be $c = P_1 V_1 = 30$ kJ, and the work done in (b) is then $_1W_2 = -30\ln 2$ kJ $= -20.79$ kJ.

For (c), the ideal gas relation may be written $P = mRT/V$, and the work equation becomes

$$_1W_2 = -\int_1^2 mRT\frac{dV}{V} = -mRT\ln\frac{V_2}{V_1} = -mRT\ln 2$$

Now, m can be determined because $m = P_1 V_1/(RT)$ in state 1; substituting, we find

$$_1W_2 = -P_1 V_1 \ln 2$$

which is the same result as for (b).

Comments

The work done in (a) is different from that found in (b) and (c) because the path was different even though the initial and final volumes were the same in all cases. The results for (b) and (c) are the same because the pressure dependence on volume for a constant-temperature ideal gas is the inverse proportionality assumed in (b). Note that the temperature of the ideal gas did not enter into the problem explicitly, nor did the mass of the system. Both were eliminated with the other given information. Also, since the system volume increased in all cases, work was done by the system on the surroundings and was therefore negative.

Example Problem 2.3

An ideal gas is contained within a cylinder with a movable piston as in Example Problem 2.2. Evaluate the work per unit mass by the gas for the path indicated in the System Diagram on the next page. Take P and T as known at states 1 and 3.

Solution

This problem is evaluated by two difference approaches.

Approach 1

The area under the path on a P-v diagram represents the work done by the system. The process on the P-T diagram is redrawn on a P-v diagram. The constant P process from 1 to 2 is again a horizontal line, but

System Diagram

State	Properties			Process
	P	T	v	
1	P_1	T_1	v_1	P = constant = P_1
2	P_1	T_3		T = constant = T_3
3	P_3	T_3	v_3	

the constant T process from 2 to 3 is given by $P = RT/v$. The work per unit mass is

$$_1w_3 = -\int_1^3 P\,dv = -\int_1^2 P\,dv - \int_2^3 P\,dv$$

$$= -P_1(v_2 - v_1) - \int_2^3 \frac{RT}{v}\,dv$$

$$= -P_1(v_2 - v_1) - RT_3 \ln \frac{v_3}{v_2}$$

This can be expressed as a function of the endpoints by noting that $v_2 = RT_2/P_2$; but $T_2 = T_3$ and $P_2 = P_1$, so $v_2 = RT_3/P_1$. Therefore, with $v = RT/P$,

$$_1w_3 = -P_1\left(\frac{RT_3}{P_1} - \frac{RT_1}{P_1}\right) - RT_3 \ln \frac{RT_3/P_3}{RT_3/P_1}$$

$$= -RT_3\left(1 - \frac{T_1}{T_3} + \ln \frac{P_1}{P_3}\right)$$

Approach 2

Integrate along the original path by expressing dv in terms of dT and dP. With $v = RT/P$,

$$dv = \left(\frac{\partial v}{\partial T}\right)_P dT + \left(\frac{\partial v}{\partial P}\right)_T dP$$

$$= \frac{R}{P} dT - \frac{RT}{P^2} dP$$

Then

$$_1W_3 = -\int_1^3 P\,dv = -\int_1^3 \left(R\,dT - \frac{RT}{P}\,dP\right)$$

Separating the integral into its elements yields

$$_1W_3 = -\int_1^2 \left(R\,dT - \frac{RT}{P}\,dP\right) - \int_2^3 \left(R\,dT - \frac{RT}{P}\,dP\right)$$

In the process from 1 to 2, $P = P_1$, and for the process from 2 to 3, $T = T_3$; thus

$$_1W_3 = -R\int_1^2 dT + RT_3 \int_2^3 \frac{dP}{P}$$

$$= -R(T_2 - T_1) + RT_3 \ln\frac{P_3}{P_2}$$

With $T_2 = T_3$ and $P_2 = P_1$,

$$_1W_3 = -RT_3\left(1 - \frac{T_1}{T_3} + \ln\frac{P_1}{P_3}\right)$$

Comments

The result is the same regardless of the approach. Work is a line integral, as indicated by the second approach; yet it can also be represented as an area integration on a P-v diagram. Also, for an ideal gas, constant-pressure work between states 1 and 2 results in $P_1(V_2 - V_1)$, and isothermal work between states 1 and 2 yields $RT_1 \ln (P_1/P_2)$. What is the value for constant-volume work? (It must be zero.)

Example Problem 2.4

A piston encloses a gas within a cylinder and is restrained by a linear spring as shown. The initial pressure and volume are 150 kPa and 0.001 m³, respectively. The spring touches the piston but exerts no force at the initial position. The gas is heated until the volume is tripled, and the pressure is 1000 kPa.

(a) Draw the P-V diagram for the process.

(b) Calculate the work done by the gas.

(c) What is the work done against the piston and spring?

Diagrams

Solution

The state and process information is given in the problem statement. The states are specified by pressure and volume, and the process is governed by the linear spring. Before the state and process table is given, the process is considered. The force relation for a linear spring is given as

$$F_{spring} = k_s d$$

where k_s is the spring constant and d is the distance displaced from the position of zero force. The pressure exerted *on* the gas by the piston and spring is the pressure exerted *by* the gas on the piston and spring (since the process is in quasi-equilibrium). Thus, the pressure in the gas is

$$P = P_{piston} + P_{spring}$$

where P_{piston} results from the mass and area of the piston, the gravitational acceleration, and atmospheric pressure. Since neither the piston mass nor g changes during the process, P_{piston} is a constant. The piston area is given as A, so the gas pressure is

$$P = P_{piston} + \frac{F_{spring}}{A} = P_{piston} + k_s \frac{d}{A}$$

Expressing the distance displaced in terms of the volume of the gas yields

$$d = \frac{V - V_1}{A}$$

since the spring exerts no force on the gas initially, that is, $F = k_s d = (k_s/A)(V - V_1) = 0$ when $V = V_1$. Therefore, the pressure-volume relation for the process is

$$P = P_{piston} + \frac{k_s}{A^2}(V - V_1)$$

The state and process chart is as follows:

State	P, kPa	V, m³	Process
1	150	0.001	$P = P_{piston} + \dfrac{k_s}{A^2}(V - V_1)$
2	1000	0.003	

The three questions are now considered.

(a) This straight-line process on a P-V diagram is indicated in the diagram. The slope of this line is k_s/A^2.

(b) The values for the constants in the pressure-volume relation can be obtained by either writing the equation of a straight line directly from the P-V diagram or, equivalently, using the state chart to obtain the numerical values. The pressure expression *at* state 1 is

$$P_1 = P_{piston} + \frac{k_s}{A^2}(0) = P_{piston} = 150 \text{ kPa}$$

The expression *at* state 2 is

$$P_2 = P_{piston} + \frac{k_s}{A^2}(V_2 - V_1)$$

$$= 150 + \frac{k_s}{A^2}(0.002 \text{ m}^3) = 1000 \text{ kPa}$$

Solving for k_s/A^2 yields

$$\frac{k_s}{A^2} = 425{,}000 \text{ kPa/m}^3 = 425{,}000 \text{ kN/m}^5$$

Thus, the presure-volume relation is

$$P = 150 + 425{,}000(V - 0.001) \qquad \text{kPa}$$

where V is in cubic meters.

The work is given by

$$_1W_2 = -\int_1^2 P \, dV$$

$$= -\int_1^2 \left[P_{piston} + \frac{k_s}{A^2}(V - V_1) \right] dV$$

$$_1W_2 = -P_{piston}(V_2 - V_1) - \frac{k_s}{2A^2}(V_2 - V_1)^2$$

$$= -150(0.002) - \frac{425,000}{2}(0.002)^2$$

$$= -0.3 - 0.85$$
$$= -1.15 \text{ kN} \cdot \text{m} = -1.15 \text{ kJ}$$

This is simply the area under the process on the P-V diagram (with a minus sign).

(c) The work required to raise the piston without the spring is

$$_1W_{2, piston} = -\int_1^2 P_{piston} \, dV = -P_{piston}(V_2 - V_1)$$

$$= -150(0.002) = -0.3 \text{ kJ}$$

The work done against the spring is

$$_1W_{2, spring} = -\int_1^2 P_{spring} \, dV = -\int_1^2 \frac{k_s}{A^2}(V - V_1) \, dV$$

$$= \frac{-k_s}{A^2}\frac{(V_2 - V_1)^2}{2} = -0.85 \text{ kJ}$$

These contributions sum to the total work done by the gas (-1.15 kJ). The spring contribution represents 73.9 percent of the total work.

Comments

The process is given by the linear spring equation which implies a linear relation between gas pressure and system volume. Note that no force is exerted by the spring initially. This fact specifies the location for the distance displaced d. If the spring had exerted a force initially, then $d = 0$ would be not at $V = V_1$ but at a system volume $V < V_1$. The pressure-volume relation is still linear, but the constants in the P-V relation would be different.

The work for the process is given as the area under the process on the P-V diagram. The spring work is represented by the triangular area above $P = P_1 = 150$ kPa. The rectangular area below $P = 150$ kPa represents the work done to displace the piston.

Polytropic Process

Many practical applications undergo processes described by a specific relationship between pressure and volume. The piston-cylinder type of problem within an internal combustion engine can often be described by an experimentally determined pressure-volume relation. A process that is particularly important is the *polytropic process*, which is described by

$$PV^n = \text{constant} \tag{2.18}$$

where n is a specified constant. Between state 1 and state 2, the polytropic process is also represented as

$$\frac{P_1}{P_2} = \left(\frac{V_2}{V_1}\right)^n \tag{2.19}$$

The $P\,dV$ work for a polytropic process is

$$_1W_2 = -\int_1^2 P\,dV = \begin{cases} \dfrac{P_1 V_1 - P_2 V_2}{1 - n} & \text{for } n \neq 1 \\[2ex] P_1 V_1 \ln \dfrac{V_1}{V_2} & \text{for } n = 1 \end{cases} \tag{2.20}$$

Generally, the value of n is known from other process information or is obtained empirically from data.

The process described by $n = 0$ is a constant-pressure one, and this process is a horizontal line on a P-V diagram. A constant-volume process is given by $n = \pm\infty$ and is a vertical line on a P-V diagram. Intermediate values of n, both positive and negative, are possible.

Example Problem 2.5

The combustion products in an internal combustion engine undergo an expansion process from an initial volume of 0.0002 m³ and initial pressure of 13 MPa. The final volume is 0.0013 m³. The experimental data for the polytropic process indicate that $n = 1.45$. Evaluate the work performed by the expanding gas.

Diagrams

Solution

The state and process information is given:

State	P, MPa	V, m³	Process
1	13	0.0002	$PV^{1.45} = \text{constant}$
2	—	0.0013	

The work for the process is given in Eq. (2.20) as

$$_1W_2 = \frac{P_1V_1 - P_2V_2}{1 - n}$$

The value of P_2 is obtained from the fact that the process is polytropic, and Eq. (2.19) yields

$$P_2 = P_1\left(\frac{V_1}{V_2}\right)^n = (13)\left(\frac{0.0002}{0.0013}\right)^{1.45} = 0.86 \text{ MPa}$$

The work is

$$_1W_2 = \frac{13(0.0002) - 0.86(0.0013)}{-0.45} = -3.29 \text{ kJ}$$

The work is performed by the gas, so the value is negative.

Comments

First note that the system is a control mass, so the problem could be considered on a specific volume basis, that is, $Pv^n = \text{constant}$ and $_1w_2 = (P_1v_1 - P_2v_2)/(1 - n)$. This is a convenient form for some problems. Also, the value of n is given in this problem, but an alternate problem statement could give P_1, V_1, P_2, and V_2 for the polytropic process. Then n could be evaluated from the P-V data.

Work on Solids

A force applied to a solid can alter the length of the solid. The normal stress acting to extend the solid is denoted by σ_z in the direction z. The work *on* the solid is

$$\delta W = \mathbf{F} \cdot d\mathbf{s} = \sigma_z A_0 \, dz \tag{2.21}$$

where A_0 is the initial cross-sectional area. No minus sign is required since the system (solid) restrains the motion; and if dz is positive (extension), work is done on the system. Alternatively this can be expressed in terms of the strain. The strain ε_z is the deformation per unit length $d\varepsilon_z = dz/L_0$, so

$$\delta W = \sigma_z A_0 L_0\, d\varepsilon_z = \sigma_z V_0\, d\varepsilon_z \tag{2.22}$$

where V_0 is the initial volume. This is integrated between two points to yield

$$_1W_2 = \int_1^2 \sigma_z V_0\, d\varepsilon_z \tag{2.23}$$

Example Problem 2.6

Calculate the work required to stretch an unstressed steel wire from 1 to 1.001 m. Young's modulus is $E = \sigma_z/\varepsilon_z = 2 \times 10^{11}$ N/m^2, and the cross-sectional area is $A_0 = 3 \times 10^{-6}$ m^2.

Solution
Equation (2.22) gives

$$\delta W = \sigma_z A_0 L_0\, d\varepsilon_z$$

Note that Young's modulus yields $\sigma_z = E\varepsilon_z$. Then

$$\delta W = E A_0 L_0 \varepsilon_z\, d\varepsilon_z$$

Integrating from the initial unstressed length to the final length gives

$$_1W_2 = \int_{\varepsilon_{z_1}=0}^{\varepsilon_{z_2}} E A_0 L_0 \varepsilon_z\, d\varepsilon_z$$

Since A_0 and L_0 represent the initial values and E is a constant,

$$_1W_2 = \frac{E A_0 L_0}{2}\, \varepsilon_{z_2}^2$$

Substituting in the numerical values gives

$$_1W_2 = \left[\frac{1}{2}(2 \times 10^{11})\frac{N}{m^2}\right](3 \times 10^{-6}\ m^2)(1\ m)\left(\frac{0.001}{1}\right)^2 = 0.3\ J$$

Work to Stretch a Film
This type of work arises in stretching a two-dimensional sheet such as a soap film on a wire frame. The force is the surface tension times the length of the line where the force is acting, L. The surface tension is denoted by σ N/m, so that the force is σL. Work performed *by* the system composed of the film is

$$\delta W = \sigma L\, dz \tag{2.24}$$

The sign results from exactly the same ideas as in the case of extending a solid. The change in the surface area of the film as it is being stretched is $dA = L\,dz$, so

$$\delta W = \sigma\,dA \qquad (2.25)$$

and the work *on* the film for a finite change is

$$_1W_2 = \int_1^2 \sigma\,dA \qquad (2.26)$$

Work by Electric Energy

We have indicated that a possible work mode is in the form of electric energy flowing across the system boundary. The potential difference **E** results in a transfer of charge $d\mathbf{Q}$. The process yields work by the system as

$$\delta W = \mathbf{E}\,d\mathbf{Q} \qquad (2.27)$$

for this quasi-equilibrium process. With the current i represented as $i = d\mathbf{Q}/dt$, where t is time,

$$\delta W = \mathbf{E}i\,dt \qquad (2.28)$$

so the work done is

$$_1W_2 = \int_1^2 \mathbf{E}i\,dt \qquad (2.29)$$

Example Problem 2.7

A battery is charged with a battery charger. The charger operates for 1 h at 12 V and a current of 25 A. Calculate the work done on the battery.

Solution

The work is given by Eq. (2.29) as

$$W = \int_1^2 \mathbf{E}i\,dt$$

The potential difference and current are taken to be constants for the charging period, so

$$_1W_2 = \mathbf{E}i(t_2 - t_1)$$

Noting that $1\text{ V} = 1\text{ kg} \cdot \text{m}^2/(\text{s}^2 \cdot \text{C})$ and $1\text{ A} = 1\text{ C/s}$, we see that

$$_1W_2 = [12\text{ kg} \cdot \text{m}^2/(\text{s}^2 \cdot \text{C})](25\text{ C/s})(1\text{ h})(3600\text{ s/h})$$
$$= 1.08 \times 10^{+6}\text{ J} = 1.08\text{ MJ}$$

Comment
The work is positive, indicating that the charger does work on the battery.

General Work Expressions

A number of mechanisms for energy transfer in the form of work have been discussed. This list is not complete, and other work modes exist for other quasi-equilibrium processes. The common feature of all the work modes is that the work is the product of a generalized force, which is an intensive property, and a generalized displacement, which is an extensive or specific intensive property. This can be represented mathematically as

$$\delta W_i = \mathbf{F}_i \, d\mathbf{X}_i \tag{2.30}$$

where δW_i = work mode i
$\mathbf{F}_i$ = generalized force i
$\mathbf{X}_i$ = generalized displacement i

A general system could have many possible work modes, so a general work expression is

$$\delta W_{\text{gen}} = \sum_i \delta W_i = \sum_i \mathbf{F}_i \, d\mathbf{X}_i$$
$$= -P \, dV + \sigma_x V \, d\varepsilon_x + \sigma \, dA + \mathbf{E} i \, dt + \cdots \tag{2.31}$$

2.5.2   Heat Transfer

Energy that may cross the system boundary because of a temperature difference falls into a separate category. When a can of frozen orange juice is placed in a warm room, experience shows us that the orange juice (the system) will increase in temperature until it approaches the temperature of the room. The state of the system is changing because energy from the surroundings is crossing the system boundary owing to the temperature difference between the room and the juice. This energy transfer is not work, since we cannot envision a way in which its sole effect is the raising of a weight. We thus define this disorganized form of energy that crosses the system boundary because of a temperature difference between the system and its surroundings as *heat transfer*. Heat transfer is given the symbol Q, and the heat transfer per unit mass is $q = Q/m$. The SI unit for heat transfer is, as for work, the joule; in USCS, it is the pound force-foot (lbf · ft), or, alternatively, the British thermal unit (Btu).

In this text, values for heat transfer in various problems are either specified in the problem statement or are an unknown to be calculated from the thermodynamics. The calculation of the amount of energy entering or leaving a system as it depends on various mechanisms and the temperature difference between the system and its surroundings is outside the realm of classical thermodynamics and is the subject of courses in heat transfer or transport processes.

An important distinction between heat transfer and work is the "disorganization" on the molecular level. Heat transfer is viewed as a disorganized energy transfer on a molecular level. This energy transfer results from molecular activity which is not directly useful to raise a weight. Work, however, is viewed as an organized energy transfer to raise a weight. This difference is important in understanding these energy transfers.

The sign convention for heat transfer is the same as for work; i.e., a heat transfer is positive *into* the system and negative *out of* the system. This can be remembered since a heat loss, being undesirable in many problems, is considered negative.

An important yet limiting specific process for heat transfer is the *adiabatic* process. The process permits no heat transfer. Practically, this occurs for a perfectly insulated system.

2.5.3   Power

To design practical devices, not only is it important to know how much energy is needed to drive a device or how much energy can be obtained from the device, but also we need to know the rate at which energy can be obtained. It may be possible to run a very small gasoline engine to move freight, but a much larger engine would allow the movement to be done more quickly. The rate at which energy is transferred is called the *power.* We denote the rate of various energy forms by placing a dot over the symbol; that is, the rate of doing work is

$$\dot{W} = \lim_{\Delta t \to 0} \frac{\delta W}{\Delta t} \tag{2.32}$$

and the rate of heat transfer across a boundary is

$$\dot{Q} = \lim_{\Delta t \to 0} \frac{\delta Q}{\Delta t} \tag{2.33}$$

The overdot implies the rate of transfer, not the time derivative of a system property. In SI, the unit of power is the watt (W), which is equal to an energy rate of 1 J/s.

2.6   Energy — What Is It?

To this point, we have used the term *energy* without giving it a careful definition. Energy can have many forms, and we are so used to working with the concept of energy that we have an intuitive feel for its meaning.

In mechanics, energy is often defined as the capacity for doing work. As we will see during our study of the second law of thermodynamics, such a definition is not adequate in thermodynamics.

What attributes have we discovered in our definitions of the particular energy forms of internal energy, kinetic energy, and potential energy and the energy transfers of work and heat transfer that might help with a more general definition? One thing that the energy transfers have in common is that each has the capacity to change the state of the system for which it is defined. If we do work on a system or have a heat transfer to a system, then the state of the system is changed (assuming that we do not add, subtract, or change more than one of the energy forms at a time so that their effects cancel).

Thus we can define *energy* as a system quantity that describes the thermodynamic state of the system and is changed by transfers to and from the system. Energy has many forms, and it can be used in any of these forms. In the remaining chapters of this text, we see how the properties of materials in a system are changed by the energy transfers; how energy can be transformed from one form to another; what the limitations are on these transformations; and how we can make use of the ability to transform energy to help us in designing useful devices.

Problems

2.1 A force of 3000 N is exerted uniformly over a plate of area 3 cm^2 at an angle of 30° to the normal. What is the pressure, in pascals, exerted on the plate?

2.2 The pressure of a gas in a tank is desired. A manometer is attached to the tank, and the level of the fluid in the manometer is measured. See Fig. P2.2. When the manometer is filled with mercury ($\rho = 13{,}550$ kg/m^3), the level reading is $h = 2$ m. Evaluate the absolute and gauge pressures in the tank. What would be the level reading if the fluid were water ($\rho = 1000$ kg/m^3)? Take the gravitational acceleration as 9.8 m/s^2.

Figure P2.2

2.3 A person starts an outing at sea level where the pressure is 0.101 MPa. Calculate the final pressure for each of the following excursions:

  (*a*) The person climbs to the top of Mt. Everest where the elevation is 8848 m (29,028 ft). Take the average air density between sea level and the mountain top to be 0.754 kg/m^3.

  (*b*) The person dives to the bottom of a sea where the depth is 395 m (1296 ft). Take the water density to be 1000 kg/m^3.

  (*c*) The person travels to the bottom of Death Valley

where the elevation is -86 m (-282 ft). Take the air density to be 1.30 kg/m³.
Neglect the fluid density variations and local gravity variations for each excursion.

2.4 A cylinder encloses a gas with a piston, as shown in Fig. P2.4. The area of the piston is 0.01 m². Take the atmospheric pressure to be 0.101 MPa and the local gravitational acceleration as 9.8 m/s². If the piston supports a mass of 50 kg (including the mass of the piston), what is the gas pressure? Will the gas pressure change if the gas volume is doubled?

2.5 A mass of 2 kg is moving at a speed of 3 m/s at a height of 10 m above a reference plane. What are the values of kinetic and potential energy of the mass, and what is the weight of the mass?

2.6 The measured pressure within a piston-cylinder arrangement diagramed in Prob. 2.4 is 0.5 MPa, and the piston area is 0.08 m². Determine the mass supported by the gas.

2.7 Find the work done, in kilojoules, by an ideal gas in going from state A to state C along the path shown in the P-V diagram in Fig. P2.7.

2.8 A 25-kg piston is required to travel a specific distance within the piston-cylinder arrangement shown in Fig. P2.8. The piston initially rests on the bottom stops. The gas within the cylinder is heated, and the piston moves until it touches the upper stops, as shown. The initial pressure and temperature are $P_1 = P_{atm} = 0.101$ MPa and $T_1 = 20°C$, respectively. (Neglect piston volume.)
(a) Draw the P-v diagram for the process.
(b) Calculate the work done by the gas.
(c) If the gas is an ideal gas ($PV = mRT$), determine the final temperature of the gas.

2.9 A piston travels between two stops and is restrained by a linear spring, as shown in Fig. P2.9. The cross-sectional area of the piston is 0.05 m², and the initial volume is 0.01 m³. At the top stops, the volume is 0.03 m³. The spring constant is 10 kN/m. The gas within the chamber is initially at 0.5 MPa, and the pressure required to just raise the piston from the stops is 1.0 MPa. The final pressure in the chamber is 6.5 MPa.
(a) Draw an accurate P-V diagram.
(b) Evaluate the work done by the gas.

2.10 A weightless and frictionless piston moves 10 cm against an ideal gas (i.e., a gas with properties that obey $Pv = RT$) contained in a cylinder that has an inside diameter of 15 cm. A spring is also within the cylinder, positioned as shown in Fig. P2.10. The spring exerts no force on the

Figure P2.4

Figure P2.7

T = constant

$P,$ kPa

V, m³

Figure P2.8

$A = 0.005m^2$

0.10 m

0.25 m

Neglect volume of stops

Figure P2.9

piston in the initial position but is touching the piston. The spring has a spring constant of 2×10^4 N/m. During the piston motion, the gas in the cylinder is maintained at 1 atm.

(a) How much work is done by the piston?

(b) What fraction of the work is done in compressing the gas?

(c) What fraction of the total work is done in the first 5 cm of piston travel?

2.11 The gas within a cylinder-piston arrangement undergoes a process with the gas pressure given by a polytropic process, i.e.,

$$PV^n = \text{constant}$$

where n is a specified constant. Evaluate the work done per unit mass between points A and B. [This problem seeks the derivation of Eq. (2.20).]

2.12 An aluminum rod 2 cm in diameter and 10 cm long is compressed under a load. What is the work done if the strain is 1 percent? The modulus of elasticity is 7×10^{10} N/m^2.

2.13 An ideal electric motor (100 percent conversion of electric to shaft work) operates through a lossless gear train to stretch a solid aluminum rod. The motor is operated at a power input of 1 kW for 15 s, and in that interval the rod is stretched so that its length changes by 2×10^{-1} m. The rod is 1 cm^2 in cross section and 9 m in length. What is the value of Young's modulus for the rod?

2.14 An alien fungus in the form of a thin sheet is found stretched across the jaws of a shop vise in the engineering shop. The fungus is holding itself in place by the surface tension force exerted by its contact with the vise jaws. A student finds that a force of 10 N must be applied to the end of the vise handle (with radius 0.3 m—see Fig. P2.14) to move the frictionless vise. She then turns the handle 24 times, causing the jaws to move 10 cm farther apart than they were initially. The vise jaws are 0.2 m long. What is the surface tension of the alien fungus?

2.15 An electric wet storage battery is used for powering the lights of a sailboat during a night race. The lights consist of three 10-W bulbs for the running lights plus a 20-W mast light. The race will last 6 h (if the present winds do not change). A battery trickle charger is available that provides a constant charge rate of 2 A to the 12-V battery while tied up at the dock. Assume that the charge-discharge cycle of the battery is 100 percent efficient (i.e., all energy stored can be recovered).

$P_{\text{surr}} = 0$

15 cm

Figure P2.10

Figure P2.14

0.2 m

0.3 m

$F = 10$ N

(a) For how long should the trickle charger be used to ensure lights for the entire race?

(b) How much will the energy stored in the battery change during the charging process?

2.16 An elevator has a mass of 1000 kg. It is at the top of a 60-m-high shaft when the cable snaps. The air in the elevator shaft is at 27°C and remains at that temperature as the elevator car falls. None of the air leaks past the car during its descent. The shaft is 2.5 m² in cross section, and the air in the shaft is initially at 1 atm. See Fig. P2.16.

(a) Sketch the P-v diagram for the air in the shaft, labeling the initial and final states with numerical values.

(b) What will the equilibrium position be of the elevator car?

(c) How much work is done by the elevator car in reaching the equilibrium position?

(d) Are any assumptions necessary to reach your answer other than those given?

Figure P2.16

2.17 An ideal frictionless isothermal compressor takes air at a pressure of 1 atm and compresses it to 10 atm at a temperature of 30°C. Each revolution of the compressor handles 1 kg of air, and the compressor operates at 200 rpm. What minimum-size electric motor should be ordered to run the compressor?

2.18 A 12-V storage battery for an aircraft is completely discharged. It is then placed on a quick-charger, and 10 A is fed to the battery for 1 h. The battery, which has a mass of 10 kg, is then loaded aboard an aircraft for delivery. The aircraft flies at 700 km/h at an altitude of 2000 m.

(a) What is the change in the total energy E of the battery from its discharged state on the ground to its state aboard the aircraft during flight?

(b) What percentage of the energy change is due to the various energy forms (kinetic, potential, and stored)?

(c) How much work was done during the charging process?

2.19 The total differential of P dV work could be written as $\delta W = -P\, dV = -P\, dV + 0\, dP$. Show that work is an inexact differential.

2.20 A 1-kg mass is moved from the floor to a position on a staircase 10 m above the floor. The new position is reached by the following paths:

(a) The mass is lifted directly upward by a frictionless pulley in a block-and-tackle system.

(b) The mass is moved on a frictionless dolly to an elevator, which is also frictionless. The elevator moves to the height necessary, and then the dolly is used to move the mass back to the final position.

In these two cases, find the work done in moving the mass to the final position. Discuss whether the work, which is a path function, depended on the path in these two cases, and if not, what characteristics were present that made work independent of path.

3

Properties of Common Substances

When you can measure what you are speaking about, and express it in numbers, you know something about it; but when you cannot measure it, when you cannot express it in numbers, your knowledge is of a meager and unsatisfactory kind: It may be the beginning of knowledge, but you have scarcely, in your thoughts, advanced to the stage of science.
Lord Kelvin

Cross-section of a 3600-rpm, tandem-compound, two-flow, reheat steam turbine, 300,000- to 350,000-kW rating range. *(General Electric.)*

3.1 Introduction

In Chap. 2, certain common properties were defined so that the concepts of state, process, and energy could be discussed. However, other properties of a system are useful in thermodynamics in addition to those defined in Chap. 2. In this chapter, some new properties are introduced, and the behavior of various properties is discussed for a particular class of substances. Still other properties are introduced in later chapters as the need arises; however, the concepts discussed here apply to all other properties.

3.2 State Postulate — Applications to Property Relations

In Chap. 2, we noted that properties could be classified as either *intensive* or *extensive*. Further, properties are defined only for a system that is in thermodynamic equilibrium. Study of a large number of substances and their behaviors during different processes reveals that the more substances present in the system and the more ways that energy can be exchanged between the system and its surroundings, the more properties are necessary to describe the stable equilibrium state of the system.

We limit discussion in this chapter to *pure, simple, compressible substances.* The words *simple compressible* imply that the only work mode considered is the $P\,dV$ work form. *Pure* implies substances composed of a single chemical species (such as H_2O, ammonia, and nitrogen), and so we exclude all mixtures of two or more substances (such as salt water, whiskey, and humid air). The exception is air, which is a nonreacting mixture of N_2, O_2, and other molecules and can be treated in many practical cases as a pure substance. In addition, the substances are taken to be of fixed molecular composition throughout our study; that is, no chemical reactions are taking place. In later chapters, we study mixtures and systems in which the species can undergo chemical transformations.

For a pure simple compressible substance, repeated observations and experiments show that two independent properties are necessary and sufficient to establish the stable equilibrium state of a system. This observation is not derivable from more basic theorems using classical approaches to thermodynamics; but no violations have been found, and it can be accorded the status of the other thermodynamics laws studied later.

The observed behavior of a pure simple compressible substance is summarized in the *state postulate:*

> Any two independent thermodynamic properties are sufficient to establish the stable thermodynamic state of a control mass composed of a pure simple compressible substance.

Thermodynamic properties are properties such as pressure, temperature, and volume, not geometric properties such as shape or elevation and not velocity. *Stable state* means an equilibrium state, so that the state postulate gives no information concerning nonequilibrium states. The above statement is for a control mass or a specified amount of matter. The state postulate can also be interpreted on a per unit mass basis or on an intensive property basis. Recalling that extensive properties can be expressed in an alternate intensive form, we see that the state postulate can be stated in a very useful form:

> Any two independent, intensive, thermodynamic properties are sufficient to establish the stable thermodynamic state of a pure simple compressible substance.

There are a number of important exceptions to these forms of the state postulate. These statements apply only to simple compressible substances, so the only work mode possible is $P\ dV$. If there are other quasi-equilibrium work modes, then the number of independent thermodynamic properties needed to determine the thermodynamic state is 1 plus the number of work modes. Also, for a control mass that contains more than one substance or that has chemically reacting components, we expect it is necessary to specify more than two independent properties to establish the state. This is explored more fully in Chaps. 9 and 10.

The state postulate indicates the information needed to specify the state of the substance in a particular system. The properties considered thus far are P, v, T, and u, and more are presented in later sections. The state postulate says that if two of these properties are independent and known, then all others are uniquely specified. If T and v are known for a pure compressible substance, then P and u have unique known values. Mathematically this is given as

$$P = P(T, v) \qquad (3.1)$$

and

$$u = u(T, v) \qquad (3.2)$$

The choice of the independent variables in the expressions is arbitrary as long as they are independent, so that equivalent alternate statements are

$$P = P(u, T) \qquad \text{or} \qquad P = P(u, v) \qquad (3.3)$$

and

$$u = u(P, T) \qquad \text{or} \qquad u = u(P, v) \qquad (3.4)$$

Since thermodynamics deals with energy and its transformation, Eqs. (3.2) and (3.4) are very important for thermodynamic analysis.

3.3 Simple Compressible Substances

Simple compressible substances are used in many engineering systems including power plants, many refrigeration systems, and thermal distribution systems using water or steam to transport energy. In addition, internal and external combustion engines can be usefully studied by assuming that they operate with simple compressible substances as the working fluid, even though in practice this is not the case. Finally, some nonreacting mixtures of pure substances, for example, dry air, can be treated as pure substances with little error, which permits considerable practical extension and application of the property relations that we develop for pure substances.

3.3.1 Liquid Phases

Suppose that, having set the values of the two independent variables, we wish, for a particular pure simple substance, to determine the *value* of a dependent variable. How could this be done, and how could the results be best presented?

As an illustration, we take a very familiar substance, cold water, and place it in a cylinder. Then we remove all air from the water (remember, we are limited to a pure substance), and we place a piston in contact with the water surface (Fig. 3.1). The water is considered to be the control mass, and we can measure the pressure in the control mass with a pressure gauge and use a thermometer to measure the control mass temperature. We have thus set the two independent properties—temperature and specific volume (the latter is found by measuring the total volume contained by the cylinder-piston arrangement and dividing by the mass of the water placed in the system). The dependent property (pressure) that results at the set values of specific volume and temperature is found from the pressure gauge. All three properties, P, v, and T, are thus measurable in this state.

Suppose now that we want to change the state of the control mass. We could perform many types of processes. For example, let us submerge the piston-cylinder arrangement within a large bath of fluid at a fixed temperature $T_1 = 100°C$ that is equal to the control mass temperature. The water and surroundings will be in thermal equilibrium at the temperature of these surroundings. We impose an initial force on the piston sufficient to give the desired specific volume of v_A. Now we increase the volume of the control mass by moving the piston upward, and we measure the resulting pressure by reading the gauge attached to the cylinder wall. If we take care to increase the volume slowly, the water will stay at constant temperature. If the control mass temperature should change infinitesimally during the process, there will be a heat transfer from the control mass to or from the surrounding bath, and the control mass will come back to thermal equilibrium with the bath at the bath temperature.

Figure 3.1 **Experimental apparatus for property measurement.**

The independent variables during this process are T and v, and we are measuring the value of the dependent variable P that exists at the given T and v. Thus, we are determining the states given by $P = P(T, v)$. If we plot the path of this process on a graph of pressure versus volume, it will appear as shown in Fig. 3.2, where A denotes the initial state and B the final state and the process between the states proceeds along an *isotherm,* or constant-temperature line, which passes through a series of states between A and B that are each in thermodynamic equilibrium. Such a process that passes through a series of equilibrium states has been defined as a *quasi-equilibrium* process.

Note in Fig. 3.2 the very great change in water pressure for a very small change in specific volume. The explanation is that liquid water is very nearly *incompressible,* and it is often assumed for convenience that the value of v is constant for liquid water (and most other liquids) as the pressure is changed at constant temperature.

3.3.2 Saturation and Phases

As the volume of the system is increased further, we find that after the control mass reaches a particular volume corresponding to state C (Fig. 3.3), *the pressure of the control mass remains constant.* We can change the volume by moving the piston, but the piston can be stopped at a position corresponding to any volume along the line CD, and the measured pressure remains at the same value found at point C; that is, the pressure in the control mass is the same at each state along CD. Observation of the water in the control mass at some point between C and D shows that some of the liquid present at point C has changed into vapor, thus filling the increased volume. As the volume is further increased, more of the liquid is con-

Figure 3.2 Constant-temperature expansion of water.

Figure 3.3 Isothermal expansion in the two-phase region.

verted to vapor, until point D is reached, where all the liquid has just disappeared and all the water is in the form of vapor.

Note that, for water, the specific volume increases dramatically during the vaporization process, and we are now forced to plot the specific volume on a logarithmic scale to keep sufficient detail at small specific volumes. The isothermal expansion of the liquid, path A-B-C, is seen to be very nearly vertical on this scale.

The physical state or form of the control mass at point C is referred to as being a *saturated liquid;* and the water temperature at this state is the *boiling point,* or *saturation temperature,* at the control mass pressure. The water in states along line CD is a mixture of two physically distinct forms, liquid and vapor. These physically distinct forms for a component are called *phases,* and states along line CD are referred to as *two-phase* states. Because the control mass has given P, v, and T values at each state along line CD, each intensive property (P and T) is the same for each phase. However, the specific intensive property v, though known for the entire two-phase mixture (the volume over the mass), is *different* for each of the two phases. The specific volume of the liquid phase can be found by measurement. We can remove a small sample of the liquid phase at some state between C and D and measure the volume and mass of the sample while holding its temperature at the boiling point for that pressure; a similar measurement can be made on the vapor phase.

Let us think about these observations in light of the state postulate that two independent properties must be specified in order to define the state of the system. Along line CD, if the pressure is specified, the temperature is automatically known; i.e., the boiling point is dependent on only the pressure of the control mass and cannot be varied independently. Thus, *along the two-phase line CD, T and P are not independent properties.* From the set of P, v, and T, then, the only independent property sets that can be used to specify the state of a two-phase system are the sets P and v or T and v.

3.3.3 Quality

Each of the two phases is at the boiling point at the existing control mass pressure at every state along line CD. Thus, if we measure the specific volume of the liquid phase at *any* state along line CD, we find the same value: the specific volume of saturated water at the control mass temperature, $v_l(T)$, is equal to v_C. Similarly, the specific volume of the vapor at any state along line CD is found to be the specific volume of the saturated vapor, $v_g(T) = v_D$. The subscripts l and g denote the liquid and vapor phases, respectively. [Some references use the subscript f (fluid) for the liquid state. Because the word *fluid* is commonly used for both the liquid and vapor states, we use the subscript l for the liquid state throughout this text.]

It is relatively simple to tabulate the specific volume of saturated liquid and saturated vapor as a function of temperature. It would thus be useful to define the specific volume v of the total mass defined by the control mass at any point along the two-phase line CD in terms of the specific volumes of the individual phases v_l and v_g. To do this, we introduce a property called the quality of the two-phase mixture. The *quality* is defined as the mass of the vapor present in the two-phase mixture divided by the total mass, and it is given the symbol x. Thus

$$x = \frac{m_g}{m} \tag{3.5}$$

For a two-phase control mass, then, the quality ranges from 0 for a control mass composed of only saturated liquid to 1.0 for a control mass composed of only saturated vapor. Quality is also often expressed as a percentage. Note that quality is undefined except for a two-phase mixture of vapor and liquid.

The system volume along the two-phase line CD is

$$V = V_l + V_g \tag{3.6}$$

Dividing by the total mass gives

$$\frac{V}{m} = v = \frac{V_l}{m} + \frac{V_g}{m}$$

$$= \frac{m_l}{m}\frac{V_l}{m_l} + \frac{m_g}{m}\frac{V_g}{m_g}$$

or

$$v = (1 - x)v_l + xv_g \tag{3.7a}$$

and the specific volume of the control mass is specified in terms of tabulated properties and the quality. An alternate way of writing Eq. (3.7a) is

$$v = v_l + x(v_g - v_l) = v_l + xv_{lg} \tag{3.7b}$$

where v_{lg} is defined as the difference between the specific volumes of the saturated vapor and the saturated liquid at the control mass temperature, or $v_g - v_l$.

Because the specific volume and, as we shall see, other specific properties of the control mass are determined once x is specified at a given P or T, it is common practice to use x and one other independent property to specify the state along the two-phase line.

3.3.4 Superheated Vapor

Let us return to Fig. 3.3. Here the process of constant temperature expansion was halted at state D, where the control mass had just been completely changed to the vapor phase. Suppose that we now continue the expansion by further withdrawing the piston so that the volume of the

Figure 3.4 **Expansion into the superheat region.**

control mass is increased while the temperature is held constant by the surrounding bath. As we do so, we find that the pressure in the system decreases, and the *P-v* trace continues along path *DE* (Fig. 3.4). In the *DE* series of states, the vapor is referred to as being *superheated,* because the temperature is above the boiling point at any pressure that exists along the path. The volume can continuously be increased past v_E, and the pressure will continue to fall. The process can be continued until the restrictions on our definition of pressure or specific volume [see Eqs. (2.1) and (2.8)] are no longer valid.

In these states note that P and T are independent. Thus, unlike the saturation states, P and T can be used as two independent properties to specify the state of a superheated vapor.

3.3.5   *P-v* **Diagram**

Suppose that we now return the piston of the experimental apparatus of Fig. 3.1 to its original position, and we change the temperature of the surrounding bath to some higher value, $T_2 > T_1$. By moving the piston slowly to increase the volume of the control mass while holding the control mass temperature constant at T_2, a new isotherm is found on the *P-v* diagram. This procedure can be repeated at higher and higher temperatures, and the isotherms for each temperature can be drawn, as in Fig. 3.5.

Critical Temperature
At higher temperatures, the length of the horizontal two-phase line decreases. Finally, for water, the isotherm in the two-phase region corre-

Figure 3.5 **Isotherms for expansion at various temperatures.**

sponding to a temperature $T = 374.14°C$ has zero length. That is, no discernible two-phase region exists, and as the volume is increased, the water undergoes a transition from liquid to vapor without the appearance of a distinct two-phase region. Experimentally, if we watch the original liquid held at or above this temperature while we expand its volume, we cannot tell when the liquid becomes a vapor, because no interface is formed between the phases — no liquid surface can be seen. This phenomenon occurs at such a high pressure and temperature for water that it is outside our everyday experience. At the high pressures present in the region of the P-v graph where this occurs, as the specific volume of the liquid is increased along an isotherm, the pressure of the control mass decreases continuously. The liquid specific volume increases, and the liquid transitions smoothly into a gas of low specific volume, with no separation of the phases being visible, which we can envision because both "phases" have the same specific volume (and thus density) and are thus indistinguishable from each other. The same phenomenon occurs for all isotherms above the critical temperature.

The lowest temperature for which the isotherm shows no phase transition is called the *critical temperature* T_{cr}. Values of the critical temperature for various substances are given in Table C.4.

It is common to call a vapor that is above the critical temperature a *gas* and to retain the word *vapor* for the phase that could exist in equilibrium with its liquid phase at the vapor temperature if the pressure were increased sufficiently.

Figure 3.6 The *P-v* liquid-vapor phase diagram for water.

P-v **Liquid-Vapor Phase Diagram**

If all the points on the various isotherms that show the onset of the two-phase region are joined and all the points corresponding to the appearance of 100 percent saturated vapor are joined, then the resulting diagram is shown on logarithmic coordinates in Fig. 3.6. The line joining saturated liquid points joins smoothly to the saturated vapor line at the point where the critical isotherm touches both lines.

Within the dome-shaped region are all states that contain a two-phase mixture of saturated liquid and saturated vapor. This part of the diagram is referred to as the *two-phase,* or *saturation, region.* Only within this region can the property of quality be defined and used.

To the left of the dome are all states in which the control mass exists in only the liquid phase. In this latter region, the liquid is at a temperature below its boiling point at any *P-v* state, and states in this region are referred to as *subcooled.* (Such a definition has no meaning above the critical isotherm, since the phenomenon of boiling cannot be observed for supercritical temperatures.)

The region to the right of the dome is called the *superheat* region.

The *critical point* is the point at which the critical isotherm is tangent to the saturation curve. The *critical pressure* is the pressure at this tangent point (values are found in Table C.4). The region above the critical isotherm and critical pressure line (critical isobar) is called the *supercritical region.*

P-v-T **Surface**

It is useful to plot the data developed for water or other simple substances on *P*, *v*, *T* coordinates rather than on the *P-v* diagram. If this is done, a *surface* of equilibrium *P-v-T* states results. For water, the portion of the surface that we have described so far appears as in Fig. 3.7. The projection of the surface onto the *P-v* plane results in the *P-v* diagram shown in Fig. 3.6.

Projection onto the *P-T* **and** *T-v* **Surfaces**

The *P-v-T* surface can be projected onto the other coordinate planes as well as onto the *P-v* plane. The phase diagram on these coordinates often provides useful insight. Examination of Fig. 3.7 shows, for example, that the region of the *P-v-T* surface examined so far would project onto the *P-T* surface as shown in Fig. 3.8, and the *T-v* surface would appear as shown in Fig. 3.9. Both diagrams are useful in presenting experimental data, as we see in the next section. Note that on all these diagrams, *T* is plotted on a linear scale, while *P* and *v* are plotted logarithmically.

Other Regions on the *P-v-T* **Diagram**

We have directed attention so far to states in which only liquid, gas, vapor, or a combination of liquid and vapor phases exists. Such a treatment, of course, ignores the existence of the other common phase of pure simple substances — the solid phase. Every simple substance will solidify if the temperature is lowered sufficiently. (However, some substances

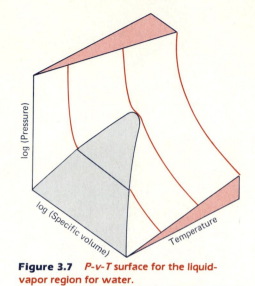

Figure 3.7 *P-v-T* surface for the liquid-vapor region for water.

T, °C

Figure 3.8 *P-T* diagram for vapor-liquid equilibrium of water.

Figure 3.9 *T-v* diagram for vapor-liquid equilibrium of water.

have solidification temperatures that are extremely low; helium, for example, freezes at temperatures below 1 K unless the control mass is at a pressure above 2 MPa.)

Let us look at the *T-v* diagram of Fig. 3.9. If we choose a new experiment as shown in Fig. 3.10, in which a fixed control mass is held at constant pressure by a weighted piston that is allowed to move, we could vary the control mass temperature by placing the experimental cylinder into a series of constant-temperature baths. Then the resulting control mass volume could be measured at each P and T by noting the resulting piston displacement. We are thus carrying out a *constant-pressure,* or *isobaric process* in which the independent variables are pressure and temperature and the dependent variable is specific volume. Thus, we are determining $v = v(T, P)$. Such a process is plotted as a horizontal line on the *P-v* diagram of Fig. 3.6. Let us choose a pressure and temperature combination that corresponds to a state in the superheated vapor region, state *A* of Fig. 3.11 ($P = 0.1$ MPa and $T = 200°$C). If we continuously lower the control mass temperature by placing the piston-cylinder arrangement containing it in a series of baths with each successive bath having a slightly lower temperature than the previous one, then path *A-B-C-D* is traced. Point *B* occurs at the saturated vapor specific volume corresponding to our constant pressure of 0.1 MPa; point *C* corresponds

Figure 3.10 Apparatus for constant-pressure experiment.

Figure 3.11 Constant-pressure process on the *T-v* diagram.

to the saturated liquid specific volume; and point *D* corresponds to a point somewhere in the subcooled liquid region.

Suppose we continue to lower the temperature of the control mass (Fig. 3.12, where the scales have been greatly expanded). At some temperature (point *E*), the system undergoes a transition to the solid phase (i.e., it freezes). This occurs at a particular temperature dependent on the pressure chosen for the experiment. Note a curious phenomenon: The freezing process occurs at constant temperature, but as the control mass freezes, the specific volume *increases,* until the control mass is completely solid at point *F*. This is not the case for all materials; it is a consequence of having chosen water for the experimental substance. (If this volume did not increase during freezing, ice would not float, since its specific volume would be less than that of the water in which it is floating!) Most substances experience a *decrease* in specific volume during the freezing process, and water is a curious exception.

If the control mass temperature is lowered still further, the specific volume changes only slightly (the specific volume of most solids is a very weak function of both pressure and temperature).

If we repeat the experiment with various initial pressures, a complete *T-v* diagram can be constructed in a manner similar to that used for the *P-v* diagram. The regions are shown and labeled in Fig. 3.13.

Figure 3.12 *T-v* diagram for the freezing of water.

Figure 3.13 *T-v* phase diagram for water.

Solid-Vapor Region

In Fig. 3.13, a large region of states exists in which ice and vapor coexist. This is generally the case below $0.01\,°C$. If large enough pressures are applied so that the specific volume of the control mass is reduced to 0.001091 m³/kg, then only the solid phase exists.

Sublimation Line

At extremely large specific volumes (which correspond to very low pressures) and low temperatures, it is possible during an expansion for a control mass to transform from a solid to a vapor without the appearance of the liquid phase. This process is called *sublimation*. The *sublimation line* marked on Fig. 3.13 indicates the boundary between the solid-vapor and superheated vapor states where this transition can occur. For example, a constant-volume heating process (a vertical line on the *T-v* diagram) at $v = 600$ m³/kg causes the control mass to undergo sublimation. The reverse of this process is responsible for the appearance of frost on cold surfaces as water vapor in the air (at very low effective water vapor pressures) deposits as solid ice on an appropriate cold surface without passing through the liquid phase. Another commonly observed example is the vaporization (sublimation) of solid carbon dioxide (dry ice), which passes directly from the solid to the vapor state as a result of heat transfer from the environment at commonly observed conditions of room pressure.

Triple Point

Note that a horizontal line (constant temperature) joins the saturation region and the solid-gas region in Fig. 3.13. This indicates that only at this particular temperature can the liquid, solid, and vapor phases coexist. This temperature is called the *triple-point temperature,* and it is 0.01°C for water. It is a convenient calibration point for thermometers. Note that this discussion is limited to a pure simple substance. However, a laboratory ice bath used for temperature calibration is usually open to the atmosphere, and the liquid-solid mixture will be saturated with dissolved air, as well as exposed to humid air rather than the vapor phase of pure water. Such an air-saturated system will not be at the triple point. The *ice point,* or temperature of an *air-saturated* ice-liquid bath in equilibrium with saturated air at standard conditions, is 0.0°C. The conditions needed to exactly maintain the ice point are more difficult to achieve in the laboratory than for a carefully constructed triple-point cell; however, for less accurate calibration, an ice bath is much more convenient to produce.

The values of the triple-point properties for various pure substances are given in Table C.1.

Figure 3.14 *P-v-T* phase diagram for a substance that expands upon freezing.

Complete *P-v-T* Surface

If the entire state surface, including solid, liquid, and vapor phases, is plotted on *P, v, T* coordinates for a substance such as water which expands upon freezing, then the diagram appears as in Fig. 3.14.

The solid phase of water can actually exist in various physically distinct forms, definable by their differing crystal structures. These structures also fall under our definition of phase, and the regions of their existence can be shown in the various diagrams.

It should not be inferred that most substances expand upon freezing. In fact, the opposite is true. For substances that contract upon freezing, the *P-v-T* phase diagram appears as shown in Fig. 3.15.

3.4   Other Thermodynamic Properties
3.4.1   Internal Energy and Enthalpy

Temperature, pressure, and specific volume are, of course, not the only properties of interest in the study of thermodynamics. For example, the property of internal energy, given the symbol U, was introduced in Chap. 2. The specific internal energy $u = U/m$ can be found for any simple compressible substance as a function of any two independent variables according to the state postulate. We could thus construct, for example, a u-P-T surface by the methods used to construct the P-v-T surface, assuming that we can invent a suitable method of measuring or determining u as a function of P and T. Then phase diagrams based on this surface could be constructed.

In many problems in thermodynamics, the group of properties $u + Pv$ occurs repeatedly. To evaluate this group at a given state, values of u and v must be found from some data source. For convenience, we can define a *new* property, called the *enthalpy* and given the symbol h, as

$$h \equiv u + Pv \tag{3.8}$$

Figure 3.15 *P-v-T* diagram for a substance that contracts during the freezing process.

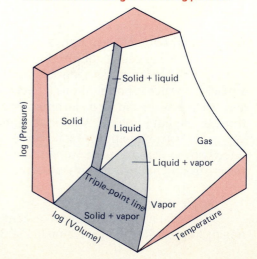

Because all the properties on the right-hand side of Eq. (3.8) can be found as a function of two independent variables, it follows that h can be tabulated as a function of the same two variables. Thus, we can evaluate and tabulate h directly at any given state, saving a great deal of effort in retrieving property data for the problems where $u + Pv$ appears.

Any specific extensive property or intensive property such as h and u can be evaluated for a state in the saturation region by relating it to the saturation properties through the quality, as was done for specific volume in Eq. (3.7b). Thus, for *any* intensive property z, we can write

$$z = z_l + x z_{lg} \tag{3.9}$$

3.4.2   Specific Heats

Suppose we wish to find the change in internal energy for a substance undergoing a process. By using Eq. (3.2), the change in internal energy is expressed [see Eq. (1.15)] as

$$du = \left(\frac{\partial u}{\partial T}\right)_v dT + \left(\frac{\partial u}{\partial v}\right)_T dv \tag{3.10}$$

Complete specification of the state is assumed by the state postulate if T and v are independent. Any two independent thermodynamic properties could have been used in Eq. (3.10), yet T and v are two very useful measurable properties. The partial derivatives in Eq. (3.10) are needed and will be important to thermodynamic analysis. The first partial derivative is particularly important and is considered here. The change in internal energy per degree of temperature change for a constant-volume process is this first partial derivative and is called the *specific heat at constant volume* (or sometimes the *specific heat capacity* at constant volume). It is given the symbol c_v and, from the above discussion, is defined mathematically as

$$c_v \equiv \left(\frac{\partial u}{\partial T}\right)_v \tag{3.11}$$

Values of c_v range from about 0.31 for argon gas to about 4.18 kJ/(kg · K) for liquid water. The values of c_v depend on the temperature of the control mass. Because the specific heat is a specific property, the property on a mass basis can be written as

$$C_v = m c_v \tag{3.12}$$

If the change in enthalpy is required, then this is considered in the same manner as the internal energy. Take the two independent variables to be P and T (this choice will be quite useful), so that

$$h = h(T, P) \tag{3.13}$$

and the change in the enthalpy is

$$dh = \left(\frac{\partial h}{\partial T}\right)_P dT + \left(\frac{\partial h}{\partial P}\right)_T dP \tag{3.14}$$

For a constant-pressure process, the first partial derivative is significant. This property is the *specific heat at constant pressure,* defined as

$$c_P = \frac{C_P}{m} \equiv \left(\frac{\partial h}{\partial T} \right)_P \tag{3.15}$$

The two specific heats defined in Eqs. (3.11) and (3.15) will prove quite useful when we study the energy flows for various thermodynamic processes. For the moment, however, note that if u or h data as a function of T at given v or P are known, then c_v and c_P can be readily calculated from the slopes of the data curves, and no additional data are needed.

3.5 Development of Property Data

Each pure substance has its distinctive P, v, T, u, h, c_v, and c_P data, and these data must usually be determined experimentally. It is possible to generate the values of all properties from some minimum set of experimental measurements, and the methods of doing this are addressed in Chap. 8. Also, the properties of some substances that have simple microscopic structure can be determined for some regions of the phase diagram without the use of direct experiment if some auxiliary spectroscopic data are available. This can be achieved by application of the methods of statistical thermodynamics (Chap. 12). Nevertheless, for most substances, careful and detailed experimental measurements are necessary to obtain the values of the minimum set of dependent properties as a function of a chosen set of independent properties.

Presentation of the experimental data relating all the properties in all regions of interest for each substance is obviously a difficult task. We need to compress as much data as possible into a small space. Many methods have been tried, and some of the most useful are described here. Others are introduced as properties are needed.

3.5.1 Graphical Data Presentation

It is very useful to be able to visualize thermodynamic processes on coordinates of various pairs of thermodynamic properties. We have done so on the P-v diagram in developing the background and definitions in this chapter. Graphical representation of processes on the phase diagrams plotted in terms of other property pairs enables visualization and analysis of many practical systems such as engines.

One very useful phase diagram can be plotted on coordinates of pressure versus enthalpy. An example appears in Fig. 3.16. On the P-h diagram, we can immediately see the change in enthalpy that occurs during any constant-pressure process, since the change in h then appears as the length of a horizontal line. Such insights are very helpful when we begin the study of thermodynamic processes and cycles. Diagrams on

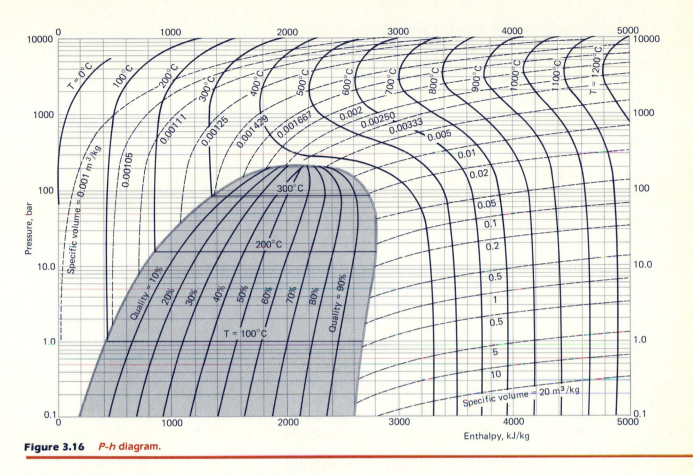

Figure 3.16   *P-h* diagram.

other property coordinates also prove useful and are discussed as other properties are introduced.

3.5.2   Equation of State

It is possible to develop mathematical relations among *P-v-T* data or other property sets for a particular substance. Such relations can be simple curve fits to experimental points or sophisticated relations based on knowledge developed from theory of the expected variation of a dependent property as a function of the independent properties. Such relations are called *equations of state*. When they are particularly simple and have good accuracy, they provide a convenient way to present *P-v-T* or other property data.

The *steam tables* compiled by the American Society of Mechanical Engineers [1] and others [2,3] are based on empirical curve fits to voluminous experimental data. The curve fits were used to generate table entries at discrete values of temperature and pressure. Computerized steam tables are generally based on the use of these equations rather than on the raw data themselves.

Simplified equations of state can often be used to describe property behavior over some limited region of states. However, the student *must* be aware of the region of validity and not use the simplified relations outside that region.

Ideal Gas Relation

Probably the most famous and the most useful equation of state is the *ideal gas relation,* which relates P, V, and T through

$$PV = n\overline{R}T \tag{3.16}$$

where n is the number of moles of the ideal gas and $\overline{R}$ is the universal gas constant [8.31441 kJ/(kmol · K) from Table B.2]. This form is often modified for engineering practice, where we usually deal in mass rather than in moles, as is common in physics and chemistry. To convert to the engineering form, note that the number of moles of gas n is equal to the mass of gas m divided by the molecular weight of the gas M. Thus, Eq. (3.16) becomes

$$Pv = \frac{\overline{R}T}{M} = \frac{\overline{R}}{M}T = RT \tag{3.17}$$

where $R \equiv \overline{R}/M$ is the particular gas constant for the particular gas under study. Thus R is simply the universal gas constant divided by the gas molecular weight (see Tables C.2 and C.4). Figure 3.17 shows the surface described by the ideal gas relation. Also shown are the projections onto the two-dimensional plots.

The microscopic description of an ideal gas must be consistent with the above macroscopic relation. This requires a number of idealizations of actual molecular phenomena. The first is that the gas is composed of a large number of very small particles. Thus, the total volume occupied by the particles is small. The second idealization is that the particles are independent of one another and are in continuous random motion. The time spent between collisions is much greater than the collision time. The final idealization is that no energy is dissipated. These concepts are helpful in interpreting the ideal gas relation and, in fact, are used in Chap. 12 to derive the ideal gas relation.

Under what conditions does this simple equation apply? The simplified microscopic description indicates that the equation does *not* apply to any substance that is in a solid, liquid, or saturation region state, or is in a state near the critical point. In the superheated and supercritical regions away from the critical point, the ideal gas relation, Eq. (3.17), is sufficiently accurate to be used for many gases that are well above their critical temperature at room conditions (CO_2, N_2, O_2, the noble gases, H_2, CO, and others). Specifically, if $P < P_{cr}$ and $T > 1.4T_{cr}$ for the particular gas, then Eq. (3.17) will be within 10 percent of the experimental values.

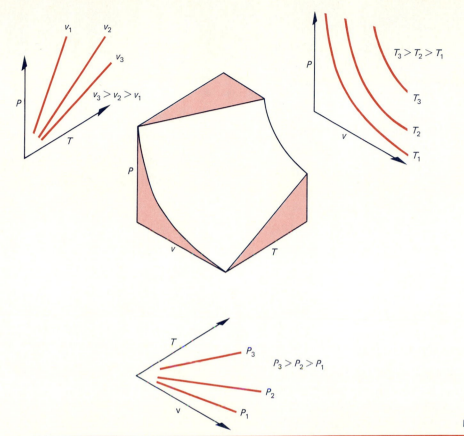

Figure 3.17   *P-v-T surface for an ideal gas.*

Additionally, if $P < 10P_{cr}$ and $T > 10T_{cr}$, then the 10 percent criterion is also met. The higher the temperature and the lower the pressure within the prescribed range, the more accurate will be the predictions based on ideal gas relations.

Other Properties of the Ideal Gas

Although the ideal gas can be defined as a substance that follows the ideal gas relation, we can infer some other property behavior for the ideal gas. Although the material in Chap. 5 is necessary for the proof, the ideal gas internal energy can be shown to be *a function of temperature only;* that is,

$$u_{\text{ideal gas}} = u(T) \tag{3.18}$$

If u for the ideal gas depends on *only* temperature, then Eqs. (3.10) and (3.11) yield

$$du = c_v \, dT \tag{3.19}$$

or

$$c_v = \left(\frac{\partial u}{\partial T}\right)_v = \frac{du}{dT} \tag{3.20}$$

From the definition of enthalpy and the use of the ideal gas relation,

$$h = u + Pv = u(T) + RT \tag{3.21}$$

and it follows that, for an ideal gas, h is also a function of temperature only. Then Eqs. (3.13), (3.14), and (3.15) yield

$$dh = c_P \, dT \tag{3.22}$$

and

$$c_P = \left(\frac{\partial h}{\partial T}\right)_P = \frac{dh}{dT} \tag{3.23}$$

Now using Eq. (3.21) in Eq. (3.23) yields

$$\frac{dh}{dT} = c_P = \frac{du}{dT} + R = c_v + R \tag{3.24}$$

or, for an ideal gas,

$$c_P - c_v = R \tag{3.25}$$

For any ideal gas, then, c_P must always be greater than c_v; and if either specific heat is known, the other is easily found from Eq. (3.25). The values for specific heats are given for many gases in Table C.2. The subscript zero implies the evaluation is at 300 K, that is, c_{P0} and c_{v0} are the values at 300 K. Also, representations for the temperature dependence are given by polynomials in Table D.1. The superscript zero in this table implies ideal gas conditions.

Example Problem 3.1

Find the change in enthalpy and the change in internal energy for nitrogen from 30 to 200°C. Take the pressure at atmospheric conditions.

Solution

The ideal gas expressions are valid since $P = 0.1$ MPa is less than $P_{cr} = 3.39$ MPa and the minimum temperature of $T = 303$ K is far greater than $T_{cr} = 126.2$ K (critical values from Table C.4). Thus

$$dh = c_P \, dT$$

and

$$h_2 - h_1 = \int_1^2 dh = \int_1^2 c_P \, dT$$

Also

$$du = c_v \, dT$$

With $c_v = c_P - R$ in Eq. (3.25),

$$du = (c_P - R) \, dT$$

Integration yields

$$u_2 - u_1 = \int_1^2 du = \int_1^2 (c_P - R) \, dT$$

$$= \int_1^2 c_P \, dT - R(T_2 - T_1)$$

This last expression can also be obtained from Eq. (3.21). The changes in h and u require the specific heat at constant pressure. One method is to assume c_P to be a constant and use this single value; the other method is to integrate the temperature dependence of the specific heat.

For a constant specific heat at constant pressure,

$$h_2 - h_1 = c_P(T_2 - T_1)$$

and

$$u_2 - u_1 = c_P(T_2 - T_1) - R(T_2 - T_1)$$

From Table C.2, $c_P = 1.0404$ kJ/(kg $\cdot$ K), so

$$h_2 - h_1 = 1.0404(200 - 30) = 176.9 \text{ kJ/kg}$$
$$u_2 - u_1 = 1.0404(200 - 30) - 0.29680(200 - 30) = 126.4 \text{ kJ/kg}$$

The temperature dependence of the specific heat is given in Table D.1 on a mole basis as

$$\bar{c}_P^0 = 28.90 - (0.1571 \times 10^{-2})(T)$$
$$+ (0.8081 \times 10^{-5})(T^2) - (2.873 \times 10^{-9})(T^3)$$

with $\bar{c}_P^0$, kJ/(kmol $\cdot$ K), and T, K. The change in enthalpy on a mole basis (the overbar indicates a unit mole basis) is

$$\bar{h}_2 - \bar{h}_1 = 28.90 \int_1^2 dT - (0.1571 \times 10^{-2}) \int_1^2 T \, dT$$

$$+ (0.8081 \times 10^{-5}) \int_1^2 T^2 \, dT - (2.873 \times 10^{-9}) \int_1^2 T^3 \, dT$$

Integration between $T_1 = 273.15 + 30 = 303.15$ K and $T_2 = 273.15 + 200 = 473.15$ K yields

$$\bar{h}_2 - \bar{h}_1 = 4990 \text{ kJ/kmol}$$

With the molecular weight of 28.016, the change in enthalpy is $h_2 - h_1 = 178.1$ kJ/kg. The change in u is evaluated directly from the above integration by combining the above equations:

$$u_2 - u_1 = \int_1^2 c_P\, dT - R(T_2 - T_1)$$

But the integral is given above, so

$$u_2 - u_1 = (h_2 - h_1) - R(T_2 - T_1)$$
$$= 178.1 - 0.29680(170) = 127.6 \text{ kJ/kg}$$

Comments

In the region of the properties of this problem, the ideal gas relation is a very accurate approximation. The temperature variation of the specific heat is relatively unimportant over the temperature range of the problem and yields a value less than 1 percent higher than the constant specific heat solution. The constant specific heat solution is given for a c_P evaluated at 300 K. A more accurate solution is obtained if the value of c_P is taken at the average of the temperatures within the needed range.

Many attempts have been made to provide an accurate $P\text{-}v\text{-}T$ relation that covers a broader range of conditions than the ideal gas relation. Such relations generally gain in accuracy at the expense of increased complexity of form and the need for more data to describe the particular gas. Some of the relations extend $P\text{-}v\text{-}T$ behavior to the liquid region. We examine two such representative equations of state.

van der Waals' Equation of State

The van der Waals equation is an attempt to modify some obvious theoretical difficulties in the ideal gas relation. In 1873, J. D. van der Waals argued that the volume of a gas should not go to zero when the pressure gets very large, as implied by the ideal gas relation, because the atoms of the gas should still take up some specific volume b. The ideal gas relation then becomes

$$v = \frac{RT}{P} + b$$

and the gas volume approaches b as P becomes very large.

In addition, the pressure measured at a surface is not the true gas pressure and is actually too small. Away from surfaces, gas molecules are exposed to short-range attractive forces from nearby gas molecules on all sides (Fig. 3.18a); near a surface, gas molecules are exposed to these forces from molecules on only one side, thus reducing the effective pressure exerted on the wall (Fig. 3.18b). Thus the gas pressure should be *increased* to a value above the measured pressure P by a certain amount. The increase in P should depend on the specific volume, since less mass and thus fewer molecules are present in a given volume to provide the attrac-

Figure 3.18 **Attractive forces between molecules. (*a*) Forces far from boundary. (*b*) Forces near boundary.**

(a)   (b)

tive forces as the specific volume increases. van der Waals proposed that the ideal gas pressure P be replaced by $P + a/v^2$ to account for these forces. The van der Waals equation then becomes

$$P = \frac{RT}{v - b} - \frac{a}{v^2} \tag{3.26}$$

The constants a and b must be determined for each gas. We extend our interpretation of Eq. (3.26) when we discuss other property relations in Chap. 8. There, it is shown that a and b are given by

$$a = \frac{27R^2T_{cr}^2}{64P_{cr}} \tag{3.27}$$

$$b = \frac{RT_{cr}}{8P_{cr}} \tag{3.28}$$

Equation (3.26) is obviously based on a physical model of the behavior of gases that is somewhat more detailed than the ideal gas model. However van der Waals' equation does not provide a precise relation between gas properties but does offer insight into the reasons for deviation from ideal gas predictions.

Benedict-Webb-Rubin Equation of State

An example of a more accurate but more complex equation that has some basis in the physical behavior of the properties of gases but is largely based on empiricism is the Benedict-Webb-Rubin equation:

$$P = \frac{RT}{v} + \left(B_0RT - A_0 - \frac{C_0}{T^2} \right)\left(\frac{1}{v^2} \right) + (bRT - a)\left(\frac{1}{v^3} \right) + \frac{\alpha a}{v^6}$$
$$+ c\frac{1 + \gamma/v^2}{v^3T^2} \exp\left(\frac{-\gamma}{v^2} \right) \tag{3.29}$$

Here, A_0, B_0, C_0, a, b, c, α, and γ are constants specific to the particular gas. With eight constants available for describing the P-v-T behavior, the range of validity of this equation of state is obviously much greater than that for the ideal gas relation. It is chiefly applied to hydrocarbons, and values of the constants are given in Table C.5.

Example Problem 3.2

Predict the specific volume of methane gas at 10 MPa and 250 K by using the ideal gas relation, the van der Waals equation, and the Benedict-Webb-Rubin (B-W-R) equation.

Solution
The ideal gas relation predicts

$$v = \frac{RT}{P} = \frac{(0.51835)(250)}{10,000} = 0.0130 \text{ m}^3/\text{kg}$$

For the van der Waals equation, the constants for methane are [from Eqs. (3.27) and (3.28)] $a = 0.8921 \text{ kJ} \cdot \text{m}^3/\text{kg}^2$ and $b = 2.669 \times 10^{-3} \text{ m}^3/\text{kg}$. If the van der Waals equation is rewritten in terms of v,

$$v^3 - v^2\left(\frac{RT}{P} + b\right) + v\left(\frac{a}{P}\right) - \frac{ab}{P} = 0$$

Solving for the positive real root gives

$$v = 0.00836 \text{ m}^3/\text{kg}$$

The B-W-R relation constants for methane are given in Table C.5. The B-W-R equation is implicit in v, as was the van der Waals equation, and must be solved by iteration. Assuming various values of v and substituting into the right-hand side of Eq. (3.29) until the right-hand side is equal to the prescribed value of P yield $v = 0.00892 \text{ m}^3/\text{kg}$.

Comments
The B-W-R result, which we would expect to be most accurate, predicts specific volume that is well below the ideal gas prediction but is 6.3 percent above the van der Waals prediction. The variations among the methods change with the particular gas and conditions chosen.

Reference 4 gives $v = 0.00858 \text{ m}^3/\text{kg}$ from experimental data. The B-W-R prediction and the van der Waals prediction are both within 4.0 percent for this case. The experimental data are probably not accurate to that degree.

The increasing accuracy of the predicted values for specific volume comes at the expense of increasing difficulty in use of the equations of state. The equation to be used must be chosen based on the required accuracy for the problem at hand.

Principle of Corresponding States

van der Waals noted that every pure substance has a critical state, and he proposed that the P-v-T data for all substances be presented in terms of the critical properties. The most easily measured of the critical properties are the critical temperature T_{cr} and the critical pressure P_{cr}. The *principle of corresponding states* declares that all substances obey the same equation of state expressed in terms of the critical properties.

If the principle is true, then we should be able to express the behavior of any simple single-phase pure substance by

$$\frac{Pv}{RT} = z(P_r, T_r) \tag{3.30}$$

where z, called the *compressibility factor,* is a function of only the reduced properties $P_r = P/P_{cr}$ and $T_r = T/T_{cr}$ of the particular substance. The value of z is a measure of the deviation of the behavior of the P-v-T data of the substance from that of an ideal gas.

If the principle of corresponding states is correct, then carefully measured P-v-T data for *any* gas could be used to obtain a plot of z as a function of P_r and T_r; that same plot should then apply to all other gases. Unfortunately, the values for z as computed by Eq. (3.30) for different substances may vary by up to 6 percent, with even greater variations near the critical point. Table C.4 shows values of z at the critical point as well as the critical properties of many substances. Charts of $z(P_r, T_r)$ have been prepared based on data for 30 gases and are presented in Fig. C.1.

In these charts are cross-plotted values of the "pseudo-reduced specific volume" v_r', defined by

$$v_r' \equiv \frac{v}{RT_{cr}/P_{cr}} = z\frac{T_r}{P_r} \tag{3.31}$$

This is *not* the true reduced specific volume $v_r = v/v_{cr}$. The critical volume is difficult to measure accurately; in addition, v_r' can be computed from only v, T_r, and P_r, and the charts remain a function of only two reduced quantities.

The principle of corresponding states has wider application than to only P-v-T data; other properties should also follow the principle. We follow up with applications to other properties in Chap. 8.

Example Problem 3.3

Find the value of the specific volume of methane at 250 K and 10 MPa by using the compressibility factor charts.

Solution

First, the critical temperature and pressure for methane are found from Table C.4 as $T_{cr} = 191.1$ K and $P_{cr} = 4.64$ MPa. The reduced properties are then $T_r = 250/191.1 = 1.31$ and $P_r = 10/4.64 = 2.16$. Figure C.1 indicates $z = 0.69$. Using Eq. (3.30) results in

$$v = \frac{zRT}{P} = \frac{0.69(0.51835)(250)}{10,000} = 0.0089 \text{ m}^3/\text{kg}$$

Comments

The compressibility factor allows relatively accurate correction for nonideal gas effects over a very wide range of states. Only the critical properties need to be known to apply the factor in determining one dependent property from any two independent properties in the collection of P, v, and T for any gas. This factor is very useful in many engineering applications where the larger amount of data necessary for the more detailed equations of state (such as that of Benedict, Webb, and Rubin) is not available.

Observe that the specific volume calculated in this example is much smaller than that predicted by the ideal gas relation, but is only 6.3 percent larger than that predicted by the van der Waals equation in Example Problem 3.2. Agreement with the B-W-R relation is nearly exact and with experimental data is within 3 percent.

Example Problem 3.4

Use the compressibility factor to determine whether the following states can be treated as an ideal gas.

(*a*) Water at 1000°C, 2 MPa
(*b*) Water at 500°C, 300 MPa
(*c*) Water at 300°C, 0.2 MPa
(*d*) Oxygen at 500°C, 100 MPa
(*e*) Oxygen at 35°C, 1.5 MPa
(*f*) Methane at 10°C, 10 MPa
(*g*) Carbon dioxide at 350°C, 1 MPa

Solution

The critical-point data given in Table C.4 are as follows:

Substance	T_{cr}, K	P_{cr}, MPa
Water	647.3	22.09
Oxygen	154.8	5.08
Methane	191.1	4.64
Carbon dioxide	304.2	7.38

To use the compressibility charts in Fig. C.1, the reduced temperature and pressure, $T_r = T/T_{cr}$ and $P_r = P/P_{cr}$, must be found. The problem statement can be expressed in terms of reduced properties as follows:

Problem Part	T_r	P_r	z
(*a*)	1.967	0.09	0.995
(*b*)	1.194	13.58	>1.5
(*c*)	0.885	0.009	0.999
(*d*)	4.995	19.69	1.35
(*e*)	1.991	0.30	0.99
(*f*)	1.482	2.16	0.84
(*g*)	2.048	0.14	0.993

The values of z have been obtained from the charts in Fig. C.1.

The compressibility factor is given as

$$z = \frac{Pv}{RT}$$

so that $z \approx 1.0$ is an ideal gas. Thus, the states given in the above table imply that (a), (c), (e), and (g) can be treated as an ideal gas. Also, (b), (d), and (f) should not be considered an ideal gas.

Comments

The compressibility chart can be used to estimate when a state or process is accurately approximated as an ideal gas. The previous recommendation of $P < P_{cr}$ and $T > 1.4 T_{cr}$ can be extended to all locations within $0.9 \leq z \leq 1.1$ if the desired accuracy is within ± 10 percent.

Incompressible Liquids and Solids

Many liquids and solids exhibit a very small change in the density or specific volume if the other properties are altered. These substances are termed *incompressible* and are represented by

$$v = \text{constant} \tag{3.32}$$

The other properties for incompressible liquids and solids are also needed. We show in Chap. 8 that the constant-volume and constant-pressure specific heats are equal for a given incompressible liquid or solid. Therefore,

$$c_P = c_v = c \tag{3.33}$$

The change in the internal energy is given by Eq. (3.10). Equation (3.32) indicates that $dv = 0$, so Eq. (3.10) yields

$$du = c_v \, dT = c \, dT \tag{3.34a}$$

and

$$u_2 - u_1 = \int_1^2 c \, dT \tag{3.34b}$$

The enthalpy is $h = u + Pv$, so

$$dh = du + P \, dv + v \, dP \tag{3.35}$$

Again, $dv = 0$ for an incompressible liquid or solid, so with Eqs. (3.34a) and (3.34b)

$$dh = c \, dT + v \, dP \tag{3.36a}$$

and

$$h_2 - h_1 = \int_1^2 c \, dT + v(P_2 - P_1) \tag{3.36b}$$

These equations are useful in determining the property changes for incompressible liquids and solids.

Example Problem 3.5

The following data are adapted from Ref. 2 for the subcooled liquid properties of water. Determine the error in the incompressible liquid approximation for the temperatures and pressures given.

Subcooled Liquid Properties for H_2O

P, MPa	T, °C	v, m³/kg	u, kJ/kg	h, kJ/kg
5	100	0.0010410	417.52	422.72
	260	0.0012749	1127.9	1134.3
10	100	0.0010385	416.12	426.50
	260	0.0012645	1121.1	1133.7
20	100	0.0010337	413.39	434.06
	260	0.0012462	1108.6	1133.5
50	100	0.0010201	405.88	456.89
	260	0.0012034	1078.1	1138.2

Solution

The incompressible fluid approximation takes $v =$ constant, so that $du = c\,dT$ and $dh = c\,dT + v\,dP$. The state is specified in the subcooled liquid region by the temperature T and pressure P. If a constant-temperature line is considered from the subcooled liquid state to the saturated liquid line, then

$$u - u_l = \int_{T_l}^{T} c\,dT = 0 \qquad \text{for an isotherm}$$

and

$$u(T, P) - u_l(T) = \int_{T_l}^{T} c\,dT = 0 \qquad \text{for an isotherm}$$

Thus,

$$u(T, P) = u_l(T)$$

Also

$$h(T, P) - h_l(T) = u(T, P) - u_l(T) + v_l[P - P_l(T)]$$

or

$$h(T, P) - h_l(T) = 0 + v_l[P - P_{sat}(T)]$$

$$h(T, P) = h_l(T) + v_l[P - P_{sat}(T)] \qquad \text{for an isotherm}$$

where v_l is evaluated at the state temperature $v = v_l(T)$.

To approximate the first line of the above table, the above equations and Table D.8 (discussed in Sec. 3.5.3) are used. At 100°C, $P_{sat}(100°C) = 101.32$ kPa, $v_l(100°C) = 0.001043$ m³/kg, $u_l(100°C) = 418.75$ kJ/kg, and $h_l(100°C) = 418.9$ kJ/kg. Thus, at the first state of $P = 5$ MPa and $T = 100°C$, the approximated values are

$$v = v_l(100°C) = 0.001043 \text{ m}^3/\text{kg}$$

$$u = u_l(100°C) = 418.75 \text{ kJ/kg}$$

$$\begin{aligned} h &= h_l(100°C) + v_l(100°C)[P - P_{sat}(100°C)] \\ &= 418.9 + 0.001043(5000 - 101.32) \\ &= 424.0 \text{ kJ/kg} \end{aligned}$$

For the second state at $P = 5$ MPa and $T = 260°C$, the approximated values are $v = v_l(260°C) = 0.001275$ m³/kg, $u = u_l(100°C) = 1128.81$ kJ/kg, and $h = 1135.2$ kJ/kg. All the values required in the above table are at two temperatures. So the approximated values of u and v are the same, and only the enthalpy must be calculated for each P and T. These calculations are carried out for all the values in the table. With the error calculated as

$$\text{Error} = \frac{\text{approximate property} - \text{tabulated property}}{\text{tabulated property}}$$

the following table is generated:

P, MPa	T, °C	v Error, %	u Error, %	h Error, %
5	100	0.192	0.295	0.303
	260	0.008	0.081	0.079
10	100	0.433	0.632	0.639
	260	0.830	0.688	0.694
20	100	0.900	1.30	1.29
	260	2.31	1.82	1.84
50	100	2.24	3.17	3.08
	260	5.95	4.70	4.78

Comments
The table indicates the errors in the incompressible approximation. Below the critical pressure, this approximation is accurate to within a few percent. Below 10 MPa, the error is less than 1 percent. The error in the specific volume is usually the largest of the property errors. Therefore, the incompressible approximation can be used in many applications with a high degree of accuracy.

3.5.3 Tabular Data

When detailed property data are to be presented for a wide range of states, so that a single equation of state is not sufficient, it is customary to present the data in tabular form. Because the commonly measured properties are temperature and pressure, usually data are presented with T and P used as the independent properties and the other properties treated as dependent. The intervals between table entries in complete tabulations are usually chosen so that linear interpolation will provide accurate values for states that fall between entries.

Use of Steam Tables

Because of the many uses of steam in industrial practice, extensive tables are available for steam properties [1–3]. Abbreviated data computed from Ref. 5 are included in Tables D.8 through D.10. These data are within 0.2 percent agreement with Ref. 1 for saturated liquid properties, within 0.15 percent for saturated vapor properties, and within 0.2 percent for superheat properties. The accuracy is generally much better than the maximum errors stated.

Example Problem 3.6

Find the value of the enthalpy of water or steam at the following states:

(a) $P = 0.2$ MPa, $T = 120.2°C$
(b) $P = 0.2$ MPa, $x = 0.59$
(c) $T = 120°C$, $v = 0.6$ m^3/kg
(d) $P = 2.5$ MPa, $T = 350°C$
(e) $T = 350°C$, $v = 0.13856$ m^3/kg
(f) $T = 375°C$, $v = 0.13$ m^3/kg
(g) $P = 2.0$ MPa, $u = 2900$ kJ/kg
(h) $P = 2.5$ MPa, $T = 100°C$

Solution

To begin, the region of the phase diagram in which the state lies must be determined. The tables are organized so that the saturation properties are given separately from the properties of superheated steam and subcooled liquid. In addition, the saturation properties are presented in two ways—with the saturation pressure given at even intervals for easy interpolation and with the saturation temperature given at even intervals for the same reason.

The region in which the prescribed state lies usually is found easily by referring to one of the two tables for saturation properties. If, at the given T or P, the given specific intensive property lies outside the range of properties that can exist for saturated liquid, saturated vapor, or their mixtures, then the state must lie in either the superheated or the subcooled region.

(a) For the state prescribed, the pressure is given at an even increment, so we first refer to the saturation property table that is presented as a function of pressure (Table D.9). An excerpt of that table is given here:

Pressure Table for Saturated Water

P_{sat}, kPa	T_{sat}, °C	Specific Volume, m³/kg			Enthalpy, kJ/kg		
		v_l	v_{lg}	v_g	h_l	h_{lg}	h_g
101.32	100.0	0.001043	1.6690	1.6700	418.8	2257.0	2675.8
125.0	106.0	0.001048	1.3697	1.3707	444.1	2240.9	2685.1
150.0	111.4	0.001053	1.1561	1.1572	466.9	2226.3	2693.2
175.0	116.1	0.001057	1.0025	1.0035	486.8	2213.4	2700.1
200.0	120.2	0.001060	0.88498	0.8860	504.5	2201.7	2706.2
225.0	124.0	0.001064	0.79229	0.7934	520.5	2191.1	2711.5

For each saturation pressure, the corresponding saturation temperature (boiling point) is given along with the values of specific volume and enthalpy for both saturated liquid and saturated vapor. The enthalpy value in the h_{lg} column is the difference in enthalpy between the saturated vapor and saturated liquid states.

For the particular state given in (a), $P = 0.2$ MPa and $T = 120.2°C$, what is the value of enthalpy? It so happens that the temperature given is exactly the saturation temperature. If the temperature were greater at this pressure, then the state would be in the superheated vapor region; if T were lower, the state would be in the subcooled liquid (or possibly solid) region. All we can say with the data given, however, is that the state is in the two-phase region. The T and P given are not independent in this case, so the enthalpy could have any value from 504.5 to 2706.2 kJ/kg, depending on the relative amounts of saturated liquid and vapor present.

(b) For this case, P and x are given. Such a state must lie in the saturation region unless the state has been misspecified, since x is defined only in the saturation region. (The only misspecification possible for x given in the range $0 < x < 1$ would be if P were greater than P_{cr}.)

To find h in this state, we use Eq. (3.9):

$$h = h_l + xh_{lg}$$

and introducing the data from the abbreviated table at 0.2 MPa, we find

$$h = 504.5 + 0.59(2201.7) = 1803.5 \text{ kJ/kg}$$

(c) For this case, we again first refer to the tables of saturation properties, to determine which region contains the specified state. In this case, however, the temperature is given, so we refer for convenience to Table D.8, a part of which is excerpted here:

Temperature Table for Saturated Water

T_{sat}, °C	P_{sat}, kPa	Specific Volume, m³/kg			Enthalpy, kJ/kg		
		v_l	v_{lg}	v_g	h_l	h_{lg}	h_g
120	198.55	0.001060	0.89100	0.8921	503.5	2202.3	2705.8
125	232.11	0.001065	0.76938	0.7704	524.8	2188.2	2712.9
130	270.15	0.001070	0.66702	0.6681	546.1	2173.8	2719.9
135	313.09	0.001075	0.58074	0.5818	567.4	2159.2	2726.6
140	361.39	0.001080	0.50739	0.5085	588.8	2144.3	2733.1

Because the specified v lies between the saturated liquid and saturated vapor values, the state does lie in the saturation region. Equation (3.7b) can be used with the table values to find the quality from

$$x = \frac{v - v_l}{v_{lg}}$$

to give

$$x = \frac{0.6 - 0.001060}{0.89100} = 0.67221$$

Now the enthalpy can be found by using the known quality:

$$h = h_l + xh_{lg} = 503.5 + (0.67221)(2202.3) = 1983.9 \text{ kJ/kg}$$

(d) For $P = 2.5$ MPa and $T = 350°C$, we check either of the saturation tables and find that for $P = 2.5$ MPa, the temperature is above the saturation temperature of 223.9°C and for $T = 350°C$, the pressure is below the saturation pressure of 16.535 MPa. In either case, the indication is that the state is in the superheat region. We then proceed to the superheated steam tables (Table D.10), a portion of which is shown now:

Properties for Superheated Steam

T, °C	v, m³/kg	h, kJ/kg
$P = 2000$ **kPa** ($T_{sat} = 212.4°C$)		
250	0.11145	2902.4
300	0.12550	3023.9
350	0.13856	3137.1
400	0.15113	3247.1
450	0.16343	3356.1
$P = 2500$ **kPa** ($T_{sat} = 223.9°C$)		
250	0.08699	2879.7
300	0.09893	3009.3
350	0.10975	3126.6
400	0.12004	3239.1
450	0.13005	3349.7

For this case, h is read directly as 3126.6 kJ/kg.

(e) Again, we check the saturation tables at the given temperature, and we find that the specific volume is greater than that for a saturated vapor at that temperature. The state is thus in the superheat region, and from the abbreviated table above the corresponding enthalpy is 3137.1 kJ/kg.

(f) This state is also found to be in the superheat region and its values to lie within the abbreviated table given above. However, neither the temperature nor the specific volume lies on a table entry. We are thus required to perform a double interpolation to find the corresponding enthalpy. This is done by first finding the values of v and h that occur at $T = 375°C$ at the pressures that bracket known values of v. Thus, at $P = 2.0$ MPa and $T = 375°C$,

$$v = 0.13856 + \frac{375 - 350}{400 - 350}(0.15113 - 0.13856) = 0.14485 \text{ m}^3/\text{kg}$$

and

$$h = 3137.1 + \frac{375 - 350}{400 - 350}(3247.1 - 3137.1) = 3192.1 \text{ m}^3/\text{kg}$$

A similar interpolation at $P = 2.5$ MPa for $T = 375°C$ gives $v = 0.11490$ m^3/kg and $h = 3182.9$ kJ/kg. Now, since we have assumed that linear interpolation is acceptable, we also assume that h lies between the known values at the two pressures in the same proportion that v lies between the two values, or

$$\frac{v - v(2 \text{ MPa})}{v(2.5 \text{ MPa}) - v(2 \text{ MPa})} = \frac{h - h(2 \text{ MPa})}{h(2.5 \text{ MPa}) - h(2 \text{ MPa})}$$

or

$$\frac{0.13 - 0.14485}{0.11490 - 0.14485} = \frac{h - 3192.1}{3182.9 - 3192.1}$$

which gives $h = 3187.5$ kJ/kg.

(g) Some tables provide internal energy as a tabulated property, but many do not. If u is not tabulated, then it must be computed from h, and this can lead to considerable tail chasing. Note first by examination of any portion of the superheat tables that both u and h are weak functions of pressure and strong functions of temperature. Thus, if P and h or P and u are given to define the state, it is usually fairly easy to find the corresponding temperature and then the other state properties (the saturation tables are first examined to determine the state region). If T and u or h are given, it is sometimes tedious to converge on the corresponding properties to good accuracy.

In the present case, we assume various temperatures and find the corresponding h and v values at $P = 2.0$ MPa. The corresponding u values are then computed from $u = h - Pv$. This effectively adds another column to the table given above, which for 2.0 MPa becomes

Properties of Superheated Steam

T, °C	v, m³/kg	u, kJ/kg	h, kJ/kg
	$P = 2000$ **kPa** ($T_{sat} = 212.4$°**C**)		
250	0.11145	2679.5	2902.4
300	0.12550	2772.9	3023.9
350	0.13856	2860.0	3137.1
400	0.15113	2944.8	3247.1

For the given $u = 2900$ kJ/kg, the temperature must lie between 350 and 400 K. Again, assuming that linear interpolation is adequate, we can determine h at the given state from

$$\frac{h - h(350°C)}{h(400°C) - h(350°C)} = \frac{u - u(350°C)}{u(400°C) - u(350°C)}$$

or

$$\frac{h - 3137.1}{3247.1 - 3137.1} = \frac{2900 - 2860.0}{2944.8 - 2860.0}$$

which gives $h = 3189.0$ kJ/kg.

(*h*) From the saturation states given in Table D.8, the saturation pressure corresponding to 100°C is 0.10132 MPa. Since the desired pressure is 2.5 MPa, the water must be in the subcooled liquid state. This state must be estimated from the saturation tables.

The saturated liquid state represents the liquid in the saturated region, and pressure and temperature are dependent. Thus the subcooled state could possibly be determined by reading the saturation value at either the pressure or the temperature of the state. The subcooled state's relation to the saturation state is interpreted by using Fig. 3.11. Constant-pressure lines on this T-v diagram are very close to the saturated liquid line for all pressures. Thus large pressure variations do not significantly alter the state. The temperature, however, accurately represents the state. Thus, the state in the subcooled region is represented by the saturated liquid value at the corresponding temperature.

The enthalpy is evaluated by assuming the liquid is an incompressible fluid, as in Example Problem 3.5. Equations (3.34) and (3.36) are valid. Considering a constant-temperature line from the saturated liquid line to the subcooled liquid state, we can evaluate the difference in enthalpy from Eq. (3.36). Integrating this expression yields

$$h - h_l = 0 + v_l(P - P_{sat})$$

$$h = h_l + v_l(P - P_{sat})$$
$$= 418.8 + 0.001043(2500 - 101.32)$$
$$= 418.8 + 2.5 = 421.3 \text{ kJ/kg}$$

Comments

We find in later chapters that it is often necessary to find the change in certain specific intensive properties between two states. To do so requires that the property values in each state be found to four or five significant figures, so that the differences will retain acceptable accuracy. Use of the tables can provide such accurate property values if tabular values are available at small increments. Complete tables [1,2,3,5] should, therefore, be used for accurate work. The abbreviated tables presented in the Appendix can introduce errors in interpolation and are presented chiefly as an aid to solving homework problems.

The major difficulty in use of the tables lies in the inevitable interpolations required for accurate solution of real problems. The use of computerized tables offers great savings in time and reduces numerical errors.

Tables for Other Substances

The exact data for the thermodynamic properties of each substance are different. Tables or other data presentations of the properties must be available for each substance if careful thermodynamic analysis is to be done. Abridged tables are shown in Tables D.8 through D.13 for the properties of steam and refrigerant 12. These tables cover the saturated liquid, two-phase, and superheat regions. The ammonia tables are Tables D.14 through D.16.

Gas Tables

As noted in Sec. 3.5.2, the internal energy and enthalpy of an ideal gas are independent of pressure and so can be presented as a function of temperature only. This makes data presentation much simpler than for substances that cannot be treated as ideal gases, such as steam. Tables of the thermodynamic properties of ideal gases based on the statistical thermodynamics approach (Chap. 12) can be prepared, and they are very accurate in comparison with experimental measurement of the properties. Such calculated properties depend on a limited amount of data that can be accurately obtained by analysis of the spectra emitted or absorbed by the gases. For monatomic gases (argon, neon, krypton, etc.) and diatomic gases (carbon monoxide, nitrogen, oxygen, hydrogen, etc.), the calculation of the properties is fairly straightforward. For more complex molecules (CO_2, H_2O in the ranges of ideality, CH_4, etc.), the calculation of the properties is more difficult. Such tables are available in Ref. 6.

In Tables D.2 through D.7, the properties of air, argon, nitrogen, oxygen, superheated water vapor, and carbon dioxide are presented under the assumption that these substances behave as ideal gases. The data are computed from Ref. 7. The values for air are computed from values for nitrogen, oxygen, and argon by assuming that air acts as an ideal mixture of ideal gases. These tables are in essentially exact agreement with the tables of Ref. 6, since both references use the fundamental statistical thermodynamics equations for their calculated properties.

Note that we have not discussed the scale of values for the quantities u, internal energy, or h, the enthalpy. We will find that the state to which we assign the value of zero is arbitrary insofar as classical thermodynamics is concerned, because only *changes* in u or h are of interest. For the gas tables, it is common practice to assign $u = 0$ kJ/kg and $h = 0$ kJ/kg at an absolute temperature of $T = 0$ K, so that u and h always have positive values.

3.5.4 Computerized Property Data Retrieval

Values for the thermodynamic properties of many substances are available in the form of computer programs that can provide the properties at any given state. This avoids the need for interpolation and/or auxiliary computation, such as when only h is given in a table and u must be calculated from $u = h - Pv$. {Such programs are included with this text for the properties of air, nitrogen, and argon (as ideal gases); for steam; and for refrigerant 12. The programs are abridged forms of more general programs that cover many more substances and a broader range of variables. These more general programs [5,7,8] have other capabilities as well, such as the ability to carry out some combustion and cycle analyses.}

The student is urged to use the property programs available with this text in evaluating the properties necessary for carrying out the homework problems, some of which can be very tedious if only tables are available for solution. The programs should help in broadening the scope of problems that can be usefully approached. The programs provide data of accuracy equivalent to the tables of properties in the appendixes.

In this text, we have avoided using programs that carry out cycle analysis, combustion analysis, etc. The ability to understand the basis of such calculations and to determine the assumptions and limitations commonly made to facilitate them is part of the value of an engineering education. After these skills are mastered, *then* fully computerized solution becomes a valuable tool.

3.6 Remarks

All the properties defined in classical thermodynamics are limited to a control mass in thermodynamic equilibrium. That is, the control mass is in mechanical, chemical, and thermal equilibrium, and has no gradients across it, so that the properties are the same in each portion of the control mass that we examine. The control mass, in other words, is at a *uniform* state. However, most problems of interest in engineering applications are not in a uniform state in thermodynamic equilibrium. How do we justify the use of temperature, for example, for a control mass where there is heat transfer occurring, since heat transfer, by definition, requires the presence of a temperature difference between the control mass and the surroundings, and the control mass is therefore not in thermal equilibrium?

Classical thermodynamics does not really answer this question in a satisfactory way, except to note that the use of the properties that we have defined (and will define) leads to the results that we expect from application of the thermodynamics relations *even in cases that are quite far from thermodynamic equilibrium.* We must take another viewpoint, that of statistical thermodynamics as introduced in Chap. 12, to show *why* we can extend the use of properties defined only for the case of thermodynamic equilibrium to cases of nonequilibrium.

There are indeed cases in which the idea of assigning a single temperature to a control mass that is far from thermal equilibrium is simply no longer valid. In the plasma necessary to sustain a fusion reaction, for example, the conditions are so far from thermodynamic equilibrium, even in the smallest control mass taken from the plasma, that the equilibrium properties do not apply, and a more detailed microscopic analysis must be invoked. However, for the vast majority of engineering problems, we can assume that application of the property information developed from the classical approach is valid.

Problems

Problems marked with a star are lengthy; thus, computerized steam tables or an accurate Mollier chart should be available if these problems arc to be completely worked.

3.1 Find P_1/P_2 for the states $v_1 = 50$ m³/kg, $T_1 = 500°C$ and $v_2 = 10$ m³/kg, $T_2 = 1200°C$ for ethane, using (a) the ideal gas relation, (b) the van der Waals equation, (c) the Benedict-Webb-Rubin equation, and (d) the compressibility factor charts.

3.2 Can the following fluids in the specified states be treated as ideal gases?
(a) Air at 0.1 MPa, 20°C
(b) Air at 13 MPa, 900°C
(c) Methane at 2 MPa, 1000°C
(d) Water at 0.1 MPa, 20°C
(e) Water at 0.01 MPa, 30°C
(f) Refrigerant 12 at 1 MPa, 50°C

3.3 Determine whether the following processes for the specified gas can be treated as ideal gas processes. If the process is an ideal gas process, then evaluate $u_2 - u_1$ and $h_2 - h_1$, assuming constant specific heats.

(a) Isothermal process (T = constant) for water from 1200 K and 2 MPa to 10 MPa
(b) Isometric process (v = constant) for refrigerant 12 from 405 K and 4.0 MPa to 470 K
(c) Isobaric process (P = constant) for nitrogen from 250 K and 17.0 MPa to 280 K

3.4 Find the specific volume of sulfur dioxide at T = 480 K and P = 15 MPa.

3.5 Nitrogen gas is originally at P = 200 atm and T = 252.4 K. It is cooled at constant volume to T = 189.3 K. What is the pressure at the lower temperature?

3.6 Plot the T_r = 1.30 isotherm for *n*-butane on a plot of compressibility factor versus P_r, using the Benedict-Webb-Rubin equation to find z.

3.7 A virial equation of state has the form

$$\frac{P}{T} = \frac{R}{v}\left[1 + \frac{B(T)}{v} + \frac{C(T)}{v^2} + \cdots\right]$$

where $B(T)$, $C(T)$, etc., are functions of temperature only and are called the first virial coefficient, the second virial coefficient, and so on. Find expressions for the virial coefficients for gases that obey (a) the ideal gas relation, (b) the van der Waals equation, and (c) the Benedict-Webb-Rubin equation.

3.8 The numerical values of the properties x, y, and z are given. Linearly interpolate and obtain the specified property.
(a) Given:

x	y
110	1.2074
115	1.0351

Find y for x = 112.3.
(b) Given:

$x = 10$		$x = 50$	
y	z	y	z
150	19.513	150	3.8894
200	21.820	200	4.3561

Find z for x = 27 and y = 150.
(c) Given: Same data as in (b). Find z for x = 32 and y = 181.

3.9 Indicate whether the following states for water are in the liquid, saturation, or superheated region. Specify the quality of the states in the saturation region.

State	P, kPa	T, °C	v, m³/kg
1	1700	200	—
2	1200	—	0.0010
3	—	75	3.0
4	500	202	—
5	350	—	0.005

3.10 Find the value of internal energy for the substances in the states indicated.
(a) Water at 0.4 MPa, 725°C
(b) Water at 3.0 MPa, 0.01 m³/kg
(c) Refrigerant 12 at 130°C, 125 kPa
(d) Water at 1.0 MPa, 100°C

3.11 Indicate whether the following states for refrigerant 12 are in the subcooled, saturation, or superheated region. The reference values for u and h are those used for tables D.11 through D.13.

State	P, kPa	T, °C	v, m³/kg	u, kJ/kg	h, kJ/kg
1	2000	—	—	—	220
2	—	100	0.04	—	—
3	—	—	0.02	325	—
4	2000	—	—	—	300
5	600	200	—	—	—
6	—	—	0.30	—	365

3.12 Find the following properties for steam from the steam tables. Show all interpolation calculations. Compare the results with those from computerized tables.
(a) $u(T = 400°C, P = 1500 \text{ kPa})$
(b) $h(T = 200°C, P = 2000 \text{ kPa})$
(c) $T(h = 2100 \text{ kJ/kg}, P = 6900 \text{ kPa})$
(d) $x(P = 500 \text{ kPa}, h = 2000 \text{ kJ/kg})$
(e) $h(u = 2000 \text{ kJ/kg}, x = 0.65)$
(f) $h(u = 3100 \text{ kJ/kg}, v = 0.1 \text{ m}^3/\text{kg})$

3.13 Find the change in the value of enthalpy h for water (steam) between an initial state where $P = 100$ kPa and $v = 1.6500$ m³/kg and a final state where $P = 3000$ kPa and $T = 525°C$.

3.14 In certain ranges of pressure and temperature, steam does act as an ideal gas. Using the steam tables from either App. D or computerized tables, plot an accurate graph of pressure versus specific volume for pressure in the range of 0.01 to 0.3 times the critical pressure (Table

C.4) for 200, 300, and 400°C. On the same graph, plot P versus v for the same temperatures, assuming that P and T are independent variables and v is calculated by using the ideal gas relation. Discuss your results. How do the results compare if T and v are chosen as the independent variables and P is calculated from the ideal gas relation? Use multicycle log-log paper for your plots.

3.15 Calculate the value of c_P for air at 100 and 1500°C, using only the enthalpy data from Table D.2. Compare with the results from the polynomial expressions for c_P from Table D.1.

3.16 Calculate c_P and c_v for steam at $T = 400$°C, $P = 200$ kPa, using data for h and u from the steam tables of App. D or computerized steam tables. Check whether $c_P - c_v = R$, which would be an indication that steam acts as an ideal gas in this state.

3.17 Calculate the change in enthalpy for nitrogen between $T_1 = 20$°C and $T_2 = 1500$°C, using the data in Table D.4. Compare the result to that found by using the polynomial expressions for c_P from Table D.1 and to that found by using a constant value of c_P based on the mean temperature between T_1 and T_2.

*3.18 Using the data from Prob. 3.14 and actual P-v-T data from the steam tables or the computerized tables, compute the compressibility factor z for steam in the same ranges of P, v, and T as for Prob. 3.14. Plot the result as z versus P with T as a parameter. On the same graph, plot z from Fig. C.1 for the same ranges of the variables. Discuss the comparison between the results. What type of graph paper did you choose for plotting the results? Why?

3.19 Find the values of u and h for water at $T = 100$°C and $P = 300$ kPa.

3.20 Steam is at $T = 600$°C and $P = 2000$ kPa. Find the change in the enthalpy of the steam when
(a) T is held constant and P is changed to 100 MPa.
(b) P is held constant and T is changed to 800°C.
Discuss the result in terms of the relative sensitivity of h to P or T. What would you conclude about the relative magnitudes of the partial derivatives of the internal energy, $(\partial u/\partial T)_P$ and $(\partial u/\partial P)_T$? Examine Fig. D.2, and discuss whether the results for the state chosen in this problem are representative of the behavior of steam in all regions of the P-h diagram.

3.21 A sealed volume of 2.0 cm³ contains saturated water. The container is heated from an initial pressure of

0.10 MPa until the vessel contains a single phase. Determine the final state of the water if the vessel contains (*a*) 0.10 g and (*b*) 1.0 g.

3.22 Nitrogen is contained in a bottle. Initially, the nitrogen is at 42 atm and 130 K. The bottle has a volume of 0.02 m³. The bottle is then heated until the nitrogen reaches 202 K.

(*a*) Can the nitrogen be treated as an ideal gas? Why or why not?

(*b*) What is the pressure of the nitrogen after heating?

(*c*) What is the mass of the nitrogen in the bottle?

3.23 Water (0.5 kg) contained within a constant-volume container ($V = 0.01$ m³) is heated. The initial pressure is 100 kPa, and the water is heated until the water exists in the saturated vapor state. Plot the quality and volume occupied by the liquid versus pressure for this process (that is, x and V_l versus P).

3.24 A piston encloses 2 kg of water in an initial volume of 0.02 m³ (see Fig. P3.24). The initial temperature is 50°C. The piston leaves the stops at a pressure of 1.0 MPa. The water is heated from its initial state to a final temperature of 200°C.

(*a*) Plot the process on a *P-v* diagram.

(*b*) Evaluate the work done by the H₂O.

(*c*) What volume is occupied by the liquid at the initial state and when the piston leaves the stops?

3.25 Refrigerant 12 is heated isothermally from an initial state at 200 kPa and 200°C to a final pressure of 800 kPa. Evaluate the work done per unit mass. Also, evaluate the work done per unit mass, assuming refrigerant 12 to be an ideal gas.

3.26 A piston encloses 1 kg of saturated water which is initially 10.0 percent liquid on a mass basis. The initial pressure is 1.0 MPa. The water is heated in the system shown in Fig. P3.26 until the water is at the saturated vapor state. The volume at the top stops is 0.19 m³. Determine the work done by the water and the final temperature and pressure.

3.27 A piston of mass equal to 900 kg and area equal to 9.8×10^{-3} m² is supported against stops (Fig. P3.27) by water at the saturated vapor state and 20 MPa (state 1). The water is cooled until half the container is liquid and half is vapor by volume (state 3). Determine the work per unit mass for the process ($_1w_3$). Draw the process on the *P-v* diagram.

3.28 Heat transfer takes place isothermally to ammonia be-

Figure P3.24

Figure P3.26

Figure P3.27

tween the saturated vapor state at $P = 500$ kPa and a final pressure of 100 kPa. Draw an accurate P-v diagram of the process, and evaluate the work done per unit mass. Numerical or graphical integration is required.

References

1. *Steam Tables,* American Society of Mechanical Engineers, New York, 1967.
2. J. H. Keenan, F. G. Keyes, P. G. Hill, and J. G. Moore, *Steam Tables, English Version,* Wiley, New York, 1969.
3. L. Haar, J. S. Gallagher, and G. S. Kell, *NBS/NRC Steam Tables,* Hemisphere, New York, 1984.
4. Richard B. Stewart and Victor J. Johnson, *A Compendium of the Properties of Materials at Low Temperatures (Phase II),* National Bureau of Standards Cryogenic Engineering Lab. Rep. WADD G0-56, Part IV, U.S Air Force Systems Command, Wright-Patterson Air Force Base, Ohio, December 1961.
5. *STEAMCALC, A Computer Program for the Calculation of the Thermodynamic Properties of Steam,* Wiley, New York, 1984.
6. J. H. Keenan, J. Chao, and J. Kaye, *Gas Tables (English Units),* 2d ed., Wiley, New York, 1980.
7. *GASPROPS, A Computer Program for the Calculation of the Thermodynamic Properties of Ideal Gases and Combustion Products.* Wiley, New York, 1984.
8. *REFRIG, A Computer Program for the Thermodynamic Properties of Common Refrigerants,* Wiley, New York, 1984.

4

First Law of Thermodynamics

Energy is an Eternal Delight.
William Blake

900-kW gas turbine with eight-stage compressor, annular combustor, and three-stage turbine. Turbine normally operates at 22,300 rpm. *(Solar Turbines, Inc.)*

4.1 Introduction

Conservation relations form the foundation of thermodynamics. The concept of a conserved quantity is important in many areas of science and is based on common experiences. The conservation of energy is discussed briefly in Chap. 1 and is formally presented as the first law of thermodynamics in this chapter. Conservation of energy and the first law of thermodynamics are different names for the same principle. Other conservation relations include the conservation of momentum and the conservation of mass.

Momentum conservation was probably first encountered in an elementary physics course. Mass conservation was probably also presented at the same time, and it is one conservation principle that we all take for granted. The water entering one end of a garden hose is expected to exit from the other end. This type of physical expectation or "physical feeling" for a phenomenon can be expressed in conservation statements or laws. This chapter focuses on the first law of thermodynamics, or conservation-of-energy principle. The relationship to other conserved quantities is also discussed.

A clear picture of the relationship between the states of the system and energy transfer is required. For a simple compressible substance, two independent intensive properties specify the state. The system changes from one state to another by an energy transfer in the form of work and/or heat transfer. This energy transfer must cross a system boundary. The amount of energy transfer depends on the path; i.e., work depends on the specific path followed in performing the work. The state of a substance depends not on how the state is reached, but on its properties only.

This chapter is essential to understanding thermodynamics. The first law of thermodynamics as well as the principle of conservation of mass are presented. The fundamental principles are first presented for a control mass and are subsequently transformed for use in a control volume.

4.2 Conservation Principles and the First Law of Thermodynamics

All the conservation statements in this section are presented for a control mass. Thus, we are considering the mass within a boundary surface which can undergo translation, rotation, and deformation, but no mass crosses the boundary. The substance contained within a cylinder with a sliding piston is a typical example. Also, a fluid within a balloon is a control mass which can be used to visualize the following statements.

4.2.1 Conservation of Mass

We have chosen to make the fundamental conservation statements about a control mass so that no mass crosses the system boundary. Therefore, this conservation principle is as follows:

The mass of a control mass never changes.

This is expressed mathematically as

$$dm = 0 \tag{4.1}$$

or

$$m = \text{constant} \tag{4.2}$$

Since the mass within the boundary is a sum of all the elements, this is also given as

$$dm = d\left(\int_m dm\right) = 0 \tag{4.3}$$

With the specific volume and density related as $\rho = 1/v$, Eq. (4.3) can also be represented as a volume integral

$$d\left(\int_V \rho \, dV\right) = d\left(\int_V \frac{dV}{v}\right) = 0 \tag{4.4}$$

where V represents the volume occupied by the control mass.

The conservation of mass is also expressed on a rate basis by considering the time interval dt. This principle is then

$$\frac{dm}{dt} = \frac{d}{dt}\left(\int_m dm\right) = \frac{d}{dt}\left(\int_V \rho \, dV\right) = 0 \tag{4.5}$$

The expressions for mass in Eqs. (4.3) and (4.4) have been used to obtain the various forms in Eq. (4.5).

All the above equations are exactly the same, and the differences are in the notation used. This statement of conservation of mass is only a reiteration of our definition of a control mass. The usefulness of conservation of mass becomes apparent in Sec. 4.3.

The conservation-of-mass statement has been employed previously. The problem which considers a control mass within a cylinder with a sliding piston has always made use of a known quantity of mass. If water as a liquid is contained within a piston-cylinder arrangement and the volume is increased, the mass will be constant as the liquid changes to a vapor.

4.2.2   First Law of Thermodynamics

This law is alternatively termed the *conservation of energy*. This energy balance is stated as follows:

> A change of the total energy (kinetic, potential, and internal) is equal to the rate of work done on the control mass plus the heat transfer to the control mass.

This is expressed mathematically as

$$dE = \delta Q + \delta W \tag{4.6}$$

The sign convention is that all energy transfer *into* the system is positive (Fig. 4.1). There is no reference to a particular path. Thus, the change in energy of the control mass from state 1 to state 2 is equal to the heat transfer to the control mass for any path plus the work done on the control mass, again, along any path. Between these two states, Eq. (4.6) is integrated to obtain

$$E_2 - E_1 = {}_1Q_2 + {}_1W_2 \tag{4.7}$$

The total energy contained by the control mass is the sum of the contribution of all the elements, so the total energy is represented as an integral over all the mass. This is given as

$$E = \int_m e \, dm \tag{4.8}$$

With the density and specific volume, this is given as a volume integral:

$$E = \int_V e\rho \, dV = \int_V e \frac{dV}{v} \tag{4.9}$$

Thus the final representation is

$$d\left(\int_V e\rho \, dV \right) = \delta Q + \delta W \tag{4.10}$$

The total energy in all the above equations is the sum of the internal, kinetic, and potential energy. These are included explicitly to yield

$$dU + d(\text{KE}) + d(\text{PE}) = \delta Q + \delta W \tag{4.11}$$

Integrating between states 1 and 2 results in

$$(U + \text{KE} + \text{PE})_2 - (U + \text{KE} + \text{PE})_1 = {}_1Q_2 + {}_1W_2 \tag{4.12}$$

Substituting in the expressions for kinetic and potential energy [see Eq. (2.11)] gives the conservation of energy as

$$(U + \tfrac{1}{2}m\mathbf{V}^2 + mgZ)_2 - (U + \tfrac{1}{2}m\mathbf{V}^2 + mgZ)_1 = {}_1Q_2 + {}_1W_2 \tag{4.13}$$

If the mass within the volume of the control mass is uniform in

$+\delta W$

$+\delta Q$

System boundary B

Figure 4.1   **Sign convention.**

space, then these equations are alternatively expressed on a specific basis or unit-mass basis as

$$d(me) = \delta(mq) + \delta(mw) \tag{4.14}$$

But the mass is constant in a control mass, so

$$de = \delta q + \delta w \tag{4.15}$$

For a process between states 1 and 2, the conservation-of-energy relation becomes

$$e_2 - e_1 = {}_1q_2 + {}_1w_2 \tag{4.16}$$

With the internal, kinetic, and potential energy expressions this is

$$(u + \tfrac{1}{2}\mathbf{V}^2 + gZ)_2 - (u + \tfrac{1}{2}\mathbf{V}^2 + gZ)_1 = {}_1q_2 + {}_1w_2 \tag{4.17}$$

As in Eq. (4.13), the left-hand side of Eq. (4.17) is an integration of an exact differential which represents the change in properties of the control mass. This change in total energy is independent of the path between the states. The right-hand sides of these equations are integrals of inexact differentials and generally depend on the particular path.

Example Problem 4.1

A frictionless piston is raised by slowly heating the gas contained in the cylinder. The initial pressure is 0.2 MPa, and the initial volume is 1 m³. The heat transfer to the tank is 2000 kJ, and the final volume is 2.0 m³. Determine the change in internal energy per unit mass of the fluid.

Diagrams

Solution

The gas is taken as a control mass, as indicated. This problem does not require property evaluation for a particular substance since all states are given. The work is calculated for the constant-pressure process, and the heat transfer is specified. The change in total energy is evaluated from the first law of thermodynamics, and the internal energy is obtained directly, since the changes in kinetic and potential energy are negligible. The states are indicated in this table:

State	P, MPa	V, m³	U, kJ	Process
1	0.2	1.0	—	P = constant
2	0.2	2.0	—	

The work done by the system is

$$_1W_2 = -\int_1^2 P\,dV = -P_1(V_2 - V_1)$$
$$= (-0.2\ \text{MPa})[(2.0 - 1.0)\ \text{m}^3] = -0.2\ \text{MJ}$$

The first law yields

$$U_2 - U_1 = {_1Q_2} + {_1W_2} = 2.0 - 0.2 = 1.8\ \text{MJ}$$

Comments

The internal energy of state 2 is 1.8 MJ greater than the internal energy of state 1.

The kinetic energy is related to the velocity of the system, and the velocities at states 1 and 2 are zero, so the kinetic energy contribution is also zero. The change in potential energy has been neglected, and from the stated information, the change in potential energy can only be estimated. The change in height of the gas center of mass depends on the specific geometric configuration, but it is probably no more than a few meters. The densities of various gases can be found from the appendixes to be of the order of a few kilograms per cubic meter. With the gravitational acceleration as 9.8 m/s², the change in potential energy ($mg\,\Delta Z$) is of the order of 0.1 kJ. This is negligible in the above calculation.

Example Problem 4.2

Evaluate the heat transfer necessary to evaporate water and to warm the vapor in the piston-cylinder experiment shown. The process is a quasi-equilibrium process. The initial state 1 is a saturated liquid at 1.0 MPa, and the evaporation process ends at state 2 on the saturated vapor line. Subsequent heat transfer takes the water to 600°C at state 3. Find $_1q_2$, $_2q_3$, and $_1q_3$.

Diagrams

Solution

The control mass undergoes a constant-pressure process, as indicated on the system diagram. The following table gives the information. States 1 and 2 are known, since both P and quality are given; v and u are found in Tables D.9 and D.10. State 3 is specified by P_3 and T_3.

State	P, MPa	T, °C	v, m³/kg	u, kJ/kg	x	Process
1	1.0†	—	(0.001127)	(761.11)	0.0	P = constant
2	1.0	—	(0.1943)	(2581.9)	1.0	P = constant
3	1.0	600	(0.4010)	(3292.8)	—	

† Values not in parentheses determine the state.

States 1 and 2 are obtained from the saturated steam tables, and state 3 is obtained from the superheated vapor tables.

The kinetic and potential energy changes are neglected, so the first law yields

$$e_2 - e_1 = u_2 - u_1 = {_1}q_2 + {_1}w_2$$

Thus

$$_1q_2 = u_2 - u_1 - {_1}w_2$$

The work done for a constant-pressure process is

$$_1w_2 = -\int_1^2 P \, dv = -P_1(v_2 - v_1)$$

Note that this is a special case of the polytropic process ($n = 0$) given in Eq. (2.20). For the process from 1 to 2,

$$_1w_2 = -1.0(0.1943 - 0.001127) \text{ MJ/kg} = -193.17 \text{ kJ/kg}$$

Thus the heat transfer is

$$_1q_2 = 2581.9 - 761.11 + 193.17 = 2014.0 \text{ kJ/kg}$$

Repeating for the process between 2 and 3 yields

$$_2w_3 = -206.7 \text{ kJ/kg} \qquad \text{and} \qquad _2q_3 = 917.6 \text{ kJ/kg}$$

Thus the total heat transfer is

$$_1q_3 = {_1}q_2 + {_2}q_3 = 2931.6 \text{ kJ/kg}$$

Because $_1q_3$ is positive, this heat transfer is *from* the surroundings *to* the system. Note that work was done *by* the system *on* the surroundings as indicated by the negative sign on $_1w_2$ and $_2w_3$.

Comments

The work performed by the control mass in each segment of this constant-pressure process is nearly equal, yet the heat transfer for evaporation (state 1 to state 2) is 68.7 percent of the total heat transfer.

The above analysis could also be expressed in terms of the enthalpy. With constant-pressure work given as

$$_1w_2 = -P_1(v_2 - v_1) = P_1v_1 - P_2v_2$$

and the first law given as

$$u_2 - u_1 = {_1}q_2 + {_1}w_2$$

we have

$$_1q_2 = u_2 - u_1 - {_1}w_2 = u_2 - u_1 + P_2v_2 - P_1v_1$$
$$= (u_2 + P_2v_2) - (u_1 + P_1v_1)$$

The enthalpy is defined as $h = u + Pv$, so

$$_1q_2 = h_2 - h_1$$

In a control mass, this expression is valid *only* for a *constant-pressure* work process. Reading the enthalpy from Tables D.9 and D.10 yields

$_1q_2 = h_g - h_l = h_{lg} = 2014.0 \text{ kJ/kg}$

and from state 2 to state 3

$_2q_3 = 3693.8 - 2776.3 = 917.5 \text{ kJ/kg}$

These are the values obtained above to within the accuracy of the tables.

The conservation-of-energy expression has been given for a process between states 1 and 2. If the control mass undergoes a complete cycle that takes the control mass from state 1 back to state 1, then Eq. (4.6) yields

$$\oint dE = \oint \delta Q + \oint \delta W \tag{4.18}$$

But total energy is a property, and according to the mathematical attributes of a property (Sec. 1.7.4), $\oint dE = 0$, so

$$\oint \delta Q = -\oint \delta W \tag{4.19}$$

Thus, the net heat transferred to a control mass undergoing a cycle is equal to the negative of the net work done. This is the starting point for some presentations of the first law of thermodynamics.

The conservation of energy can also be presented on a rate basis. For a time interval Δt, the expression is

$$\frac{dE}{dt} = \frac{\delta Q}{\Delta t} + \frac{\delta W}{\Delta t} \tag{4.20}$$

With the definitions presented in Sec. 2.5.3, this can be written as

$$\frac{dE}{dt} = \dot{Q} + \dot{W} \tag{4.21}$$

The energy within the volume is given in Eq. (4.9), so this can also be expressed as

$$\frac{d}{dt}\left(\int_V e\rho \, dV\right) = \dot{Q} + \dot{W} \tag{4.22}$$

This form is useful in the consideration of control volume formulations.

Example Problem 4.3

A control mass undergoes a three-process power cycle with air as the working fluid. The fluid undergoes an isothermal compression, a constant-pressure heating, and a polytropic process given by $Pv^k = $ constant, where $k = c_P/c_v$. Determine q and w for each individual process. Assume the specific heats are constant and $P_1 = 0.1$ MPa, $T_1 = 20°C$, and $P_2 = 0.6$ MPa.

Diagram

Solution

First, the ideal gas approximation must be checked for air for the conditions of this process. The most critical check occurs at the lowest temperature and largest pressure. This occurs at state 2, as can be seen. With the critical properties given from Table C.4 as $T_{cr} = 132.5$ K and $P_{cr} = 3.77$ MPa, then $T_r = 293.15/132.5 = 2.21$ and $P_r = 0.6/3.77 = 0.159$. The compressibility factor from Fig. C.1 yields

$$z \approx 1.0$$

Thus the ideal gas approximation is very accurate.

The state information is given:

State	P, MPa	T, °C	v, m³/kg	u, kJ/kg	h, kJ/kg	Process
1	0.1	20	—	—	—	$T = $ constant
2	0.6	20	—	—	—	$P = $ constant
3	0.6	—	—	—	—	$Pv^k = $ constant
1	0.1	20	—	—	—	

For air, $R = 0.28700$ kJ/(kg · K) and $k = 1.400$ (Table C.2) thus

$$v_1 = \frac{RT_1}{P_1} = \frac{(0.28700)(293.15)}{100} = 0.84134 \text{ m}^3/\text{kg}$$

Also,

$$v_2 = \frac{RT_2}{P_2} = 0.14022 \text{ m}^3/\text{kg}$$

The state table indicates that both state 1 and state 2 are completely specified, but state 3 requires one more property. This is evaluated from the process information between states 1 and 3. Thus

$$P_1 v_1^k = P_3 v_3^k$$

or

$$v_3 = v_1 \left(\frac{P_1}{P_3}\right)^{1/k}$$

Thus

$$v_3 = (0.84134)(\tfrac{1}{6})^{1/1.4} = 0.23396 \text{ m}^3/\text{kg}$$

and state 3 is then specified. Also $T_3 = P_3v_3/R = 489.11$ K.

The work is evaluated from Eq. (2.17) as

$$_1w_2 = -\int_1^2 P \, dv$$

For the process from 1 to 2, $P = RT/v$, so for this isothermal process [see Eq. (2.20)]

$$_1w_2 = -\int_1^2 \frac{RT}{v} \, dv = -RT_2 \ln \frac{v_2}{v_1} = +150.75 \text{ kJ/kg}$$

For the process from 2 to 3, $P =$ constant and

$$_2w_3 = -P_2(v_3 - v_2) = -56.24 \text{ kJ/kg}$$

For the process from 3 to 1, $Pv^k =$ constant $= P_1v_1^k = P_3v_3^k$, so

$$_3w_1 = -\int_3^1 P \, dv = -\text{const} \int_3^1 \frac{dv}{v^k}$$

$$= -\text{const} \left(\frac{v_1^{1-k} - v_3^{1-k}}{1 - k}\right)$$

Using $P_1v_1^k = P_3v_3^k =$ constant yields [see Eq. (2.20)]

$$_3w_1 = \frac{P_1v_1 - P_3v_3}{k - 1} = -140.61 \text{ kJ/kg}$$

The heat transfer for each process is evaluated from the first law with the work expressions given above. Conservation of energy for no kinetic or potential energy changes [Eq. (4.17)] states that

$$u_2 - u_1 = {}_1q_2 + {}_1w_2$$

so

$$_1q_2 = u_2 - u_1 - {}_1w_2$$

The change in internal energy is evaluated from Eq. (3.19) as

$$u_2 - u_1 = \int_1^2 c_v \, dT = c_v(T_2 - T_1)$$

Thus

$$_1q_2 = c_v(T_2 - T_1) - {}_1w_2$$

For the isothermal process from state 1 to state 2

$$_1q_2 = 0 - {_1}w_2 = -150.75 \text{ kJ/kg}$$

With $c_v = 0.7180 \text{ kJ/(kg} \cdot \text{K)}$, the constant-pressure process from state 2 to state 3 yields

$$_2q_3 = 0.7180(489.11 - 293.15) + 56.24 = 196.94 \text{ kJ/kg}$$

The process from state 3 to state 1 yields

$$_3q_1 = 0.7180(293.15 - 489.11) + 140.61 = -0.09 \approx 0 \text{ kJ/kg}$$

The process described by $Pv^k = $ constant is shown in Chap. 5 to be an adiabatic process, so the finite value is a result of property uncertainty.

In summary,

State	w, kJ/kg	q, kJ/kg
1 to 2	150.75	-150.75
2 to 3	-56.24	196.94
3 to 1	-140.61	-0.09
Total	-46.10	46.10

This satisfies Eq. (4.19), or the heat transfer is equal to the negative of the work done for a cyclic process.

Comments

The constant-pressure process can also be considered in terms of enthalpy. The heat transfer from state 2 to state 3 is

$$\begin{aligned}_2q_3 &= u_3 - u_2 - {_2}w_3 \\ &= u_3 - u_2 + P_2(v_3 - v_2) \\ &= (u_3 + P_3v_3) - (u_2 + P_2v_2) \\ &= h_3 - h_2\end{aligned}$$

With Eq. (3.22),

$$_2q_3 = c_P(T_3 - T_2) = 1.005(489.11 - 293.15) = 196.94 \text{ kJ/kg}$$

This is the value obtained above when four significant figures are used for c_P and c_v.

4.2.3 Other Conservation Relations

Two conservation statements can also be made for momentum: linear momentum and the moment of momentum. The conservation of linear momentum has been used previously in obtaining relations for pressure measurement. These principles form the foundation for solid mechanics and fluid mechanics. They are not covered in this text. They are mentioned here because the form of these two conservation relations is very similar to that of the statements of conservation of mass and conservation of energy. The conservation of linear momentum is outside the realm of classical thermodynamics and is studied in a course on fluid mechanics or transport phenomena. Yet this equation is necessary to evaluate the detailed velocity distributions within flowing systems. This text focuses on thermodynamics, so any needed velocity distributions are specified; thus, the solution to the linear momentum equation is not required. This point is similar to that made on heat transfer (see Sec. 2.5.2).

4.3   Control Volume Formulation

Many systems are most conveniently analyzed as a system fixed in space which permits mass to cross the boundaries. Such a system is defined to be a *control volume* (CV). The boundary surface B can move, and mass passes through the boundary at the crossing boundaries. Most flow systems including pumps, turbines, heat exchangers, and valves are analyzed with control volumes.

The conservation statements for mass and energy are presented for a control mass (CM) in Sec. 4.2. The mass under consideration is always the same, and the conservation statement specifies what is necessary to change the property in question. An alternate, yet equivalent, approach considers a control volume and specifies what is necessary to change the property in the control volume. The difference is that the changes in the properties resulting from flows at the inlets and outlets must be included.

This section restates the conservation principles presented in Sec. 4.2 for a control volume. The conservation principles are exactly the same; yet since mass crosses the boundary, extra contributions must be included. The mathematical justification for the extra terms is presented in App. F.

4.3.1   Conservation of Mass

The mass flowing into a specified control volume must equal the mass leaving the control volume; otherwise, the mass within the control volume must change. When a garden hose is initially turned on, there is no water leaving the hose since water is filling the volume of the hose. This is stated as follows:

> The change of mass within a control volume is equal to the mass entering minus the mass leaving.

This is simply a "bookkeeping" of the mass within the control volume at any time.

Let the symbol $\dot{m}$ stand for the mass flow rate. All flow streams entering the control volume across the system boundary are added to get the total rate at which mass enters the system:

$$\dot{m}_{in} = \sum_{in} \dot{m} \tag{4.23}$$

Here the subscript *in* denotes mass flow at the inlets made up of contributions from the multiple inlet streams, each with individual mass flow rate $\dot{m}$. Similarly for the outlet streams, the mass flow rate at the outlets is

$$\dot{m}_{out} = \sum_{out} \dot{m} \tag{4.24}$$

where the subscript *out* denotes outlets.

Now the mass balance on the control volume (CV) becomes simply

$$\frac{\partial m_{CV}}{\partial t} = \dot{m}_{in} - \dot{m}_{out} = \sum_{in} \dot{m} - \sum_{out} \dot{m} \tag{4.25}$$

This states that the rate of change of mass within the control volume is equal to the mass that enters the control volume per unit time minus the mass that leaves the control volume per unit time. In short, what stays within is equal to what goes in minus what comes out. This is shown schematically in Fig. 4.2.

A more detailed view of each term in Eq. (4.25) takes into consideration every element within the control volume and each element at the crossing boundary. The mass within the control volume is expressed as

$$m_{CV} = \int_{m_{CV}} dm = \int_{CV} \rho \, dV = \int_{CV} \frac{dV}{v} \tag{4.26}$$

$$\dot{m}_{in} = \dot{m}_1 + \dot{m}_2$$
$$\dot{m}_{out} = \dot{m}_3 + \dot{m}_4$$

$$\frac{\partial m_{CV}}{\partial t} = \dot{m}_{in} - \dot{m}_{out}$$

Figure 4.2 Conservation of mass for a control volume.

This is the same term as in the control mass formulation given in Eq. (4.4).

The boundary terms are expressed in terms of the local velocity and area elements. Consider an individual area element dA with a mass flow rate per unit area across this area element of $d\dot{m}/dA$. Then

$$\dot{m} = \int_A \frac{d\dot{m}}{dA}\, dA$$

The mass flow rate per unit area is also expressed in terms of the velocity. The volume swept out as the fluid particle moves (Fig. 4.3) is

$$\Delta L \cos\theta\, dA \tag{4.27}$$

For a time increment Δt, the volume flow rate across the boundary area element is

$$\frac{\Delta L \cos\theta\, dA}{\Delta t} \tag{4.28}$$

but the ratio of ΔL to Δt as the time increment goes to zero is the magnitude of the velocity:

$$\lim_{\Delta t \to 0} \frac{\Delta L \cos\theta}{\Delta t} = |\mathbf{V} \cdot \mathbf{n}| \tag{4.29}$$

Velocity is a vector, so the dot product with the outward normal from the crossing boundary gives the normal velocity as $\mathbf{V}\cos\theta$. Thus, the volume flow rate is

$$\lim_{\Delta t \to 0} \frac{\Delta L}{\Delta t} \cos\theta\, dA = |\mathbf{V} \cdot \mathbf{n}|\, dA \tag{4.30}$$

The mass flow rate per unit area of the crossing boundary is

$$d\dot{m} = \rho |\mathbf{V} \cdot \mathbf{n}|\, dA \tag{4.31}$$

For the boundary B with the crossing boundary at the inlet denoted A_{in} and the crossing boundary at the outlet denoted A_{out}, we have

$$\dot{m}_{\text{in}} = \int_{A_{\text{in}}} \rho |\mathbf{V} \cdot \mathbf{n}|\, dA = \int_{A_{\text{in}}} d\dot{m} \tag{4.32}$$

and

$$\dot{m}_{\text{out}} = \int_{A_{\text{out}}} \rho |\mathbf{V} \cdot \mathbf{n}|\, dA = \int_{A_{\text{out}}} d\dot{m} \tag{4.33}$$

The conservation-of-mass statement is then

$$\frac{\partial}{\partial t}\left(\int_{\text{CV}} \rho\, dV \right) = \int_{A_{\text{in}}} \rho |\mathbf{V} \cdot \mathbf{n}|\, dA - \int_{A_{\text{out}}} \rho |\mathbf{V} \cdot \mathbf{n}|\, dA \tag{4.34a}$$

For multiple inlet and outlet ports, this becomes

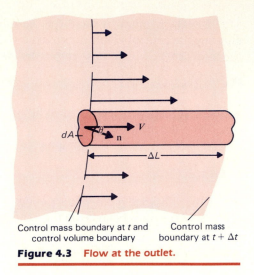

Control mass boundary at t and control volume boundary

Control mass boundary at $t + \Delta t$

Figure 4.3   Flow at the outlet.

$$\frac{\partial}{\partial t}\left(\int_{CV} \rho \, dV\right) = \sum_{in} \int_{A_{in}} d\dot{m} - \sum_{out} \int_{A_{out}} d\dot{m} \qquad (4.34b)$$

There are other important forms for these equations as well as a number of useful idealizations. The particular expression employed depends on the problem considered. These are addressed in detail in Sec. 4.4. One special case considers no mass entering or leaving the control volume. The conservation of mass reduces to

$$\frac{dm_{CV}}{dt} = 0 \qquad (4.35)$$

or, by integrating with respect to time,

$$m_{CV} = \text{constant} \qquad (4.36)$$

which we have assumed all along for such a system. This is exactly the expression for a control mass [Eq. (4.2)].

4.3.2   Conservation of Energy

The mass flowing into the control volume carries energy with it, and the same is true for the outlets from the control volume. The net gain in energy from the inlets and outlets must equal the change in energy within the control volume plus the heat transfer and the work done. The work term requires special attention since the fluid moving into and out of the control volume is doing work. This is the conservation statement:

> The change in energy within the control volume is equal to the net energy transported to the control volume plus the heat transferred and the work done to the control volume.

This is shown schematically in Fig. 4.4.

Let $\dot{E}$, termed the *energy flux,* denote the energy transported with the fluid at the inlets and outlets. The overdot indicates the rate of flow of energy as in the mass flow rate. The energy within the control volume is denoted by E_{CV}, so the conservation of energy is

$$\frac{\partial E_{CV}}{\partial t} = \dot{E}_{net} + \dot{Q}_{CV} + \dot{W} \qquad (4.37)$$

or, in terms of inlets and outlets,

$$\frac{\partial E_{CV}}{\partial t} = \dot{E}_{in} - \dot{E}_{out} + \dot{Q}_{CV} + \dot{W} \qquad (4.38)$$

Each term requires detailed consideration.

The energy change within the control volume is given by

$$\frac{\partial E_{CV}}{\partial t} = \frac{\partial}{\partial t}\left(\int_{CV} \rho e \, dV\right) \qquad (4.39)$$

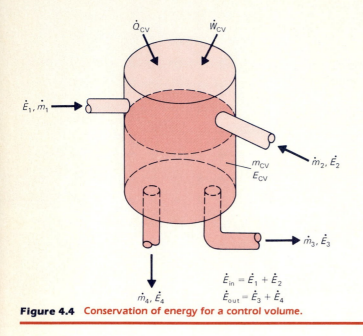

$$\dot{E}_{in} = \dot{E}_1 + \dot{E}_2$$
$$\dot{E}_{out} = \dot{E}_3 + \dot{E}_4$$

Figure 4.4 **Conservation of energy for a control volume.**

This represents the change of total energy within the control volume in terms of the sum of the contributions of the volume elements. This is often called the *energy storage* term since any accumulation of energy within the control volume is included in this expression.

The mass flow transports energy into and out of the control volume. These contributions are denoted by $\dot{E}_{in}$ and $\dot{E}_{out}$, respectively. The development of the mass flow rate term for the conservation of mass is extended to include the energy transport. Equation (4.31) represents the mass flow rate per unit area element of the boundary. The fluid carries energy across each element, so the energy transported with the fluid element is given by Eq. (4.31) multiplied by the specific total energy, or

$$d\dot{E} = e\rho|\mathbf{V} \cdot \mathbf{n}| \, dA = e \, d\dot{m} \tag{4.40}$$

The sum, or integral, over all the inlet and outlet boundaries is

$$\dot{E}_{in} = \int_{A_{in}} e\rho|\mathbf{V} \cdot \mathbf{n}| \, dA = \int_{A_{in}} e \, d\dot{m} \tag{4.41}$$

and

$$\dot{E}_{out} = \int_{A_{out}} e\rho|\mathbf{V} \cdot \mathbf{n}| \, dA = \int_{A_{out}} e \, d\dot{m} \tag{4.42}$$

Therefore, the net energy transported into the control volume by fluid motion is

$$\dot{E}_{net} = \int_{A_{in}} e\rho |\mathbf{V} \cdot \mathbf{n}|\, dA - \int_{A_{out}} e\rho |\mathbf{V} \cdot \mathbf{n}|\, dA$$

$$= \int_{A_{in}} e\, d\dot{m} - \int_{A_{out}} e\, d\dot{m} \tag{4.43}$$

Equation (4.37) is given with Eqs. (4.39) and (4.43) as

$$\frac{\partial}{\partial t}\left(\int_{CV} \rho e\, dV\right) = \int_{A_{in}} e\, d\dot{m} - \int_{A_{out}} e\, d\dot{m} + \dot{Q}_{CV} + \dot{W} \tag{4.44}$$

For multiple inlets and multiple outlets this is

$$\frac{\partial}{\partial t}\left(\int_{CV} \rho e\, dV\right) = \sum_{in} \int_{A_{in}} e\, d\dot{m} - \sum_{out} \int_{A_{out}} e\, d\dot{m} + \dot{Q}_{CV} + \dot{W} \tag{4.45}$$

Heat Transfer Contribution

The heat transfer to the control volume is the energy transport across any portion of the boundary. This energy transfer is the same as considered for the control mass. It generally occurs across the inside boundary of the control volume.

Work Contribution

The work term in the conservation of energy includes the general formulation contained in Sec. 2.5.1. The contributions of $P\, dV$ work, electrical work, and other work forms which transport energy across the boundary are incorporated into the energy balance. Three additional forms of work were not included in Sec. 2.5.1 but must be included in the work contribution when fluid motion is permitted. The general definition for work is

$$\delta W = \mathbf{F} \cdot d\mathbf{s}$$

The rate of doing work is

$$\dot{W} = \lim_{\Delta t \to 0} \frac{\delta W}{\Delta t} = \lim_{\Delta t \to 0} \frac{\mathbf{F} \cdot d\mathbf{s}}{\Delta t} \tag{4.46}$$

With the velocity defined as

$$\mathbf{V} = \lim_{\Delta t \to 0} \frac{d\mathbf{s}}{\Delta t} \tag{4.47}$$

the rate of doing work on a fluid element is

$$\dot{W} = \mathbf{F} \cdot \mathbf{V} \tag{4.48}$$

Shear work results from fluid flowing over the boundary where a force is exerted on it by the boundary element dA. This force is expressed in terms of the shear stress as $\tau\, dA$. Thus, the work done by the fluid is the shear work and is

$$\delta \dot{W}_{shear} = \tau\, dA \cdot \mathbf{V} \tag{4.49}$$

The shear work done on the entire boundary, B, is

$$\dot{W}_{shear} = \int_B \delta\dot{W}_{shear} = \int_B \boldsymbol{\tau} \cdot \mathbf{V}\, dA \tag{4.50}$$

The particular choice of the boundary prescribes the contribution of this term. At the inside boundaries which coincide with a solid surface, the fluid "sticks" to the walls, so the velocity is zero. Therefore, the shear work contribution for an inside boundary is zero. At inlets and outlets to the control volume, the flow is generally normal to the area element. Since the normal component of the shear stress is zero, the shear work is zero for this case. If a control volume includes a shaft crossing the boundary and interacts with the fluid through a shearing motion, this contribution is included under a different label, *shaft work* (see the next paragraph) and is not included here. Other contributions to the shear work term are possible but are rarely significant, so they are not included.

Shaft work is the contribution resulting from the rotation of a paddle wheel or turbine rotor within the control volume. The shaft crosses the boundary and transfers energy between the control volume and surroundings. This shaft work term is truly a result of shear forces acting on the paddle wheel, yet the measurable response from the control volume is in the rotation of the shaft. This is a directly usable work mode. This work contribution is termed *shaft work* $\dot{W}_{shaft}$. This work is usually the desired result in a work calculation because it is directly usable.

Flow work is related to the normal force at the boundary. The normal force includes the contribution from the pressure and viscous effects, yet the pressure force is the dominant contribution in most applications. At the inside boundary, the net work is zero since the wall counterbalances the fluid force. At the crossing boundary, the pressure force is $-P\, dA$, and it acts on an area element which is perpendicular to the normal velocity component $\mathbf{V} \cdot \mathbf{n}$. Thus, the flow work for the crossing boundaries is

$$\dot{W}_{flow} = -\int_B P\mathbf{V} \cdot \mathbf{n}\, dA \tag{4.51}$$

With the mass flow rate expression in Eqs. (4.31) through (4.33) and multiplying by $\rho v = 1$, we find

$$\dot{W}_{flow} = -\int_B Pv\rho\mathbf{V} \cdot \mathbf{n}\, dA = \int_{A_{in}} Pv\, d\dot{m} - \int_{A_{out}} Pv\, d\dot{m} \tag{4.52}$$

In summary, the rate of work done in the control volume formulation is the sum of the shear, shaft, and flow work plus the general forms discussed in Sec. 2.5.1. Since the shear work is typically zero for the reasons just discussed, the control volume work is

$$\dot{W} = \dot{W}_{shear}^{\,0} + \dot{W}_{shaft} + \dot{W}_{gen} + \dot{W}_{flow}$$

$$\dot{W} = \dot{W}_{shaft} + \dot{W}_{gen} + \int_{A_{in}} Pv \, d\dot{m} - \int_{A_{out}} Pv \, d\dot{m} \qquad (4.53a)$$

The $\dot{W}_{gen}$ term includes the $P \, dV$ work, electrical work, etc. from Eq. (2.31), and it is combined with $\dot{W}_{shaft}$ and expressed in the following equations as $\dot{W}_{CV}$. Thus,

$$\dot{W} = \dot{W}_{shaft} + \dot{W}_{gen} + \int_{A_{in}} Pv \, d\dot{m} - \int_{A_{out}} Pv \, d\dot{m}$$

$$= \dot{W}_{CV} + \int_{A_{in}} Pv \, d\dot{m} - \int_{A_{out}} Pv \, d\dot{m} \qquad (4.53b)$$

Conservation-of-Energy Expression

The conservation of energy for a control volume is obtained from Eq. (4.37) with Eqs. (4.39), (4.43), and (4.53b). These substitutions yield

$$\frac{\partial}{\partial t}\left(\int_{CV} \rho e \, dV\right) = \int_{A_{in}} e \, d\dot{m} + \int_{A_{in}} Pv \, d\dot{m} + \dot{Q}_{CV} + \dot{W}_{CV}$$

$$- \int_{A_{out}} e \, d\dot{m} - \int_{A_{out}} Pv \, d\dot{m} \quad (4.54)$$

Combining the integrals and rearranging yield

$$\frac{\partial}{\partial t}\left(\int_{CV} \rho e \, dV\right) = \int_{A_{in}} (e + Pv) \, d\dot{m} + \dot{Q}_{CV} + \dot{W}_{CV}$$

$$- \int_{A_{out}} (e + Pv) \, d\dot{m} \quad (4.55)$$

This conservation statement says that the change of energy within the control volume is equal to the entering energy contributions minus the leaving energy contributions. The energy balance is also expressed in terms of multiple inlet streams and multiple outlet streams as

$$\frac{\partial}{\partial t}\left(\int_{CV} \rho e \, dV\right) = \sum_{in} \int_{A_{in}} (e + Pv) \, d\dot{m} + \dot{Q}_{CV} + \dot{W}_{CV}$$

$$- \sum_{out} \int_{A_{out}} (e + Pv) \, d\dot{m} \quad (4.56)$$

The results for conservation of mass in Eq. (4.34) compare with the conservation of energy in Eq. (4.56). Both describe the change in the property within the control volume—the property is mass in Eq. (4.34) and energy in Eq. (4.56). These changes are equal to the incoming contributions minus the outgoing contributions. In the case of the conservation of mass, the contributions are mass flow rates. For conservation of energy with the sign convention indicated in Fig. 4.1, the incoming contribution is heat transfer, energy flux, flow work, and work, and the outgoing contribution is the energy flux and flow work.

A very important term appears in the flux terms. It is the sum of the specific total energy and the product Pv. The Pv term results from the flow work performed as fluid enters and leaves the control volume. The total energy is the sum of the internal, kinetic, and potential energy, so

$$e + Pv = \frac{E}{m} + Pv = u + \tfrac{1}{2}\mathbf{V}^2 + gZ + Pv \tag{4.57}$$

Enthalpy is defined as $h = u + Pv$, so

$$e + Pv = h + \tfrac{1}{2}\mathbf{V}^2 + gZ \tag{4.58}$$

This indicates another important application for enthalpy. Equation (4.56) is written as

$$\frac{\partial}{\partial t}\left(\int_{CV} \rho e \, dV\right) = \sum_{in} \int_{A_{in}} (h + \tfrac{1}{2}\mathbf{V}^2 + gZ)\, d\dot{m} + \dot{Q}_{CV} + \dot{W}_{CV}$$

$$- \sum_{out} \int_{A_{out}} (h + \tfrac{1}{2}\mathbf{V}^2 + gZ)\, d\dot{m} \tag{4.59}$$

where $e = u + \tfrac{1}{2}\mathbf{V}^2 + gZ$. In short, what stays within is equal to what goes in minus what goes out.

As in the case of conservation of mass, a number of important idealizations are of practical significance. The first consideration is at the inlet and outlet streams. The others involve the control volume. These are addressed in detail in Sec. 4.4. For the special case of no mass entering or leaving the control volume, all $\dot{m}$'s are zero, and Eq. (4.59) reduces to the control mass form:

$$\frac{d}{dt}\left(\int_{CV} \rho e \, dV\right) = \frac{dE_{CV}}{dt} = \dot{Q}_{CV} + \dot{W}_{CV} \tag{4.60}$$

Since $\dot{W}_{gen}$ is included in the work term, the same expression results as previously given for a control mass [Eq. (4.22)]. If there is no energy transfer, the result is

$$\frac{dE_{CV}}{dt} = 0 \tag{4.61}$$

which is the isolated system result.

4.4 Control Volume Analysis

The analysis of a system requires detailed consideration of the inlets and outlets to the control volume as well as the state within the control volume. The spatial considerations are addressed at both the inlets and outlets and the volume itself. *Uniform properties* are spatially constant but may vary with time. At inlets and outlets this means that the state across the crossing boundary is described at any time by single values of the properties. Uniform properties within the control volume indicate that all elements within the control volume have the same value for the properties at any time. *Nonuniform properties* vary spatially. The time variation must also be addressed at the inlets and outlets as well as within the control volume. *Steady state* flow conditions at the inlets and outlets mean that there is no variation of the properties with time. The density, velocity, specific internal energy, etc. have constant values for all times. Steady state properties within the control volume indicate that there is no accumulation within the control volume that causes the properties within the control volume to vary with time. *Unsteady state* conditions imply time variation either at the inlets and outlets or within the control volume. In summary, uniform and nonuniform describe the spatial variation while steady state and unsteady state describe the time variation.

4.4.1 Inlet and Outlet Considerations

The mass flow rate for an inlet or outlet is given by the form [Eq. (4.32) or (4.33)]

$$\dot{m} = \int_A d\dot{m} = \int_A \rho |\mathbf{V} \cdot \mathbf{n}| \, dA \tag{4.62}$$

where all the quantities in the integrand can vary with time and space. Thus, this is the general nonuniform, unsteady form for an inlet or outlet. All considerations of problems should start with this expression. Applying appropriate idealizations reduces this term to a form useful for the particular problem. A few common idealizations are addressed here.

If the *velocity and density (or specific volume) are uniform across the inlets and outlets and the velocity is normal to the ports*, then $|\mathbf{V} \cdot \mathbf{n}| = \mathbf{V}$ and the mass flow rates in Eq. (4.62) reduce to

$$\dot{m}_{\text{in}} = (\rho \mathbf{V} A)_{\text{in}} = \left(\frac{\mathbf{V} A}{v}\right)_{\text{in}} \tag{4.63}$$

$$\dot{m}_{\text{out}} = (\rho \mathbf{V} A)_{\text{out}} = \left(\frac{\mathbf{V} A}{v}\right)_{\text{out}} \tag{4.64}$$

for a single inlet and outlet. These expressions can be used if the stated information is interpreted as average values. Many problems specify only the inlet and outlet pipe sizes and velocities, so Eqs. (4.63) and (4.64) are directly applicable.

Instrumentation for flow measurements is often calibrated to read *volumetric flow rate* $\dot{V}$. Because conservation of mass in its simplest form is written in terms of mass flow rate $\dot{m}$, it is convenient to have relations between $\dot{m}$ and $\dot{V}$. The volumetric flow rate is easily converted to mass flow rate if the density or specific volume is uniform at the inlets and outlets. Dividing by the specific volume of the flowing material at the port yields

$$\dot{m} = \frac{\dot{V}}{v} \tag{4.65}$$

Unsteady considerations at inlets and outlets result from the time variation of the mass flow rate, usually through a time variation of the fluid velocity. These conditions generally require the summation of all time increments or an integration of the inlet and outlet mass flow rates in time. The mass transported into or out of the control volume from t_1 to t_2 is

$$\int_{t_1}^{t_2} \dot{m} \, dt = m \tag{4.66}$$

The unsubscripted mass without an overdot on the right-hand side represents the total mass crossing the control volume for the time interval $t_2 - t_1$.

The inlet and outlet terms for conservation of energy [see Eq. (4.56)] are

$$\int_A (e + Pv) \, d\dot{m} = \int_A (e + Pv)\rho|\mathbf{V} \cdot \mathbf{n}| \, dA \tag{4.67}$$

where $e = u + \frac{1}{2}\mathbf{V}^2 + gZ$. This general form incorporates nonuniform properties and unsteady time variations. Again, simplifying idealizations are useful as in the case of the mass flow rate.

If the *properties (h and v) and velocity are uniform* at the inlets and outlets, then the crossing boundary terms are

$$(h + \tfrac{1}{2}\mathbf{V}^2 + gZ)\dot{m} \tag{4.68}$$

which can also be combined with Eqs. (4.63) and (4.64) for $\dot{m}$. This yields

$$(h + \tfrac{1}{2}\mathbf{V}^2 + gZ)_{\text{in}} (\rho\mathbf{V}A)_{\text{in}} \tag{4.69}$$

and

$$(h + \tfrac{1}{2}\mathbf{V}^2 + gZ)_{\text{out}} (\rho\mathbf{V}A)_{\text{out}} \tag{4.70}$$

When the *time variation* must be included, time integration of Eq. (4.67) is often required. This contribution takes the form

$$\int_{t_1}^{t_2} \int_A (e + Pv) \, d\dot{m} \, dt \tag{4.71}$$

which represents all the contributions of $(e + Pv) \, d\dot{m}$ for $t_2 - t_1$. If the inlet or outlet has *uniform, steady state* properties ($e + Pv = $ constant), then this term is

$$(e + Pv)m = (h + \tfrac{1}{2}\mathbf{V}^2 + gZ)m \tag{4.72}$$

Equation (4.72) requires $e + Pv$ to be uniform and steady state, but $d\dot{m}$ could be nonuniform as well as unsteady.

TABLE 4.1 Mass and Energy Terms for a Control Volume†

	Nonuniform and Unsteady	Uniform	Unsteady Time Integration
Inlets and outlets	$\dot{m} = \int_A d\dot{m} = \int_A \rho\|\mathbf{V} \cdot \mathbf{n}\| dA$	$\dot{m} = \rho\mathbf{V}A$	$\int_{t_1}^{t_2} \dot{m}\, dt = m$
	$\int_A (e + Pv)\, d\dot{m}$	$(e + Pv)\dot{m}$	$\int_{t_1}^{t_2} \int_A (e + Pv)\, d\dot{m}\, dt$
	$= \int_A (e + Pv)\rho\|\mathbf{V} \cdot \mathbf{n}\| dA$		$= (e + Pv)m$ with uniform and steady state properties (not necessarily $\dot{m}$)
Within the control volume	$m_{\mathrm{CV}} = \int_{\mathrm{CV}} \rho\, dV$	$m_{\mathrm{CV}} = (\rho V)_{\mathrm{CV}}$	$\int_{t_1}^{t_2} \dfrac{\partial m_{\mathrm{CV}}}{\partial t}\, dt = m_2 - m_1$
	$E_{\mathrm{CV}} = \int_{\mathrm{CV}} \rho e\, dV$	$E_{\mathrm{CV}} = (\rho e V)_{\mathrm{CV}} = (me)_{\mathrm{CV}}$	$\int_{t_1}^{t_2} \dfrac{\partial E_{\mathrm{CV}}}{\partial t}\, dt = E_2 - E_1$ $= (em)_2 - (em)_1$ with uniform properties

† Here $e = u + \frac{1}{2}\mathbf{V}^2 + gZ$.

All the above forms along with the required idealizations are shown in Table 4.1.

4.4.2 Considerations Within the Control Volume

The idealizations addressed at the inlets and outlets must also be considered for the terms describing the properties within the control volume. The general form for the mass within the control volume [Eq. (4.26)] is

$$m_{\mathrm{CV}} = \int_{\mathrm{CV}} \rho\, dV \tag{4.73}$$

This includes nonuniform property variations.

If the *density (or specific volume) is uniform* throughout the control volume, then

$$m_{\mathrm{CV}} = (\rho V)_{\mathrm{CV}} \tag{4.74}$$

Also, the properties within the control volume vary with time in some applications. This requires the integration of the conservation of mass over the time interval considered. The variation of mass with time or *unsteady* contribution yields

$$\int_{t_1}^{t_2} \frac{\partial m_{\mathrm{CV}}}{\partial t}\, dt = m_2 - m_1 \tag{4.75}$$

The right-hand side of Eq. (4.75) represents the mass within the control volume at time t_2 minus the mass within the control volume at t_1.

The energy within the control volume is given by

$$E_{\mathrm{CV}} = \int_{\mathrm{CV}} \rho e\, dV \tag{4.76}$$

where $e = u + \frac{1}{2}\mathbf{V}^2 + gZ$. This term represents the energy contained within the control volume in its most general form, including possible nonuniform and unsteady variations.

The total energy and density within the control volume might be idealized as *uniform* in space. This yields

$$E_{CV} = (\rho eV)_{CV} = (me)_{CV} \tag{4.77}$$

where Eq. (4.74) has been used. This is an important expression since only one value describes the complete control volume energy.

Start-up and filling problems describe the accumulation of mass and energy within the control volume. This *unsteady* consideration results in an energy storage term. The time integration is generally required, and the needed expression is

$$\int_{t_1}^{t_2} \frac{\partial E_{CV}}{\partial t}\, dt = E_2 - E_1 \tag{4.78}$$

where E_2 and E_1 are the energies within the control volume at t_2 and t_1, respectively. If it is also assumed that the properties are *uniform* throughout the volume at any instant, then

$$\int_{t_1}^{t_2} \frac{\partial E_{CV}}{\partial t}\, dt = (em)_2 - (em)_1 \tag{4.79}$$

All the above expressions with the appropriate idealizations are given in Table 4.1.

The inlets and outlets are considered separately from the properties within the control volume since the terms have very different meanings. They also result in different forms for the various idealizations. Table 4.1 summarizes the above assumptions, but not all possible conditions have been addressed. Two types of problems that occur very often in engineering analysis combine the idealizations given above, and are considered next. Both assume uniform properties.

4.4.3 Steady State Analysis

The analysis of the long-term operation of a particular component or complete system eliminates the need for describing the initial start-up period. This permits consideration of the steady state operation where no mass or energy accumulates within the control volume. Thus

$$\frac{\partial m_{CV}}{\partial t} = 0 \tag{4.80}$$

$$\frac{\partial E_{CV}}{\partial t} = 0 \tag{4.81}$$

The conservation of mass and of energy in Eqs. (4.34) and (4.55) are

$$0 = \int_{A_{in}} \rho |V \cdot n| \, dA - \int_{A_{out}} \rho |V \cdot n| \, dA \tag{4.82}$$

or

$$\int_{A_{in}} dm = \int_{A_{out}} dm \tag{4.83}$$

and

$$0 = \int_{A_{in}} (e + Pv) \, dm + \dot{Q}_{CV} + \dot{W}_{CV} - \int_{A_{out}} (e + Pv) \, dm \tag{4.84}$$

where $e = u + \frac{1}{2}V^2 + gZ$. Assuming uniform properties at the inlets and outlets yields (see Table 4.1)

$$\dot{m}_{in} = \dot{m}_{out} \tag{4.85}$$

and

$$(h + \tfrac{1}{2}V^2 + gZ)_{in}\dot{m}_{in} + \dot{Q}_{CV} + \dot{W}_{CV} = (h + \tfrac{1}{2}V^2 + gZ)_{out}\dot{m}_{out} \tag{4.86}$$

These expressions are the basis of much of the cycle component analysis presented in the following chapters. The subscripts *in* and *out* represent all the inlets and outlets. In short, both these expressions state, "what goes in equals what comes out."

4.4.4 Unsteady State Analysis

Time variation occurs in filling or discharging problems as well as in start-up and shutdown considerations. Unsteady state analysis is needed when there is accumulation or storage within the control volume and/or time variation at the inlets and outlets. Integrating the conservation statements in Eqs. (4.25) and (4.55) from t_1 to t_2 and using Table 4.1 yield

$$\int_{t_1}^{t_2} \frac{\partial m_{CV}}{\partial t} \, dt = \int_{t_1}^{t_2} \dot{m}_{in} \, dt - \int_{t_1}^{t_2} \dot{m}_{out} \, dt \tag{4.87}$$

or

$$m_2 - m_1 = m_{in} - m_{out} \tag{4.88}$$

and

$$\int_{t_1}^{t_2} \frac{\partial}{\partial t}\left(\int_{CV} e\rho \, dV \right) dt = \int_{t_1}^{t_2} \int_{A_{in}} (e + Pv) \, dm \, dt + \int_{t_1}^{t_2} \dot{Q}_{CV} \, dt$$
$$+ \int_{t_1}^{t_2} \dot{W}_{CV} \, dt - \int_{t_1}^{t_2} \int_{A_{out}} (e + Pv) \, dm \, dt \tag{4.89}$$

If the properties are considered to be uniform (at the inlets and outlets and within the control volume) and the inlets and outlets are steady, then

Table 4.1 yields the simplified expressions. The heat transfer and work terms are

$$\int_{t_1}^{t_2} \dot{Q}\, dt = {}_1Q_2 \tag{4.90}$$

and

$$\int_{t_1}^{t_2} \dot{W}\, dt = {}_1W_2 \tag{4.91}$$

where the right-hand sides represent the heat transferred and work done for $t_2 - t_1$.

This results in the conservation of energy as

$$(em)_2 - (em)_1 \equiv (u + \tfrac{1}{2}\mathbf{V}^2 + gZ)_2 m_2 - (u + \tfrac{1}{2}\mathbf{V}^2 + gZ)_1 m_1$$

$$= (h + \tfrac{1}{2}\mathbf{V}^2 + gZ)_{in} m_{in} + {}_1Q_2 + {}_1W_2$$

$$- (h + \tfrac{1}{2}\mathbf{V}^2 + gZ)_{out} m_{out} \tag{4.92}$$

This equation includes the unsteady storage term within the control volume and all inlet and outlet flow terms (terms with subscripts *in* and *out*). In short, the conservation of energy states that the change of energy within equals energy in minus energy out.

These two idealizations for conservation of mass and energy are summarized in Table 4.2.

4.5   Control Volume Applications

There are three major subdivisions to thermodynamic analysis.

1. *Control volume location*
   The control volume must be indicated on the system drawing. Carefully distinguish between the crossing boundary and inside boundary. Denote the energy transfers. The boundary must be drawn so that the transfer process of interest (mass flow, work, heat transfer) crosses the boundary if the solution for that quantity is sought. The conservation statements connect the desired result to other quantities. These other quantities must be known if the desired quantity is to be evaluated.
2. *Conservation statements*
   The conservation of mass and the conservation of energy are applied to the control volume. The first point to consider is the time variation of the problem. Is unsteady or steady state analysis appropriate? The second point is the spatial variation of the inlets and outlets as well as within the control volume. Are the properties uniform within the control volume or at the inlets and outlets? All idealizations must be carefully considered.

3. *Properties*

Two independent properties for a simple compressible substance are required to determine the state. Tabular information or equations of state are needed to obtain the results.

These three steps are many times done simultaneously or iteratively depending on the problem.

The results presented in Secs. 4.3 and 4.4 are applied to typical engineering components in this section. They are subdivided into four subsections depending on whether the component requires unsteady state or steady state analysis and whether there is a work interaction or not. The starting equations for all the applications considered are given in Table 4.2.

4.5.1 Steady State Work Applications

Turbines, compressors, pumps, and fans are typical steady state devices that have a work interaction with the surroundings. All these devices generally have a single inlet and a single outlet. Also the change in potential energy from inlet to outlet is usually very small. Thus, Table 4.2 yields

Mass:

$$\dot{m}_{in} = \dot{m}_{out}$$

Energy:

$$(h + \tfrac{1}{2}\mathbf{V}^2)_{in}\dot{m}_{in} + \dot{Q}_{CV} + \dot{W}_{CV} = (h + \tfrac{1}{2}\mathbf{V}^2)_{out}\dot{m}_{out}$$

or

$$(h + \tfrac{1}{2}\mathbf{V}^2)_{in} + q_{CV} + w_{CV} = (h + \tfrac{1}{2}\mathbf{V}^2)_{out} \tag{4.93}$$

A turbine is a power-producing component which does work through the rotation of a shaft. The high-pressure fluid expands to a low

---

TABLE 4.2 Special Cases for Control Volume Analysis

Steady State with Uniform Properties

Mass: $\sum\limits_{in} \dot{m}_{in} = \sum\limits_{out} \dot{m}_{out}$

Energy: $\sum\limits_{in} (h + \tfrac{1}{2}\mathbf{V}^2 + gZ)\dot{m} + \dot{Q}_{CV} + \dot{W}_{CV} = \sum\limits_{out} (h + \tfrac{1}{2}\mathbf{V}^2 + gZ)\dot{m}$

Unsteady State with Uniform Properties and Steady State Port Properties

Mass: $m_2 - m_1 = m_{in} - m_{out}$

Energy: $(u + \tfrac{1}{2}\mathbf{V}^2 + gZ)_2 m_2 - (u + \tfrac{1}{2}\mathbf{V}^2 + gZ)_1 m_1$

$\qquad\qquad = \sum\limits_{in} (h + \tfrac{1}{2}\mathbf{V}^2 + gZ)m + {}_1Q_2 + {}_1W_2 - \sum\limits_{out} (h + \tfrac{1}{2}\mathbf{V}^2 + gZ)m$

Figure 4.5 Steady state work applications. (*a*) Turbine; (*b*) compressor; (*c*) fan; (*d*) pump.

pressure, doing work against the turbine blades. The kinetic and potential energy changes are usually, but not always, small. The heat transfer to the surroundings is also generally small compared to the work produced. A schematic of a turbine is shown in Fig. 4.5*a*.

Compressors require a work input to produce a high pressure at the outlet from the low pressure at the inlet. They are usually used for a gaseous working fluid. Fans, on the other hand, are employed to move fluid and alter velocities rather than produce large pressure changes. Pumps also require work input and are generally associated with liquids. All three devices require work input and typically have small heat transfers. Schematic representations of these steady state work applications are shown in Fig. 4.5*b*, *c*, and *d*.

Example Problem 4.4

A steam turbine is designed to have a power output of 9 MW for a mass flow rate of 17 kg/s. The inlet state is 3 MPa, 450°C, and 200 m/s, and the outlet state is 0.5 MPa, saturated vapor, and 80 m/s. What is the heat transfer for this turbine? See diagrams on next page.

Solution
The conservation statements for a single inlet and a single outlet are

Mass:
$$\frac{\partial}{\partial t}\left(\int_{CV} \rho \, dV\right) = \int_{A_{in}} d\dot{m} - \int_{A_{out}} d\dot{m}$$

Energy:
$$\frac{\partial}{\partial t}\left(\int_{CV} \rho e \, dV\right) = \int_{A_{in}} (h + \tfrac{1}{2}\mathbf{V}^2 + gZ) \, d\dot{m} + \dot{Q}_{CV} + \dot{W}_{CV}$$

$$- \int_{A_{out}} (h + \tfrac{1}{2}\mathbf{V}^2 + gZ) \, d\dot{m}$$

Diagrams

All statements in the problem indicate steady operation, so the storage terms (within the control volume terms) are zero. Also there is no time variation at the ports, so the analysis is steady state. Only single values of properties are given for the inlet and outlet, so the ports are assumed to be uniform and the terms are given by Eq. (4.68). Thus the governing equations are

Mass:

$$\dot{m}_{in} = \dot{m}_{out}$$

Energy:

$$(h + \tfrac{1}{2}\mathbf{V}^2 + gZ)_{in}\dot{m}_{in} + \dot{Q}_{CV} + \dot{W}_{CV} = (h + \tfrac{1}{2}\mathbf{V}^2 + gZ)_{out}\dot{m}_{out}$$

They are the same as those given in Table 4.2 for a single inlet and a single outlet. Assuming the potential energy change is small compared with the other terms, we find

$$\dot{Q}_{CV} = -\dot{W}_{CV} + \dot{m}(h_{out} - h_{in}) + \frac{\dot{m}}{2}(\mathbf{V}^2_{out} - \mathbf{V}^2_{in})$$

The property information is given from Tables D.9 and D.10:

State	P, MPa	T, °C	x	h, kJ/kg
In	3.0	450	—	(3343.1)
Out	0.5	—	1.0	(2746.6)

Substituting numerical values yields

$$\dot{Q}_{CV} = 9000 + 17(2747.6 - 3343.1) + \tfrac{17}{2}(80^2 - 200^2)\tfrac{1}{1000}$$
$$= 9000 - 10{,}123.5 - 285.6$$
$$= -1409.1 \text{ kW}$$

Note that the change in the kinetic energy term requires the conversion from watts to kilowatts when the mass flow rate is in kilograms per second and the velocity is in meters per second.

Comments

First, the rate of heat transfer is from the turbine to the surroundings. The temperature of the steam is greater than that of the surroundings, so this is expected. The heat transfer is 1409.1/9000 = 15.6 percent of the work from the turbine. Also, the kinetic energy contribution is only 285.6/10,123.5 = 2.8 percent of the enthalpy change and is often neglected.

Example Problem 4.5

A compressor is required to provide 0.5 MPa and 200°C air for a stationary power plant. The air intake is at 0.1 MPa and 20°C. The outlet velocity is 25 m/s. Evaluate the work per unit mass needed for an adiabatic compressor.

Diagram

Solution
The state information is as follows:

State	P, MPa	T, °C	h, kJ/kg
In	0.1	20	
Out	0.5	200	

Thus, two independent properties are known for each state, and the other information can be obtained.

The conservation statements are given in Eq. (4.93) as

Mass:

$$\dot{m}_{in} = \dot{m}_{out}$$

Energy:

$$(h + \tfrac{1}{2}\mathbf{V}^2)_{in} + q_{CV} + w_{CV} = (h + \tfrac{1}{2}\mathbf{V}^2)_{out}$$

The inlet area to a compressor is often very large, so the velocity at the inlet is approximately zero for a fixed $\dot{m} = \rho A \mathbf{V}$. The control surface could be placed at another location in the compressor inlet with a smaller area, but then the inlet velocity would be needed. Also, the compressor is adiabatic, so $q = 0$. Thus,

$$w = (h_{out} - h_{in}) + \tfrac{1}{2}\mathbf{V}_{out}^2$$

The change in enthalpy for air is determined by treating air as an ideal gas (this is accurate, which can be determined from the generalized chart in Fig. C.1). For an ideal gas [Eq. (3.22)]

$$dh = c_P \, dT$$

So if c_P is constant, Table C.2 yields

$$w = c_P(T_{out} - T_{in}) + \tfrac{1}{2}\mathbf{V}_{out}^2$$

$$= (1.0052)(200 - 20) + \frac{1}{2}\frac{(25)^2}{1000}$$

$$= 180.94 + 0.31 = +181.25 \text{ kJ/kg}$$

Comments
The outlet kinetic energy contribution to the work is ~ 0.2 percent for this example. A more accurate approach includes the temperature dependence of the specific heat. Table D.1 gives

$$\bar{c}_P^o = 28.11 + (0.1967 \times 10^{-2})(T) + (0.4802 \times 10^{-5})(T^2)$$
$$- (1.966 \times 10^{-9})(T^3) \quad \text{kJ/(kmol} \cdot \text{K)}$$

Then

$$w = \int_{in}^{out} c_P \, dT + \tfrac{1}{2}\mathbf{V}_{out}^2$$

yields

$$w = +183.07 + 0.31 = +183.38 \text{ kJ/kg}$$

The difference between the constant c_P solution and variable c_P solution is very small since the temperature dependence of c_P near 20°C indicates that c_P is almost constant in this temperature range.

4.5.2 Steady State Flow Applications

Heat exchangers, condensers, steam generators, diffusers and nozzles, throttling valves, and pipes are devices that operate at steady state and do not produce or consume work. Most experience very small changes in potential energy, so

Mass:

$$\dot{m}_{\text{in}} = \dot{m}_{\text{out}}$$

Energy:

$$(h + \tfrac{1}{2}\mathbf{V}^2)_{\text{in}}\dot{m}_{\text{in}} + \dot{Q}_{\text{CV}} = (h + \tfrac{1}{2}\mathbf{V}^2)_{\text{out}}\dot{m}_{\text{out}} \tag{4.94}$$

The number of inlets and outlets depends on the device as well as on the location of the control volume.

A heat exchanger is used to transfer energy from one fluid to another. There is no work interaction with the surroundings, and the changes in potential and kinetic energies are typically small. The heat transferred is computed from the enthalpy difference, and the analysis is very dependent on the choice of the control volume. Figure 4.6a indicates two possible control volumes, and the specific one used depends on the desired and known information. For control volume A,

$$q = h_{1,\,\text{out}} - h_{1,\,\text{in}} \tag{4.95}$$

and for control volume B,

$$\dot{m}_1 h_{1,\,\text{in}} + \dot{m}_2 h_{2,\,\text{in}} = \dot{m}_1 h_{1,\,\text{out}} + \dot{m}_2 h_{2,\,\text{out}} \tag{4.96}$$

A steam generator and condenser are types of heat exchangers that perform a specific task. A steam generator uses an energy source such as the burning of pulverized fuel or a nuclear reactor to create a high-temperature fluid. There is heat transfer from this fluid to water which passes through tube banks to form a superheated vapor. This is diagramed in Fig. 4.6b. A condenser circulates cold water to condense a high-quality mixture to the liquid state. For the steam generator and condenser, the first law reduces to

$$q = h_{\text{out}} - h_{\text{in}} \tag{4.97}$$

For a condenser, it is useful in some situations to have the control volume as shown in Fig. 4.6a (control volume B).

Figure 4.6   *Steady state flow applications. (a) Heat exchanger; (b) steam generator; (c) nozzle and diffuser; (d) throttling valve; (e) pipe.*

Diffusers and nozzles are used to control the velocity of the fluid. A diffuser decreases the velocity while increasing the pressure, and the opposite is true for a nozzle. These devices are simply area-changing devices, as shown in Fig. 4.6c. Since kinetic energy is typically important and there is typically a single inlet and outlet, Eq. (4.94) reduces to

$$q = (h_{\text{out}} - h_{\text{in}}) + \tfrac{1}{2}(\mathbf{V}_{\text{out}}^2 - \mathbf{V}_{\text{in}}^2) \tag{4.98}$$

where q might also be negligible.

A throttling valve is simply a flow constriction, as shown in Fig. 4.6d. This reduces the pressure but performs no work. Generally, the heat transfer is small. If a control volume sufficiently far from the actual constriction is chosen, then the change in kinetic energy is small. Equation (4.94) yields

$$h_{in} = h_{out} \tag{4.99}$$

This is a very important result which is applied to any valve.

One final steady state flow device is a simple pipe. Assuming that potential energy changes are important and that there is no work or heat transfer, we see from Table 4.2 that

$$(h + \tfrac{1}{2}\mathbf{V}^2 + gZ)_{in} = (h + \tfrac{1}{2}\mathbf{V}^2 + gZ)_{out}$$

or

$$(h_{in} - h_{out}) + \tfrac{1}{2}(\mathbf{V}_{in}^2 - \mathbf{V}_{out}^2) + g(Z_{in} - Z_{out}) = 0 \tag{4.100}$$

If the fluid is incompressible and isothermal, then Eq. (3.36a) yields

$$dh = c\,dT + v\,dP = v\,dP$$

so

$$\frac{P_{in} - P_{out}}{\rho} + \tfrac{1}{2}(\mathbf{V}_{in}^2 - \mathbf{V}_{out}^2) + g(Z_{in} - Z_{out}) = 0 \tag{4.101}$$

This is termed *Bernoulli's equation*. This derivation of Bernoulli's equation is *not* general since Bernoulli's equation is generally derived from Newton's second law. It is presented here to show the connection with fluid mechanics.

Example Problem 4.6

A high-velocity water jet is proposed for use as a cutting device with possible applications in metal, concrete, and plastic. A sapphire nozzle with dimensions shown in the diagram uses liquid water at a mass flow rate of 0.0065 kg/s. The water discharges to 0.1 MPa. Assume the flow is isothermal, and determine the pressure of the water source needed to operate this device.

Diagram

$D = 0.015$ m

$D = 1 \times 10^{-4}$ m

Solution

The properties are evaluated by assuming the liquid is incompressible. This process is idealized as adiabatic, isothermal, and steady state. The inlet and outlet are assumed to have uniform properties; thus conservation of mass yields

$$\dot{m} = \rho A \mathbf{V} \qquad \text{or} \qquad \mathbf{V} = \frac{\dot{m}}{\rho A} = \frac{v\dot{m}}{A}$$

Equation (4.101), neglecting changes in potential energy, yields

$$0 = v(P_{out} - P_{in}) + \frac{\dot{m}^2 v^2}{2}\left[\left(\frac{1}{A_{out}}\right)^2 - \left(\frac{1}{A_{in}}\right)^2\right]$$

With $A = \pi D^2/4$,

$$P_{in} = P_{out} + \frac{8\dot{m}^2 v}{\pi^2}\left(\frac{1}{D_{out}^4} - \frac{1}{D_{in}^4}\right)$$

With numerical values, taking $v = 0.001004$ m³/kg at $T = 30°C$, we have

$$P_{in} = 100 + \frac{8(0.0065)^2(0.001004)}{\pi^2(1000)}\left[\frac{1}{(1 \times 10^{-4})^4} - \frac{1}{(0.015)^4}\right]$$

$$= 100 + 343{,}800 \text{ kPa} = 343.9 \text{ MPa}$$

Comments

The required pressure is very high, so the incompressible assumptions might introduce some error. Also, the dependence of the inlet pressure on the nozzle diameter controls the water outlet speed. For a nozzle diameter of 1×10^{-4} m, the outlet speed is

$$\mathbf{V} = \frac{v\dot{m}}{A} = \frac{(0.001004)(0.0065)}{\pi(10^{-8})/4} = 831 \text{ m/s}$$

This very large speed is required for this application.

4.5.3   Unsteady State Work Applications

Short-operation turbines and the start-up of systems require an unsteady analysis. Another type of unsteady work problem involves the increase of fluid within a tank required to move a piston. This type of problem is diagramed in Fig. 4.7a. Conservation of mass and energy from Table 4.2 for this type of problem yields (without changes in the kinetic and potential energy)

Mass:

$$m_2 - m_1 = m_{in} - m_{out} \tag{4.102a}$$

Energy:

$$u_2 m_2 - u_1 m_1 = h_{in}m_{in} + {}_1Q_2 + {}_1W_2 - h_{out}m_{out} \tag{4.102b}$$

Remember that the state within the control volume has been assumed to be uniform at any instant.

(a)

(b)

Figure 4.7   Unsteady state applications. (*a*) Unsteady state work application; (*b*) unsteady state flow application.

Example Problem 4.7

An insulated 0.25-m³ cylinder is used to displace a load through the motion of a piston, as shown in the diagram. The piston has negligible volume and is initially at the bottom of the cylinder. The valve to a steam line (1.5 MPa and 350°C) is opened to raise the load, and it is closed when the piston reaches the top of the cylinder. The total work done is -50 kJ. What is the final pressure in the cylinder if the final mass is 0.25 kg?

Diagram

Solution

The states are indicated in the table:

State	P, MPa	T, °C	v, m³/kg	u, kJ/kg	h, kJ/kg
1	—	—	0	—	—
2	—	—	1		
In	1.5	350	—		

The control volume is indicated in the diagram, and it includes the valve. The initial state within the control volume is not required, since there is no fluid within the control volume. The inlet is completely specified, but the final state needs one more property. The control volume is placed around the steam under the piston and enlarges with time to move the piston. The cylinder is insulated, so there is no heat transfer during filling. All properties are taken as uniform. Table 4.2 yields

$$m_2 - m_1 = m_{in} - m_{out}$$

but $m_1 = 0$ and $m_{out} = 0$, so

$$m_2 = m_{in}$$

Thus from Table 4.2, conservation of energy states that

$$u_2 m_2 = h_{in} m_{in} + {}_1W_2$$

Eliminating m_{in} with conservation of mass yields

$$u_2 m_2 = h_{in} m_2 + {}_1W_2$$

or, by dividing through by m_2,

$$u_2 = h_{in} + {}_1w_2$$

$$= h_{in} - \frac{50}{0.25}$$

$$= h_{in} - 200 \qquad kJ/kg$$

Table D.10 yields $h_{in} = 3147.2$ kJ/kg, so $u_2 = 2947.2$ kJ/kg. The final state is specified by u and v. Interpolation yields $v = 1.0150$ m^3/kg for $P = 300$ kPa and $u = 2947.2$ kJ/kg and $v = 0.76097$ m^3/kg for $P = 400$ kPa and $u = 2947.2$ kJ/kg. Thus linearly interpolating between the u values yields

$$P_2 = 306 \text{ kPa}$$

Also, $T_2 = 389.5°C$.

Comments

If the control volume had been chosen without including the valve, the enthalpy on the tank side of the valve would be needed in the energy balance. However, the enthalpy on the pipe side is known. To relate these two enthalpies, a control volume around the valve is needed as shown in the margin. This is an unsteady flow problem without mass and energy accumulation. Since no work or heat transfer crosses the control volume boundary, Table 4.2 reduces to

Mass:

$$m_{in} = m_{out}$$

Energy:

$$0 = h_{in}m_{in} - h_{out}m_{out}$$

The energy equation yields

$$h_{in} = h_{out}$$

even if the inlet flow is unsteady. This is the connection between enthalpies necessary to complete the problem if the control volume is chosen around the tank.

Observe that the final steam temperature in the tank is *greater* than that in the steam line. Why should this be so?

Out

In

4.5.4   Unsteady State Flow Applications

Charging and discharging of tanks without work are important applications. The purpose is to add or withdraw fluid from a tank. The discharge process is diagramed in Fig. 4.7b. The kinetic and potential energy changes are usually small, so conservation of mass and energy for the diagramed problem are

Mass:

$$m_2 - m_1 = -m_{out} \tag{4.103a}$$

Energy:

$$u_2 m_2 - u_1 m_1 = {_1}Q_2 - h_{out} m_{out} \tag{4.103b}$$

The energy equation [Eq. (4.103b)] contains the assumption that h_{out} is steady. If h_{out} varies with time, then the final term in Eq. (4.103b) must be replaced by a term of the form in Eq. (4.71).

Example Problem 4.8

Water flows through a pipe with a 5-cm inside diameter (ID) at a velocity of 0.5 m/s into an initially empty 0.2-m³ drum (approximately a 55-gal drum). How long will it take to fill the drum? What is the volumetric flow rate of the water through the pipe?

Solution

The water is assumed to have a uniform velocity profile in the pipe, so the velocity given is used as the fluid velocity. An alternative interpretation is to assume that the velocity stated is the average velocity in the pipe. The specific volume of water at atmospheric conditions is $v = 0.001$ m³/kg. The mass flow rate is

$$\dot{m} = \frac{\mathbf{V}A}{v} = \frac{0.5\pi(0.025)^2}{0.001} = 0.982 \text{ kg/s}$$

The volumetric flow rate in the pipe is

$$\dot{V} = \dot{m}v = (0.982)(0.001) = 0.982 \times 10^{-3} \text{ m}^3/\text{s}$$

The conservation of mass in Eq. (4.34) yields

$$\frac{\partial}{\partial t}\left(\int_{CV} \rho \, dV\right) = \int_{A_{in}} d\dot{m} - \int_{A_{out}} d\dot{m}$$

This is an unsteady filling problem which requires an integration over time. The inlet and outlet terms are given by Eq. (4.66), and the control volume term is given by Eq. (4.75). The integrated conservation-of-mass expression is $m_2 - m_1 = m_{in} - m_{out}$, which is the same as that given in Table 4.2. Since $m_{out} = 0$,

$$m_2 - m_1 = m_{in}$$

But the flow is steady, so $m_{in} = \dot{m}_{in}(t_2 - t_1)$ and

$$m_2 - m_1 = \dot{m}_{in}(t_2 - t_1)$$

If the drum is initially empty at $t_1 = 0$, then $m_1 = 0$. At t_2 the drum is full, so $m_2 = V/v = 0.2/0.001 = 200$ kg. Then the conservation of mass yields

$$200 \text{ kg} - 0 \text{ kg} = (0.982 \text{ kg/s})(t_2 \text{ s})$$

Solving for t_2 yields

$$t_2 = \frac{200}{0.982} = 204 \text{ s} = 3.39 \text{ min}$$

Example Problem 4.9

A tank of volume V is to be filled with an ideal gas. Initially the tank is at P_1 and T_1. The port is regulated with a valve, and the port properties are constant at T_{in}. The process is adiabatic. Determine the final temperature T_2 to obtain P_2.

Diagram

Solution

The properties are those of an ideal gas, as given in Sec. 3.5.2. The control volume is chosen as the tank and the valve. The properties are uniform and steady at the inlet. The properties within the control volume are unsteady, but the properties are assumed uniform throughout the volume at any instant. The conservation statements are (Table 4.2)

Mass:

$$m_2 - m_1 = m_{in}$$

Energy:

$$m_2 u_2 - m_1 u_1 = h_{in} m_{in}$$

since there is no work or heat transfer. The inlet mass flow rate is not specified, so it is eliminated from the above equations, and

$$m_2 u_2 - m_1 u_1 = h_{in}(m_2 - m_1)$$

or

$$m_2(u_2 - h_{in}) = m_1(u_1 - h_{in})$$

This is the expression needed to evaluate the final temperature through $u_2 m_2$.

The properties are determined for the ideal gas, by assuming constant specific heats:

$$PV = mRT \quad \text{or} \quad m = \frac{PV}{RT}$$

and

$$du = c_v \, dT$$

With $h = u + Pv = u + RT$ and $c_P - c_v = R$, we have

$$u_2 - h_{in} = u_2 - u_{in} - RT_{in} = c_v(T_2 - T_{in}) - RT_{in} = c_v T_2 - c_P T_{in}$$

Substituting this into the above expression and performing the same manipulations for $u_1 - h_{in}$ yield

$$\frac{P_2 V}{RT_2}(c_v T_2 - c_P T_{in}) = \frac{P_1 V}{RT_1}(c_v T_1 - c_P T_{in})$$

With V and R as constants and $k = c_P/c_v$,

$$P_2 - P_2 k \frac{T_{in}}{T_2} = P_1 - P_1 k \frac{T_{in}}{T_1}$$

so

$$T_2 = \frac{1}{(P_2 - P_1)/(kP_2 T_{in}) + P_1/(P_2 T_1)}$$

Comments

As P_2 approaches P_1, then T_2 approaches T_1. But this is expected, since there is not much filling when P_2 is close to P_1. If P_2 is much larger than P_1, then the second term in the denominator is negligible, and P_1 is negligible compared with P_2. In this case,

$$T_2 \rightarrow kT_{in}$$

Because k is greater than unity [$k = c_P/c_v = (R + c_v)/c_v = 1 + R/c_v$], the final temperature is above the inlet value.

4.6 Other Statements of the First Law

This chapter discusses conservation of mass and energy for both a control mass and a control volume. The conservation principles are the same, but the resulting equations have different terms.

The first law of thermodynamics has many different forms, and all are equivalent. We have chosen to develop the first law from a general control mass viewpoint, as given in Eq. (4.6). We stated the first law as "energy can be neither created nor destroyed" for an isolated system in

Chap. 1. We could alternately state that the energy is constant for an isolated system. Another statement of the first law is that the work done by a system in a cyclic process is proportional to the heat transferred to the system. A different approach is to start with the idea that a perpetual-motion machine of the first kind is impossible. A perpetual-motion machine of the first kind is a continuously operating device that produces a continuous supply of energy without receiving energy input. The equivalency of these statements is shown here.

The control mass form of conservation of energy [Eq. (4.6)] is

$$dE = \delta Q + \delta W$$

This expresses the change in energy for a process. An alternative statement is that the energy is constant for an isolated system. An isolated system is one with no energy transfers with the surroundings. Thus, if $\delta Q = \delta W = 0$, then Eq. (4.6) yields

$$dE = 0$$

as stated. Another form states that an extensive property exists whose increment is equal to the work received by the system while surrounded by an adiabatic wall. For an adiabatic wall, $\delta Q = 0$, and the work received is positive by our sign convention, so

$$dE = \delta W$$

Therefore, the work is equal to the change in total energy. The total energy is an extensive property. Note that for this particular process (adiabatic), dE is as usual an exact differential; therefore, δW for this process must also be an exact differential.

Another approach is to make statements concerning cycles. The perpetual-motion machine principle states that a continuously operating device which produces a continuous supply of energy without receiving energy input (a perpetual-motion machine of the first kind) is impossible. If a device has no energy transfers, then $\delta Q = 0$ and $\delta W = 0$. Thus it is impossible to have dE equal anything but zero. Another form based on cycles states that the work done by a system in a cyclic process is proportional to the heat transfer. This equality was demonstrated in Eq. (4.19).

Problems

4.1 An ideal gas (CO_2) with molecular weight 44.01 is expanded slowly and isobarically. The gas has a mass of 0.1 kg, and the expansion occurs at 1 atm. If the expansion occurs from an initial temperature of 50°C to a final temperature of 150°C, find the average value of c_v for the gas during the process.

4.2 Water is heated in a piston-cylinder arrangement from the saturated liquid state to the saturated vapor state (Fig. P4.2). Evaluate the heat transfer per unit mass for (*a*) 0.05 MPa, (*b*) 0.10 MPa, (*c*) 0.5 MPa, and (*d*) 20 MPa.

4.3 A piston-cylinder arrangement initially contains water at $P = 20$ atm and $T = 240°C$. The piston is moved to a final position, where the volume is twice the initial volume. During the piston movement, there is heat transfer to the water in such a way that the pressure in the cylinder remains constant.

(*a*) Sketch the process on a *P-v* diagram, indicating the relative position of the saturation curve.

(*b*) Find the work done per kilogram of water during the expansion.

(*c*) Find the heat transfer per kilogram of water during the expansion.

(*d*) Find the value of the enthalpy of the water after expansion.

$_1q_2$ **Figure P4.2**

4.4 A cylinder-piston system contains 2 kg of H_2O at 150 kPa and has a volume of 0.35 m³. The piston is moved so that the final volume of the cylinder is 2.314 m³. During the piston motion from the initial to the final state, there is heat transfer to the cylinder in such a way as to hold the temperature constant.

(*a*) What is the final pressure in the cylinder?

(*b*) How much work was done by the steam?

(*c*) Evaluate the heat transfer during the process.

(*d*) Sketch the process on *P-v* and *P-h* diagrams, labeling the initial state 1 and the final state 2.

4.5 One kilogram of nitrogen fills a piston-cylinder system. The piston is weightless and frictionless and separates the nitrogen from the surroundings, which are at 1 atm. The initial volume in the cylinder is 1 m³ at 1 atm. See Fig. P4.5. Heat transfer to the nitrogen occurs until the volume is doubled. Assuming the specific heat of nitrogen can be taken as constant during the process, evaluate the heat transfer to the nitrogen during the process.

4.6 One kilogram of water is contained in a sealed kettle with a volume of 3 m³ at $T = 20°C$. An electric immersion heater inside the kettle is turned on and remains on until the kettle is filled with saturated vapor. Electricity costs $0.15 per kilowatthour.

(*a*) Draw a sketch defining the system you choose for analysis. Also, sketch a *P-v* diagram of the process undergone by that system. (Show the saturation dome on your sketch.) Note any necessary assumptions.

(*b*) Find the final pressure in the kettle.

(*c*) Find the cost of the electricity necessary to carry out the process.

$P_\infty = 1$ atm

N_2

Q

Figure P4.5

4.7 Develop an equation for the heat transfer to an ideal gas during a constant-pressure expansion in terms of the initial pressure P_1, the initial and final volumes V_1 and V_2, and c_P (assumed temperature-independent) and R for the gas.

4.8 Develop an equation for the change in internal energy U of an ideal gas during a constant-pressure expansion in terms of the initial pressure P_1, the initial and final volumes V_1 and V_2, and c_v and R for the gas.

4.9 In Prob. 2.16, an elevator is allowed to fall to an equilibrium position. For that problem, determine the heat transfer from the air in the elevator shaft in order for the assumption of an isothermal process to be justified.

4.10 Two blocks A and B are initially at 100 and 500°C, respectively (Fig. P4.10). They are brought together and isolated from the surroundings. Determine the final equilibrium temperature of the blocks. Block A is aluminum [$c_P = 0.900$ kJ/(kg · K)] with $m_A = 0.5$ kg, and block B is copper [$c_P = 0.386$ kJ/(kg · K)] with $m_B = 1.0$ kg.

4.11 Refrigerant 12 is initially contained within the cylinder shown at 20°C and 0.7 MPa (Fig. P4.11). There is a heat transfer to the refrigerant 12 until the volume is 21 times the initial value. The piston is locked in place, and then the fluid is cooled to 20°C. Determine the heat transfer per unit mass for this process.

4.12 The cylinder shown in Fig. P4.12 contains 0.1 kg of water at 0.150 MPa. The volume initially enclosed is 0.005 m³. The spring touches the piston but exerts no force at this initial state. The piston rises owing to a heat transfer to the water until the water is in the saturated vapor state. The spring constant is 150 kN/m, and the piston area is 0.02 m². Determine the final temperature and pressure in the water, and evaluate the heat transfer.

4.13 Water (1 kg) at 0.2 MPa is initially enclosed within a volume of 0.10 m³, and the piston rests on the stops (Fig. P4.13). The piston will move when the pressure is 1.0 MPa. A total heat transfer of 2500 kJ is added to the water. Determine the work done and the final state of the water.

4.14 Steam occupies one-half of the partitioned insulated chamber (Fig. P4.14). The initial pressure and temperature of the steam are 0.3 MPa and 350°C. The other half of the chamber contains a vacuum. The partition is removed, and the steam fills the entire volume at the end of the process. Determine the final temperature of the steam.

4.15 A sealed container [$c = 1.0$ kJ/(kg · K), $\rho = 2500$ kg/m³] containing water at 0.10 MPa and $x = 0.10$ is unknow-

Figure P4.10

Figure P4.11

Figure P4.12

Figure P4.13

Figure P4.14

Partition

ingly put into a high-temperature furnace. The volume occupied by the water is 0.03 m³, and the volume of the container material is 0.005 m³. Calculate the heat transfer (to the container and water together) necessary for the water to reach 3.0 MPa.

4.16 Two kilograms of H_2O are contained within a piston-cylinder arrangement. See Fig. P4.16. The initial temperature is 105°C, and there are equal masses of liquid and vapor initially. The system is heated to a position where the piston is locked, and then the system is cooled to the saturated vapor state at $T = 50$°C. Draw the process on a P-v diagram, and evaluate the work done during the process.

Figure P4.16

4.17 A piston and weights of total mass equal to 400 kg and area of 0.0098 m² contain 5.0 kg of H_2O at 500°C. See Fig. P4.17. The water is cooled until the volume is 1.0 m³, where the piston rests on stops. Then the H_2O is cooled while the piston remains on the stops to a state that could support a piston and weights of mass equal to 200 kg. At this point the stops are removed, and weights are removed to yield 200 kg. The system is then cooled to the saturated liquid state. Draw the process on a P-v diagram, and evaluate the work done.

Figure P4.17

Figure P4.18

4.18 Water (1.0 kg) is enclosed in a cylinder with a movable piston (cross-sectional area of 1.0 m²). See Fig. P4.18. Initially, the piston rests on stops, and the water is at 0.1 MPa and $x = 0.63875$. The water is heated, and the piston rises after a heat transfer of 902.96 kJ to the water. There is an additional heat transfer until $T = 750$°C. Find the total heat transfer and work done. Also, plot the process on the P-v diagram.

Figure P4.19

4.19 Consider the system diagrammed in Fig. P4.19 which is composed of air (subscript a) and H_2 (subscript H) separated by a movable adiabatic piston. The initial data are $V_a = 1$ m³, $T_a = 25$°C, $P_a = 0.15$ MPa, $P_H = 0.15$ MPa, and $m_H = 0.10$ kg. There is a total heat transfer of 80 kJ to the air resulting in a final air temperature of $T_a = 45$°C. Find the temperature *increase* in the H_2 and the work done on the H_2. Assume both gases to be ideal.

Movable adiabatic piston

Figure P4.20

4.20 A system consists of an adiabatic piston of cross-sectional area equal to 1.0 m² (neglect piston mass and volume), a linear spring, and air (treat the air as an ideal gas). See Fig. P4.20. A spring force of 413.15 kN is required to deflect the spring 1.0 m ($k = 413.15$ kN/m), and the spring exerts no force on the piston initially. The H_2O does work on the piston as it compresses the spring and air while the air is maintained at a constant temperature of 200°C. The initial volume of H_2O is 0.008 m³ and of air is 2.0 m³, so the total volume is 2.008 m³. Determine the work done by the H₂O as it is heated from its initial saturated liquid state at 0.1 MPa to the final saturated vapor state at 1.0 MPa.

4.21 Consider a system of carbon dioxide enclosed in an expandable vessel. The carbon dioxide undergoes a complete cycle in the following manner: For the expansion, the final value minus the initial value of the internal energy is found to be −150 kJ. For the compression process, there is 35 kJ of heat transfer from the system. Find the amount of work done on the carbon dioxide in the compression process.

4.22 A piston performs work on air within the system shown in Fig. P4.22. There is a heat transfer between the chambers, but there is no heat transfer with the surroundings. Initially, the air is at 500 kPa and 200°C, and the N₂ is initially at 1500 kPa. The initial volumes are $V_{N_2} = 0.01$ m³ and $V_{air} = 0.01$ m³. Determine the work performed, final temperature, and heat transfer between chambers if the process ends when the N₂ pressure is 1580 kPa.

Figure P4.22

4.23 A pressure cooker has a volume of 0.1 m³. When the valve on the cooker begins to release steam, it maintains the pressure in the cooker at 150 kPa. Assume 10 kg of water is initially in the cooker.

(a) What is the quality of the saturated mixture in the cooker when steam begins to escape?

(b) What is the quality of the saturated mixture in the cooker when 7 kg of steam has escaped through the valve?

(c) If the cooker were placed on a stove with an initial charge of 10 kg of water (liquid plus vapor) at 20°C and 1 atm, evaluate the heat transfer to the water between the time the stove is turned on and the pressure valve begins to release steam?

4.24 A control mass undergoes a cycle; 10 kJ of work is done by the control mass, and there is a heat transfer of 50 kJ to the control mass for the first portion of the cycle. The cycle is completed with no work being done during the last portion. What is the heat transfer on this last portion of the cycle?

4.25 A control mass undergoes the process from state 1 to state 2 shown in Fig. P4.25. The heat transfer is $_1Q_2 = 200$ kJ. The control mass returns adiabatically from state 2 to state 1 by another process. What is the work on the return process, that is, $_2W_1$?

Figure P4.25

4.26 Water flows in a 0.05-m-diameter pipe. At the inlet to a tank, the water is at 20°C and atmospheric pressure. See Fig. P4.26. Evaluate the mass flow rate $\dot{m}$ and $\int_A h\,d\dot{m}$ at the inlet for the following two velocity profiles.

(a) $\mathbf{V}_z(r) = \mathbf{V}_{z,\,max}\left[1 - \left(\dfrac{r}{r_0}\right)^2\right]$

where

$\mathbf{V}_{z,\,max} = 1$ m/s   and $r_0 = \dfrac{D_{pipe}}{2}$

(b) $\mathbf{V}_z(r) = \mathbf{V}_{z,\,max}$

Note that $dA = 2\pi r\,dr$.

Figure P4.26

4.27 A crude-oil storage tank is in the shape of a right circular cylinder. The tank is 20 m high and holds 2000 m³ when full. Crude oil has a specific volume of 0.0015 m³/kg. At 10 a.m. on June 16, the tank contains 1000 m³, and crude oil is being pumped into the tank through one pipe at a rate of 2 m³/min. Oil is being pumped out of the tank at a velocity of 1.5 m/s in another pipe of 0.15-m ID. What will the height of the oil in the tank be at noon on June 17? What volume of oil will be in the tank at that time?

4.28 A conical storage tank is oriented with its axis vertical and its apex toward the ground. It has a base 1 m in diameter and is 2 m high. Water enters the cone at a rate of 5.0 kg/s. How fast is the water rising in the cone when it reaches a height of 1 m? Take the specific volume of water as 0.001003 m³/kg.

4.29 A steam turbine is designed to operate at a mass flow rate of 1.5 kg/s. The inlet conditions are $P_1 = 2$ MPa, $T_1 = 400°C$, and $\mathbf{V}_1 = 60$ m/s; the outlet conditions are $P_2 = 0.1$ MPa, $x_2 = 0.98$, and $\mathbf{V}_2 = 150$ m/s. The change in elevation from inlet to outlet is 1 m, and the heat loss is 50 kW.
(a) Evaluate the power output of the turbine.
(b) Evaluate the power output if the changes in kinetic and potential energy are neglected.
(c) What are the diameters of the inlet pipe and exhaust pipe?

4.30 What flow rate of steam is required to produce 500 kW from an adiabatic turbine with inlet conditions of 800 kPa and 400°C and exit conditions of 6 kPa and $x = 95$ percent?

4.31 A compressor is to be purchased for the new mechanical engineering building. It must compress air from atmospheric pressure, 25°C to 10 atm, 600°C; and the outlet velocity from the compressor must not exceed 10 m/s.

Assume the compressor to be adiabatic and frictionless.
(a) How much work is required for each kilogram of air that is compressed?
(b) What power is required to drive the compressor if 2 kg/s of air is to be compressed?

4.32 An adiabatic compressor requires 10 kW to compress 0.05 kg/s of oxygen from 1 atm, 50°C to a final pressure of 5 atm. Using the ideal gas tables for all properties, find the temperature of the oxygen as it leaves the compressor.

4.33 A large pump is used to take water from a nearby lake at a rate of 1 m³/s and raise its pressure from 120 to 700 kPa so that it can be fed into a fire-safety main. Assuming the pump is adiabatic and frictionless, how much power is necessary to drive the pump?

4.34 Ammonia flows at 15 kg/min through a pipe 0.5 cm in diameter, and encounters a diffuser. Upstream of the diffuser the pressure P_A is 0.9 MPa, and the temperature T_A is 20°C. Downstream the pressure P_B is 0.3689 MPa, and the speed V_B is 100 m/s. See Fig. P4.34.
(a) Find the downstream temperature or quality. *Note:* Only one of these will appropriately fix the downstream state.
(b) Find the downstream diffuser diameter.

4.35 Steam at atmospheric pressure is used to heat air in the heat exchanger shown in Fig. P4.35. The steam enters as a saturated vapor and leaves as a saturated liquid. The air enters the heat exchanger at atmospheric pressure and 20°C and undergoes a constant-pressure heating. If the mass flow rate of air is 1 kg/s and the outlet temperature of the air is 80°C, what is the required water mass flow rate?

4.36 A fluid flows through a valve. The inlet conditions are 1.0 MPa and 400°C, and the exit pressure is 0.3 MPa. See Fig. P4.36. Evaluate the exit temperature for (a) steam and (b) air as the fluid.

4.37 An adiabatic mixing chamber is designed to obtain refrigerant 12 in the saturated vapor state at 0.1 MPa. See Fig. P4.37. Gaseous refrigerant 12 at 1.0 MPa and 150°C passes through a valve and into the chamber. Liquid refrigerant 12 enters the chamber at 0.1 MPa and −40°C. If the mass flow rate of refrigerant 12 gas is 0.080 kg/s, determine the mass flow rate of liquid.

4.38 A heat exchanger transfers energy by means of heat transfer from a hotter to a colder fluid without allowing the fluids to physically mix. Such an exchanger is used to cool liquid ethyl alcohol (ethanol) (Table C.3) from 40 to 30°C. The alcohol flows at a rate of 10 kg/s. Cooling

0.5 cm D_B

$P_A = 0.9$ MPa
$T_A = 20\,°C$

$P_B = 0.3689$ MPa
$V_B = 100$ m/s   **Figure P4.34**

1   2
                    Steam
                    Air
4   3
                    Figure P4.35

Figure P4.36

Figure P4.37

Superheated vapor

Saturated vapor

Liquid

water is available at 20°C at a flow rate of $\dot{m}_{H_2O}$. See Fig. P4.38.

(a) Consider the control volume(s) that you wish to specify in order to determine the outlet water temperature. Write the first law for all such control volumes.

(b) Prepare a plot of cooling water outlet temperature versus mass flow rate of the water $\dot{m}_{H_2O}$. Do not plot for any flow rates that predict that the water outlet temperature can exceed 40°C.

4.39 A heat exchanger is a device for transferring energy from a hotter fluid to a colder fluid. The conditions present in one such device are shown in Fig. P4.39. The hot stream is air, the cold stream is water. How many kilograms per second of cooling water are required to cool 2 kg/s of air from 150 to 100°C if the temperature change of the water is 10°C? Assume constant specific heats.

4.40 Nitrogen flows through a pipe in which a porous plug has been inserted. The pressure drop across the plug is large. See Fig. P4.40. If the pressure upstream of the plug is 5 atm, derive an expression for the change in temperature of the nitrogen across the plug. Note any assumptions that you make.

4.41 Steam is continuously removed through an orifice from a steam line that contains saturated steam at 1 MPa. The bleed steam enters a chamber kept at constant pressure by an exhaust line. The temperature and pressure of the bleed steam in the exhaust chamber are measured as 250 kPa and 135°C. See Fig. P4.41.

(a) What is the quality of the steam in the steam line? (The device described is called a *throttling calorimeter* and is used for the purpose outlined, i.e., finding the quality of steam in a supply line.)

(b) Can you think of possible limitations on the range of quality in the line that can be measured?

4.42 Steam at $T_A = 400°C$ and 10 atm enters a subsonic nozzle at a speed of $\mathbf{V}_A = 30$ m/s and leaves the nozzle at a speed of $\mathbf{V}_B = 100$ m/s. After leaving the nozzle at state B, the steam enters a turbine, where it is expanded to state C. At state C, the steam has a pressure of 0.08 atm, a quality of $x_C = 0.98$, and a speed $\mathbf{V}_C = 10$ m/s. If the mass flow rate entering the nozzle is 2 kg/s, what is the power output of the turbine?

4.43 A continuous copper sheet passes through a furnace at a velocity of 5 m/s. The sheet is 2 m × 0.05 m in cross section. It is heated from a uniform temperature at the inlet of 27°C to a uniform temperature of 977°C at the furnace outlet. The furnace is heated by electric resistance

Figure P4.38

Figure P4.39

Figure P4.40

Figure P4.41

heaters, and heat transfer from the furnace to the surroundings is negligible. How large a power supply must be provided to feed the heaters?

4.44 Steam is used in a lift as shown in Fig. P4.44. The steam source is at 1.0 MPa and 300°C. The initial state is 0.5 MPa, saturated vapor, and 0.002 m³. The steam slowly fills the volume until the volume is 0.040 m³, and the mass added is 0.15 kg. Determine the final state and the heat transfer.

Figure P4.44

4.45 A balloon is to be filled from a helium tank. The tank is large, and the tank pressure remains effectively constant while the balloon is filled. A valve between the tank and the balloon controls the helium flow to a constant rate of 0.05 kg/s. The pressure in the balloon varies during the filling process. The temperature of the helium entering the balloon is 20°C. See Fig. P4.45.

(a) How much work is done in filling the balloon to a final pressure of 500 kPa, where the final radius is 2 m if the balloon is adiabatic? (It takes 8 min to fill the balloon.)

(b) What is the final temperature in the balloon?

4.46 A steam line with very large capacity contains saturated vapor at $P = 1$ MPa. The line is connected through a valve to an evacuated and insulated tank with a volume of 5 m³. At $t = 0$, the valve is cracked open, and steam flows into the tank until the pressure in the tank equalizes with that in the steam line. The valve is immediately shut. At the moment the valve is shut, what is the temperature of the steam in the tank?

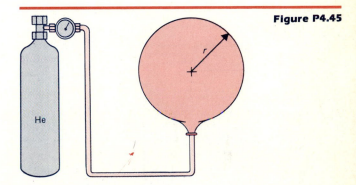

Figure P4.45

4.47 Air at 1 atm and 20°C initially fills a bottle of 0.1-m³ volume. The bottle is attached to an air line that provides air at 20°C and 50 atm, and the bottle is "charged" to a pressure of 50 atm. There is heat transfer from the bottle, so the air within is held at 20°C throughout the process. Identify an appropriate control volume. What is the heat transfer for your control volume during this process?

4.48 Find the temperature increase ΔT that results during the adiabatic filling of an evacuated tank. The gases enter the tank from a line at 1.0 MPa and 20°C. The process stops when the pressure in the tank is equal to the line pressure. Consider (a) air, (b) ethane, and (c) argon.

4.49 A tank initially contains H_2O at 0.05 MPa and 100°C and is filled from a steam line at 0.60 MPa and 200°C. The tank is filled to a level of 90 percent liquid on a volume basis. The 1.0-m³ tank is maintained at 100°C during the process. Evaluate the required heat transfer to keep the tank isothermal.

4.50 A steam pipeline carries saturated vapor at pressure P. An evacuated and insulated 3-m³ tank is connected to the pipeline through a valve. At $t = 0$ the valve is opened, and steam flows into the tank. When the tank pressure just reaches the pipeline pressure P, the temperature of the steam in the tank is measured to be 250°C. What is the pressure in the pipeline? (Give your answer within a 100-kPa span, for example, $1000 < P < 1100$ kPa.)

4.51 A power cycle uses refrigerant 12 as the working fluid. It operates between pressures of 60 and 600 kPa, and the states at the process endpoints are shown on the P-h diagram. The quality at state 3 is $x_3 = 0.98$. See Fig. P4.51.

Construct a state table for this heat engine, and find the net work output divided by the heat transfer to the working fluid, that is, $|w_{net}|/_1q_2$. (Use Tables D.11 to D.13 or computerized tables for all properties.)

Figure P4.51

4.52 A refrigeration cycle operates between the same two pressures as the heat engine of Prob. 4.51. The P-h diagram is shown in Fig. P4.52. The quality $x_3 = 0.8$. Construct a state table for this refrigeration cycle.

(a) Find the value of $_4q_3 /_3w_2$.

(b) Find the rate at which refrigerant 12 must circulate in kilograms per second to provide 3 tons of cooling capacity. (Note: 1 ton = 12,000 Btu/h. See App. B for conversion factors.)

Figure P4.52

5

Entropy and the Second Law of Thermodynamics

In any attempt to bridge the domains of experience belonging to the spiritual and physical sides of our nature, time occupies the key position.

A. S. Eddington, *The Nature of the Physical World*

Plate-type immersion cooler for heat transfer from hot engine oil to the water in an engine cooling system. *(Harrison Radiator Division, General Motors.)*

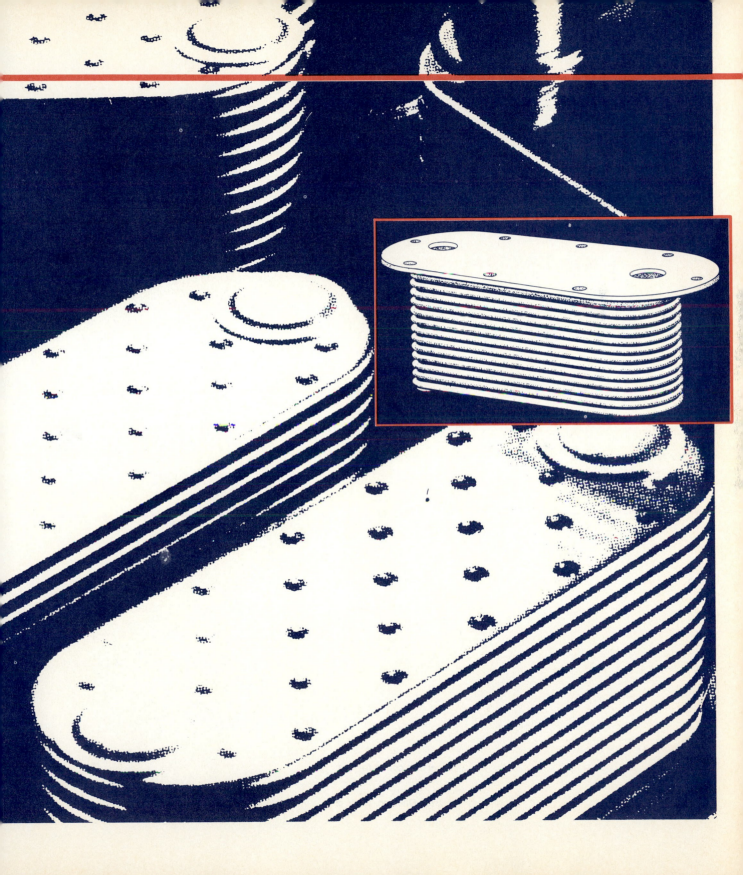

5.1 Introduction

The first law of thermodynamics, like the conservation-of-mass principle, describes the conservation of a particular property. The first law expresses the manner in which a system's energy is altered by energy transfer across the system boundary and mass transport into and out of the system. The first law is a strict accounting procedure that describes the changes in the system's energy. The conservation of mass quantifies the changes in the system's mass. Both conservation statements directly relate the change in a system's property to transfers at a boundary. The second law of thermodynamics also relates a system property to energy transfer at a boundary, but the relation simply specifies the direction of change. Like all other physical laws used in classical thermodynamics, the second law cannot be proved but is a statement of observed phenomena.

5.1.1 Physical Observations

Two examples are presented to demonstrate the inability of the first law of thermodynamics and the conservation-of-mass statement to completely describe systems. Both examples consider isolated systems, so the heat transfer and work across the system boundary are zero. Figure 5.1a presents an isolated chamber with steam on one side of a partition and a vacuum on the other side. This is the initial state of the system and is indicated as state 1. The partition is suddenly removed, and the system reaches equilibrium in the final state where the steam fills the entire chamber. This final state is denoted as state 2. The conservation of mass and first law are presented in the figure, indicating that the transformation is a constant-mass and constant-internal-energy process. Clearly, these conservation statements are satisfied, and the process is physically reasonable. In Fig. 5.1b, the steam completely fills the chamber in its initial state 1. The system transforms itself into two subsystems with the steam residing in one-half of the chamber and a perfect vacuum in the other half. A partition is inserted, and this final state is state 2. This case is also described by the constant-mass and constant-internal-energy process, as indicated in Fig. 5.1. Again the principles of conservation of mass and energy are satisfied; yet, from our experience, the process in Fig. 5.1b is impossible. Regardless of the time allowed for this process, it will never occur. The inability of the first law and conservation of mass to explain this impossibility indicates the need for another fundamental law.

Figure 5.2 shows an isolated system containing two blocks of material labeled A and B. In Fig. 5.2a block A is at a high temperature T_H, and

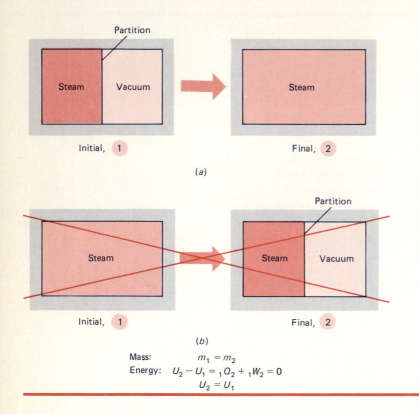

Initial, 1   Final, 2

(a)

(b)

Mass: $m_1 = m_2$
Energy: $U_2 - U_1 = {}_1Q_2 + {}_1W_2 = 0$
$U_2 = U_1$

Figure 5.1   Expansion into a vacuum for an isolated system.

block B is at a low temperature T_L. They are initially brought together and isolated from all surroundings, as indicated by state 1. There is a subsequent heat transfer between block A and block B until they reach a common intermediate equilibrium temperature T. This final state is denoted as state 2. The statements of conservation of mass and energy are presented in Fig. 5.2, indicating that this is a constant-mass and constant-internal-energy process. The temperature of the hotter block is expected to decrease, and the temperature of the colder block is expected to increase to the equilibrium temperature. The process satisfies the conservation statements as well as our physical expectations. In Fig. 5.2b, the two blocks are initially brought together at the same temperature, and the final state results when block A is at the high temperature T_H and block B is at the low temperature T_L. Again the conservation statements given in Fig. 5.2 are satisfied, yet our experience leads us to expect that this is an impossible process.

    These are only a few examples that demonstrate that there is a fundamental principle, the second law of thermodynamics, which is not described by the principles of conservation of mass and energy. These examples present the two key features of the second law of thermodynamics. First, the second law prescribes the *direction* for the processes; second, some thermodynamic characteristic of the system, a system *prop-*

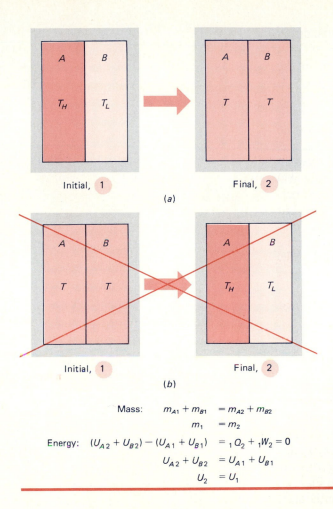

$$\text{Mass:} \quad m_{A1} + m_{B1} = m_{A2} + m_{B2}$$
$$m_1 = m_2$$
$$\text{Energy:} \quad (U_{A2} + U_{B2}) - (U_{A1} + U_{B1}) = {}_1Q_2 + {}_1W_2 = 0$$
$$U_{A2} + U_{B2} = U_{A1} + U_{B1}$$
$$U_2 = U_1$$

Figure 5.2 Heat transfer between two blocks within an isolated system.

erty, exists and is always changing in the specified direction. These features are discussed below.

The example in Fig. 5.1 indicates that the system (Fig. 5.1*a*) tends to proceed to a more random state and that we expect the reverse process (Fig. 5.1*b*), which proceeds to a more organized state (that is, a state with more individual control masses, each with different intensive properties), to be impossible. Thus, there is a specific direction to the process. The direction appears to be related to the disorder, randomness, or uncertainty of the microscopic scale of the system. There is an inability of the system to self-organize. The example in Fig. 5.2 also demonstrates that the organized, or structured, state of a separate high-temperature material and low-temperature material in state 1 tends toward a more disorganized, or less structured, state of a single homogeneous temperature in state 2. Again the reverse process is not possible. Thus, the second law needs to describe the direction for processes, and the direction is described in terms of a system property that characterizes the system's randomness, disorganization, or uncertainty.

Observe that the randomness, disorganization, or uncertainty in the two systems of Figs. 5.1 and 5.2 was changed in different ways. In Fig. 5.1, the state of the steam was changed by rearranging the positions of the molecules of the steam. Because more positions are available for a given molecule after the expansion than before it, our uncertainty as to the position of a given molecule is greater following the expansion. In Fig. 5.2, the state of the system was changed by allowing heat transfer between subsystems, with no spatial rearrangement of the system. By statistical mechanics, the microscopic disorder of a material depends on the temperature. The heat transfer process between the subsystems increases our uncertainty about the molecular disorder in the two subsystems compared with the disorder before the process. (Proving this is outside the scope of classical thermodynamics, but this discussion is meant to show where the idea of uncertainty comes from in these processes. These ideas are discussed quantitatively and in greater detail in Chap. 12.)

Thus, the disorder of the system can be changed in at least two ways: geometric rearrangement and heat transfer.

5.1.2 Increasing Disorder by Heat Transfer

The above examples focus on isolated systems, and this is the starting point for the second law presentation; yet, energy transfers are fundamental to thermodynamics and require special attention. The example in Fig. 5.2 indicates that Fig. 5.2a is the possible process and that the isolated system tends to a less structured, or more random, state. If the system is redefined to be either block A or block B, then a heat transfer across the system boundary exists between the blocks. This is diagramed in Fig. 5.3. For control mass A, the process starts from a high-temperature initial state and ends at a lower-temperature final state as a result of heat transfer from

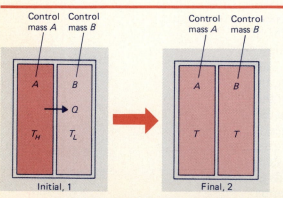

Initial, 1   Final, 2

Negative heat transfer for control mass A
Positive heat transfer for control mass B

Figure 5.3 Heat transfer between control mass A and control mass B.

this control mass to block B. Physically, block A proceeds from a high-temperature, or less organized, state (from a molecular viewpoint) to a lower-temperature, or more organized, state (on the molecular scale). This process, which tends to organize or reduce the random molecular nature of block A, is a result of heat loss from this control mass or a negative energy transfer. In Sec. 1.3, heat transfer is associated with a disorganized energy transfer, so a heat loss from the control mass is consistent with a decrease in the randomness of block A. However, control mass B receives the heat transfer from block A, and this is interpreted as positive disorganized energy transfer. Thus control mass B increases from an initial low-temperature organized state to a more random high-temperature state as a result of the heat transfer. Remember that the possible process is one where the complete isolated system tends to a more random (or uncertain) state, but the subsystems can individually experience an increase or a decrease in disorder.

This last discussion links heat transfer or disorganized energy transfer to the system property that characterizes the randomness, uncertainty, or disorder of the state. This connection between heat transfer and an uncertainty property which shows the directionality of a process is the second law of thermodynamics. The property that describes the randomness or uncertainty is called *entropy*. The direction of the process or change in state as specified by the change in entropy is tied to the direction of the heat transfer. The subsequent sections of this chapter present the macroscopic viewpoint of the second law of thermodynamics. The microscopic calculation and interpretation is presented in Chap. 12.

5.2   Entropy and the Second Law for an Isolated System

The microscopic disorder of a system is prescribed by a system property called *entropy*. The previous examples indicate that an isolated system, after removal of internal constraints, undergoes processes that are not quantified by the first law of thermodynamics. These facts are combined into the statement of the second law of thermodynamics:

> The entropy S, an extensive equilibrium property, must always increase or remain constant for an isolated system.

This is expressed mathematically as

$$dS_{isolated} \geq 0$$

or

$$(S_{final} - S_{initial})_{isolated} \geq 0 \qquad (5.1)$$

Entropy, like our other thermodynamic properties, is defined only at equilibrium states or for quasi-equilibrium processes. Equations (5.1) shows that the entropy of the final state is never less than that of the initial state for any process which an isolated system undergoes. In the examples presented in Sec. 5.1, the final state of the isolated system results after the removal of an internal constraint. The final state has the largest value of entropy for the complete isolated system. Thus, an isolated system composed of subsystems tends toward a state which increases the total entropy of the composite or isolated system. This is the maximum possible entropy subject to any constraints that are imposed.

Expressions for changes in the property S as a function of other system properties must be developed before we can apply the second law. However, some characteristics of entropy can be examined first. Entropy is an extensive property of the system, so any system composed of subsystems yields

$$S = m_A s_A + m_B s_B + \cdots = \sum_i m_i s_i \tag{5.2}$$

where s is the specific entropy or entropy per unit mass and the subscript denotes the subsystem. The second law, expressed as an integral over all the elements of an isolated system, is

$$dS_{isolated} = d\left(\int_m s \, dm \right)_{isolated} = d\left(\int_V \rho s \, dV \right)_{isolated} \geq 0 \tag{5.3}$$

This property is specified by the state of the system, or it is used to specify the state of the system. As prescribed by the state postulate for a simple compressible substance in Chap. 3, the entropy is completely specified by two independent properties of the system; or it can be used as an independent property, combined with another property, to specify all other properties. Entropy is a measure of the molecular disorder of the substance. Larger values of entropy imply larger disorder or uncertainty, and lower values imply more microscopically organized states. The units of entropy S are kilojoules per kelvin.

A key difference between the second law and the first law is the inequality sign. The first law specified a direct relationship between energy and energy transfer while the second law indicates only the direction for the change in molecular disorder or uncertainty. An alternate, yet equivalent, presentation which is quite useful is to define the term *entropy production* or *entropy generation* S_{gen} and to eliminate the inequality sign:

$$(dS - \delta S_{gen})_{isolated} = 0 \tag{5.4}$$

or

$$(S_{final} - S_{gen} - S_{initial})_{isolated} = 0 \tag{5.5}$$

Here δS_{gen} is the entropy generated during a change in system state and is always positive or zero. Equation (5.4) is exactly the same as Eq. (5.1), where δS_{gen} represents the "amount of inequality" as prescribed by the second law. The negative sign preceding the entropy generation is necessary to "reduce" the change in entropy to zero. As is demonstrated in subsequent sections, δS_{gen} is a path-dependent function or inexact differential unlike dS, which represents the change in a property and is independent of path or an exact differential. [The relationship between an exact differential and inexact differential in Eq. (5.4) should not be surprising since the first law connects the exact differential dU and the inexact differentials δQ and δW. From Sec. 4.6, for an adiabatic process, $dU = \delta W$.]

The second law is also expressed on a rate basis by considering the time interval dt. This law is then

$$\left.\frac{dS}{dt}\right|_{\text{isolated}} = \frac{d}{dt}\left(\int_V \rho s\, dV\right)_{\text{isolated}} \geq 0 \tag{5.6}$$

or

$$\left.\frac{dS}{dt}\right|_{\text{isolated}} - \dot{S}_{\text{gen}} = 0 \tag{5.7}$$

where the overdot on $\dot{S}_{\text{gen}}$ indicates the rate of entropy generation.

The above expressions are the starting point for further considerations of the second law. The change in entropy of a control mass and control volume are developed from the isolated system. Before these expressions are presented, the special case of an equality versus an inequality in Eq. (5.1), or zero entropy generation versus finite entropy generation in Eq. (5.4), must be discussed. Also, the thermodynamic definitions of temperature and pressure are required.

5.3   Reversible and Irreversible Processes

The concepts of reversible and irreversible processes play a fundamental role in thermodynamics, and the reduction of irreversibilities is a goal of many analyses. To better understand these concepts, consider the isolated system diagramed in Fig. 5.4. The isolated system is composed of a number of subsystems, four in this example, labeled A, B, C, and D. The subsystems might communicate thermally or mechanically, or they might even transfer mass across their boundaries (except for the outside boundary of the isolated system). Therefore, the total system is isolated when all four subsystems are considered. But if the system is taken as one subsystem, then the other three are viewed as surroundings to the chosen subsystem. If subsystem A has energy transfers in the form of heat transfer and work with the other subsystems, then subsystem A is a control mass and subsystems B, C, and D are the surroundings to control mass A. Reversible and irreversible processes are strictly applicable for only isolated systems and must be interpreted for subsystems (such as subsystem A) with appropriate terminology. This leads to a set of subcategories of reversible and irreversible processes.

The second law puts a direction on all isolated system processes:

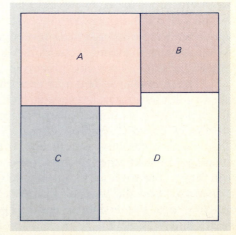

Figure 5.4   An isolated system composed of four subsystems.

$S_{final} \geq S_{initial}$. Thus there is a forward and backward direction to all processes. When $S_{final} > S_{initial}$, the forward direction is defined and the backward direction is impossible, as stated by the second law. Such a process is termed *irreversible*. When $S_{final} = S_{initial}$, then the forward and backward processes are equally possible, and the process is termed *reversible*. The inequality sign is the key distinguishing feature of reversible and irreversible processes. Alternatively, a reversible process produces no entropy generation, so S_{gen} is zero; and an irreversible process has a net entropy generation, so S_{gen} is finite and positive.

The previous discussion pertains to an isolated system which could be composed of subsystems. Reversible and irreversible processes are defined in terms of the *composite* entropy of all the subsystems as required by the second law. These processes can also be referred to as *totally reversible* and *totally irreversible*.

If a particular subsystem is considered as the system, say subsystem A in Fig. 5.4, then the concepts of internally reversible or irreversible and externally reversible or irreversible are needed. An *internally reversible process* considers the system (not surroundings), and it can be reversed at any point and proceeds in reverse through the same states. Note that this requires that all transfers with the surroundings be reversible, but the surroundings do not necessarily undergo a reversible process. An *internally irreversible* process has irreversibilities within the boundaries. An *externally reversible process* (or *externally irreversible process*) pertains to the surroundings of the system. In our example, the surroundings are subsystems B, C, and D, so an externally reversible process (or externally irreversible process) specifies the particular process for these subsystems. The restriction of $S_{gen} \geq 0$ for irreversible or reversible processes is valid for subsystems, but the specific definitions must await the control mass formulation in Sec. 5.6.

Physically, a totally reversible process can be reversed at any point, and the system and surroundings (making up the isolated system) proceed backwards along the same process. The totally reversible or simply reversible process must include both the system *and* surroundings. This differs significantly from an internally reversible process, which views only the system as reversible and excludes the surroundings from consideration. Clearly the totally reversible process is more difficult to achieve.

The reversible process is an idealization, and all actual processes are irreversible and approach reversible processes only in special cases. Figure 5.5 shows some examples of irreversible and corresponding reversible processes. The first example considers the expansion of a substance against a piston and inelastic spring. As the substance expands, energy is dissipated by the frictional work during the interaction of the piston and wall. The disorder within the piston and wall has increased, so the entropy is increased and the process is irreversible. Friction is a very common contributor to irreversible processes. The inelastic spring also dissipates energy and increases the disorder. This dissipation also contributes to the

IRREVERSIBLE   REVERSIBLE

Piston motion

Expansion

Heat transfer

Figure 5.5   Irreversible and reversible processes.

irreversible process. If the ideal case of frictionless motion and an elastic spring are considered, then the process can be reversed at any point in the process and restored to its original state. Thus this process is reversible. The unrestrained expansion of steam into a vacuum increases the disorder and is clearly irreversible, whereas a controlled, slow expansion does work against the surroundings, which can be reversed to achieve a reversible process. The heat transfer across a finite temperature difference is irreversible since the disorder of the entire system increases. Only for the very idealized case of heat transfer across an infinitesimal temperature difference can a reversible heat transfer be obtained. This case is considered in greater detail in succeeding sections, but it should be viewed as an ideal heat transfer similar to the ideal work of a constant-pressure piston motion.

In each of the above cases, the viewpoint can be a subsystem denoted by subsystem A in each case. It is possible for subsystem A to be internally reversible while the irreversibilities occur exterior to the system boundary. Specifically the steam in the subsystem A undergoing the unrestrained gas expansion (Fig. 5.5) could be proceeding through an internally reversible process, but the process for the complete isolated system is irreversible. This case indicates that this is an externally irreversible process. Similar subsets of reversible and irreversible processes are possible for the other examples.

The main features of irreversible processes are that there are dissipative effects which are not directly useful and that the processes can proceed through nonequilibrium states. These both tend to increase the molecular disorder, or entropy, of the systems. Both contributions make processes irreversible. However, the reversible process progresses through equilibrium states, and there are no dissipative effects. Therefore, reversible processes are idealizations of the real situations.

A clear understanding of the relationship of reversible and irreversible processes with the quasi-equilibrium process presented in Sec. 2.2.3 is important. The previous definition of a quasi-equilibrium process pertained to only the system and indicated nothing about the surroundings. *A reversible process must specify reversibility for both the system and the surroundings.* Therefore, a reversible process is always a quasi-equilibrium process since both the system and surroundings undergo reversible processes, and thus pass through a series of states. Also, all internally reversible processes are quasi-equilibrium processes. Conversely, a quasi-equilibrium process is an internally reversible process since these both pertain to the system only. A quasi-equilibrium process is not necessarily a reversible process (totally reversible), since irreversibilities can occur outside the system boundaries.

5.4 Temperature and Pressure Definitions

Temperature and pressure are discussed in Sec. 2.3, yet formal definitions were lacking. The development to this point has relied on our intuitive concepts of these properties. Now specific definitions are possible. The entropy property requires the definition of both temperature and pressure. In this section we state the definitions for temperature and pressure and show that these definitions are consistent with our concepts of these properties [1,2].

The definitions of temperature and pressure are developed from the differential form of the system energy. As indicated by the state postulate, the energy of a pure simple compressible substance is specified by any two independent properties. Thus, $E = E(S, V)$, or if we exclude the potential and kinetic energies, $U = U(S, V)$. The differential of this property, an exact differential, is

$$dU = \left(\frac{\partial U}{\partial S}\right)_V dS + \left(\frac{\partial U}{\partial V}\right)_S dV \tag{5.8}$$

The two partial derivatives in this expression are intensive parameters since the ratio of two extensive parameters is an intensive parameter.

5.4.1 Temperature

Temperature is defined in classical thermodynamics as the first partial derivative in Eq. (5.8)

$$T \equiv \left(\frac{\partial U}{\partial S}\right)_V \tag{5.9}$$

This is a rather abstract definition for temperature and is appropriate only if it satisfies our intuitive ideas of temperature. The two intuitive concepts that must be demonstrated from the above definition of temperature are thermal equilibrium as stated by the zeroth law (Sec. 2.3.3) and our perceptions of hot and cold. Both are demonstrated with the two-block examples shown in Fig. 5.2.

Figure 5.2 diagrams an isolated system of a high-temperature block A and a low-temperature block B which are brought into contact and then isolated from the surroundings (state 1). The final state 2 results after the blocks are in thermal equilibrium at T. The first law states

$$dU = d(U_A + U_B) = 0 \quad \text{or} \quad dU_A = -dU_B \tag{5.10}$$

The second law yields

$$dS = d(S_A + S_B) \geq 0 \tag{5.11}$$

The entropy change is positive as the system undergoes the irreversible heat transfer process until the final equilibrium state is reached. Then the entropy change is equal to zero when infinitesimal deviations from the final equilibrium state are considered. For thermal equilibrium in an isolated system, the change in entropy is

$$dS = d(S_A + S_B) = 0 \tag{5.12}$$

The volumes are taken as constants without loss of generality. Thus, $S = S(U, V)$ reduces to $S = S(U)$, and

$$dS = dS_A + dS_B = \left(\frac{\partial S_A}{\partial U_A}\right)_{V_A} dU_A + \left(\frac{\partial S_B}{\partial U_B}\right)_{V_B} dU_B = 0 \tag{5.13}$$

Applying the first law [Eq. (5.10)], we obtain

$$dS = \left[\left(\frac{\partial S_A}{\partial U_A}\right)_{V_A} - \left(\frac{\partial S_B}{\partial U_B}\right)_{V_B}\right] dU_A = 0 \tag{5.14}$$

But the partial derivatives are defined as temperature in Eq. (5.9), so Eq. (5.14) is

$$dS = \left(\frac{1}{T_A} - \frac{1}{T_B}\right) dU_A = 0 \tag{5.15}$$

Thermal equilibrium results when $dS = 0$, so this requires

$$T_A = T_B \tag{5.16}$$

which is the expected requirement for thermal equilibrium. Thus the temperature definition in Eq. (5.9) together with the second law satisfies the expected thermal equilibrium concept.

As the two blocks proceed toward the final thermal equilibrium state in Fig. 5.2, the second law and the temperature definition combine to describe the change in entropy, similar to Eq. (5.15),

$$dS = \left(\frac{1}{T_A} - \frac{1}{T_B}\right) dU_A \geq 0 \tag{5.17}$$

From Fig. 5.3, the first law for control mass A yields

$$dU_A = \delta Q_A + \delta W_A \tag{5.18}$$

but the block does not do any work, so

$$dU_A = \delta Q_A \tag{5.19}$$

Substituting Eq. (5.19) into Eq. (5.17) yields

$$dS = \left(\frac{1}{T_A} - \frac{1}{T_B}\right) \delta Q_A \geq 0 \tag{5.20}$$

This example was assumed to initially have

$$T_H = T_A > T_B = T_L \tag{5.21}$$

So with the second law requiring $dS \geq 0$, the heat transfer is

$$\delta Q_A \leq 0 \tag{5.22}$$

Thus, the heat transfer is from the higher-temperature block $T_A = T_H$ to the lower-temperature block $T_B = T_L$. This is exactly as expected and as presented in Sec. 5.1.

The temperature definition in Eq. (5.9) is consistent with our intuitive concepts of temperature. The actual measurement of temperature and the construction of the temperature scale are deferred to Sec. 5.12.

5.4.2 Pressure

The other partial derivative in Eq. (5.8) is used to define pressure. The formal definition is

$$P \equiv -\left(\frac{\partial U}{\partial V}\right)_S \tag{5.23}$$

Like the previous definition for temperature, this rather abstract defini-

Figure 5.6 Two subsystems that communicate thermally and mechanically.

tion for pressure must satisfy our previous concepts. The two intuitive concepts of pressure that must be demonstrated are mechanical equilibrium at equal pressures and the equivalence of the defined pressure with the previous force-related pressure [Eq. (2.1)].

The system considered is an isolated system composed of two subsystems which are separated by a movable piston. The two subsystems also communicate thermally. Figure 5.6 indicates the initial state, denoted 1, and the final state, denoted 2. The first law is given as

$$dU = d(U_A + U_B) = 0 \qquad \text{or} \qquad dU_A = -dU_B \tag{5.24}$$

and the total volume of the chamber is constant, so

$$dV = d(V_A + V_B) = 0 \qquad \text{or} \qquad dV_A = -dV_B \tag{5.25}$$

As in the thermal equilibrium example above, the final equilibrium state is prescribed from the second law as

$$dS = d(S_A + S_B) = 0 \tag{5.26}$$

The state postulate yields $S = S(U, V)$, so

$$dS = \left(\frac{\partial S_A}{\partial U_A}\right)_{V_A} dU_A + \left(\frac{\partial S_A}{\partial V_A}\right)_{U_A} dV_A$$
$$+ \left(\frac{\partial S_B}{\partial U_B}\right)_{V_B} dU_B + \left(\frac{\partial S_B}{\partial V_B}\right)_{U_B} dV_B \tag{5.27}$$

With Eqs. (5.24) and (5.25),

$$dS = \left[\left(\frac{\partial S_A}{\partial U_A}\right)_{V_A} - \left(\frac{\partial S_B}{\partial U_B}\right)_{V_B}\right] dU_A$$
$$+ \left[\left(\frac{\partial S_A}{\partial V_A}\right)_{U_A} - \left(\frac{\partial S_B}{\partial V_B}\right)_{U_B}\right] dV_A \tag{5.28}$$

The first two partial derivatives in Eq. (5.28) are temperatures, as defined in Eq. (5.9). Equation (1.16) is used to eliminate the third and fourth partial derivatives by noting

$$\left(\frac{\partial S}{\partial V}\right)_U \left(\frac{\partial V}{\partial U}\right)_S \left(\frac{\partial U}{\partial S}\right)_V = -1 \tag{5.29}$$

Thus

$$\left(\frac{\partial S}{\partial V}\right)_U = \frac{-1}{(\partial V/\partial U)_S (\partial U/\partial S)_V} \tag{5.30}$$

and Eq. (1.17) yields

$$\left(\frac{\partial S}{\partial V}\right)_U = -\frac{(\partial U/\partial V)_S}{(\partial U/\partial S)_V} \tag{5.31}$$

Thus the definitions for temperature and pressure give

$$\left(\frac{\partial S}{\partial V}\right)_U = \frac{P}{T} \tag{5.32}$$

and Eq. (5.28) is

$$dS = \left(\frac{1}{T_A} - \frac{1}{T_B}\right) dU_A + \left(\frac{P_A}{T_A} - \frac{P_B}{T_B}\right) dV_A \tag{5.33}$$

Since $dS = 0$ at thermodynamic equilibrium and dU_A and dV_A are independent, each of the parenthetical expressions must be zero, or

$$\frac{1}{T_A} - \frac{1}{T_B} = 0 \quad \text{or} \quad T_A = T_B \tag{5.34}$$

$$\frac{P_A}{T_A} - \frac{P_B}{T_B} = 0 \quad \text{or} \quad P_A = P_B \tag{5.35}$$

This last expression is the required equality for mechanical equilibrium.

The definitions for temperature and pressure given in Eqs. (5.9) and (5.23), respectively, are substituted into Eq. (5.8) to yield

$$dU = T\,dS - P\,dV \tag{5.36}$$

This is an extremely important relation in thermodynamics and is used extensively. It is known as the *Gibbs equation* and is considered in Sec. 5.5. It applies generally to a control mass. The first law for a control mass is

$$dU = \delta Q + \delta W \tag{5.37}$$

For an adiabatic reversible process, Eq. (5.36) yields

$$dU = -P\,dV \tag{5.38}$$

and Eq. (5.37) yields

$$dU = \delta W \tag{5.39}$$

Comparing these last expressions indicates that the previous force definition of pressure in Eq. (2.1) which is used in the work relation in Eq. (2.16) is in agreement with the pressure definition in Eq. (5.23).

Thus, the definitions for temperature and pressure in terms of partial derivatives of internal energy [Eqs. (5.9) and (5.23)] are consistent with the expected behavior of these properties. The next task is to present values and relations for entropy determination.

5.5 Entropy — The Property

Entropy is a property that is specified for every equilibrium state of a substance. Entropy represents the disorder, or uncertainty, of the microscopic scale, yet macroscopically it is used as all other properties. Thus, the relations presented in Chap. 2 for other properties are interpreted here for this property. The state postulate for a pure simple compressible substance states that two independent intensive properties specify the state. Therefore, entropy is an additional property that can be used to specify a state. As with all previous properties, data for entropy can be found in tabular, graphical, equation, and computer retrieval forms. The steam tables provide tabular information for the entropy of steam, and the interpolation methods described in Chap. 3 for obtaining v, u, and h also apply for determining s. Within the saturation region, the entropy is evaluated with the quality as

$$s = s_l + x s_{lg} \tag{5.40}$$

The tables of thermodynamic properties present numerical values for entropy in App. D for various substances.

The graphical presentation of property information is used extensively in thermodynamic analysis. The analysis to this point in the text focuses on the P-v diagram. The P-v diagram is very important to simple compressible work calculation since the graphical area under the process representation is the work done. Much of the thermodynamic analysis performed in the subsequent sections uses the T-s and h-s diagrams. Typical T-s and h-s diagrams are shown in Fig. 5.7 (specific substances are presented in App. D). The lines of constant volume, pressure, and enthalpy are shown on the T-s diagram. The h-s diagram is called the *Mollier chart,* and the constant-volume, constant-pressure, and constant-temperature lines are indicated. For ideal gases with temperature-independent c_P, the T-s and h-s diagrams are similar, since h is directly proportional to T.

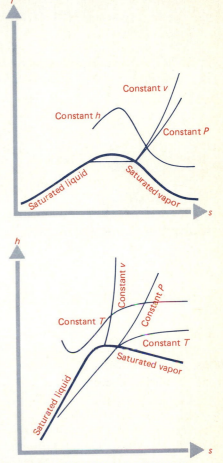

Figure 5.7 The *T-s* and *h-s* diagrams.

Example Problem 5.1

Evaluate the entropy (per unit mass) for water and refrigerant 12 at the following states with $P = 100$ kPa: saturated liquid, saturated vapor,

$T = 150°C$, and $T = 200°C$. Use Tables D.8, D.9, and D.10 for water and Tables D.11, D.12, and D.13 for refrigerant 12.

Solution

All the states are specified, and the properties are given in the indicated tables. Thus, in kilojoules per kilogram-kelvin,

	s_l	s_g	s_{150}	s_{200}
Water	1.3020	7.3598	7.6146	7.8347
Refrigerant 12	0.89454	1.5749	1.9273	2.0075

Comments

Comparisons between the absolute values of s for different substances are not possible since the reference state for each is different. The entropy increases from left to right in the above table, which satisfies the physical concepts of increasing molecular disorder. The saturated liquid is more ordered than the saturated vapor state. The superheated states are more molecularly disordered than the saturation states.

Example Problem 5.2

Steam occupies one-half of the partitioned insulated chamber shown. The initial pressure and temperature of the steam are 0.3 MPa and 350°C. The other half of the chamber is a vacuum. The partition is removed, and the steam fills the entire volume at the end of the process. Determine the final temperature of the steam and the increase in entropy (entropy generation) of the process.

Diagrams

Solution

The initial state is completely specified, and the final volume is known:

State	P, MPa	T, °C	V, m³	u, kJ/kg	s, kJ/(kg · K)
1	0.3	350	V_1		
2			$2V_1$		

The control mass is chosen to be the steam in the entire chamber, so the mass is constant and

First law: $\quad U_1 = U_2$

Second law: $\quad S_2 - S_1 \geq 0 \quad$ or $\quad S_{gen} = S_2 - S_1 \geq 0$

or, on a unit mass basis,

First law: $\quad u_1 = u_2$

Second law: $\quad s_2 - s_1 \geq 0 \quad$ or $\quad s_{gen} = s_2 - s_1 \geq 0$

Thus the first law specifies the second needed property for state 2. Table D.10 yields $v_1 = 0.9535$ m³/kg, $u_1 = 2884.4$ kJ/kg, and $s_1 = 7.8717$ kJ/(kg · K). The final state is specified by $v_2 = 2v_1 = 1.9070$ m³/kg (since the mass is constant) and $u_2 = u_1 = 2884.4$ kJ/kg. Interpolating first on u, then on v from Table D.10 yields

$$T_2 = 348.8°C \qquad P_2 = 0.167 \text{ MPa} \qquad s_2 = 8.1642 \text{ kJ/(kg · K)}$$

Therefore, the entropy generation is

$$S_{gen} = s_2 - s_1 = 0.2925 \text{ kJ/(kg · K)}$$

and the process is clearly irreversible.

Comments

The process is almost isothermal since T_2 is only a fraction of 1 percent lower than T_1. Checking the compressibility chart reveals that the states are well approximated by the ideal gas equation of state. A constant-internal-energy process is also isothermal since

$$du = c_v \, dT$$

Therefore, $T_1 \simeq T_2$ as evaluated in this example is expected. This fact helps in determining the final state from the property tables.

5.5.1 Entropy Relations

In Sec. 5.4 we define the equilibrium properties of temperature and pressure. The state postulate yields $U = U(S, V)$, so the change in internal energy combined with the temperature and pressure definitions yields

$$dU = T\,dS - P\,dV \tag{5.36}$$

This is the Gibbs equation which relates equilibrium thermodynamic properties. Thus this expression depends not on the path between any two states but on only the substance. A companion expression is obtained from enthalpy as

$$dH = d(U + PV) = dU + P\,dV + V\,dP \tag{5.41}$$

Substituting Eq. (5.36) for dU yields

$$dH = T\,dS + V\,dP \tag{5.42}$$

These expressions are also represented on an intensive basis as

$$du = T\,ds - P\,dv \tag{5.43}$$

$$dh = T\,ds + v\,dP \tag{5.44}$$

The changes in entropy are obtained directly from these equations as

$$dS = \frac{dU}{T} + \frac{P}{T}\,dV \tag{5.45}$$

$$dS = \frac{dH}{T} - \frac{V}{T}\,dP \tag{5.46}$$

or

$$ds = \frac{du}{T} + \frac{P}{T}\,dv \tag{5.47}$$

$$ds = \frac{dh}{T} - \frac{v}{T}\,dP \tag{5.48}$$

Example Problem 5.3

Consider the vaporization of water at 100 kPa. Using Eqs. (5.47) and (5.48), evaluate the change in entropy per unit mass (i.e., the change in entropy from the saturated liquid to the saturated vapor states s_{lg}) from the tabular values of u, h, P, T, and v.

Solution

The constant-pressure evaporation of water is also at constant temperature. Equation (5.47) is integrated from the saturated liquid state to the saturated vapor state. Taking P and T as constants, we have

$$\int_{s_l}^{s_g} ds = \int_{u_l}^{u_g} \frac{du}{T} + \int_{v_l}^{v_g} \frac{P}{T}\,dv$$

$$s_g - s_l = \frac{1}{T_{sat}}(u_g - u_l) + \frac{P_{sat}}{T_{sat}}(v_g - v_l)$$

$$s_{lg} = \frac{u_{lg}}{T_{\text{sat}}} + \frac{P_{\text{sat}}}{T_{\text{sat}}} v_{lg}$$

$$= \frac{2089.0}{372.75} + \frac{100}{372.75}(1.6898) = 6.0576 \text{ kJ/(kg} \cdot \text{K)}$$

Since the evaporation is at constant pressure, Eq. (5.48) yields

$$\int_{s_l}^{s_g} ds = \int_{h_l}^{h_g} \frac{dh}{T} - \int_{P_{\text{sat}}}^{P_{\text{sat}}} \frac{v}{T} dP = \int_{h_l}^{h_g} \frac{dh}{T}$$

$$s_g - s_l = \frac{h_g - h_l}{T_{\text{sat}}} = \frac{h_{lg}}{T_{\text{sat}}} = \frac{2258}{372.75} = 6.0577 \text{ kJ/(kg} \cdot \text{K)}$$

These results are to be compared with the tabulated value of 6.0578 kJ/(kg · K).

Comments

Equations (5.47) and (5.48) are viewed as required relations between thermodynamic properties. They describe the property surfaces and depend on only the state.

Equations of state are presented in Chap. 3 to mathematically relate the properties of a particular substance. They are shown in Sec. 3.5.2 to be very useful for particular substances. The changes in entropy for the ideal gas and incompressible fluid idealizations are considered here.

5.5.2 Ideal Gas Relations

The equation of state for an ideal gas is $Pv = RT$, and as stated in Sec. 3.2, the change in internal energy is a function of temperature only. This is demonstrated by rewriting Eq. (5.47) as

$$ds = \frac{du}{T} + \frac{R}{v} dv \tag{5.49}$$

where $P/T = R/v$ for an ideal gas. Since ds is an exact differential, Eq. (1.31) yields

$$\left[\frac{\partial}{\partial v} \left(\frac{1}{T} \right) \right]_u = \left[\frac{\partial}{\partial u} \left(\frac{R}{v} \right) \right]_v \tag{5.50}$$

But the right-hand side of this expression is zero, so

$$\left(\frac{\partial T}{\partial v} \right)_u = 0 \tag{5.51}$$

Therefore, temperature is independent of the specific volume for a constant internal energy, or temperature is a function of only internal energy.

Thus internal energy is a function of temperature only, or $u = u(T$ only), and Eqs. (3.19) and (3.22) yield

$$du = c_v\, dT \qquad \text{and} \qquad dh = c_P\, dT$$

These relations and the equation of state are combined with the Gibbs equation in Eqs. (5.47) and (5.48) to yield

$$ds = c_v \frac{dT}{T} + R \frac{dv}{v} \qquad (5.52)$$

$$ds = c_P \frac{dT}{T} - R \frac{dP}{P} \qquad (5.53)$$

Finite changes in entropy are evaluated by integrating these last equations to yield

$$s_2 - s_1 = \int_1^2 c_v \frac{dT}{T} + R \ln \frac{v_2}{v_1} \qquad (5.54)$$

$$s_2 - s_1 = \int_1^2 c_P \frac{dT}{T} - R \ln \frac{P_2}{P_1} \qquad (5.55)$$

Either of the previous expressions for the change in entropy is used depending on the information that is known. Unlike the internal energy and enthalpy for an ideal gas which require only temperature for their evaluation, entropy requires two properties [see Eq. (5.52) or (5.53)]. Completion of the evaluation of the change in entropy requires the functional form of the specific heat variation with temperature.

The specific heats are often accurately approximated by a single average value. Thus, for *constant specific heats*,

$$s_2 - s_1 = c_v \ln \frac{T_2}{T_1} + R \ln \frac{v_2}{v_1} \qquad (5.56)$$

$$s_2 - s_1 = c_P \ln \frac{T_2}{T_1} - R \ln \frac{P_2}{P_1} \qquad (5.57)$$

Table C.2 presents specific heats at 300 K for various gases.

More accurate values for changes in entropy are obtained if the temperature variation of the specific heats is used in the integration in Eqs. (5.54) and (5.55). The temperature dependence of the specific heat at constant pressure is given in Table D.1 for various gases. These expressions are substituted into the integrals, and the values for the change are obtained. Note that $c_P - c_v = R$, so only the c_P integration needs to be performed. When the temperature dependence of c_P is included, the results are termed *variable specific heat* solutions.

A few differences occur in the determination of s from the gas tables, Tables D.2 through D.7, as compared with the properties discussed in Chap. 3. Because these tables are based on an ideal gas relation, the values for u and h are functions of T only. The tables are thus presented as a function of the *single* property T, unlike the steam tables (Tables D.8 through D.10) which require *two* independent properties to define a state. However, for the entropy of an ideal gas, Eqs. (5.54) and (5.55) show that entropy is not a function of T only, but that a second independent property (usually v or P) must be specified. However, by choosing P as the second independent variable, the value of s is obtained by writing Eq. (5.55) as

$$s - s_{ref} = \int_{T_{ref}}^{T} c_P \frac{dT}{T} - R \ln \frac{P}{P_{ref}} = s_0(T) - R \ln \frac{P}{P_{ref}} \qquad (5.58)$$

where

$$s_0(T) \equiv \int_{T_{ref}}^{T} c_P \frac{dT}{T}$$

In this form, s is determined from summing $s_0(T)$, which is a function of T only, with the second term, which is a function of P only. The subscript 0 in $s_0(T)$ denotes that this is the temperature-dependent portion of the entropy of an ideal gas and so is *not* a reference value. In the gas tables, $s_0(T)$, often given the symbol ϕ, is tabulated, and the value of $s - s_{ref}$ is simply obtained by subtracting the pressure term $R \ln (P/P_{ref})$ from $s_0(T)$. The difference in entropy for two states is

$$s_2 - s_1 = s_0(T_2) - s_0(T_1) - R \ln \frac{P_2}{P_{ref}} + R \ln \frac{P_1}{P_{ref}} \qquad (5.59)$$

or

$$s_2 - s_1 = s_0(T_2) - s_0(T_1) - R \ln \frac{P_2}{P_1} \qquad (5.60)$$

The reference state is chosen at absolute zero temperature and atmospheric pressure.

5.5.3 Incompressible Fluid or Solid Relations

The equation of state for an incompressible fluid or solid, as stated in Eq. (3.32), is $v = $ constant. Equation (3.34a) states that

$$du = c \, dT$$

This expression is substituted into Eq. (5.47) to give the change in entropy

$$ds = c \frac{dT}{T} + \frac{P}{T} \, dv \qquad (5.61)$$

but $v = $ constant, so

$$ds = c \frac{dT}{T} \qquad (5.62)$$

Finite changes in entropy are

$$S_2 - S_1 = \int_1^2 c\, \frac{dT}{T}$$ (5.63)

If the specific heat is approximated by a constant, then

$$S_2 - S_1 = c \ln \frac{T_2}{T_1}$$ (5.64)

Otherwise, the temperature integration must be performed. With $dh = du + P\,dv + v\,dP$ and Eq. (5.48), the same result is obtained.

Example Problem 5.4

Two blocks A and B which are initially at 100 and 500°C, respectively, are brought into contact and isolated from the surroundings. They are allowed to reach a final state of thermal equilibrium. Determine the entropy change of each block and of the isolated system. Block A is aluminum [$c_P = 0.900$ kJ/(kg · K)] with $m_A = 0.5$ kg, and block B is copper [$c_P = 0.386$ kJ/(kg · K)] with $m_B = 1.0$ kg.

Solution

The diagram of the system is as presented in Fig. 5.2. The blocks are incompressible, so only temperature is required to specify the states. The initial temperatures are specified, and the final temperature is unknown. The first and second laws for the isolated system are

First law: $dU = \delta Q + \delta W = 0$

$$U_2 - U_1 = 0$$

$$U_{2A} + U_{2B} = U_{1A} + U_{1B}$$

Second law: $dS \geq 0$ or $dS - \delta S_{\text{gen}} = 0$

$$S_{\text{gen}} = S_2 - S_1 = (S_{2A} + S_{2B}) - (S_{1A} + S_{1B})$$

or

$$S_{\text{gen}} = (S_{2A} - S_{1A}) + (S_{2B} - S_{1B})$$

The properties are obtained from Eqs. (3.34a) and (5.64) as

$$U_2 - U_1 = mc(T_2 - T_1)$$

$$S_2 - S_1 = mc \ln \frac{T_2}{T_1}$$

The first law is used to determine the final state. Rearranging the first law expression above yields

$$U_{2A} - U_{1A} = U_{1B} - U_{2B}$$

$$m_A c_A (T_{2A} - T_{1A}) = m_B c_B (T_{1B} - T_{2B})$$

With $T_{2A} = T_{2B} = T_2$, then

$$T_2 = \frac{m_A c_A T_{1A} + m_B c_B T_{1B}}{m_A c_A + m_B c_B} = 284.7°C$$

The entropy generated is given above as

$$S_{gen} = m_A c_A \ln \frac{T_2}{T_{1A}} + m_B c_B \ln \frac{T_2}{T_{1B}}$$

$$= 0.1809 - 0.1260 = 0.0549 \text{ kJ/K}$$

Thus there is an increase in the entropy of block A of $+0.1809$ kJ/K and a decrease in the entropy of block B of -0.1260 kJ/K. The entropy generation of the isolated system is greater than zero, as stated by the second law. Therefore, this is an irreversible process.

Comments

The final temperature is a weighted average of the initial values as specified by the first law. The final temperature, with the material properties, determines the entropy generated within each block. The entropy generated must be zero or positive for the isolated system, but each block undergoes different directional changes. Block A is increasing in temperature; so it is increasing in molecular disorder, and the entropy increases. Block B is decreasing in temperature, or it is becoming more molecularly ordered, and the entropy decreases. Since a heat transfer from block B to block A is associated with a transfer of disorganization, these increases and decreases are expected.

Also, if the initial temperature of block B is closer to that of block A, say $T_{1B} = 110°C$, then the entropy generation is ($T_2 = 104.6°C$)

$$S_{gen} = 0.00553 - 0.00546 = 0.00007 \text{ kJ/K}$$

Thus, the entropy changes for each block are much smaller than for the case of large temperature difference. The process is irreversible; but since the initial block temperatures are closer, it is approaching a reversible process.

5.6 Control Mass Formulation

The second law is presented for an isolated system, and as indicated by irreversible and reversible processes, the isolated system processes are directional. This directionality of the entropy change for an isolated system is a fundamental and important aspect of the second law. But the isolated system is somewhat abstract, and the primary interest of much of thermodynamic analysis lies in control mass and control volume systems. Therefore, the second law must be transformed to these viewpoints. This section focuses on the control mass formulation, and Sec. 5.7 considers the control volume formulation.

Many of the previous examples in this chapter consider an isolated system as a set of subsystems. One subsystem is considered the control mass, and the others compose the surroundings. From this viewpoint, a general control mass undergoes a process with arbitrary energy transfers with the surroundings, and the control mass and surroundings form the isolated system. The change in entropy for the control mass is desired. Thus the surroundings are idealized and specified, so the only unspecified entropy changes are those associated with the control mass.

A control mass exchanges energy with the surroundings by work and heat transfer. Therefore, two types of idealized subsystems of the surroundings are conceived: a *reversible work reservoir* and a *reversible heat transfer reservoir*. Both reservoirs are reversible, indicating that they proceed slowly through equilibrium states and can be restored to the original states at any point in the process. Both are control masses and are termed *reservoir* to indicate a very large system that is unchanging during work or heat transfer.

The reversible heat transfer reservoir participates in heat transfer only (not in work transfers). Therefore, this reservoir is a constant-volume control mass. All the heat transfers to the reversible heat transfer reservoir transform this system through equilibrium states which are specified by a uniform temperature. The heat transfers can be to or from this reservoir. From the Gibbs equation [Eq. (5.36)] for constant-volume processes,

$$dS = \frac{dU}{T} \tag{5.65}$$

and the first law for a control mass that does no work is

$$dU = \delta Q \tag{5.66}$$

Thus, for the reversible heat transfer reservoir, the entropy change is

$$dS = \frac{\delta Q}{T} \tag{5.67}$$

where δQ is positive *into* the reversible heat transfer reservoir. The temperature is uniform throughout this reservoir.

The reversible work reservoir participates in only work interactions (no heat transfers), and for a compressible system, the work is a result of boundary expansion. The work, either positive or negative, is a quasi-equilibrium process. The Gibbs equation for the reversible work reservoir is given in Eq. (5.36) as

$$dS = \frac{dU}{T} + \frac{P}{T} dV$$

but the first law with only $P\,dV$ work and no heat transfer is

$$dU = \delta\cancel{Q}^{0} + \delta W = -P\,dV \tag{5.68}$$

Thus, substituting Eq. (5.68) into Eq. (5.36) yields

$$dS = -\frac{P\,dV}{T} + \frac{P}{T}\,dV = 0 \tag{5.69}$$

so the reversible work reservoir has no entropy change.

In summary, the control mass form of the second law is obtained by considering a general control mass that undergoes both work interactions with reversible work reservoirs and heat transfers with reversible heat transfer reservoirs. The composite system of control mass, reversible work reservoirs, and reversible heat transfer reservoirs forms an isolated system. The work reservoirs have no entropy change, and the heat transfer reservoir's entropy change is $dS = \delta Q/T$. A general control mass is shown in Fig. 5.8, where there could be an arbitrary number of work and heat transfer reservoirs.

The second law for the isolated system (control mass plus the reservoirs) is

$$dS_{\text{isolated}} \geq 0 \tag{5.70}$$

but

$$S_{\text{isolated}} = S_{\text{CM}} + \sum_{i}\cancel{S}^{0}_{\substack{\text{reversible}\\\text{work}\\\text{reservoirs}}} + \sum_{i} S_{\substack{\text{reversible}\\\text{heat transfer}\\\text{reservoirs}}} \tag{5.71}$$

The entropy change for each reversible heat transfer reservoir is $dS_i = \delta Q_i/T_i$, where δQ_i is positive into the reservoir and T_i is the reservoir temperature and so is the temperature of the boundary between the reversible heat transfer reservoir and the control mass. Thus, in terms of the

Isolated system

Figure 5.8   *Control mass, reversible heat transfer reservoirs, and reversible work reservoirs within an isolated system.*

control mass, the heat transfer to the control mass from each reservoir is

$$\delta Q_{\text{CM},\,i} = -\delta Q_{\substack{\text{reversible} \\ \text{heat transfer} \\ \text{reservoir},\, i}} \qquad (5.72)$$

Thus, substituting Eqs. (5.71) and (5.72) into (5.70) yields

$$dS_{\text{isolated}} = dS_{\text{CM}} - \sum_i \left(\frac{\delta Q_i}{T_i}\right)_{\text{CM}} \geq 0 \qquad (5.73)$$

or

$$dS_{\text{CM}} \geq \sum_i \left(\frac{\delta Q_i}{T_i}\right)_{\text{CM}} \qquad (5.74)$$

since the change in entropy of all reversible work reservoirs is zero. Note especially that the temperature in the second law is the boundary temperature interfacing the control mass and heat transfer reservoirs. Thus this is the control mass form of the second law:

> The change in entropy S, an extensive property of the control mass, is greater than or equal to the sum of the heat transfers divided by the corresponding absolute temperatures of the boundary.

Note that dS_{CM} can be positive or negative since the heat transfer can be positive or negative. Recall that heat transfer is a microscopically disorganized energy transfer, so positive heat transfer increases microscopic disorder and increases entropy while negative heat transfer decreases microscopic disorder and decreases entropy. Also, there is no contribution to entropy change by work interactions. Since work is an organized energy transfer mechanism and entropy describes microscopic disorder, this is appropriate.

Equation (5.74) is the control mass form for the second law. The inequality applies to irreversible processes. Since the development is in terms of reversible surroundings, the irreversibilities are within the control mass; so this type of process is termed *internally irreversible*. The equality applies to an internally reversible process. The inequality is eliminated if the concept of entropy generation is used. The development is as above, but Eq. (5.4) is the starting point. Thus

$$(dS - \delta S_{\text{gen}})_{\text{isolated}} = 0 \qquad (5.75)$$

and with Eqs. (5.67), (5.69), (5.71), and (5.72),

$$dS_{\text{CM}} - \sum_i \left(\frac{\delta Q_i}{T_i}\right)_{\text{CM}} - \delta S_{\text{gen}} = 0 \qquad (5.76)$$

or

$$\delta S_{\text{gen}} = dS_{\text{CM}} - \sum_i \left(\frac{\delta Q_i}{T_i}\right)_{\text{CM}} \geq 0 \qquad (5.77)$$

This is the entropy generation by the process.

The control mass forms of the second law are given in Eq. (5.74) or (5.77). They can also be expressed as

$$dS_{CM} = d\left(\int_V \rho s \, dV\right) \geq \sum_i \left(\frac{\delta Q_i}{T_i}\right)_{CM} \tag{5.78}$$

If the control mass has a uniform distribution of mass, then

$$d(ms)_{CM} \geq \sum_i \left[\frac{\delta(mq)_i}{T_i}\right]_{CM} \tag{5.79}$$

or

$$ds_{CM} \geq \sum_i \left(\frac{\delta q_i}{T_i}\right)_{CM} \tag{5.80}$$

In terms of entropy generation, this is

$$\delta S_{gen} = d\left(\int_V \rho s \, dV\right) - \sum_i \left(\frac{\delta Q_i}{T_i}\right)_{CM} \geq 0 \tag{5.81}$$

and

$$\delta s_{gen} = ds_{CM} - \sum_i \left(\frac{\delta q_i}{T_i}\right)_{CM} \geq 0 \tag{5.82}$$

The heat transfers are relative to the control mass, and the T_i corresponds to the temperature at boundary i of the control mass where the heat transfer occurs. The sum includes all the possible heat transfers with the surroundings. All real processes are irreversible to some extent; so the inequality sign is applicable, and the entropy generation is greater than zero. Only for the idealization of a reversible process is the equal sign appropriate, and $\delta s_{gen} = 0$. Equations (5.78) and (5.81) are the forms of the second law that are the starting point for all subsequent analysis. The importance of these expressions cannot be overemphasized. The control mass forms of the second law are expressed on a rate basis as

$$\left.\frac{dS}{dt}\right|_{CM} = \frac{d}{dt}\left(\int_V \rho s \, dV\right)_{CM} \geq \sum_i \left(\frac{\dot{Q}_i}{T_i}\right)_{CM} \tag{5.83}$$

or

$$\dot{S}_{gen} = \frac{d}{dt}\left(\int_V \rho s \, dV\right)_{CM} - \sum_i \left(\frac{\dot{Q}_i}{T_i}\right)_{CM} \geq 0 \tag{5.84}$$

The second law connects the system property, entropy, to the heat transfer. If the *process is reversible,* the entropy generation is zero, and

$$dS_{CM, \, rev} = \sum_i \left(\frac{\delta Q_i}{T_i}\right)_{CM, \, rev} \tag{5.85}$$

For a single heat transfer

Figure 5.9 Heat transfer and work represented as areas on *T-s* and *P-v* diagrams, respectively.

$$dS_{\text{CM, rev}} = \left(\frac{\delta Q}{T}\right)_{\text{CM, rev}} \tag{5.86}$$

or, since the reversible heat transfer occurs across an infinitesimal temperature difference,

$$\delta Q_{\text{rev}} = T\, dS \tag{5.87}$$

This shows the importance of the *T-s* diagram to thermodynamic analysis. The heat transfer for a specified reversible process between states 1 and 2 is

$$_1Q_{2,\text{rev}} = \int_1^2 T\, dS \tag{5.88}$$

or

$$_1q_{2,\text{rev}} = \int_1^2 T\, ds \tag{5.89}$$

Thus the area under the process curve on a *T-s* diagram is the heat transfer per unit mass. This is similar to the work representation on a *P-v* diagram. This is shown schematically in Fig. 5.9.

Example Problem 5.5

Calculate the heat transfers required to evaporate water at 0.1 and 1.0 MPa. Assume the process to be reversible.

Solution

All the states are known. and Table D.9 yields

State	P, MPa	T, °C	x	s, kJ/(kg · K)
$1a$	0.1†	(99.6)	0	(1.3020)
$1b$	1.0	(179.9)	0	(2.1367)
$2a$	0.1	(99.6)	1.0	(7.3598)
$2b$	1.0	(179.9)	1.0	(6.5827)

† Entries *not* in parentheses determine the state.

The system is chosen as the water within a container which is isothermal for the constant-pressure evaporation. The second law for a reversible process is

$$ds = \frac{\delta q_{\text{rev}}}{T} \quad \text{or} \quad \delta q_{\text{rev}} = T\, ds$$

Integration for an isothermal process yields

$$_1 q_{2,\text{rev}} = T_{\text{sat}}(s_2 - s_1) = T_{\text{sat}} s_{lg}$$

Using absolute temperature results in

$$_1 q_{2,\text{rev}} = \begin{cases} 2258.0 \text{ kJ/kg} & \text{for } P = 0.1 \text{ MPa} \\ 2014.3 \text{ kJ/kg} & \text{for } P = 1.0 \text{ MPa} \end{cases}$$

Comments

The area on the T-s diagram is shown in the diagram. The s_{lg} is very different in each case, but since the temperatures are also different, the heat transfers are comparable.

An alternative approach to this problem starts with the first law for the control mass as

$$du = \delta q + \delta w$$

with $\delta w = -P\, dv$,

$$\delta q = du + P\, dv$$

which for a constant-pressure process is

$$\delta q = dh$$

so

$$_1 q_2 = h_2 - h_1 = h_{lg}$$

From Table D.9,

$$_1 q_2 = \begin{cases} 2258 \text{ kJ/kg} & \text{for } P = 0.1 \text{ MPa} \\ 2014 \text{ kJ/kg} & \text{for } P = 1.0 \text{ MPa} \end{cases}$$

which agrees with the above analysis.

Diagram

Example Problem 5.6

Water is transformed from the saturated vapor state to the saturated liquid state at $P = 0.1$ MPa. This process occurs by a heat transfer across the container wall to the environment (heat transfer reservoir) at T_0. Calculate the entropy generation for the container per unit mass of water for the heat transfer with $T_0 = 25°C$ and $T_0 = 99°C$.

Diagram

Solution

The heat transfer required to condense the water at 0.1 MPa is the negative of that evaluated in Example Problem 5.5 for evaporation. The water was considered as the control mass, and the second law was applied to that example. The process was assumed to be internally reversible in that example. The required heat transfer per unit mass is calculated as

$q = 2258.0$ kJ/kg

and the water is isothermal at $T_{sat} = 99.6°C$. This is the heat transfer across the container wall in the present example.

The entropy generation for the heat transfer with the environment focuses on the wall as the system with T_{sat} at the inside surface and T_0 at the outside surface. The process is assumed to be at steady state, so q at the inside surface is the same as q at the outside surface. The second law is applied as given in Eq. (5.84),

$$\dot{S}_{gen} = \frac{d}{dt}\left(\int_V \rho s \, dV\right)_{CM} - \sum_i \left(\frac{\dot{Q}_i}{T_i}\right)_{CM} \geq 0$$

The entropy change within the control mass (the wall) is zero. The temperature varies spatially across the wall, so the local value of the entropy varies spatially. But the values do not change in time, since steady state is assumed; so the change in entropy within the control mass is zero.

Integrating the second law with respect to time yields

$$S_{gen.} = -\sum_{i=1}^{2} \left(\frac{Q_i}{T_i}\right)_{CM} \geq 0$$

The heat transfer into the wall, Q, must be equal to the heat transfer from the water, $m_w q$, where m_w is the mass of the water. The heat transfer into the wall is positive at T_{sat}, and the heat transfer to the environment is negative at T_0, so

$$S_{gen} = -\frac{m_w q}{T_{sat}} + \frac{m_w q}{T_0} = m_w q \left(\frac{1}{T_0} - \frac{1}{T_{sat}} \right) \geq 0$$

Thus, with $T_0 = 25°C$

$$S_{gen} = m_w(2258)\left(\frac{1}{298.15} - \frac{1}{372.75} \right)$$

$$= m_w(1.516) \quad kJ/K$$

and with $T_0 = 99°C$

$$S_{gen} = m_w(2258)\left(\frac{1}{372.15} - \frac{1}{372.75} \right)$$

$$= m_w(0.009767) \quad kJ/K$$

Comments

The entropy generation is greatly reduced as T_0 is increased from 25 to 99°C. Thus the heat transfer is approaching the reversible heat transfer idealization. If the environment is $T_0 = T_{sat} - dT$, then

$$\frac{S_{gen}}{m_w} = q \left(\frac{1}{T_{sat} - dT} - \frac{1}{T_{sat}} \right)$$

$$= q \frac{T_{sat} - T_{sat} + dT}{(T_{sat} - dT)(T_{sat})}$$

$$= q \frac{dT}{(T_{sat} - dT)(T_{sat})}$$

Thus, as the infinitesimal temperature difference is reduced to zero, the heat transfer becomes reversible, or

$$S_{gen} \rightarrow 0 \quad \text{as } dT \rightarrow 0$$

5.7   Control Volume Formulation and Analysis

Control volume analysis permits mass to cross the boundary. This mass transfer carries the property entropy into and out of the volume. Therefore, the change in entropy within the control volume is modified by the mass transport. The control mass form of the second law states that the change in entropy within the boundary is related to the quantity dQ/T at the boundary. Thus, this is the control volume statement of the second law:

The change in entropy within the control volume minus the net entropy transported into the control volume is greater than or equal to the sum of the heat transfers divided by the corresponding boundary absolute temperatures.

Let $\dot{S}$ denote the rate at which entropy is transported with the fluid at the inlets and outlets, where the overdot indicates the rate of flow of entropy. The entropy within the control volume is denoted by S_{CV}, so the second law is

$$\frac{\partial S_{CV}}{\partial t} - \dot{S}_{net} \geq \sum_i \left(\frac{\dot{Q}_i}{T_i}\right)_{CV} \tag{5.90}$$

Or, with the net rate of flow of entropy equal to the inlet minus the outlet contributions,

$$\frac{\partial S_{CV}}{\partial t} - \dot{S}_{in} + \dot{S}_{out} \geq \sum_i \left(\frac{\dot{Q}_i}{T_i}\right)_{CV} \tag{5.91}$$

The terms on the left-hand side of the inequality require detailed consideration.

The change in entropy within the control volume is given as

$$\frac{\partial S_{CV}}{\partial t} = \frac{\partial}{\partial t}\left(\int_{CV} \rho s \, dV\right) \tag{5.92}$$

The mass transport contributions into and out of the control volume, denoted $\dot{S}_{in}$ and $\dot{S}_{out}$, respectively, follow the development presented in the control volume formulation of conservation of mass and energy. The fluid element carries entropy across each boundary element, so the transported entropy is the mass flow rate in Eq. (4.31) multiplied by the entropy per unit mass, or

$$d\dot{S} = s\rho|\mathbf{V} \cdot \mathbf{n}| \, dA = s \, d\dot{m} \tag{5.93}$$

The integrals over all the inlet and outlet boundaries are

$$\dot{S}_{in} = \int_{A_{in}} s\rho|\mathbf{V} \cdot \mathbf{n}| \, dA = \int_{A_{in}} s \, d\dot{m} \tag{5.94}$$

and

$$\dot{S}_{out} = \int_{A_{out}} s\rho|\mathbf{V} \cdot \mathbf{n}| \, dA = \int_{A_{out}} s \, d\dot{m} \tag{5.95}$$

Thus the net transport is

$$\dot{S}_{net} = \int_{A_{in}} s\rho|\mathbf{V} \cdot \mathbf{n}| \, dA - \int_{A_{out}} s\rho|\mathbf{V} \cdot \mathbf{n}| \, dA$$

$$= \int_{A_{in}} s \, d\dot{m} - \int_{A_{out}} s \, d\dot{m} \tag{5.96}$$

The control volume form of the second law is given as

$$\frac{\partial}{\partial t}\left(\int_{CV} \rho s \, dV\right) - \int_{A_{in}} s \, d\dot{m} + \int_{A_{out}} s \, d\dot{m} \geq \sum_i \left(\frac{\dot{Q}_i}{T_i}\right)_{CV} \tag{5.97}$$

For multiple inlets and outlets, this is

$$\frac{\partial}{\partial t}\left(\int_{CV} \rho s \, dV\right) - \sum_{in} \int_{A_{in}} s \, d\dot{m} + \sum_{out} \int_{A_{out}} s \, d\dot{m} \geq \sum_i \left(\frac{\dot{Q}_i}{T_i}\right)_{CV} \tag{5.98}$$

The left-hand side of this expression should be compared with Eqs. (F.16) and (F.17) in App. F: the Reynolds transport theorem could have been used to develop these expressions. Equation (5.97) is the general control volume form of the second law. The first term is the entropy storage term within the control volume. The next two terms are the entropy transported by mass transport into and out of the control volume. The right-hand side represents the entropy transport via heat transfer. The equivalent expression corresponding to Eq. (5.97) from using the entropy generation concept is

$$\dot{S}_{gen} = \frac{\partial}{\partial t}\left(\int_{CV} \rho s \, dV\right) - \int_{A_{in}} s \, d\dot{m} + \int_{A_{out}} s \, d\dot{m} - \sum_i \left(\frac{\dot{Q}_i}{T_i}\right)_{CV} \geq 0 \tag{5.99}$$

Before we proceed to the idealizations at the crossing boundaries and within the control volume, note that the above expressions [Eqs. (5.97) and (5.99)] reduce to the special cases considered in previous sections. For the case of no mass transport into or out of the control volume, these expressions reduce to

$$\frac{d}{dt}\left(\int_{CV} \rho s \, dV\right) = \frac{dS_{CV}}{dt} \geq \sum_i \left(\frac{\dot{Q}_i}{T_i}\right)_{CV}$$

$$\dot{S}_{gen} = \frac{d}{dt}\left(\int_{CV} \rho s \, dV\right) - \sum_i \left(\frac{\dot{Q}_i}{T_i}\right)_{CV} \geq 0$$

These are the same expressions given in Eqs. (5.83) and (5.84). If the system is isolated so that there are no interactions with the surroundings, the heat transfer is zero, and

$$\frac{d}{dt}\left(\int_{CV} \rho s \, dV\right) = \frac{dS_{CV}}{dt} \geq 0$$

$$\dot{S}_{gen} = \frac{d}{dt}\left(\int_{CV} \rho s \, dV\right) \geq 0$$

These expressions are equivalent to Eqs. (5.6) and (5.7).

A summary of the governing equations for the control mass formulation and the control volume formulation is given in Table 5.1. For convenience, the conservation of mass and energy are also given.

TABLE 5.1   General Conservation Relations

	Control Mass	Control Volume
Mass	$\dfrac{d}{dt}\left(\displaystyle\int_V \rho\,dV\right)=0$	$\dfrac{\partial}{\partial t}\left(\displaystyle\int_{CV} \rho\,dV\right)=\displaystyle\sum_{in}\dot m-\sum_{out}\dot m$
First law	$\dfrac{d}{dt}\left(\displaystyle\int_V e\rho\,dV\right)=\dot Q+\dot W$	$\dfrac{\partial}{\partial t}\left(\displaystyle\int_{CV} e\rho\,dV\right)=\displaystyle\sum_{in}\int_{A_{in}}(e+Pv)\,d\dot m$ $+\dot Q_{CV}+\dot W_{CV}-\displaystyle\sum_{out}\int_{A_{out}}(e+Pv)\,d\dot m$
Second law	$\dfrac{d}{dt}\left(\displaystyle\int_V s\rho\,dV\right)\ge\displaystyle\sum_i\left(\dfrac{\dot Q_i}{T_i}\right)$ or $\dot S_{gen}=\dfrac{\partial}{\partial t}\left(\displaystyle\int_V s\rho\,dV\right)$ $-\displaystyle\sum_i\left(\dfrac{\dot Q_i}{T_i}\right)\ge 0$	$\dfrac{\partial}{\partial t}\left(\displaystyle\int_{CV} s\rho\,dV\right)+\displaystyle\sum_{out}\int_{A_{out}}s\,d\dot m$ $-\displaystyle\sum_{in}\int_{A_{in}}s\,d\dot m\ge\displaystyle\sum_i\left(\dfrac{\dot Q_i}{T_i}\right)_{CV}$ or $\dot S_{gen}=\dfrac{\partial}{\partial t}\left(\displaystyle\int_{CV} s\rho\,dV\right)$ $+\displaystyle\sum_{out}\int_{A_{out}}s\,d\dot m-\displaystyle\sum_{in}\int_{A_{in}}s\,d\dot m$ $-\displaystyle\sum_i\left(\dfrac{\dot Q_i}{T_i}\right)_{CV}\ge 0$

Here $e = u + \tfrac{1}{2}\mathbf{V}^2 + gZ$.

5.7.1   Spatial and Time Variation Idealizations

The analysis of a system requires detailed consideration of the crossing boundaries and the state within the control volume. *Uniform state* and *nonuniform state* describe spatial variation, while *steady state* and *unsteady state* describe time variation. These terms describe both the crossing boundaries and the state within the control volume. The idealizations parallel the development of the energy equations. Table 5.2 presents the idealizations for the first and second laws as well as the conservation of mass (this is an expanded version of Table 4.1). Recall that the uniform state idealization eliminates the spatial integration, and that the unsteady state analysis generally requires a time integration.

   The two classifications of problems that often occur in engineering analysis are the steady state and unsteady state idealizations with uniform states at the crossing boundaries and within the control volume. For *steady state analysis* there is no time variation, so Eq. (5.98) reduces to

$$\sum_{out}\int_{A_{out}}s\,d\dot m-\sum_{in}\int_{A_{in}}s\,d\dot m\ge\sum_i\left(\frac{\dot Q_i}{T_i}\right)_{CV} \tag{5.100}$$

With uniform properties at the inlets and outlets, this results in

$$\sum_{out}s\dot m-\sum_{in}s\dot m\ge\sum_i\left(\frac{\dot Q_i}{T_i}\right)_{CV} \tag{5.101}$$

TABLE 5.2   Control Volume Relations

	Nonuniform and Unsteady	Uniform	Unsteady Time Integration
Inlets and outlets	$\dot{m} = \int_A d\dot{m} = \int_A \rho \lvert \mathbf{V} \cdot \mathbf{n}\rvert \, dA$	$\dot{m} = \rho \mathbf{V} A$	$\int_{t_1}^{t_2} \dot{m}\, dt = m$
	$\int_A (e + Pv)\, d\dot{m} = \int_A (e + Pv)\rho \lvert \mathbf{V} \cdot \mathbf{n}\rvert\, dA$	$(e + Pv)\dot{m}$	$\int_{t_1}^{t_2} \int_A (e + Pv)\, d\dot{m}\, dt = (e + Pv)m$ with uniform and steady state properties
	$\int_A s\, d\dot{m} = \int_A s\rho \lvert \mathbf{V} \cdot \mathbf{n}\rvert\, dA$	$s\dot{m}$	$\int_{t_1}^{t_2} \int_A s\, d\dot{m}\, dt = sm$ with uniform and steady state properties
Within the control volume	$m_{CV} = \int_{CV} \rho\, dV$	$m_{CV} = (\rho V)_{CV}$	$\int_{t_1}^{t_2} \dfrac{\partial m_{CV}}{\partial t}\, dt = m_2 - m_1$
	$E_{CV} = \int_{CV} \rho e\, dV$	$E_{CV} = (\rho e V)_{CV} = (me)_{CV}$	$\int_{t_1}^{t_2} \dfrac{\partial E_{CV}}{\partial t}\, dt = E_2 - E_1 = (em)_2 - (em)_1$ with uniform properties
	$S_{CV} = \int_{CV} \rho s\, dV$	$S_{CV} = (\rho s V)_{CV} = (ms)_{CV}$	$\int_{t_1}^{t_2} \dfrac{\partial S_{CV}}{\partial t}\, dt = S_2 - S_1 = (sm)_2 - (sm)_1$ with uniform properties

The entropy generation form for steady analysis is

$$\dot{S}_{\text{gen}} = \sum_{\text{out}} s\dot{m} - \sum_{\text{in}} s\dot{m} - \sum_i \left(\frac{\dot{Q}_i}{T_i}\right)_{CV} \geq 0 \qquad (5.102)$$

The *unsteady state analysis* requires the time integration of Eq. (5.98). With uniform properties, this integration yields

$$s_2 m_2 - s_1 m_1 - \sum_{\text{in}} sm + \sum_{\text{out}} sm \geq \int_{t_1}^{t_2} \sum_i \left(\frac{\dot{Q}_i}{T_i}\right)_{CV} dt \qquad (5.103)$$

The time integration of Eq. (5.99) with Table 5.2 yields

$$_1 S_{\text{gen}, 2} = s_2 m_2 - s_1 m_1 + \sum_{\text{out}} sm - \sum_{\text{in}} sm$$

$$- \int_{t_1}^{t_2} \sum_i \left(\frac{\dot{Q}_i}{T_i}\right)_{CV} dt \geq 0 \quad (5.104)$$

These equations [Eqs. (5.101) through (5.104)] are summarized, along with the corresponding first law and conservation of mass, in Table 5.3.

5.7.2   Applications

The application of the second law to specific problems requires the same general approach as the first law analysis. A review of the general statements in Sec. 4.5 is appropriate at this point. A careful denotation of the system and boundary transfers is important. Now the second law is used

TABLE 5.3 Relations for Special Cases

Steady State with Uniform Properties

Mass:
$$\sum_{\text{in}} \dot{m} = \sum_{\text{out}} \dot{m}$$

First law:
$$\sum_{\text{in}} (h + \tfrac{1}{2}\mathbf{V}^2 + gZ)\dot{m} + \dot{Q}_{\text{CV}} + \dot{W}_{\text{CV}} = \sum_{\text{out}} (h + \tfrac{1}{2}\mathbf{V}^2 + gZ)\dot{m}$$

Second law:
$$\sum_{\text{out}} s\dot{m} - \sum_{\text{in}} s\dot{m} \geq \sum_{i} \left(\frac{\dot{Q}_i}{T_i}\right)_{\text{CV}}$$

or

$$\dot{S}_{\text{gen}} = \sum_{\text{out}} s\dot{m} - \sum_{\text{in}} s\dot{m} - \sum_{i} \left(\frac{\dot{Q}_i}{T_i}\right)_{\text{CV}} \geq 0$$

Unsteady State with Uniform Properties and Steady State Port Properties

Mass:
$$m_2 - m_1 = \sum_{\text{in}} m - \sum_{\text{out}} m$$

First law:
$$(u + \tfrac{1}{2}\mathbf{V}^2 + gZ)_2 m_2 - (u + \tfrac{1}{2}\mathbf{V}^2 + gZ)_1 m_1 = \sum_{\text{in}} (h + \tfrac{1}{2}\mathbf{V}^2 + gZ)m$$
$$+ {}_1Q_2 + {}_1W_2 - \sum_{\text{out}} (h + \tfrac{1}{2}\mathbf{V}^2 + gZ)m$$

Second law:
$$s_2 m_2 - s_1 m_1 + \sum_{\text{out}} sm - \sum_{\text{in}} sm \geq \int_{t_1}^{t_2} \sum_i \left(\frac{\dot{Q}_i}{T_i}\right)_{\text{CV}} dt$$

or

$${}_1 S_{\text{gen}, 2} = s_2 m_2 - s_1 m_1 + \sum_{\text{out}} sm - \sum_{\text{in}} sm - \int_{t_1}^{t_2} \sum_i \left(\frac{\dot{Q}_i}{T_i}\right)_{\text{CV}} dt \geq 0$$

as another governing relation. Properties available include the new property of entropy. A steady state application is presented in Example Problem 5.7, and unsteady applications are seen in Example Problems 5.8 and 5.9.

Example Problem 5.7

A steam turbine inlet state is given as 6 MPa and 500°C, and the outlet pressure is 10 kPa. Determine the work output per unit mass if the process is reversible and adiabatic. Evaluate the work output per unit mass if the output quality is 0.90 and 1.0 and the turbine is adiabatic. Compare these outputs to the reversible adiabatic case, and calculate the entropy generation.

Solution
The state information given above and Tables D.9 and D.10 yield

Diagram

State	P, kPa	T, °C	h, kJ/kg	x	s, kJ/(kg · K)
1	6000	500	(3421.3)	—	(s_1 = 6.8793)
2a	10		(2179.3)	(0.8305)	s_1
2b	10		(2345.7)	0.9	(7.4009)
2c	10		(2585.0)	1.0	(8.1511)

The first law for an adiabatic turbine is

$$-w = h_{in} - h_{out} = h_1 - h_2$$

The second law is

$$S_{gen} = S_{out} - S_{in} - \left(\frac{q}{T}\right)_{CV} \geq 0$$

but the turbine is adiabatic, so

$$S_{gen} = s_2 - s_1 \geq 0$$

The states are all specified, so the enthalpies are obtained. The results for w and s_{gen} are as follows:

Process	w, kJ/kg	$\dfrac{w}{_1w_{2a}}$, %	s_{gen}, kJ/(kg · K)
1-2a	−1242.0	100.0	0.0
1-2b	−1075.6	86.6	0.5216
1-2c	−836.3	67.3	1.2718

Comments

The turbine is adiabatic, so the loss in work and the increase in the entropy generation are a result of the irreversibilities in the process. The percentage drop in the work output from the reversible case increases as the output moves toward the saturated vapor line.

The outlet state of the turbine is slightly "wet." Some power plants provide saturated steam to the turbine, and the turbines must be designed to remove even greater amounts of condensed liquids. This liquid must be removed before it passes through the turbine blades. This is done by including in the turbine centrifugal liquid removal stages that spin the liquid to the outer casing.

Example Problem 5.8

An initially evacuated tank of volume V is filled with an ideal gas. The inlet is controlled with the valve, and the inlet properties are constants at T_{in} and P_{in}. Determine the final tank temperature T_2 when the tank pressure is P_2. Evaluate the entropy generation for the filling process. The process is adiabatic.

Diagram

Solution

The state information is represented as follows:

State	P	T	m
In	P_{in}	T_{in}	
1	—	—	0.0
2	P_2		

The initial state is eliminated because nothing exists in the tank. The governing equations are used to specify the final state. The governing equations are

Mass:

$$m_2 - \cancel{m_1}^{0} = m_{in}$$

First law:

$$m_2 u_2 = h_{in} m_{in}$$

Second law:

$$_1 S_{gen, 2} = m_2 s_2 - m_{in} s_{in} \geq 0$$

By eliminating m_{in}, the first and second laws are

$$m_2 u_2 = h_{in} m_2$$

and

$$_1 S_{gen, 2} = m_2 s_2 - m_2 s_{in} = m_2 (s_2 - s_{in}) \geq 0$$

Thus

$$u_2 = h_{in}$$

and

$$_1 S_{gen, 2} = s_2 - s_{in} \geq 0$$

The properties for an ideal gas give $h = u + Pv = u + RT$, so the first law is

$$u_2 = u_{in} + RT_{in}$$

Then

$$du = c_v \, dT$$

and with $c_P = c_v + R$,

$$c_v T_2 = c_v T_{in} + RT_{in} = c_P T_{in}$$

Thus

$$T_2 = \frac{c_P}{c_v} T_{in} = kT_{in}$$

The change in entropy is given in Eq. (5.53) as

$$ds = c_P \frac{dT}{T} - \frac{R}{P} dP$$

Therefore

$$_1S_{gen, 2} = c_P \ln \frac{T_2}{T_{in}} - R \ln \frac{P_2}{P_{in}}$$

or

$$_1S_{gen, 2} = c_P \ln k - R \ln \frac{P_2}{P_{in}}$$

Comments

The final temperature is always greater than the inlet temperature since $k > 1$. The entropy generation is always positive. This is seen from the last equation. Rearranging yields

$$_1S_{gen, 2} = c_P \ln k + R \ln \frac{P_{in}}{P_2}$$

With $k > 1$ and $P_{in} \geq P_2$ for a filling process, the entropy generation is greater than zero. The filling process with an ideal gas is always irreversible.

Example Problem 5.9

A 0.01-m^3 chamber contains refrigerant 12 at $30\,°C$. Initially, one-tenth of the total volume is liquid, and the remainder is vapor. The chamber is filled by a refrigerant 12 source which is at $P_{in} = 1.0$ MPa and $50\,°C$. The tank is maintained at $30\,°C$ by a heat transfer with the surroundings at $30\,°C$. The process ends when the tank is filled entirely with liquid. Determine the heat transfer and entropy generation during this process.

Diagram

Solution

The necessary information is given to determine the states. The initial liquid volume and vapor volume give the quality of that state, since

$$x = \frac{m_g}{m} = \frac{m_g}{m_g + m_l} = \frac{1}{1 + m_l/m_g} = \frac{1}{1 + V_l v_g/(V_g v_l)}$$

The state information is tabulated as follows:

State	P, MPa	T, °C	x	u, kJ/kg	h, kJ/kg	s, kJ/(kg · K)
In	1.0	50	—	—	(374.45)	(1.5620)
1	—	30	0.22846	(255.05)	—	(1.2006)
2	—	30	0	(228.07)	—	(1.0988)

The conservation statements for this unsteady filling problem are as follows:

Mass:

$$m_2 - m_1 = m_{in}$$

First law:

$$u_2 m_2 - u_1 m_1 = h_{in} m_{in} + {}_1Q_2$$

Second law:

$${}_1S_{gen, 2} = s_2 m_2 - s_1 m_1 - s_{in} m_{in} - \int_{t_1}^{t_2} \frac{\dot{Q}}{T} \, dt \geq 0$$

The volume is given by $V = 0.01$ m^3, so the initial mass and final mass are obtained with the specific volumes. Therefore, $m_1 = 1.6750$ kg and $m_2 = 12.9232$ kg, and conservation of mass yields $m_{in} = 11.2482$ kg. The heat transfer is evaluated from the first law, which is rearranged as

$$\begin{aligned}
{}_1Q_2 &= u_2 m_2 - u_1 m_1 - h_{in} m_{in} \\
&= 228.07(12.9232) - 255.05(1.6750) - 374.45(11.2482) \\
&= -1691.7 \text{ kJ}
\end{aligned}$$

This is the heat loss *to* the surroundings.

The entropy generation is determined from the second law. The temperature T_0 at the boundary is constant at 30°C, so the second law becomes

$${}_1S_{gen, 2} = s_2 m_2 - s_1 m_1 - s_{in} m_{in} - \frac{{}_1Q_2}{T_0}$$

Substituting the values yields

$$\begin{aligned}
{}_1S_{gen, 2} &= 1.0988(12.9232) - 1.2006(1.6750) - 1.5620(11.2482) \\
&\quad - \left(\frac{-1691.7}{303.15} \right) \\
&= 0.1997 \text{ kJ/K}
\end{aligned}$$

Comments
    The refrigerant 12 and the surroundings are at the same temperature. Therefore, the chamber wall that contains refrigerant 12 is at the same constant temperature. The above analysis can be interpreted as including or not including the chamber wall (since $m_{wall}\, du_{wall}$ and $m_{wall}\, ds_{wall}$ are zero). If there is a temperature drop across the wall so that refrigerant 12 is at 30°C and the surroundings are at 25°C, then the entropy generation expression which includes the entire system of refrigerant 12 and the chamber wall requires an idealization. The heat transfer is idealized as being steady, so the change in internal energy and the entropy of the wall are zero (see Example Problem 5.6 for a similar approximation). Then the entropy generation for the refrigerant 12 and chamber is the same as the above expression with the temperature of the surroundings replaced by $273.15 + 25 = 298.15$ K. The entropy generation is 0.2933 kJ/K for this case.
    How would the above analysis be altered if the filling process were initially adiabatic to the point where the valve was closed and then cooled as a control mass to the same final state? The states are all the same, and the governing equations are the same. Thus, the answers are the same.

5.8 Isentropic Process

An *isentropic* process is a constant-entropy process. If a control mass undergoes a process which is both reversible and adiabatic, then the second law [Eq. (5.74)] specifies the entropy change to be zero. Steady state and reversible flow through an adiabatic control volume also have no entropy change from inlet to outlet [Eq. (5.99)]. Both examples are isentropic processes. Although an isentropic process might be an idealization of an actual process, this process serves as a limiting process, for particular applications.

    The comparison between the isentropic processes and the adiabatic but irreversible processes is shown in Fig. 5.10. The processes denoted by solid lines are the reversible processes, and the dashed lines indicate the irreversible processes. The irreversible processes proceed from the initial state of the process to larger values of entropy. The process for state 1 to state 2 is an expansion process while the process from state 3 to state 4 is a compression process.

5.8.1 Isentropic Process for an Ideal Gas

The entropy change for an ideal gas is presented in Eqs. (5.54) and (5.55). For an isentropic process, $s_2 - s_1 = 0$, so these expressions become

$$0 = \int_1^2 c_v \frac{dT}{T} + R \ln \frac{v_2}{v_1} \qquad (5.105)$$

Figure 5.10 Process comparisons (1 to 2 and 3 to 4 are isentropic).

$$0 = \int_1^2 c_P \frac{dT}{T} - R \ln \frac{P_2}{P_1} \qquad (5.106)$$

The specific results depend on the approximation made for the temperature dependence of the specific heats.

Assuming that the specific heats are accurately approximated by constant values eliminates the integrals in the above equations. The *constant specific heat* equations are

$$0 = c_v \ln \frac{T_2}{T_1} + R \ln \frac{v_2}{v_1} \qquad (5.107)$$

$$0 = c_P \ln \frac{T_2}{T_1} - R \ln \frac{P_2}{P_1} \qquad (5.108)$$

Dividing by the specific heats and taking the exponential of each expression yield

$$\left(\frac{T_2}{T_1}\right)_s = \left(\frac{v_2}{v_1}\right)_s^{-R/c_v} = \left(\frac{v_1}{v_2}\right)_s^{R/c_v} \tag{5.109}$$

$$\left(\frac{T_2}{T_1}\right)_s = \left(\frac{P_2}{P_1}\right)_s^{R/c_P} \tag{5.110}$$

where the subscript indicates that the process occurs at constant entropy. The power on each expression is rewritten in terms of $k = c_P/c_v$ by noting that $c_P - c_v = R$, so $R/c_v = k - 1$ and $R/c_P = (k - 1)/k$. Thus, Eqs. (5.109) and (5.110) become

$$\left(\frac{T_2}{T_1}\right)_s = \left(\frac{v_1}{v_2}\right)_s^{k-1} \tag{5.111}$$

$$\left(\frac{T_2}{T_1}\right)_s = \left(\frac{P_2}{P_1}\right)_s^{(k-1)/k} \tag{5.112}$$

A relation between pressure and volume is obtained by eliminating the ratio of temperatures in the above expressions. This yields

$$\left(\frac{P_2}{P_1}\right)_s = \left(\frac{v_1}{v_2}\right)_s^{k} \tag{5.113}$$

These last three equations are specific relations that are used for an ideal gas undergoing an isentropic process if the specific heats are taken as constants. These expressions are summarized in Table 5.4.

   If the specific heats cannot be approximated as constants, then the temperature dependence of the specific heats must be included. The *variable specific heat* solution for an ideal gas undergoing an isentropic process is obtained from Eq. (5.60). For an isentropic process, this expression yields

$$s_2 - s_1 = 0 = s_0(T_2) - s_0(T_1) - R \ln \frac{P_2}{P_1} \tag{5.114}$$

where the first terms are a function of T only and are obtained from Tables D.2 to D.7. Since $s_0(T)$ is available only in table form, the direct solution for the constant-entropy process using the above expression could require a lengthy iterative procedure. To avoid this difficulty, Tables D.2 through D.7 include two additional entries at each state, denoted P_0 and v_0. These entries are useful for isentropic processes.

   Equation (5.114) is rearranged to yield

$$\ln \frac{P_2}{P_1} = \frac{s_0(T_2) - s_0(T_1)}{R} \tag{5.115}$$

The pressure ratio between two states connected by an isentropic process for an ideal gas depends on only the $s_0(T)$ values of the two states. Therefore, this pressure ratio is a *function of absolute temperature* only. The

TABLE 5.4 Ideal Gas Relations

$$Pv = RT$$
$$c_P - c_v = R$$

Variable Specific Heats	Constant Specific Heats

$$u_2 - u_1 = \int_1^2 c_v \, dT \qquad\qquad u_2 - u_1 = c_v(T_2 - T_1)$$

$$h_2 - h_1 = \int_1^2 c_P \, dT \qquad\qquad h_2 - h_1 = c_P(T_2 - T_1)$$

$$s_2 - s_1 = \int_1^2 c_v \frac{dT}{T} + R \ln \frac{v_2}{v_1} \qquad\qquad s_2 - s_1 = c_v \ln \frac{T_2}{T_1} + R \ln \frac{v_2}{v_1}$$

$$s_2 - s_1 = \int_1^2 c_P \frac{dT}{T} - R \ln \frac{P_2}{P_1} \qquad\qquad s_2 - s_1 = c_P \ln \frac{T_2}{T_1} - R \ln \frac{P_2}{P_1}$$

$$s_2 - s_1 = s_0(T_2) - s_0(T_1) - R \ln \frac{P_2}{P_1}$$

For isentropic processes, use

$$\left(\frac{P_{0,1}}{P_{0,2}}\right)_s = \left(\frac{P_1}{P_2}\right)_s$$

$$\left(\frac{v_{0,1}}{v_{0,2}}\right)_s = \left(\frac{v_1}{v_2}\right)_s$$

in Tables D.2 to D.7.

For isentropic processes: $\Delta S = 0$

reversible and adiabatic

$$\left(\frac{T_2}{T_1}\right)_s = \left(\frac{v_1}{v_2}\right)_s^{k-1}$$

$$\left(\frac{T_2}{T_1}\right)_s = \left(\frac{P_2}{P_1}\right)_s^{(k-1)/k}$$

$$\left(\frac{P_2}{P_1}\right)_s = \left(\frac{v_1}{v_2}\right)_s^{k}$$

values of $s_0(T)$ are integrals of the specific heat at constant pressure from the reference state to the state in question. The difference in entropy between two states is evaluated by taking the difference in $s_0(T)$ values and including the pressure term [Eq. (5.114)]. A similar procedure is used in the evaluation of the pressure ratio in Eq. (5.115). The value of the pressure ratio that occurs if an isentropic process is carried out between the reference state and the desired state is denoted $P_0 = P/P_{ref}$. From the definition of $s_0(T)$ below Eq. (5.58), $s_0(T_{ref}) = 0$, so Eq. (5.115) yields

$$\ln P_0 = \ln \frac{P}{P_{ref}} = \frac{s_0(T) - 0}{R} = \frac{s_0(T)}{R} \tag{5.116}$$

Thus, this ratio is a function of temperature only, and it can be tabulated. The specific volume is useful in some applications, and the corresponding ratio is obtained from the ideal gas relation as

$$v_0 = \frac{v}{v_{ref}} = \frac{RT\,P_{ref}}{PR\,T_{ref}} = \frac{T\,P_{ref}}{T_{ref}\,P} \tag{5.117}$$

This is a function of temperature only and is tabulated also.

The specific values for $s_0(T)$, P_0, and v_0 are given in Tables D.2 to D.7, and they all depend on temperature only. The use of $s_0(T)$ is discussed in Sec. 5.5. Note that differences in $s_0(T)$ are needed in Eq. (5.60) so that the specific choice of the reference value is eliminated for pure compressible substances; however, the tables are based on $s_0(T = 0 \text{ K})$ as the reference value for use with mixtures or when chemical reactions occur (Chaps. 9 through 12). And P_0 and v_0 are used as ratios only, so the reference state is eliminated. For an isentropic process between a state 1 and another state 2, the ratio is

$$\left(\frac{P_1}{P_2}\right)_s = \left(\frac{P_1}{P_{\text{ref}}}\frac{P_{\text{ref}}}{P_2}\right)_s = \left(\frac{P_{0,1}}{P_{0,2}}\right)_s \qquad (5.118)$$

Thus, for an ideal gas undergoing an isentropic process from a specified state to a desired pressure, the final pressure is evaluated by the pressure ratio in Eq. (5.118). Similarly, the specific volume ratio is

$$\left(\frac{v_1}{v_2}\right)_s = \left(\frac{v_{0,1}}{v_{0,2}}\right)_s \qquad (5.119)$$

Note that entries in Tables D.2 to D.7 for P_0 and v_0 are used for an isentropic process only. The following examples demonstrate the use of these tables.

The tabulated values of P_0 and v_0 in App. D are used in ratio form as given in Eqs. (5.118) and (5.119). Therefore, their absolute values are unimportant. The values presented in the tables have been adjusted by a constant to obtain reasonable magnitudes for presentation. This does not affect the use of P_0 and v_0.

Example Problem 5.10

Air is to be compressed isentropically from 1 atm and 20°C to a final pressure of 20 atm. What is the temperature of the air leaving the compressor?

Solution

If air is assumed to have constant specific heat, then Table 5.4 can be used directly to give

$$\left(\frac{T_2}{T_1}\right)_s = \left(\frac{P_2}{P_1}\right)_s^{(k-1)/k} = (20)^{0.4/1.4} = 2.354$$

or

$$T_2 = 2.354(20 + 273.15) \text{ K} = 690.1 \text{ K} = 416.9°\text{C}$$

Using the gas tables for air (Table D.2), which do not assume constant specific heat, we go to the entry for 20°C, which by interpolation gives $P_0 = 1.2991$. Since the ratios P_2/P_1 and $P_{0,2}/P_{0,1}$ must be equal [see Eq. (5.118)], we note that in state 2

$$P_{0,2} = 1.2991 \times 20 = 25.982$$

Because P_0 depends on T only, we search the table for the corresponding T; again, using linear interpolation, we find $T_2 = 404.9\,°C$.

Comments

We find a difference of 12 degrees between the constant specific heat result and the result obtained by using the tables. Note that the tables themselves are generated from statistical thermodynamic relations based on models of ideal gases with a few constants obtained from spectroscopic measurements. Comparison with experimental data for real gases shows the gas tables to be accurate within a few percent over the temperature range presented, unless the pressures are extremely high. However, the values for v_0 and P_0 are highly nonlinear with T, and linear interpolation can lead to significant error. The computerized tables or more complete tables are recommended for high accuracy.

5.8.2 Isentropic Process for an Incompressible Fluid or Solid

The entropy change for an incompressible fluid or solid is given in Eq. (5.62) as

$$ds = c\,\frac{dT}{T} \tag{5.120}$$

For an isentropic process, $ds = 0$, so $dT = 0$. Thus, an isentropic process is an isothermal process for an incompressible fluid or solid. Also, the internal energy is given [Eq. (3.34a)] as

$$du = c\,dT \tag{5.121}$$

TABLE 5.5 Incompressible Fluid or Solid Relations

$$v = \text{constant}$$
$$c_P = c_v = c$$

Variable Specific Heats	Constant Specific Heats
$u_2 - u_1 = \displaystyle\int_1^2 c\,dT$	$u_2 - u_1 = c(T_2 - T_1)$
$h_2 - h_1 = \displaystyle\int_1^2 c\,dT + v(P_2 - P_1)$	$h_2 - h_1 = c(T_2 - T_1) + v(P_2 - P_1)$
$s_2 - s_1 = \displaystyle\int_1^2 c\,\frac{dT}{T}$	$s_2 - s_1 = c\ln\dfrac{T_2}{T_1}$
For isentropic processes:	For isentropic processes:
$\quad T_2 = T_1$	$\quad T_2 = T_1$
$\quad u_2 = u_1$	$\quad u_2 = u_1$
$h_2 - h_1 = v(P_2 - P_1)$	$h_2 - h_1 = v(P_2 - P_1)$

So $du = 0$ for an isentropic process. The change in enthalpy is

$$dh = du + P\, dv + v\, dP \tag{5.122}$$

But $du = 0$ for this process and $dv = 0$ for incompressible fluids or solids, so

$$dh = v\, dP \tag{5.123}$$

This last expression is integrated to yield

$$h_2 - h_1 = v(P_2 - P_1) \tag{5.124}$$

since $v = $ constant. This last expression is particularly useful in adiabatic work considerations of liquid pumps.

Applicable relations for incompressible fluids or solids are summarized in Table 5.5.

5.9   Special Considerations

A few applications require special consideration. These specific applications are quite important and are discussed here. The first is the work done in a steady state flow problem, and the second is the unsteady discharging problem. The starting points are the differential forms of conservation of mass, first law and second law. The expressions in Table 5.1 are rewritten with Table 5.2 by considering a differential time element dt and differential area element dA. All terms are multiplied by dt, and the terms with an overdot become differentials. Thus,

Mass:

$$dm_{CV} = dm_{in} - dm_{out} \tag{5.125}$$

First law:

$$dE_{CV} = (e + Pv)_{in}\, dm_{in} + \delta Q_{CV} + \delta W_{CV} - (e + Pv)_{out}\, dm_{out} \tag{5.126}$$

Second law:

$$dS_{CV} + s_{out}\, dm_{out} - s_{in}\, dm_{in} \geq \left(\frac{\delta Q}{T}\right)_{CV}$$

or

$$\delta S_{gen} = dS_{CV} + s_{out}\, dm_{out} - s_{in}\, dm_{in} - \left(\frac{\delta Q}{T}\right)_{CV} \geq 0 \tag{5.127}$$

The summation signs have not been included for the inlets, outlets, or heat transfers.

The *steady state work application* involves a single inlet and outlet. The changes in properties within the control volume are zero, because the application is steady state. The process is also considered to be reversible, and the changes in potential and kinetic energy are neglected. Thus, the governing expressions are

Mass:

$$dm_{in} = dm_{out} \tag{5.128}$$

First law:

$$h_{in}\, dm_{in} + \delta Q_{CV, rev} = h_{out}\, dm_{out} - \delta W_{CV} \tag{5.129}$$

Second law:

$$s_{out}\, dm_{out} - s_{in}\, dm_{in} = \left(\frac{\delta Q}{T}\right)_{CV, rev} \tag{5.130}$$

Dividing the first law and second law by the conservation of mass yields

$$h_{in} + \delta q_{CV, rev} = h_{out} - \delta w_{CV} \tag{5.131}$$

and

$$s_{out} - s_{in} = \left(\frac{\delta q}{T}\right)_{CV, rev} \tag{5.132}$$

Expressing the changes in enthalpy and entropy as differentials for a control volume yields

$$\delta q_{CV, rev} = dh - \delta w_{CV} \tag{5.133}$$

$$ds = \left(\frac{\delta q}{T}\right)_{CV, rev} \tag{5.134}$$

Eliminating the heat transfer between Eqs. (5.133) and (5.134) yields

$$T\, ds = dh - \delta w_{CV} \tag{5.135}$$

This expression is compared to Eq. (5.44), which is rearranged as

$$T\, ds = dh - v\, dP \tag{5.136}$$

Thus

$$\delta w_{CV} = +v\, dP \tag{5.137}$$

or

$$w_{CV} = \int_{1}^{2} v\, dP \tag{5.138}$$

This expression gives the shaft work within a control volume analysis as an integral of $v\, dP$. The only assumption is that the device is *reversible* and, therefore, is *not* restricted to isothermal or adiabatic processes.

It is convenient to summarize the work expressions for reversible shaft work in steady state control volume analysis, $\delta w_{CV} = v\, dP$, and the work resulting from boundary movement, $\delta w = -P\, dv$. The shaft work requires the reversible idealization as well as steady state conditions. It is applicable for control volume analysis only. The boundary movement work is valid for both a control volume and a control mass, but the

TABLE 5.6 **Work Relations for Polytropic Processes**

$Pv^n = $ constant

where $n = 0$, constant pressure

$n = 1$, constant temperature for an ideal gas

$n = k$, constant entropy for an ideal gas

$n \rightarrow \infty$, constant volume

$v\,dP$	$-P\,dv$
${}_{in}w_{CVout} = \int_{in}^{out} v\,dP$	${}_1w_2 = -\int_1^2 P\,dv$
${}_{in}w_{CVout} = \begin{cases} \dfrac{n}{n-1}[(Pv)_{out} - (Pv)_{in}] & n \neq 1 \\[2mm] (Pv)_{out}\ln\dfrac{P_{out}}{P_{in}} & n = 1 \end{cases}$	${}_1w_2 = \begin{cases} \dfrac{1}{1-n}[(Pv)_1 - (Pv)_2] & n \neq 1 \\[2mm] (Pv)_1\ln\dfrac{v_1}{v_2} & n = 1 \end{cases}$

boundary of the control surface must move. The evaluation of the work between two states requires the integral of these expressions and, therefore, the dependence of v on P or P on v. Table 5.6 gives the results for both $v\,dP$ and $-P\,dv$ work for a polytropic process.

Example Problem 5.11

Evaluate the work per unit mass to (*a*) pump liquid water from 0.1 MPa and 25°C to 1.5 MPa and (*b*) compress steam isothermally from 0.1 MPa and 300°C to 1.5 MPa. Both processes are reversible and operating under steady state conditions.

   Solution
   (*a*) Liquid water is incompressible, so Table 5.6 for $n \rightarrow \infty$ yields

$$w = v(P_2 - P_1)$$

The specific volume at 25°C is the saturated liquid value of $v = 0.001003$ m³/kg, so

$$w = (0.001003)(1500 - 100) = 1.404 \text{ kJ/kg}$$

   (*b*) Steam at a reduced temperature of $T_r = 573.15$ K/647.29 K = 0.885 and $P_r = 1.5$ MPa/22.09 MPa = 0.0679 is approximated as an ideal gas (Fig. C.1). At lower pressures, the approximation is more accurate. Table 5.6 gives the work relation with $n = 1$ ($Pv = RT = $ constant for an isothermal process). Thus

$$w = P_1v_1 \ln\frac{P_2}{P_1} = RT\ln\frac{P_2}{P_1} = (0.2870)(573.15)\ln\frac{1.5}{0.1} = 445.5 \text{ kJ/kg}$$

A second important application is the discharge of a constant-volume vessel which is an *unsteady state application*. The different aspect of this problem is that the exit state, which is often the same as the state within the control volume, changes with time and so cannot be assumed constant. If there is only one exit and there are no inlets, then the discharge is [see Eqs. (5.125) through (5.127)]

Mass:

$$dm_{CV} = -dm_{out} \tag{5.139}$$

First law:

$$dE_{CV} = \delta Q_{CV} - (e + Pv)_{out}\, dm_{out} \tag{5.140}$$

Second law:

$$dS_{CV} + s_{out}\, dm_{out} \geq \left(\frac{\delta Q}{T}\right)_{CV} \tag{5.141}$$

The changes in kinetic and potential energy are neglected, and the changes in control volume properties are expressed in their intensive form. Thus the first and second laws are

$$d(um)_{CV} = m_{CV}\, du_{CV} + u_{CV}\, dm_{CV} = \delta Q_{CV} + h_{out}\, dm_{CV} \tag{5.142}$$

and

$$d(sm)_{CV} - s_{out}\, dm_{CV} = s_{CV}\, dm_{CV} + m_{CV}\, ds_{CV}$$
$$- s_{out}\, dm_{CV} \geq \left(\frac{\delta Q}{T}\right)_{CV} \tag{5.143}$$

where the conservation of mass has been used. The properties at the outlet and within the control volume vary with time, but if they are equal at any instant, then $u_{CV} = u_{out}$ and $s_{CV} = s_{out}$,

$$m_{CV}\, du_{CV} = \delta Q_{CV} + (Pv)_{CV}\, dm_{CV} \tag{5.144}$$

$$m_{CV}\, ds_{CV} \geq \left(\frac{\delta Q}{T}\right)_{CV} \tag{5.145}$$

These expressions govern the *discharge of a single-phase substance* from a vessel including the heat transfer contributions. For a two-phase substance, the enthalpy of the outgoing material is not generally equal to the enthalpy of the control volume and the analysis must start from the first law.

Consider the discharge from a constant-volume, adiabatic tank which is described by Eq. (5.144) and the Gibbs equation [Eq. (5.43)]. Equation (5.144), for an adiabatic process, is

$$m_{CV}\, du_{CV} = (Pv)_{CV}\, dm_{CV} \qquad (5.146)$$

Since the volume is constant, $d(m_{CV}v_{CV}) = 0$ or $v_{CV}\, dm_{CV} = -m_{CV}\, dv_{CV}$. Thus

$$m_{CV}\, du_{CV} = -m_{CV} P_{CV}\, dv_{CV} \qquad (5.147)$$

or

$$du_{CV} = -P_{CV}\, dv_{CV} \qquad (5.148)$$

This is the Gibbs equation given in Eq. (5.43) when

$$ds_{CV} = 0 \qquad (5.149)$$

Thus the constant-volume discharge from an adiabatic tank is a constant-entropy process. However, we have assumed that the outlet entropy is equal to the entropy of the uniform mass within the control volume; this has just been shown to result in an internally reversible process with respect to the control volume. This is true in all cases and does *not* require the assumption of a reversible process. This is important in discharge calculations since a property is known at every point for the process.

Example Problem 5.12

An insulated 1.0-m³ tank contains steam at 2.0 MPa and 500°C. Vapor is withdrawn from the top of the tank for use in a related experiment. What are the final temperature and the mass withdrawn if the final pressure is 0.2 MPa?

Diagrams

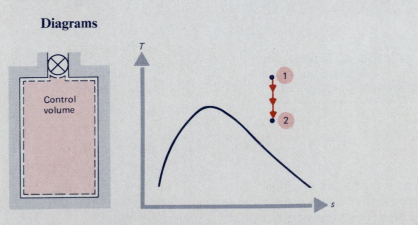

Solution

The initial state is specified by pressure and temperature. The final state is specified by constant entropy and pressure. Thus

State	P, MPa	T, °C	v, m³/kg	s, kJ/(kg · K)
1	2.0	500	(0.17556)	(7.4286)
2	0.2	(182.7)	(1.0385)	$s_2 = s_1$

The states and process are specified so that the remaining properties are obtained from Table D.10. These properties are entered into the above table.

The final temperature is given by interpolation as 182.7°C. The mass withdrawn is found through the change in specific volume as

$$m_{\text{out}} = m_1 - m_2 = \left(\frac{V}{v}\right)_1 - \left(\frac{V}{v}\right)_2$$

$$= V\left(\frac{1}{v_1} - \frac{1}{v_2}\right) = 1\left(\frac{1}{0.17556} - \frac{1}{1.0385}\right) = 4.73 \text{ kg}$$

Comments

The solution to this process is straightforward if $s_2 = s_1$ is used. An alternative approach to the solution, which is required in the analysis of the discharge of two-phase substances, starts with the first law. Equation (5.148) is

$$du = -P \, dv$$

for an adiabatic tank. This expression would have to be incremented through the process by using Table D.9. Specifically, for two intermediate states a and b,

$$u_b - u_a = P_{\text{sat}}(v_a - v_b)$$

where v_a and u_a are known from the previous step and u_b and v_b are evaluated through the quality, that is, $u_b = u_{lb} + x_b u_{lgb}$ and $v_b = v_{lb} + x_b v_{lgb}$. Thus the quality is obtained and the solution is completed.

5.10 Component Efficiencies

In many components the adiabatic performance is the desired situation. A steam turbine seeks to produce work given a high-temperature and high-pressure inlet state and an outlet pressure. Any energy transfer from the turbine as heat transfer is a loss and reduces the work output. A compressor that takes a fluid from the inlet state to the desired outlet pressure requires a work input. A heat loss from a desired adiabatic compressor requires a larger work input. (Some compressors operate almost isothermally, so there is a heat transfer. These compressors require special consideration.) The inefficiencies in these components result from irreversible processes.

The component efficiencies are used to compare the actual process to the adiabatic and reversible processes. The reduction of the irreversibilities of a component process is desired to increase the efficiency of the process. Therefore, the component efficiency is the ratio of the actual to the isentropic result, with the limiting value being unity. The isentropic device is taken as the standard for comparison of real adiabatic operation.

5.10.1 Turbine Efficiency

The efficiency of a turbine is a comparison of the actual work produced with the work produced for an isentropic process. The inlet to the turbine is at a specified state, and the exit is at a specified pressure. The exhaust pressure is the outlet condition, since factors such as the available cooling water temperature to the condenser usually set the outlet pressure for a given application. The two processes compared in turbine efficiency are shown in Fig. 5.11a. The actual process and isentropic process start at the same state. The exit state is different in each case, but the exit states for the actual and isentropic processes are at the same pressure. Denoting the actual work by w_a and the isentropic work by w_s, we define the turbine efficiency as

$$\eta_t \equiv \frac{w_a}{w_s} \tag{5.150}$$

The reduction of irreversibilities moves the actual process exit state toward the isentropic exit state. Applying the first law for steady state, adiabatic flow with uniform properties at the inlet and exit, we find for the turbine efficiency

$$\eta_t = \frac{(h_{in} - h_{out})_a}{(h_{in} - h_{out})_s} \tag{5.151}$$

where the changes in kinetic and potential energy are negligible. Figure 5.11a shows the process for a steam turbine, but the definition is valid for a gas turbine also. Actual turbines typically provide efficiencies in the range of 60 to 80 percent, with very large turbines approaching efficiencies of 90 percent.

5.10.2 Compressors and Pumps

In a compressor, the comparison is again made between the actual and the isentropic case. The desired performance of the compressor is to produce the exit pressure with the minimum work. Assuming that the actual process is adiabatic, we see that the irreversibilities require a larger work input. The compressor efficiency is

$$\eta_c \equiv \frac{w_s}{w_a} \tag{5.152}$$

The inlet state is the same for each process, and both processes exhaust to the same outlet pressure. The first law for adiabatic steady state flow yields

$$\eta_c = \frac{(h_{in} - h_{out})_s}{(h_{in} - h_{out})_a} \tag{5.153}$$

where the changes in kinetic and potential energies are not significant. The isentropic work is less than the actual work input, and the limit is a compressor efficiency of unity. The comparison of these two processes is shown in Fig. 5.11*b*.

Compressors typically have efficiencies of near 60 percent when they are operated near design conditions. Because compressors operate efficiently if the volume of gas being compressed is maintained at a minimum, it is often desirable to provide cooling within the compressor to minimize the expansion of the gas owing to temperature increases. For cooled compressors, a different definition for the efficiency is used where the isothermal rather than the isentropic compressor is used for comparison.

Pumps provide an interesting special case. For most situations involving pumps, the fluid flowing through the pump can be considered to be *incompressible* — that is, v = constant. The work for a pump is conveniently expressed from Eq. (5.138) as

$$w_s = \int_1^2 v \, dP = v(P_2 - P_1) \tag{5.154}$$

which is valid for a *reversible* pump or other device through which an *incompressible* fluid is flowing. [This form of *Bernoulli's equation,* which is of fundamental importance in fluid mechanics, is extended from Eq. (4.101) to account for a pump in the line. It is also valid for an isothermal pump.] Therefore, the pump efficiency is

$$\eta_p = \frac{w_s}{w_a} = \frac{v(P_2 - P_1)}{w_a} \tag{5.155}$$

5.10.3 Nozzles

Another component efficiency is defined for a nozzle. A nozzle increases the velocity as a result of a pressure decrease. This component is generally adiabatic. The efficiency compares the actual exit kinetic energy to the kinetic energy at the exit for an isentropic process. The inlet state for each process and the exhaust pressure are the same for the compared processes. The nozzle efficiency is

$$\eta_N \equiv \frac{(KE)_a}{(KE)_s} = \frac{\mathbf{V}_a^2/2}{\mathbf{V}_s^2/2} \tag{5.156}$$

The component efficiencies compare two processes that have the same inlet state. The exit state is at the same pressure for the actual and ideal processes.

5.10.4 Control Mass Efficiency

The expansion or work stroke in an internal combustion engine requires a control mass analysis. An efficiency for this process or other piston-cylinder process compares the actual work output to the isentropic work to the same final pressure. This efficiency is expressed in terms of internal energy as

$$\eta_{CM} = \frac{w_a}{w_s} = \frac{(u_2 - u_1)_a}{(u_2 - u_1)_s} \qquad (5.157)$$

The initial state is the same in this comparison, and kinetic and potential energy changes are neglected.

Example Problem 5.13

A steam turbine of efficiency η_t has an inlet state of 350°C and 4.0 MPa. The mixture that emerges from the turbine exit is condensed in the condenser at 0.1 MPa. The process is adiabatic, and the mass flow rate is 0.5 kg/s. Determine the percentage of the initial mass that is liquid at the turbine exhaust and the turbine power as the turbine efficiency degrades from $\eta_t = 95$ to 80 to 60 percent.

Diagrams

Solution

The initial state is specified. The outlet state is determined from the turbine efficiency and the outlet pressure. The turbine efficiency compares the isentropic process to the actual process, so that η_t and the isentropic path specify the actual outlet state. For an adiabatic turbine, Eq. (5.151) yields

$$\eta_t = \frac{(h_1 - h_2)_a}{(h_1 - h_2)_s}$$

or

$$h_{2a} = h_1 - \eta_t(h_1 - h_2)_s$$

Thus, with state 2_s specified, the above expression gives the actual outlet state (with $P_2 = 0.1$ MPa).

The state information is as follows:

State	P, MPa	T, °C	x	h, kJ/kg	s, kJ/(kg · K)
1	4.0	350	—	(3093.4)	(6.5835)
2_s	0.1	—	(0.8719)	(2385.9)	$s_{2s} = s_1$
2_{95}	0.1	—	(0.8875)	2421.3	—
2_{80}	0.1	—	(0.9345)	2527.4	—
2_{60}	0.1	—	(0.9972)	2668.9	—

The percentage of the initial mass that is liquid at state 2 is ($\dot{m}_1 = \dot{m}_2 = \dot{m}$)

$$\frac{\dot{m}_{l2}}{\dot{m}} = \frac{\dot{m}_2 - \dot{m}_{g2}}{\dot{m}} = 1 - x$$

and the quality is given in the above table. The power output for this adiabatic turbine is

$$\dot{W} = \dot{m}(h_2 - h_1)_a = \dot{m}\eta_t(h_2 - h_1)_s$$

The table above is used to evaluate the percentage of the mass that is liquid and the power output. Thus

Turbine Efficiency, %	Mass Liquid, %	Power, kW
100	12.81	−353.8
95	11.25	−336.1
80	6.55	−283.0
60	0.28	−212.3

Comments

The reversible process is given by $\eta_t = 1.0$ and indicates the maximum work that is possible from the stated inlet to the outlet pressure. All other efficiencies must produce less. The percentage of liquid in the turbine exhaust is very dependent on the turbine efficiency. If the turbine is designed to operate at $\eta_t = 0.95$, then the condenser must handle 11.25 percent liquid.

If the water leaving the condenser is saturated liquid, an energy balance for the condenser requires a heat transfer of $\dot{Q} = \dot{m}(h_{out} - h_{in}) = \dot{m}(h_l - h_{2a})$. The transfer from the water changes from -984.3 kW for $\eta_t = 1.0$ to -1125.8 kW for $\eta_t = 0.6$. Therefore, a degradation of the turbine performance requires changes in the other cycle component performance.

5.11 Cyclic Processes and the Carnot Cycle

Processes that return to their initial state are called *cyclic processes*. The individual processes that make up the elements of the cyclic process vary, and they depend on the particular application. An ideal steam power cycle is composed of constant-pressure heat transfer processes (to the working fluid in the steam generator and from the working fluid in the condenser) and adiabatic work processes (work addition by the pump and work output by the turbine). The idealized spark-ignition engine is composed of adiabatic and constant-volume processes. The fuel and air are compressed adiabatically, and the subsequent combustion is idealized as a constant-volume heating. The hot gases expand adiabatically, doing work. Then the exhaust gases are further expanded at constant volume. In these idealized examples, the processes are usually assumed to be reversible. These examples (and there are many others) indicate that a cyclic process is composed of many different individual processes, and the combination depends on the application.

These example cycles have a common feature — they operate between two limiting temperatures. The high temperature results from the combustion process in the steam generator or within the cylinder. The low temperature results from the cooling process. The characteristics of these two-temperature cycles are shown from a general viewpoint as a high-temperature heat transfer reservoir or source at T_H and a low-temperature heat transfer reservoir at T_L. The cycle operating between these two temperatures is arbitrary. The general power cycle or engine is shown schematically in Fig. 5.12.

The first law for an arbitrary cycle states

$$-\oint \delta W = \oint \delta Q \tag{5.158}$$

which is valid for an arbitrary collection of processes and for both reversible and irreversible cycles. For the two heat transfer cycles shown in Fig. 5.12,

$$|W| = \left|\oint \delta W\right| = |Q_H| - |Q_L| \tag{5.159}$$

Figure 5.12 *Engine representation.*

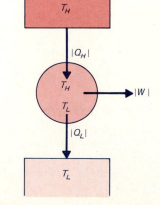

The absolute value signs are used to indicate magnitudes, and the sign is explicitly indicated to show the direction of the heat transfer. The second law for a cycle [Eq. (5.74)] states

$$\oint dS_{CM} = 0 \geq \oint \sum_i \left(\frac{\delta Q_i}{T_i}\right)_{CM} \tag{5.160}$$

where the zero results from the consideration of a cycle. Equations (5.158) and (5.160) are general equations for cycles. These expressions lead to a very important statement about cycles operating between two heat transfer reservoirs. For reversible heat transfers with the two heat transfer reservoirs, the second law is

$$0 \geq \frac{|Q_H|}{T_H} - \frac{|Q_L|}{T_L} \tag{5.161a}$$

or

$$\frac{|Q_L|}{|Q_H|} \geq \frac{T_L}{T_H} \tag{5.161b}$$

This last expression could also be obtained from the entropy generation expression in Eq. (5.77).

The *efficiency* of a cycle η is defined as

$$\eta \equiv \frac{\text{desired output}}{\text{required input}} \tag{5.162}$$

This efficiency should not be confused with the component efficiency defined in Sec. 5.10. The cycle efficiency compares the complete cycle's desired output to required input, whereas the component efficiency considers a process (not a cycle) and compares actual to isentropic paths. A power cycle or engine as diagrammed in Fig. 5.12 has a work output $|W|$ and an input from the high-temperature reservoir $|Q_H|$. Thus, this efficiency is

$$\eta = \frac{|W|}{|Q_H|} \tag{5.163}$$

Equation (5.159) yields

$$\eta = 1 - \frac{|Q_L|}{|Q_H|} \tag{5.164}$$

The ratio of heat transfers is eliminated with Eq. (5.161b) to yield

$$\eta \leq 1 - \frac{T_L}{T_H} \tag{5.165}$$

where the equality applies to a reversible cycle and the inequality applies to an irreversible cycle. Thus

$$\eta_{\text{irr}} < \eta_{\text{rev}} = 1 - \frac{T_L}{T_H} \tag{5.166}$$

Figure 5.13 Carnot cycle.

or the maximum efficiency of a heat engine operating between two heat transfer reservoirs occurs for a reversible cycle. This statement applies regardless of the specific processes of the cycle, but they must be reversible.

A very important reversible cycle that uses two heat transfer reservoirs is shown in Fig. 5.13a and b. The two specific examples undergo the same processes but operate with different components. The cycle includes isothermal energy exchange with the two heat transfer reservoirs, and the other two processes take place adiabatically. All processes are reversible. The heat transfers occur across infinitesimal temperature differences, so the temperature within the system is the same as the source or sink. The other two processes are reversible and adiabatic, so they are isentropic. The state diagram in Fig. 5.13c shows the following four processes:

1 to 2   Isothermal heat transfer at T_H
2 to 3   Isentropic expansion process
3 to 4   Isothermal heat transfer at T_L
4 to 1   Isentropic compression process

This cycle is called a *Carnot cycle*.

A control mass undergoing a Carnot cycle is shown in Fig. 5.13a. There is a heat transfer to the control mass from state 1 to state 2 at T_H, and some expansion work is obtained. The process from state 2 to state 3 is an adiabatic expansion to the temperature T_L. Then there is a heat transfer from the control mass from state 3 to state 4 which occurs reversibly at T_L. The final process takes the control mass adiabatically from state 4 back to state 1. An alternate situation is a power plant which includes four control volume elements with the working fluid undergoing the processes as it moves through the elements (Fig. 5.13b). There is an isothermal heat transfer to the fluid at the high temperature in the boiler from state 1 to state 2. The fluid is expanded adiabatically through the turbine from state 2 to 3 to obtain a work output. The condenser is used for the isothermal heat transfer from state 3 to 4 at T_L. The adiabatic compression of the fluid to state 4 is achieved by the pump. Both types are represented by the general power cycle diagram shown in Fig. 5.12.

The above analysis considers a power cycle or engine used to produce a net output of work. A refrigerator is a cyclic device that seeks a heat transfer from a low-temperature region and requires work input. This device is shown schematically in Fig. 5.14. The efficiency for a refrigerator is termed the *coefficient of performance* (COP) and is given as

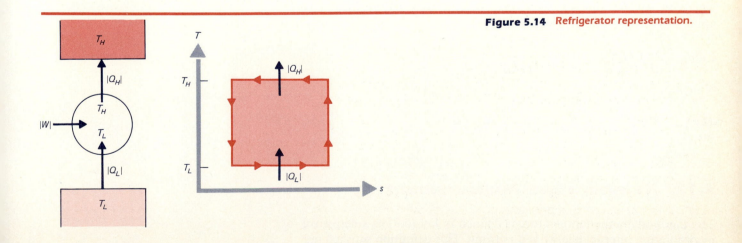

Figure 5.14   Refrigerator representation.

$$COP = \frac{|Q_L|}{|W|} \tag{5.167}$$

The first law states

$$-|W| = |Q_L| - |Q_H| \tag{5.168}$$

or

$$COP = \frac{|Q_L|}{|Q_H| - |Q_L|} = \frac{1}{|Q_H|/|Q_L| - 1} \tag{5.169}$$

The second law yields

$$0 \geq \frac{|Q_L|}{T_L} - \frac{|Q_H|}{T_H} \quad \text{or} \quad \frac{|Q_H|}{|Q_L|} \geq \frac{T_H}{T_L} \tag{5.170}$$

Thus

$$COP \leq \frac{1}{T_H/T_L - 1} \tag{5.171}$$

The equality applies to the reversible refrigerator and the inequality to the irreversible refrigerator. Thus

$$COP_{irr} < COP_{rev} = \frac{1}{T_H/T_L - 1} \tag{5.172}$$

Note that the COP theoretically ranges from 0 to large values > 1. Further consideration is given to these cycles in Chap. 6.

A heat pump (HP) operates as indicated in Fig. 5.14, but it seeks a heat transfer to a high-temperature region so the desired output is $|Q_H|$. The above analysis yields

$$COP_{HP} = \frac{|Q_H|}{|W|} = \frac{|Q_H|}{|Q_H| - |Q_L|} = \frac{1}{1 - |Q_L|/|Q_H|} \leq \frac{1}{1 - T_L/T_H} \tag{5.173}$$

where the equality sign applies to the reversible cycle and the inequality applies to the irreversible cycle. Therefore,

$$COP_{HP, irr} < COP_{HP, rev} = \frac{1}{1 - T_L/T_H} \tag{5.174}$$

Note that

$$COP_{HP} - COP = \frac{|Q_H|}{|Q_H| - |Q_L|} - \frac{|Q_L|}{|Q_H| - |Q_L|} = 1 \tag{5.175}$$

5.12   Temperature Measurement

The thermodynamic temperature is defined in Eq. (5.9) as a derivative incorporating internal energy and entropy. This definition satisfied our expectations of thermal equilibrium and our perceptions of hot and cold. This definition of thermodynamic temperature has led, through some analysis, to Eq. (5.161). For a reversible engine operating between two heat transfer reservoirs, this expression is

$$\frac{|Q_L|}{|Q_H|} = \frac{T_L}{T_H} \tag{5.176}$$

This expression is valid for an arbitrary reversible engine and is independent of the material. This expression, in principle, supplies a method to measure thermodynamic temperature.

The approach is to operate a reversible engine between a specified temperature reservoir and another temperature reservoir where the temperature is desired. Since the heat transfers are measurable, in principle, the unknown temperature is obtained. The Kelvin scale is based on the assigned value of the triple point of water as 273.16 K. The measurement of the ratio of heat transfers for a cycle operating between this temperature and the desired temperature yields the desired temperature. Therefore, any temperature is assigned by this reversible engine.

The practical approach to temperature measurement is to use an ideal gas thermometer. This thermometer is based on the ideal gas equation of state

$$Pv = RT \tag{5.177}$$

If a specified amount of gas occupies a specified volume, then

$$\frac{P}{T} = \text{constant} \tag{5.178}$$

Therefore, the measurement of pressure yields the temperature for a fixed-volume thermometer containing an ideal gas. Again, assigning 273.16 K to the triple point of water and measuring the pressure in this ideal gas thermometer yield the value for the constant in Eq. (5.178). Subsequent pressure measurements with this ideal gas thermometer yield the desired temperatures. Note that this method is based on the validity of the ideal gas equation of state. The verification of a specific gas obeying this ideal gas equation of state requires the measurement of temperature, which is possible with the reversible engine described above.

5.13   Other Statements of the Second Law

The original development of classical thermodynamics took into consideration engines and cyclic machines since the development of thermodynamics, like other disciplines, is based on physical observations. A short history of this development is presented in App. A. In this text we state the laws directly and subsequently demonstrate their use. In this section we present a few of the other forms for the second law to show the correspondence with the present approach. An understanding of the other forms is useful to a clear interpretation of the present approach. Also, the other statements of the second law serve as good examples of thermodynamic analysis.

Other statements of the second law are probably as numerous as the number of other authors. No attempt is made to present all possible statements. Also, complete proofs of other statements of the second law from the present approach are not developed. The comparison of the various approaches is presented and discussed.

A number of variations of the second law deal with the impossibility of transforming a single heat transfer to work. This is the form referred to as the *Kelvin-Planck statement* [3]:

> It is impossible to construct an engine that, operating continuously, will produce no effect other than the extraction of heat from a single reservoir and the performance of an equivalent amount of work.

This statement is demonstrated by considering the cycle expression ("operating continuously" implies a cycle) in Sec. 5.11. The first law for a cycle with a single high-temperature reservoir is

$$|W| = |Q_H|$$

where the work is negative for the engine output and the heat transfer is positive into the engine. The second law for these same conditions is

$$0 \geq \frac{|Q_H|}{T_H}$$

This expression indicates that the heat transfer must be negative, or out of the engine. Therefore, there is a contradiction, and this leads to the impossibility. Note that it *is* possible to produce work from a heat transfer if the system does not operate in a cycle.

Another statement that is similar to the Kelvin-Planck form is

> It is impossible to construct a machine which would work periodically and continuously and perform work at the expense of heat extracted from only one source.

This was formulated by Planck and is rephrased above from Kestin [4]. Another similar form is

> A perpetual-motion machine of the second kind (PMM2) is impossible.

where a PMM2 is a system which operates cyclically, delivering work while exchanging energy, as a heat transfer with a single reservoir of uniform temperature. This statement is from Keenan [5].

An alternate approach considers the impossibility of a heat transfer from a low-temperature reservoir to a high-temperature reservoir. This second law statement is referred to as the *Clausius statement* [3]:

> It is impossible to construct a device that, operating continuously, will

produce no effect other than the transfer of heat from a cooler to a hotter body.

The first law for a cycle that does not perform work, that accepts a heat transfer from a low-temperature body $|Q_L|$, and that has a heat transfer to a high-temperature body $|Q_H|$ is

$$0 = |Q_L| - |Q_H|$$

or

$$|Q_H| = |Q_L|$$

The second law for a cycle is

$$0 \geq \frac{|Q_L|}{T_L} - \frac{|Q_H|}{T_H}$$

or

$$\frac{|Q_L|}{T_L} \leq \frac{|Q_H|}{T_H}$$

But the first law requires that $|Q_L| = |Q_H|$, so the second law requires

$$T_L \geq T_H$$

This contradicts the statement that T_L is lower than T_H and leads to the impossibility of such a device.

These examples are second law forms that show the historical approach to classical thermodynamics. There are many other examples that are not presented here.

5.14 Summary

The basic elements of classical thermodynamics are now complete. The first and second laws are presented for both the control mass and control volume. These laws combined with the state postulate and conservation of mass form the basis of thermodynamic analysis. The governing principles are summarized in Table 5.1. Some of the more common idealizations are presented in Tables 5.2 and 5.3. These expressions are applied to a clearly defined system indicating the boundary transfers. The properties required for the analysis have also been presented. The properties are contained within the appendixes. The ideal gas relations and incompressible fluid or solid relations are summarized in Tables 5.4 and 5.5, respectively. Also, the Gibbs equation, Eq. (5.43), and Eq. (5.44) are important property relations. Be sure to separate the thermodynamic analysis from the property evaluation.

In the following chapters we apply the thermodynamic principles and properties to various systems and processes. New definitions are presented, but they are based on the basic elements delineated here.

Problems

5.1 Indicate whether the entropy of the control mass has increased, decreased, or remained constant when the following processes are complete. Explain your answer.

(a) A frictionless piston slowly compresses a gas in an adiabatic cylinder.

(b) A frictionless piston slowly compresses an isothermal gas in a diabatic (permits heat transfer) cylinder.

(c) A piston that fits tightly in a cylinder slowly compresses an isothermal gas in a diabatic cylinder.

(d) A cake of ice floating in water that is very near freezing slowly increases in mass until all the water is frozen. (Take the control mass as the ice plus the water.)

(e) One kilogram of putty is dropped onto a floor tile and sticks without bouncing.

(f)* A bead with a hole through it slides down a slack wire hung between two hooks that are at equal heights and comes to rest at the lowest point on the wire. Take the bead as the control mass.

5.2 For the following processes, consider the total entropy generated in the control mass and in the surroundings. Describe the processes that change the entropy and the sign of each change, and show that $S_{gen} \geq 0$ in each case.

(a) A frictionless piston slowly compresses a gas in an adiabatic cylinder.

(b) A frictionless piston slowly compresses an isothermal gas in a diabatic (permits heat transfer) cylinder.

(c) A piston that fits tightly in a cylinder slowly compresses an isothermal gas in a diabatic cylinder.

(d) A cake of ice floating in water that is very near the freezing temperature slowly increases in mass until all the water is frozen. (Take the control mass as the ice plus the water.)

(e) One kilogram of putty is dropped onto a floor tile and sticks without bouncing.

(f) A bead with a hole through it slides down a slack wire hung between two hooks that are at equal heights and comes to rest at the lowest point on the wire. Take the bead as the control mass.

5.3 Two reservoirs A and B contained within an isolated system are at temperatures T_A and T_B, respectively, and communicate thermally with each other. Prove that the entropy of the isolated system increases for this process regardless of whether $T_A > T_B$ or $T_B > T_A$.

5.4 A control mass in initial state 1 undergoes a process that results in final state 2. Because entropy is a property, we know that $(\Delta S)_{CM} = S_2 - S_1$ is the same regardless of the process that results in the change of state. Describe the entropy change of the *surroundings* $(\Delta S)_{surr}$ in terms of $(\Delta S)_{CM}$ if the process is (a) heat transfer from the control mass to the surroundings, (b) heat transfer from the surroundings to the control mass, (c) adiabatic work done on the control mass with friction, and (d) adiabatic work done by the control mass without friction.

5.5 At $T = 200°C$ and $P = 500$ kPa, find the specific entropy of (a) nitrogen, using the gas tables (Table D.4); (b) steam, using the steam tables; (c) steam, using the T-s diagram (Fig. D.1); (d) steam, using the P-h diagram (Fig. D.2); and (e) steam, using the Mollier diagram (Fig. D.3).

5.6 Refrigerant 12 expands isothermally from an initial state at $P = 300$ kPa, $T = 40°C$ to a final state at $P = 150$ kPa. Find the change in specific entropy of refrigerant 12 that occurs in this process, using the tables.

5.7 Calculate the change in entropy for air between the states $P_1 = 1$ atm, $T_1 = 50°C$ and $P_2 = 10$ atm, $T_2 = 1000°C$, using Eq. (5.55) and the expression for variable specific heat from Table D.1. Compare your result with that from using Table D.2.

5.8 A massless container of fixed volume encloses 1 kg of nitrogen at 1000 K. A second similar container encloses 3 kg of argon at $T_a < 1000$ K. Both containers are isolated from the surroundings. The containers are brought together, and there is heat transfer until the containers are in thermal equilibrium. Plot a curve of S_{gen}, the entropy generation during this process, versus T_a. (Here T_a is in the range $100 \leq T_a \leq 1000$ K.) Assume constant specific heats for argon and nitrogen.

5.9 A reversible adiabatic process occurs for steam between an initial state of $P_1 = 15$ kPa, $v_1 = 10.0$ m³/kg and a final state with $P_2 = 100$ kPa. What is the temperature change for this process?

5.10 A piston compresses an ideal gas in a cylinder from a state P_1, v_1 to a final pressure P_2.
(a) Derive an expression for the ratio of work done if the process is isentropic, w_s, to the work done if the process is isothermal, w_T, in terms of T, v, and k only.
(b) Sketch the two processes on a P-v diagram and a T-s diagram. Label all states.

5.11 A piston compresses an ideal gas in a cylinder from a state P_1, v_1 to a final temperature T_2.
(a) Derive an expression for the ratio of the work done if

the process is isentropic, w_s, to the work done if the process is isobaric (constant-pressure), w_P, in terms of k only.

(b) Sketch the two processes on P-v and T-s diagrams. Label all states.

5.12 A closed copper vessel has a volume of 2 m³. In the vessel is 4 kg of water (liquid plus vapor) at a pressure of 200 kPa. The surroundings are at 160°C. There is a heat transfer to the vessel until all the water is just evaporated.

(a) What is the final pressure in the vessel?

(b) What is the entropy change of the water in the vessel during the heat transfer process?

(c) What is the entropy change of the surroundings due to the heat transfer process?

(d) What is the entropy generation due to the heat transfer process?

5.13 A glass of ice water initially contains 50 percent ice and 50 percent water by mass. The ice water undergoes heat transfer with the surroundings, which are at T_0, until the water in the glass reaches very nearly a temperature T_0.

(a) If $T_0 = 27°C$ and the entropy generation is $S_{gen} = 3$ kJ/K, what is the total mass of water plus ice in the glass during the process?

(b) At what value of T_0 would $S_{gen} = 0$ for the process described? (Prove your result.)

The enthalpy of melting *(latent heat of fusion)* of ice is 335 kJ/kg.

5.14 A reversible isothermal process occurs in which a fixed amount of refrigerant 12 is evaporated and heated from an initial state with $T_1 = 40°C$, $h_1 = 300$ kJ/kg to a final state at $P_2 = 200$ kPa. How much heat transfer occurs during this process?

5.15 A 1-m³ tank is half filled with refrigerant 12 liquid and half filled with refrigerant 12 vapor. The initial pressure is 700 kPa. There is a heat transfer until one-half of the original liquid by mass is evaporated. An automatic valve allows saturated vapor to escape at constant tank pressure. Determine the required heat transfer.

5.16 Refrigerant 12 is reversibly heated isothermally from an initial state at 200 kPa and 200°C to a final pressure of 800 kPa in a piston-cylinder arrangement. Evaluate the work done per unit mass.

5.17 Five kilograms of saturated refrigerant 12 (liquid plus vapor) are in a 0.491-m³ tank. A valve is opened, and the tank is heated so that refrigerant 12 vapor escapes at constant pressure until 1 kg of saturated vapor just fills the tank.

(a) Sketch the T-s diagram for this process, and label the states.

(b) What is the initial pressure in the tank?

(c) What is the entropy change of the control volume $(\Delta S)_{CV}$ if the control volume is taken as the tank interior?

5.18 Three kilograms of saturated water (liquid plus vapor) are in a tank of volume 6.2006 m³. A valve is opened, and there is a heat transfer to the tank in such a way that water vapor escapes at constant temperature until 1 kg of saturated vapor just fills the tank.

(a) Sketch the T-s diagram for the material in the tank during this process, and label the states.

(b) What is the initial quality of the water in the tank?

(c) What is the entropy change of the control volume $(\Delta S)_{CV}$ if the control volume is taken as the tank interior?

5.19 A bladder initially contains 7 kg of refrigerant 12 (liquid plus vapor) at 100 kPa, and the volume is 1 m³. The bladder (which expands at constant internal pressure) is heated until the refrigerant 12 is all saturated vapor. What is the change in the entropy of refrigerant 12?

5.20 Neon is in an insulated container at an initial state defined by P_1 and T_1. Work is done until the system comes to a final state P_2 and T_2. One of the end states of the process has $P = 1$ atm and $T = 25°C$. The other has $P = 2$ atm and $T = 139.7°C$. Which state is the initial state? (Prove your result.)

5.21 Consider two possible expansion processes for steam (see Fig. P5.21). Steam is initially at 0.60 MPa and 400°C. The steam is partitioned into a chamber with the other side of the partition being a vacuum. The partition is removed, and the steam fills the complete chamber. Determine the temperature and pressure of the final state and the entropy generated if (a) $V_s = 0.1V_v$ and (b) $V_s = 10V_v$.

5.22 A reversible process for a piston-cylinder control mass is shown on the P-v and T-s diagrams in Fig. P5.22. Evaluate the change in internal energy per unit mass for the process.

5.23 The inlet conditions to a steam turbine are $P = 6000$ kPa, $T = 400°C$. The outlet conditions are $P = 8$ kPa and $x = 0.904$. What is the turbine efficiency?

5.24 A space-solar power plant uses large solar mirrors to concentrate solar energy onto a boiler, which produces steam at $P = 600$ kPa, $T = 400°C$. The steam enters a revers-

Adiabatic chamber   **Figure P5.21**

Figure P5.22

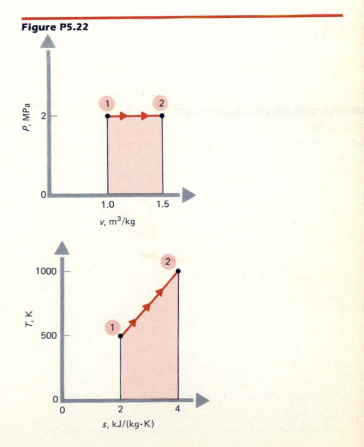

ible adiabatic turbine at a flow rate of 4 kg/s. The turbine exit pressure is 8 kPa.

(a) What is the quality of steam leaving the turbine?

(b) What is the turbine power output?

5.25 The inlet state to an adiabatic steam turbine is 2.0 MPa and 450°C, and the outlet pressure is 0.030 MPa. The turbine is rated so that the actual work output divided by the work output for reversible operation (between inlet state and given outlet pressure) is 0.85. Specify the outlet state (either x_2 or T_2 depending on the state location), and evaluate the actual work output.

5.26 A compressor is used to compress steam in a steady flow process from 150 kPa, 120°C to 1.5 MPa, 200°C. The work input is measured to be 485.4 kJ/kg. Kinetic and potential energy changes are negligible. The environment is at 27°C.

(a) Find the magnitude and direction of any heat transfer.

(b) Find the entropy change of the fluid flowing through the compressor.

(c) Find the total entropy change for the overall process.

5.27 Hydrogen enters a turbine at 1000 kPa, 400°C and exhausts at 200 kPa. Neglecting kinetic and potential energy changes and assuming an isentropic turbine, find (a) the exhaust temperature and (b) the work output.

5.28 Air at 1 atm, 27°C is flowing in a pipe. An adiabatic porous plug in the pipe causes the air pressure to drop to 0.1 atm. What is the change in specific entropy of the air across the plug?

5.29 A small swimming pool pump is specified by the manufacturer to have an efficiency of 83 percent. The flow rate through the pump is 20 kg/min. The pump must increase the pressure of the inlet water by 200 kPa. What is the required size for the electric motor to drive the pump for the following cases?

(a) The inlet and outlet pipes to the pump each have 5-cm ID.

(b) The inlet pipe has a 2.5-cm ID, the outlet has a 5-cm ID.

(c) The inlet pipe has a 5-cm ID, the outlet has a 2.5-cm ID.

5.30 Water flows reversibly through the piping system shown in Fig. P5.30. How large an electric motor (efficiency = 90 percent) should be selected to drive the pump (efficiency = 70 percent)?

5.31 An adiabatic air compressor has an efficiency of 75 percent and is required to compress 1 kg/s of ambient air

Figure P5.30

($P_1 = 1$ atm, $T_1 = 27°C$) to a final pressure of 5 atm. How much power input is required?

5.32 Steam flows reversibly into an adiabatic nozzle at 15 MPa and 500°C. The steam leaves at 400°C. If the inlet velocity is 100 m/s, calculate the (*a*) work done, (*b*) heat transferred, (*c*) entropy change of the steam, (*d*) exit pressure, and (*e*) exit velocity. Indicate any assumptions.

5.33 An incompressible fluid flows through a throttling valve with $P_1 > P_2$. See Fig. P5.33. Develop the explicit relationship for the entropy generation for this process in terms of P_1, T_1, P_2, v, and c.

Figure P5.33

Figure P5.34

5.34 Argon gas is available from a large-capacity supply line in a chemical plant at $P = 500$ kPa, $T = 27°C$. See Fig. P5.34. An evacuated 2-m³ tank is to be filled adiabatically from the supply line.
(*a*) What is the temperature of argon in the tank when it reaches the supply line pressure?
(*b*) How much entropy was generated during the process?

5.35 The supply line described in Prob. 5.34 is connected to an identical tank, which initially contains argon at 100 kPa and 27°C.
(*a*) What is the temperature of the argon in the tank when it reaches the supply line pressure?
(*b*) How much entropy was generated during the process?

5.36 An evacuated tank has a volume of 0.5 m³. It is connected through a large valve to an air line in which air is available at 70°C, 5 MPa. The valve is opened, allowing air to flow adiabatically into the tank. The valve is closed when the tank pressure reaches 700 kPa. The tank is then allowed to sit until it comes to thermal equilibrium at room temperature ($T_{room} = 20°C$). What is the final pressure inside the tank?

5.37 A tank initially contains very little water (assume $m_1 = u_1 = s_1 = 0$). The tank is attached to a steam line which is at 0.8 MPa and 400°C. The valve connecting the tank and steam line is opened until the tank pressure reaches 0.8 MPa, and then the valve is closed. During the process there is a heat transfer of 1000 kJ *from* the tank to the surroundings (surroundings are at 300 K) for each kilogram in the tank at the end of the process ($_1Q_2/m_2$). Determine the final state in the tank (give P and T or P and x, as appropriate) and the entropy generation per unit mass (S_{gen}/m_2).

5.38 The tank shown in Fig. P5.38 initially contains 0.3 kg of an ideal gas, N_2, at 530 K and an adiabatic, volumeless,

Figure P5.38

weightless, and frictionless piston. The 0.25-m³ tank is connected to an infinite source of ideal gas, air, at 530 K and 0.7 MPa. Air enters the tank until the temperature of the air is 560 K, and the final mass of air is 0.15 kg. Calculate the entropy generation of the process, assuming there is 8.6 kJ of heat transfer *from* the N_2 to the surroundings at 25°C.

5.39 A chamber of volume 0.001 m³ is attached to a steam line as shown in Fig. P5.39. The steam line state is constant at $P_{in} = 0.600$ MPa and $x_{in} = 0.97$. A small portion of the fluid from the line is diverted into a valve, through the chamber, and out a second valve. The chamber is insulated so the process is adiabatic.
  (*a*) The mass flow rate into the chamber is equal to the mass flow rate out, and the chamber pressure is 0.050 MPa. Determine the fluid temperature within the chamber and the entropy generation.
  (*b*) Then the outlet valve is quickly closed, and the filling process terminates when the chamber pressure is equal to the steam line pressure. The mass addition to the chamber is 0.002 kg. Determine the fluid temperature within the chamber and the entropy generation.

Figure P5.39

5.40 A bottle of pressurized nitrogen gas is sitting in the laboratory at 27°C and an initial pressure of 500 kPa and contains an initial mass of nitrogen of 1 kg. The valve on the tank is left open, and nitrogen escapes into the laboratory until the tank pressure reaches 100 kPa.
  (*a*) Define the system or control volume, and write the first and second laws for it. Note any assumptions used to simplify the equations.
  (*b*) Find the entropy generation for the process.

5.41 A tank with 10-m³ volume contains a gas with a temperature independent c_P of 0.3 kJ/(kg · K) and a gas constant of $R = 0.0857$ kJ/(kg · K). The tank is initially at $T = 327°C$, $P = 10$ atm, and $s = 6$ kJ/(kg · K) at this condition. The tank is discharged to a final pressure of 1 atm. Treat the gas as ideal.
  (*a*) What is the final temperature of the gas in the tank?
  (*b*) What is the change in total entropy of the gas in the tank, ΔS_{CV}?

5.42 A stainless-steel tank has a volume of 0.07 m³. It initially contains pure nitrogen at 300°C and 600 kPa. At a particular time, a low-flying aircraft knocks the valve from the outlet of the tank, and nitrogen discharges adiabatically from the tank until the tank pressure just reaches 100 kPa.
  (*a*) What is the temperature of the nitrogen remaining in the tank just as the tank pressure reaches 100 kPa?

(b) How much nitrogen was lost from the tank during the transient discharge?

(c) What is the change in total entropy ΔS_{CV} inside the tank during the discharge process?

(d) How much entropy is generated during the process?

5.43 An insulated air storage tank initially is at 1.5 MPa and 30°C. The valve on the 0.4-m³ tank is opened, and the air exhausts until the tank pressure is 0.5 MPa. What is the tank temperature at this state? How much mass is left in the tank? (Evaluate these quantities with both constant and variable specific heats.)

5.44 Water is flowing through a pump. It enters the pump at a pressure of 500 kPa, and the actual work done on the water passing through the pump is 6 kJ/kg. The water leaves the pump at 5 MPa and 30°C. The pump efficiency is 78 percent. What is the temperature of the water entering the pump?

5.45 Tests are run on a newly designed steam turbine. At an inlet pressure of 60 bars and an inlet temperature of 430°C, the turbine exhausts to a pressure of 0.10 bar. The exit quality from the turbine is measured to be 0.90.

(a) What is the enthalpy of the steam leaving the turbine?

(b) What is the value of the turbine work?

(c) What is the turbine efficiency?

5.46 A turbine with an efficiency of 80 percent operates at the same inlet conditions and outlet pressure as that described in Prob. 5.45.

(a) What is the quality of steam leaving the turbine?

(b) What is the turbine work?

(c) What is the entropy generation for this turbine?

(d) What is the percentage change in quality and turbine work in comparison with an isentropic turbine operating at the same conditions?

5.47 Air enters an adiabatic compressor at 20°C and 0.1 MPa. The outlet pressure is 10 times the inlet pressure. If the steady state compressor efficiency is 95 percent, determine the outlet temperature and the actual work per unit mass required, assuming (a) constant specific heats and (b) variable specific heats.

5.48 An adiabatic air compressor has an efficiency of 75 percent and is required to compress 1 kg/s of ambient air ($P_1 = 1$ atm, $T_1 = 27°C$) to a final pressure of 5 atm. How much power input is required?

5.49 An adiabatic air compressor has an efficiency of 75 percent. Air enters the compressor at 27°C, 1 atm and exits at 3 atm. What is the temperature of the air at the compressor exit?

5.50 Nitrogen is compressed adiabatically at a rate of 40 m³/min (inlet conditions) in an axial-flow compressor from 1 to 5 MPa. The temperature of the nitrogen entering the compressor is 40°C. The velocity of nitrogen entering the compressor is 100 m/s, and that leaving is 10 m/s. The compressor has an efficiency of 75 percent. Calculate the shaft power required to drive the compressor.

5.51 Steam enters a turbine at an inlet temperature of 900°C and a pressure of 3.0 MPa. The outlet pressure is 7.5 kPa. The mass flow rate of the steam is 2 kg/s. The inlet and outlet velocities are small.
(a) If the turbine is isentropic, what is its power output?
(b) If the turbine efficiency is 85 percent, what is the power output?
(c) What is the outlet enthalpy of the steam from the real turbine?
(d) Explain in physical terms why the enthalpy of the steam leaving the real turbine differs from that of the isentropic turbine (if it does).

5.52 A steam turbine provides a measured work output of 4.0 MW when a mass flow of 5 kg/s flows through it. See Fig. P5.52. The outlet steam is at 25°C and has a quality of 90 percent and a velocity of 100 m/s. The inlet steam pressure to the turbine is 2 MPa. What are (a) the actual inlet steam temperature and (b) the turbine efficiency? State any necessary assumptions.

5.53 An air turbine drives a water pump with the conditions shown in Fig. P5.53. The air undergoes an isentropic process ($k = 1.4$) in going through the turbine. The pump has an efficiency of 70 percent. What is the inlet temperature T_{in} of the air?

5.54 A steam turbine with 70 percent efficiency drives an air compressor with an isentropic efficiency of 78 percent, as shown in Fig. P5.54. What mass flow rate $\dot{m}$ of steam, in kilograms per hour, must be supplied to the turbine?

5.55 A steam turbine (efficiency 72 percent) drives an air compressor (efficiency 80 percent) as shown in Fig. P5.55. What mass flow rate $\dot{m}$ of air, in kilograms per hour, can be compressed?

5.56 A two-stage air compressor is compressing 2 kg/s at the conditions shown in Fig. P5.56. The first stage is operated as an adiabatic compressor; the second stage is operated isothermally. The actual power required to drive the compressor is 650 kW.
(a) What would T_{out} be if each stage of the compressor were reversible?

$\dot{m} = 5$ kg/s

$P = 2$ MPa

$\hat{W}_a \approx \hat{m}\,(\hat{h}_{out} - \hat{h}_{in}) + \frac{\dot{m}}{2}\,V^2$

solve for $\hat{h}_{in}$, T_{in}

$\eta_T = \frac{\hat{W}_a}{\hat{W}_s}$ $\Delta S = 0$ for N_S

$-S_1 =$

$\dot{W}_a = -4.0$ MW

25 °C
$x = 0.9$
$V = 100$ m/s   **Figure P5.52**

Air →

$\dot{m}_a = 1$ kg/s $P_{out} = 1$ MPa
T_{in}

Turbine   Pump

Water

$T_{out} = 30°C$ $\dot{m} = 5$ kg/s
$P_{in} = 0.1$ MPa   **Figure P5.53**

$P = 5$ atm   **Figure P5.54**
$\dot{m}_{out} = 0.1$ kg/s

$\dot{m}$

$P = 2$ MPa
$T = 350$ °C

Turbine   Compressor

$P = 8$ kPa $P = 1$ atm
$T = 27$ °C

$P_{in} = 2$ MPa   **Figure P5.55** $P_{out} = 5$ atm

$T_{in} = 350$ °C
$\dot{m}_{st} = 0.1$ kg/s

$T_{s,12} = T_{in}\left(\frac{P_{out}}{P_{in}}\right)^{k-1/k}$

$\Delta \hat{h}_{air} = C_p\,(T_{2,5} - T_{in})$

$\dot{W}$

Air compressor

Turbine

$P_{out} = 8$ kPa

$P_{in} = 1$ atm
$T_{in} = 27$ °C

$W_t = \dot{m}_{st}\,\eta_T\,(\hat{h}_{out} - \hat{h}_{in})$

$W_t = -W_c = \dot{m}_{air}\,\dfrac{(\hat{h}_{out,s} - \hat{h}_{in})}{\eta_c}$ comp

(b) What power input would be required to drive the compressor if each stage were reversible?

(c) If T_{out} for the actual compressor were 250°C, how much power would be required to drive the *second* stage?

5.57 Two adiabatic compressors are used to take inlet air at $P_1 = 0.1$ MPa and $T_1 = 300$ K to an outlet state of $P_4 = 0.7$ MPa and $T_4 = 600$ K. An adiabatic valve connects the compressors. Both compressor efficiencies have the same value of $\eta_c = 0.70$. If $P_2 = 0.3$ MPa, evaluate the work per unit mass for each compressor and the entropy generation per unit mass for this process. Determine T_3 and P_3. Draw the process on the T-s diagram. Assume air is an ideal gas.

5.58 Two valves are used to control an adiabatic steady state turbine. The air (treat it as an ideal gas) enters the first valve at $P_1 = 1000$ kPa and $T_1 = 800$ K. The turbine exhausts to a pressure of $P_3 = 140$ kPa. The outlet of the second valve is at $P_4 = 100$ kPa and $T_4 = 500$ K. The turbine efficiency is $\eta_t = 0.90$. Determine work per unit mass of the turbine, the entropy generation per unit mass for the process from 1 to 4, and the entropy generation for the turbine alone. Draw the process on a T-s diagram.

5.59 The system shown in Fig. P5.59 is used to separate liquid water and water vapor while recovering some work from the steam. Vapor first enters a turbine at 350°C and 4.0 MPa. The mixture emerges from the turbine and enters a separator where saturated liquid and saturated vapor at 0.10 MPa are obtained as separated streams. The flow rate at state 1 is 0.5 kg/s in this adiabatic process.

(a) Find the percentage of initial mass flow converted to liquid water if the turbine produces 300 kW.

(b) Explain what happens if the turbine does *not* produce any work.

5.60 An inventor makes the following claims. Determine whether the claims are valid and explain why or why not. $\eta_{rev} = 1 - \frac{T_L}{T_H} = .8$

.69 < .8

(a) A flame at 1500 K is used as a heat source, and the low-temperature reservoir is at 300 K. The inventor indicates that 69 percent of the heat transfer to the cyclic device from the flame is returned as work.

(b) A building receives a heat transfer of 50,000 kJ/h from a heat pump. The inside temperature is maintained at 21°C, and the surroundings are at −1°C. The inventor claims a work input of 7000 kJ/h is required. $COP = \frac{1}{1 - \frac{T_L}{T_H}} = 13.4$ $COP \eta_t$ $\frac{|Q_l|}{W} = 7.1$

(c) An engine operates between 1000 and 400 K with a heat transfer into the engine of 550 kW. The inventor states that the heat transfer to the low-temperature reservoir is 250 kW, and the work output is 250 kW. $Q_H = \dot{W} + Q_L$

$P = 1$ atm T_{out}

$T = 27$ °C $P = 10$ atm

$\dot{W}$ 1st stage   2nd stage

$P = 5$ atm   **Figure P5.56**

Figure P5.57

4   W_B B   3 W_A A   1   2

Figure P5.58

1   2   −w   Turbine   3   4

Figure P5.59

$\dot{W}_t = -300 kW$ $h_1 = 3093.4$ $\dot{W}_t = \dot{m}(h_2 - h_1)$

1   Turbine   −$\dot{W}$ $x_2 = \frac{h_2 - h_{f,2}}{h_{fg,2}}$

$\dot{m}_4 = \dot{m}_1 - \dot{m}_3$

$\frac{\dot{m}_4}{\dot{m}_1} = \frac{\dot{m}_1 - \dot{m}_3}{\dot{m}_1}$

3   2   Separator   4

Gas   Liquid

5.61 A cyclic device has a single work input $|W|$ and a single heat transfer $|Q_H|$ to a high-temperature reservoir at T_H. Is the device possible? Why or why not?

5.62 Solar thermal power plants use solar energy to increase the temperature of some working fluid to T_H. This fluid can be used to power a heat engine. Low-cost flat-plate solar collectors can provide working fluids at temperatures in the range of about $30 \leq T_H \leq 100°C$ (concentrating collectors can provide much higher temperatures). If cooling water is available at 27°C, plot the efficiency of a reversible heat engine powered by flat-plate collectors versus T_H, for T_H in the range given for flat-plate collectors.

5.63 An inventor brings you a device. She claims that it will collect solar energy at 120°C and that each 1 m² of the collecting surface will provide power from the attached heat engine at a rate of 0.5 kW/m². You know that the solar energy on a clear day near noon in your area rarely exceeds 1 kW/m².
(a) Does the proposed device violate the first law?
(b) Does it violate the second law?
(c) Could the device operate as advertised if a reservoir were available at a low enough temperature?

5.64 A Carnot engine receives a heat transfer from a boiler at $T_H = 1000$ K, and it rejects a heat transfer of 20 kJ to a low-temperature reservoir at T_L. What is the value of the product $Q_H T_L$ for this engine?

5.65 A Carnot engine is used in a nuclear power plant. It receives 1500 MW of power as a heat transfer from a source at 327°C, and it rejects thermal waste to a nearby river at 27°C. The river temperature rises by 3°C because of this power rejection by the plant.
(a) What is the mass flow rate of the river?
(b) What is the efficiency of the power plant?
(c) What is the power output of the plant?

5.66 A Carnot engine provides power to a plant that stamps out engineering students. The engine receives a heat transfer from a steam line that contains saturated vapor at $P = 250$ kPa and rejects a heat transfer to saturated refrigerant 12 at $P = 10$ kPa. See Fig. P5.66. If the engine receives 10 kW from the steam line and it requires 5 kJ to stamp out one student, how many students can be stamped out per second?

Figure P5.66

Saturated steam, $P = 250$ kPa

$\dot{Q}_H = 10$ kW

$-\dot{W}$

$\dot{Q}_C$

Saturated R-12, $P = 10$ kPa

5.67 A reversible heat engine has a work output of 500 MW when receiving a heat transfer of 1000 MW from a heat transfer reservoir. The engine rejects a heat transfer at a constant temperature of 27°C to a flowing river. The average river temperature increases by 2.0°C.

(*a*) What is the mass flow rate of the river?

(*b*) What is the temperature of the heat transfer reservoir providing energy to the heat engine?

(*c*) What is the efficiency of the heat engine?

(*d*) If the heat engine were irreversible, would the temperature *change* of the river be greater or smaller? Explain.

5.68 A reversible heat engine receives energy from a solar collector at 80°C and rejects a heat transfer to the surroundings at 25°C. See Fig. P5.68. The solar collector converts 50 percent of the incident solar energy to usable thermal energy. If 1 kW of solar energy strikes each 1 m² of the collector, what collector area is necessary to provide 5 kW of power output from the heat engine?

Figure P5.68

5.69 A farmer buys a small heat engine from Carnot Products Co., Ottumwa, Iowa, to drive an irrigation pump. The engine takes a heat transfer from a nearby solar pond that is kept at $T_H = 73$°C. The pump is 70 percent efficient and pumps 1 kg/s of water at 27°C from a well. The water undergoes a 2-MPa pressure change in the pump. The water leaving the pump is used as the reservoir for the heat rejection from the so-called Carnot engine.

(*a*) Sketch the system.

(*b*) Calculate the rate of heat transfer from the solar pond.

(*c*) State your assumptions.

References

1. H. B. Callen, *Thermodynamics and an Introduction to Thermostatics,* 2 ed., Wiley, New York, 1985.
2. W. C. Reynolds and H. C. Perkins, *Engineering Thermodynamics,* McGraw-Hill, New York, 1977.
3. S. L. Soo, *Thermodynamics of Engineering Science,* Prentice-Hall, Englewood Cliffs, N.J., 1958.
4. J. Kestin, *A Course in Thermodynamics,* vol. 1, Blaisdell, New York, 1966.
5. J. H. Keenan, *Thermodynamics,* Wiley, New York, 1941.

6

Thermodynamic Cycles and Common Energy Systems

When the Waters were dried an' the earth did appear, . . .
The Lord He created the Engineer
Rudyard Kipling

Cross-section of aircraft gas turbine engine (fan-jet). The large fan at the inlet drives air around the compressor-combustor-turbine assembly, where it is mixed with the hot exhaust gases from the turbine stage. The mixed gases are then ejected from the engine. *(United Technologies' Pratt and Whitney.)*

In this chapter, we examine in some detail the common thermodynamic cycles used for the conversion of heat transfer to work, and we discuss the hardware that can be used to implement these cycles. We present the limitations imposed on the ideal thermodynamic cycles by real hardware and examine cycles for producing refrigeration and air conditioning.

6.1   Heat Engine Cycles

Devices for converting heat transfer to work are called *heat engines*. In Chap. 5, the Carnot cycle was used to demonstrate the limitations that the second law places on the efficiency of all heat engine cycles that operate between the same heat addition and rejection temperatures. Although the Carnot cycle is useful in determining such ideal behavior, it is not a practical cycle to use in the design of a heat engine. It is not possible to design hardware that will allow a heat transfer to a working fluid at a constant temperature in a reversible process over a finite time. To do so would require that the temperature difference between the system and its surroundings be infinitesimally small; this has the unfortunate side effect of making the energy transfer require an extremely long time. If we wish to operate a Carnot engine at a reasonable power level, then many cycles per unit time must be carried out; and if this is done, the engine operation is very far from the reversible ideal. The same difficulty arises in the heat rejection process, which for the true Carnot cycle must also be isothermal and reversible.

Because of these difficulties, other cycles have been developed which have *theoretical* efficiencies the same as or lower than that of the Carnot cycle. These cycles usually operate with nonisothermal heat transfers but have higher *actual* thermodynamic efficiencies when implemented in practical engines than are attainable from engines operating with near-isothermal high transfer.

There are other reasons for developing heat engines that do not operate on the Carnot cycle. These include the characteristics of the energy source available to drive the heat engine, the characteristics of the working fluid chosen for the cycle, material limitations in the hardware, requirements for steady or varying work output, and other practical considerations. For these reasons and because of the difficulty of constructing an efficient heat engine that actually operates on an approximation to the Carnot cycle, a number of other heat engine cycles have been proposed and engines built that operate on these cycles.

In analyzing cycles, the states at the beginning and end of each process making up the cycle must be known, as must the path of the process connecting the states. As we have seen, definition of the states requires that two independent properties be known if the working fluid is a pure substance; then all other properties can be determined at that state. Thus, if the state at a cycle point is an unsaturated fluid, then *any* two properties will define the state of the system. However, if the fluid is saturated, then two properties are again required. But at most one of these can be an intensive property from the set of *P* and *T*, since these intensive properties are not independent for a saturated system. In other words, for a saturated system, either one extensive or specific intensive and one intensive property from the set of *P* and *T* must be specified, or two extensive (or specific intensive) properties must be specified to define the state. Cycle analysis then depends on applying our knowledge of some of the end states along with the first and second laws to the processes connecting the end states; next the state of the system throughout the cycle can be calculated. The work done and heat transferred during the cycle can be found, and the cycle efficiency determined from these values. This approach is illustrated for various cycles.

Heat engine cycles can be categorized in a number of ways. Here we examine two categories: (1) cycles in which the energy source (often from combustion of a fuel, but also including sources such as solar, nuclear, and geothermal) is provided externally to the working fluid and is transferred across the system boundary in the form of a heat transfer (an *external heat source* engine) and (2) engines in which a fuel is burned within the system boundary and the energy thus released is used to increase the temperature of the working fluid (an *internal source or internal combustion* engine). This categorization is not without flaw, since some of the cycles we examine, such as the Brayton cycle, are operated as both internal and external heat source engines.

6.1.1 Cycle Analysis Methodology

For each cycle to be analyzed for efficiency, a certain minimum amount of information must be given, so that the state at each point in the cycle can be determined. Then the efficiency and other information about the cycle behavior can be found. To carefully lay out the given information, it is useful to tabulate the properties at convenient state points in the cycle. It then becomes obvious which points have the necessary minimum of two properties known and which points require further analysis to determine the state. Unknown properties at a cycle point usually may be found from the first or second law relationships that describe the process connecting an unknown state point to an adjacent point with known properties. Steady-state operation is always assumed for cycle analysis.

Example Problem 6.1

A system operates on the cycle shown with air as the working fluid. The processes making up the cycle are reversible and are noted on the figure. Find the cycle efficiency.

Solution

First, a general table is made of the known properties at points 1 through 4:

Cycle Diagram

State	P, kPa	T, K	v, m³/kg	h, kJ/kg	s, kJ/(kg · K)
1	1000	—	0.15	—	—
2	—	—	0.6	—	—
3	10	—	—	—	—
4	—	—	—	—	—

These are all the direct entries that can be made in the table. Only state 1 is completely defined, and all properties in row 1 are fixed (since two properties are known). Since air at moderate conditions is the working fluid, the ideal gas law may be used to find T_1 immediately. So

$$T_1 = \frac{(Pv)_1}{R} = \frac{150 \text{ kPa} \cdot \text{m}^3/\text{kg}}{R} = 523 \text{ K}$$

Because the process between states 1 and 2 is isothermal, $T_2 = T_1$, and these entries may be made in the table. Two properties, T_2 and v_2, are now known at state point 2, so all properties at point 2 can be found. From the ideal gas relation, for example,

$$P_2 = \frac{RT_2}{v_2} = 250 \text{ kPa}$$

The process connecting points 2 and 3 is adiabatic and reversible. The process 2-3 is, therefore, isentropic, and $s_3 = s_2$. For an ideal gas,

$$s_3 - s_2 = c_P \ln \frac{T_3}{T_2} - R \ln \frac{P_3}{P_2} = 0$$

or (see Table 5.4)

$$T_3 = T_2 \left(\frac{P_3}{P_2}\right)_s^{(k-1)/k} = (523)\left(\frac{10}{250}\right)^{(1.4-1)/1.4} = 208 \text{ K}$$

Now T_3 and P_3 are known, and all other properties can be found for state 3.

Because the cycle we are examining is exactly the Carnot cycle, the cycle efficiency can now be found as

$$\eta_C = 1 - \frac{T_3}{T_2} = 1 - \frac{208}{523} = 0.602$$

It is left to the reader to determine P, v, and T at point 4.

Comments

A table of properties and knowledge of the processes that connect the states in the cycle have allowed us to find information about the cycle behavior in this case. The same method is used to follow through the behavior of other cycles.

6.1.2 Air-Standard Cycles

It is useful to study the efficiency of certain cycles by making simplifying assumptions about the behavior of the processes that make up the cycle. This approach allows us to separate the general behavior of the cycle from the properties of the particular working fluid in the cycle and from the irreversibilities caused by particular hardware. In addition, idealized heat transfer processes can be substituted for the combustion processes that occur in internal combustion engines. Thus, internal combustion engines can be treated as if they were heat engines. The general behavior of individual cycles and the comparison between cycles can then be demonstrated on a common basis. These idealized cycles, called *ideal air-standard cycles,* are usually analyzed as in Example Problem 6.1, by assuming that the working fluid behaves as an ideal gas with temperature-independent specific heats. In addition, the mechanical devices and heat transfer processes used in the cycle are taken to be reversible. These restrictions can be relaxed if an actual device is to be modeled more carefully.

Because we make these assumptions, the cycle efficiencies based on these analyses are greater than can be realized in practice. The particular causes of deviation from the air-standard efficiencies are pointed out for each cycle.

Note that it is not difficult to analyze the air-standard cycles by allowing for the temperature dependence of the properties of the working fluid. And this should always be done in analyzing a real cycle. The gas tables (App. D) can be used. The constant-specific-heat assumption is invoked in the analysis of air-standard cycles simply to allow comparison of different cycles.

6.2 External Heat Transfer Cycles

The arc of a balance-wheel
flows like a curved rush of swallows, come
  over a hill . . .
Things lost come again in sudden new beauty.
Look long on an engine. It is sweet to the
  eyes.
MacKnight Black

6.2.1 Carnot Cycle

The Carnot cycle is analyzed for a particular case in Example Problem 6.1 and in Chap. 5. However, a few remarks are in order to deepen our understanding of the analysis of other cycles.

The P-v diagram for the Carnot cycle is illustrated in Fig. 6.1. Let us examine the slopes of the lines on this diagram for consistency. We assume an air-standard cycle. For the isothermal processes, an ideal gas follows the relation

$$P = \frac{RT}{v} = \frac{C_1}{v} \tag{6.1}$$

Or, by taking the derivative,

$$\left(\frac{\partial P}{\partial v}\right)_T = \frac{-C_1}{v^2} \tag{6.2}$$

However, the constant $C_1 = RT$ is different for each isotherm but always positive. So the slope of the P-v curve will always be negative, and the slope will be greater for the higher-temperature isotherm at any given specific volume. For the isentropic processes, $Pv^k = C_2$, so the slopes of these curves are

$$\left(\frac{\partial P}{\partial v}\right)_s = \frac{-kC_2}{v^{1+k}} \tag{6.3}$$

which are again always negative, with the slope decreasing more rapidly with v (since $k > 1$ for all ideal gases) than for the isotherms. At point 2, the slope of the isentropic curve is greater than the slope of the isotherm by the ratio

$$\frac{(\partial P/\partial v)_s}{(\partial P/\partial v)_T} = \frac{kC_2/v^{1+k}}{C_1/v^2} \tag{6.4}$$

Evaluating the ratio at P_2 and v_2 and noting that $C_1 = RT_2 = P_2 v_2$ and $C_2 = (Pv^k)_2$, we find

$$\frac{(\partial P/\partial v)_s}{(\partial P/\partial v)_T} = k \tag{6.5}$$

or, the slope of the isentropic line at point 2 is greater than the slope of the isotherm at that point by a factor k. Similar analysis at other points leads to the shape of the Carnot cycle diagram shown in Fig. 6.1.

It is often useful to plot cycles in other coordinates than P and v. For example, the T-s diagram for the Carnot cycle plots as a rectangle (Fig. 6.2). This makes it very easy to determine the efficiency of the cycle, since the ratio q_L/q_H is available from the relations $q_H = T_1(s_2 - s_1)$ and $q_L = T_4(s_4 - s_3)$. This gives the cycle efficiency directly [Eq. (5.164)] as

$$\eta_C = 1 - \frac{|q_L|}{|q_H|} = 1 - \frac{T_L}{T_H} \tag{6.6}$$

Figure 6.1 **Pressure-volume diagram for Carnot cycle.**

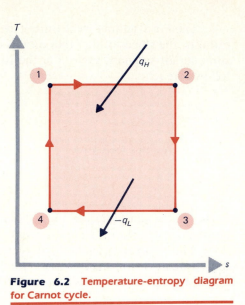

Figure 6.2 **Temperature-entropy diagram for Carnot cycle.**

Another useful plot is the *P-h* diagram, which allows the inlet and exit conditions to such equipment as compressors, turbines, and throttling valves to be displayed, aiding in the design process. The *P-h* diagram for the Carnot cycle is shown in Fig. 6.3. The isotherms are vertical because the enthalpy of an ideal gas is a function of temperature only. In general, the isentrope shape can be inferred from the definition

$$h = u + Pv \tag{6.7}$$

Taking the total derivative of $h(s, P)$ gives

$$dh = \left(\frac{\partial h}{\partial s}\right)_P ds + \left(\frac{\partial h}{\partial P}\right)_s dP \tag{6.8}$$

Equation (5.44) shows that

$$dh = T\,ds + v\,dP \tag{6.9}$$

For an isentropic process, comparison of Eqs. (6.8) and (6.9) shows that

$$\left(\frac{\partial h}{\partial P}\right)_s = v \quad \text{or} \quad \left(\frac{\partial P}{\partial h}\right)_s = \frac{1}{v} \tag{6.10}$$

Thus, the slope of the isentropes is always positive, and the slope is greater at higher pressure. As the system moves along the isentropic compression line from 4 to 1 on the *P-h* diagram, v decreases and the slope $\left(\frac{\partial P}{\partial h}\right)_s$ increases.

The *h-s*, or Mollier, diagram is also convenient for the description of cycles, particularly those using a vapor that undergoes a phase change for the working fluid. The change in enthalpy across an ideal compressor, turbine, or pump appears as a vertical line, while energy addition to

Figure 6.3 **Pressure-enthalpy diagram for Carnot cycle.**

heating or cooling devices is conveniently found as a change in vertical distance. The *h-s* diagram is explained in Sec. 6.3 on the Rankine engine. For air-standard cycles, the *h-s* diagram is similar to the *T-s* diagram because of the assumption of an ideal gas (often with temperature-independent specific heat) as the working fluid, which makes *h* directly proportional to *T* and independent of *P*.

Similar analyses can be performed on other cycles to predict the general shape of the diagrams in the various coordinates.

6.2.2 Stirling Cycle

This is the first of the practical heat engine cycles that we study. The Stirling cycle uses a compressible fluid as the working fluid, and the cycle is made up of a constant-temperature energy addition process followed by a constant-volume process, a constant-temperature energy rejection, and finally a return to the initial state by another constant-volume process, as shown in Fig. 6.4.

To accomplish this cycle requires some ingenious hardware. The cycle of operation is shown in Fig. 6.5. Two pistons *A* and *B* enclose the working gas, which is forced back and forth through a *regenerator C*. The regenerator is typically a chamber containing wire or ceramic mesh or fine metallic gauze and is used for the temporary storage of energy. Because the regenerator is within the system boundary, the energy stored and released by the regenerator during the cycle is treated not as a heat transfer, but simply as a change in the internal energy. Because the internal energy changes of the regenerator and the working fluid are always equal and opposite in sign, the changes do not affect the first law as written for the Stirling engine (i.e., the working fluid as the control mass) as a whole. The idea of regeneration is common to most external heat transfer engines.

To carry through the cycle, there is a heat transfer to the working fluid in the cylinder containing piston *A* while piston *A* is moved to the left at a rate just sufficient to hold the temperature of the gas constant. This process of constant-temperature energy addition is continued until state 2 is reached (Fig. 6.4). When state 2 is reached, both pistons are moved to the right at the same rate. Thus the gas volume is held constant and gas is forced through the gauze, which is at a lower temperature because of the previous step in the cycle. The gas is thus cooled by transferring some of its energy to the regenerator. When state 3 is reached, the cool gas is in the cylinder containing piston *B*. Piston *B* is then moved to the left while piston *A* is held in position. At the same time, there is a heat transfer from the gas at such a rate as to hold the gas temperature constant during compression. When state 4 is reached (at the same volume as state 1), both pistons are moved to the left at the same rate, again holding the gas volume constant while forcing the gas through the regenerator. The cold gas regains the energy it had previously given to the regenerator, thus

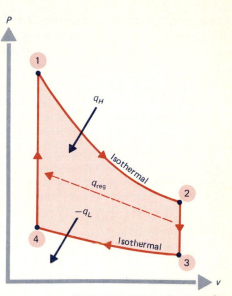

Figure 6.4 Pressure-volume diagram for Stirling cycle.

Figure 6.5 Operation of Stirling cycle.

cooling the regenerator material. The regenerator is then ready to receive energy in the next cycle, and the system is thus returned to state 1.

One interesting attribute of the Stirling engine is that the working fluid is a closed system, in that it is always contained within the engine, so no contamination need enter the lubricant. In addition, the working fluid can be chosen for its thermal properties since the inventory of working fluid is small enough that cost is not important. Furthermore, in the ideal cycle just described, since all heat transfer to the cycle takes place at constant temperature and all heat loss is also at constant temperature, the ideal Stirling cycle efficiency must equal the Carnot efficiency! This presupposes that all the processes in the cycle can be carried out reversibly, which, we noted in the beginning of the chapter, is not possible in real systems. It is, of course, possible to include compression and expansion efficiencies in the cycle if they are known.

To carry out the interrelated motion of the pistons or other devices needed to provide the processes making up the Stirling cycle, many ingenious methods have been tried. Stirling's original patent drawing is shown in Fig. 6.6, and present Stirling engine designs are no less complex.

Figures 6.7 and 6.8 show the *T-s* and *P-h* diagrams for the air-standard Stirling cycle.

Figure 6.7   *T-s* diagram for Stirling cycle.

Figure 6.6   Original patent drawing of Stirling engine.

Figure 6.8   *P-h* diagram for Stirling cycle.

Example Problem 6.2

Determine the cycle efficiency for the ideal air-standard Stirling cycle by using cycle analysis.

Solution

The cycle efficiency can be determined by computing the cycle work (Fig. 6.4) as

$$w = -\oint P \, dv = {_1}w_2 + {_3}w_4$$

$$= -\int_1^2 P \, dv - \int_3^4 P \, dv$$

$$= -RT_1 \ln \frac{v_2}{v_1} - RT_3 \ln \frac{v_4}{v_3}$$

However, $v_2 = v_3$ and $v_1 = v_4$, so

$$w = -R(T_1 - T_3) \ln \frac{v_2}{v_1}$$

The heat transfer to the cycle is

$$q_H = T_1(s_2 - s_1) = RT_1 \ln \frac{v_2}{v_1}$$

and the cycle efficiency is then

$$\eta_S = \frac{|w|}{|q_H|} = 1 - \frac{T_3}{T_1}$$

which is exactly the Carnot efficiency, as expected.

6.2.3 Ericsson Cycle

The Ericsson cycle was originally developed, as was the Stirling cycle, in an effort to find a practical external heat transfer engine that could substitute air for steam as the working fluid. The cycle is composed of two constant-pressure processes connected by constant-temperature energy addition and rejection processes (Fig. 6.9). Again, in the ideal reversible (air-standard) cycle, the efficiency must approach that of the Carnot cycle, since the Ericsson cycle operates between two constant temperatures at which energy is added and rejected. The *T-s* and *P-h* diagrams for the Ericsson cycle are shown in Figs. 6.10 and 6.11.

The Ericsson cycle was originally used in very large marine piston

Figure 6.9 *P-v diagram for Ericsson cycle.*

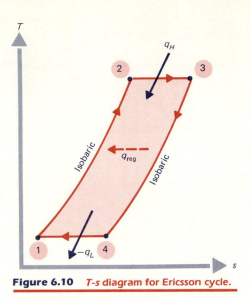

Figure 6.10 *T-s diagram for Ericsson cycle.*

engines—so large that they were impractical in competition with steam engines of similar power. However, it is possible to envision a system using a turbine compressor system employing the Ericsson cycle, as shown in Fig. 6.12. Here, the lower-temperature working fluid enters the regenerator (heat exchanger) at state 1, and energy is added to the fluid from the hotter fluid that has left the turbine. The fluid leaves the regenerator at state 2 and enters the turbine. The turbine is designed so that as the working fluid does work in the turbine, energy is added at exactly the same rate at which work is done at each point in the turbine. Thus the working fluid is kept at constant temperature. The fluid leaves the turbine at state 3 and enters the regenerator, where it transfers energy to the working fluid en route to the turbine. This energy exchange process is assumed to take place at constant pressure. The fluid exits the regenerator at state 4 and enters the compressor. The compressor is designed so that the work done at each point is balanced by the heat transfer at each point. Thus, the working fluid is maintained at constant temperature and leaves the compressor at state 1, to complete the cycle.

In practice, it is difficult to design turbines and compressors that have the internal cooling profiles necessary to achieve isothermal conditions. In addition, of course, real components are not reversible; and, for example, fluid friction in the regenerator makes the pressure vary through the regeneration process. Also, the regenerator cannot be designed in practice to bring the outlet temperature of the cold fluid up to the inlet temperature of the hot fluid, as the cycle diagram implies. This would require an infinite heat transfer surface area in the regenerator or zero resistance to heat transfer across the boundary between the fluids. Neither condition can be attained in practice. The actual efficiency will be below the Carnot efficiency for these reasons and others. However, it is possible

Figure 6.11 *P-h diagram for Ericsson cycle.*

Figure 6.12   Turbocompressor version of Ericsson cycle.

Figure 6.13   *P-v* diagram for Brayton (Joule) cycle.

to build components that approximate the Ericsson processes, as we will see.

Ericsson Cycle Efficiency

The ideal air-standard Ericsson cycle with regeneration is a reversible cycle with heat transfer processes that occur at constant temperature, and so the air-standard cycle efficiency is equal to the Carnot efficiency.

6.2.4   Brayton Cycle (External Heat Transfer)

The Brayton cycle, also called the *Joule cycle,* was developed originally for use in a piston engine with fuel injection, but its most common embodiment at present is in open- and closed-cycle turbine engines. The open-cycle engine can be used with either internal combustion or an external heat transfer, while the closed-cycle engine has an external energy source. First, we examine the closed-cycle air-standard Brayton engine.

The air-standard simple Brayton cycle is composed of constant-pressure heat transfer processes separated by isentropic expansion and compression processes, as shown in Fig. 6.13. The corresponding *T-s* and *P-h* diagrams are shown in Figs. 6.14 and 6.15.

The turbine compressor system that uses such a cycle is shown in Fig. 6.16. The working fluid in the closed cycle enters the high-temperature heat exchanger in state 1, and there is an energy addition in a constant-pressure process until the working fluid attains the higher temperature of state 2. The fluid then enters the turbine, where an isentropic expansion occurs, producing power. The fluid leaves the turbine at state 3, is further cooled at constant pressure in the low-temperature heat exchanger, and then enters the compressor at state 4. The fluid is com-

Figure 6.14   *T-s* diagram for closed Brayton cycle.

Figure 6.15 *P-h diagram for Brayton cycle.*

Figure 6.16 *Diagram of simple Brayton cycle hardware.*

pressed isentropically to state 1, and the cycle is repeated. A considerable fraction of the turbine work is used to power the compressor.

The turbine and compressor are assumed to be isentropic in the ideal air-standard Brayton cycle, rather than isothermal as in the Ericsson cycle. A turbine can be built which is nearly adiabatic, although reversibility is much more difficult to approach. Thus the Brayton cycle is easier to implement in a practical device than the Ericsson cycle. The Brayton cycle, however, provides heat transfer to the working fluid over a range of temperatures, as can be seen from the *T-s* digram. The Brayton cycle does not have the capability in its simplest form, therefore, of approaching the Carnot efficiency when it operates between the same two reservoir temperatures, unlike the Ericsson cycle.

Example Problem 6.3

Determine the efficiency of an ideal air-standard closed-cycle Brayton engine in terms of (*a*) the compressor inlet and outlet temperatures, (*b*) the turbine inlet and outlet temperatures, and (*c*) the turbine (or compressor) inlet/outlet pressure ratio.

Solution
The cycle efficiency (Fig. 6.14) is given by

$$\eta_B = \frac{|w|}{|q_H|} = 1 - \frac{|q_L|}{|q_H|} = 1 - \frac{c_P(T_3 - T_4)}{c_P(T_2 - T_1)} = 1 - \frac{T_4}{T_1}\frac{T_3/T_4 - 1}{T_2/T_1 - 1}$$

For the isentropic turbine and compressor, Table 5.4 yields

$$\frac{P_2}{P_3} = \left(\frac{T_2}{T_3}\right)^{k/(k-1)}$$

$$\frac{P_1}{P_4} = \left(\frac{T_1}{T_4}\right)^{k/(k-1)}$$

However, $P_2/P_3 = P_1/P_4$, so it follows that

$$\frac{T_2}{T_3} = \frac{T_1}{T_4}$$

or

$$\frac{T_2}{T_1} = \frac{T_3}{T_4}$$

Substituting the latter two ratios into the efficiency equation yields

$$\eta_B = 1 - \frac{T_4}{T_1} = 1 - \frac{T_3}{T_2} \qquad (6.11a)$$

which provides the answers to (a) and (b). For (c), substitute for the temperature ratio in terms of the pressure ratio, to obtain

$$\eta_B = 1 - \frac{1}{(P_1/P_4)^{(k-1)/k}} = 1 - \frac{1}{r_P^{(k-1)/k}} \qquad (6.11b)$$

where r_P is the pressure ratio P_1/P_4.

Comments

The efficiency of the closed-cycle Brayton engine is found to depend on the isentropic pressure ratio only. If we increase the inlet pressure to the turbine, the inlet temperature to the turbine is also increased. The inlet turbine temperature is usually limited by the material properties of the turbine blades, which places a practical upper limit on the cycle efficiency.

The closed-cycle (external heat addition) Brayton engine has received considerable attention for its use with nuclear and, more recently, high-temperature solar energy systems.

Effect of Actual Turbine and Compressor Efficiencies

Real turbines and compressors are, of course, not isentropic. For air-standard cycles, the efficiency of each device, as defined in Chap. 5, is

Figure 6.17   Brayton cycle with real compressor and turbine.

easily included in the analysis. The real compressor and turbine are taken to exit at the same pressure as the isentropic device (Brayton turbine and compressor efficiencies are usually given with respect to isentropic, not isothermal, devices). The *T-s* diagram for the Brayton cycle (with no regeneration) as modified to include a nonisentropic compressor and turbine is shown in Fig. 6.17.

Example Problem 6.4

A simple air-standard Brayton cycle operates between 1 and 5 atm. The inlet air to the compressor is at 27°C, and the turbine inlet is at 1127°C. The compressor is 65 percent efficient, and the turbine efficiency is 75 percent. What is the efficiency of this Brayton cycle?

Solution

Refer to Fig. 6.17. The outlet compressor temperature is found from the compressor efficiency [by taking $c_P \neq f(T)$] by [Eq. (5.153)]:

$$\eta_c = 0.65 = \frac{h_{1,s} - h_4}{h_{1,a} - h_4} = \frac{c_P(T_{1,s} - T_4)}{c_P(T_{1,a} - T_4)}$$

The value of $T_{1,s}$ is found from the isentropic relation

$$T_{1,s} = T_4 \left(\frac{P_1}{P_4}\right)^{(k-1)/k} = (273 + 27)(5)^{(1.4-1)/1.4} = 475 \text{ K}$$

which results in

$$T_{1,a} = T_4 + \frac{T_{1,s} - T_4}{0.65} = 300 + \frac{475 - 300}{0.65} = 569 \text{ K}$$

Similarly, for the turbine,

$$T_{3,s} = 1400 \left(\frac{1}{5}\right)^{0.4/1.4} = 884 \text{ K}$$

and

$$T_{3,a} = 1400 + 0.75(884 - 1400) = 1013 \text{ K}$$

The cycle efficiency is then

$$\eta_B = 1 - \frac{|Q_L|}{|Q_H|} = 1 - \frac{c_P(T_{3,a} - T_4)}{c_P(T_2 - T_{1,a})}$$

$$= 1 - \frac{1013 - 300}{1400 - 569} = 0.142$$

This compares with the ideal air-standard cycle efficiency of

$$\eta_{B, \text{ideal}} = 1 - \frac{1}{(P_1/P_4)^{(k-1)/k}} = 1 - \frac{1}{5^{0.4/1.4}} = 0.369$$

Comments

Obviously, if the turbine and/or compressor efficiency becomes low enough, the efficiency of the cycle can become zero (or even negative). So no net cycle work can be obtained, or work input to the cycle may be required for operation. This is not a desirable way to operate a power cycle!

Improving the Brayton Cycle Efficiency

The idea of regeneration can be incorporated into the Brayton cycle by modifying the hardware as shown in Fig. 6.18. For the simple closed Brayton cycle (Fig. 6.14), the turbine exhaust gases at state 3 may be at a higher temperature than the gases at the outlet of the compressor (state 1). It is then possible to use a portion of the rejected energy q_L to preheat the gases leaving the compressor and thereby reduce the heat transfer required to the cycle.

The T-s diagram for an ideal air-standard Brayton cycle with regeneration is shown in Fig. 6.19. Note that the gases leaving the compressor could, at best, be heated by the regenerator to $T_B = T_3$, the turbine exhaust temperature. The turbine exhaust gas will, in that case, be cooled to $T_A = T_1$. For such an ideal case, the energy added to the cycle becomes $_Bq_2$, and the work output remains unchanged. The ideal air-standard Brayton cycle efficiency with ideal regeneration then is given by

Figure 6.18 **Brayton cycle with Ideal regeneration.**

$$\eta_{B,\,\text{reg, ideal}} = 1 - \frac{|Q_L|}{|Q_H|} = \frac{|Q_H| - |Q_L|}{|Q_H|}$$

$$= \frac{c_P(T_2 - T_B) - c_P(T_A - T_4)}{c_P(T_2 - T_B)}$$

$$= \frac{(T_2 - T_3) - (T_1 - T_4)}{T_2 - T_3}$$

$$= 1 - \frac{T_1 - T_4}{T_2 - T_3} \tag{6.12}$$

If for a particular cycle $T_3 < T_1$, then regeneration by using turbine exhaust is not possible. It is also not possible to build a regenerator that will actually make $T_B = T_3$. For the real case, the compressor outlet temperature will be raised to some lower temperature $T_{B'}$ (Fig. 6.19). A *regenerator efficiency* η_{reg} can be defined as the fraction of the maximum enthalpy that is recovered by the real regenerator, or

$$\eta_{\text{reg}} = \frac{h_{B'} - h_1}{h_B - h_1} = \frac{h_3 - h_{A'}}{h_3 - h_A} \tag{6.13}$$

Or, for the constant-property air-standard cycle,

$$\eta_{\text{reg}} = \frac{T_{B'} - T_1}{T_3 - T_1} = \frac{T_3 - T_{A'}}{T_3 - T_1} \tag{6.14}$$

The cycle efficiency with the real regenerator is now

$$\eta_{B,\,\text{reg}} = 1 - \frac{|Q_L|}{|Q_H|} = 1 - \frac{c_P(T_{A'} - T_4)}{c_P(T_2 - T_{B'})}$$

$$= 1 - \frac{T_{A'} - T_4}{T_2 - T_{B'}} \tag{6.15}$$

Figure 6.19 *T-s* diagram for a Brayton cycle with Ideal regeneration.

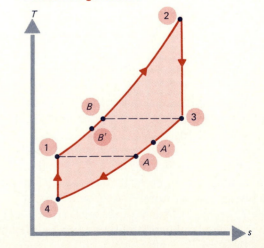

The values of $T_{A'}$ and $T_{B'}$ can be determined from known values of the regenerator efficiency.

Example Problem 6.5

Find the ideal air-standard cycle efficiency of a simple closed Brayton cycle, a Brayton cycle with an ideal regenerator, and with a regenerator with an efficiency of 70 percent. Compare the results. The cycle operates between 20 and 1000°C at a pressure ratio of 8.

Solution

For the simple Brayton cycle, Eq. (6.11b) gives

$$\eta_B = 1 - \frac{1}{(P_1/P_4)^{(k-1)/k}} = 1 - \frac{1}{(8)^{(1.4-1)/1.4}} = 0.448$$

By adding an ideal regenerator, Eq. (6.12) gives

$$\eta_{B,\text{ reg, ideal}} = 1 - \frac{T_1 - T_4}{T_2 - T_3}$$

The values of T_2 and T_4 are given. The isentropic expansion relations can be used to find the other temperatures:

$$T_1 = T_4 \left(\frac{P_1}{P_4}\right)^{(k-1)/k}$$
$$= (20 + 273)(8)^{0.4/1.4}$$
$$= 531 \text{ K} = 258°\text{C}$$

and

$$T_3 = \frac{1273}{8^{0.4/1.4}} = 703 \text{ K} = 430°\text{C}$$

The cycle efficiency is then

$$\eta_{B,\text{ reg, ideal}} = 1 - \frac{258 - 20}{1000 - 430} = 0.582$$

For a real regenerator, $T_{A'}$ and $T_{B'}$ are, from Eq. (6.14),

$$T_{A'} = T_3 - \eta_{\text{reg}}(T_3 - T_1) = 430 - 0.7(430 - 258) = 310°\text{C}$$

and

$$T_{B'} = T_1 + \eta_{\text{reg}}(T_3 - T_1) = 258 + 0.7(430 - 258) = 378°\text{C}$$

The cycle efficiency is then, from Eq. (6.15),

$$\eta_{B,\,\mathrm{reg}} = 1 - \frac{310 - 20}{1000 - 378} = 0.534$$

So the simple cycle has an efficiency of 44.8 percent, and adding an ideal regenerator to the cycle increases the efficiency to 58.2 percent. However, considering a real regenerator drops the predicted efficiency back to 53.4 percent.

Comments

Although the regenerator will improve cycle efficiency when the temperature constraints are met that allow its use, there are other real effects that we have not considered. Usually, a significant pressure drop is incurred in the regenerator, and this must be overcome at the expense of useful work output from the engine.

The Brayton cycle efficiency can be improved in other ways. For example, if a multistage compressor is used, the working fluid can be cooled between each stage of compression. If enough stages of such *intercooling* are added, the compressor approaches isothermal operation. Another option is to use a multistage turbine and heat the working fluid between each stage. Again, isothermal heat transfer can be approached. Thus, the cycle diagram will approach the case of two isothermal heat transfer processes separated by two adiabatic (regenerative) constant-pressure processes, which is the Ericsson cycle. The Ericsson cycle has the same efficiency as the Carnot efficiency when operating between the same two temperatures. The Brayton cycle can thus be made to approach the Carnot efficiency, but at the expense of increasing complexity. Of course, real turbine and compressor efficiencies cause even a complex real cycle to depart considerably from the Carnot efficiency.

6.3 Rankine Cycle

I showed him round last week, o'er all—an' at the last says
  he:
"Mister McAndrew, don't you think steam spoils romance
  at sea?"
Damned ijjit! I'd been doon that morn to see what ailed
  the throws,
Manholin', on my back—the cranks three inches off my nose.
Romance! Those first-class passengers they like it very
  well,
Printed an' bound in little books; but why don't poets
  tell?
I'm sick of all their quirks an' turns—the loves an' doves
  they dream—
Lord, send a man like Robby Burns to sing the Song o'
  Steam!
Kipling, M'Andrew's Hymn

Figure 6.20 *Idealized simple Rankine cycle.*

The simple Rankine cycle is the common cycle used in all steam power plants. The Rankine cycle was devised to make use of the characteristics of water as a working fluid and to handle the phase change between liquid and vapor. Of course, many other materials can be chosen instead of water as working fluids. The choice depends on many factors, including the need to accommodate the temperatures of heat transfer to and from the vapor and liquid states while maintaining low vapor pressures in the system.

In an idealized simple Rankine cycle (Fig. 6.20), heat transfer takes place to the working fluid in a constant-pressure process in the boiler. The P-v diagram is shown in Fig. 6.21. Liquid enters the boiler from the pump at a relatively low temperature (state 1) and is heated to saturation along line 1-2. The liquid experiences a slight increase in volume owing to thermal expansion. At point 2, the saturation temperature is reached. Since the boiler is operated at effectively constant pressure, further heat transfer takes place at constant temperature. And the energy added to the working fluid goes into the heat of vaporization, producing vapor until a quality of 100 percent is reached at point 3. Then the vapor is expanded (in an assumed isentropic process for the simple cycle) through a turbine to produce work and leaves at state 4. The expansion is limited in the practical cycle by the appearance of vapor condensation in the turbine and by the saturation pressure available at the temperature of the cooling

Figure 6.21 *Diagram for simple ideal Rankine cycle.*

medium used in the condenser. If excessive condensation is allowed to occur, the condensed liquid droplets will rapidly erode the turbine blades.

At the end of the expansion process (state 4), low-pressure but fairly high-quality vapor leaves the turbine. The vapor is then condensed to a liquid as the vapor comes into contact with surfaces in the condenser that are cooled, usually by a cold-water stream. Because the condenser operates near the temperature of the cooling water, the condensation process occurs at temperatures well below the normal (atmospheric) boiling point of most working fluids. And the condenser pressure on the working fluid side is, therefore, often subatmospheric. The liquid leaves the condenser at state 5.

Figure 6.22 *T-s diagram for the ideal simple Rankine cycle.*

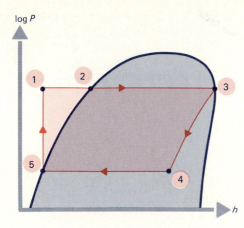

Figure 6.23   *P-h diagram for the simple ideal Rankine cycle.*

Following condensation, the liquid enters a pump. The working fluid is returned to the high pressure necessary for energy addition at the higher boiler temperature, and the cycle is repeated.

Note that the *P-v* diagram used to illustrate the Rankine cycle (Fig. 6.21) is a log-log plot. This is necessary to illustrate the cycle because the specific volume ranges over four decades over the pressure range of interest in steam cycles. However, such a plot greatly distorts the *P-v* behavior of water that would be presented on a linear *P-v* diagram. For example, on a linear plot, the saturated liquid line is essentially vertical, while the saturated vapor line is greatly curved.

The *T-s*, *P-h*, and *h-s* diagrams for the cycle are shown to scale in Figs. 6.22 through 6.24, respectively. Each plot is on a linear scale except for *P* on a *P-h* diagram (see Chap. 3). Only a portion of the *h-s* plot is shown. This allows the boiler-turbine part of the Rankine cycle to be shown in detail.

Note that this is the first cycle which cannot be treated as an air-standard cycle, since the ideal gas law is invalid for saturated vapor and liquid. Rather, the Rankine cycle efficiency is dependent not only on the states of the working fluid but also on the particular working fluid used and its properties.

Figure 6.24   *The h-s diagram for the ideal simple Rankine cycle.*

Example Problem 6.6

Find and compare the efficiency of the simple ideal Rankine cycle engine operating between 20 and 100°C for working fluids of water and refrigerant 12. Compare the results to the Carnot efficiency.

Solution

The *T-s* diagrams for each working fluid are shown. The properties at each state are found from the property tables. Note that all points are for saturated vapor or liquid states except for states 1 and 4. The second law across the ideal turbine and pump is used to fill in the entropy values at states 1 and 4.

State Diagrams

State	P, kPa	T, °C	v, m³/kg	h, kJ/kg	s, kJ/(kg · K)	x
Steam:						
1	101.32†		—	—	s_5	—
2	(101.32)	100	(0.001043)	(418.9)	(1.3062)	0.0
3	(101.32)	100	(1.6699)	(2675.8)	(7.3554)	1.0
4	(2.34)	20	—	—	s_3	—
5	(2.34)	20	(0.001002)	(83.9)	(0.2965)	0.0
Refrigerant 12:						
1	3343.2		—	—	s_5	—
2	(3343.2)	100	(0.0011129)	(314.98)	(1.3476)	0.0
3	(3343.2)	100	(0.003906)	(376.86)	(1.5134)	1.0
4	(567.0)	20	—	—	s_3	—
5	(567.0)	20	(0.0007524)	(218.89)	(1.0663)	0.0

† Entries *not* in parentheses determine the state.

To determine the work output of the cycle, the first law can be written for the ideal turbine as

$$_3w_{4,s} = h_{4,s} - h_3$$

The value of $h_{4,s}$ is not known at this point; however, since $s_{4,s} = s_3$, the quality $x_{4,s}$ can be found from

$$x_{4,s} = \frac{s_{4,s} - s_{l,4}}{s_{lg,4}}$$

and $h_{4,s}$ is given by

$$h_{4,s} = h_{l,4} + x_{4,s}h_{lg,4}$$

For water, this results in $x_{4,s} = 0.8432$, $h_{4,s} = 2154.0$ kJ/kg, and $_3w_{4,s} = -521.8$ kJ/kg.

The heat transfer to the working fluid in process 1-2-3 must be found in two parts. That in the constant-temperature portion 2-3 is easily found from

$$\begin{aligned}_2q_3 &= T_2(s_3 - s_2) \\ &= (100 + 273.15)(7.3554 - 1.3062) \\ &= 2257.3 \text{ kJ/kg}\end{aligned}$$

In portion 1-2, the first law gives

$$_1q_2 = h_2 - h_1$$

To find h_1, the work done by the ideal pump is taken from Eq. (5.138):

$$_5w_1 = v_5(P_1 - P_5) = 0.001002(101.32 - 2.34) = 0.099 \text{ kJ/kg}$$

Now the first law, written for the pump, gives

$$h_1 = h_5 + {_5w_1} = 83.9 + 0.099 = 84.0 \text{ kJ/kg}$$

and the heat transfer to the water in process 1-2 is

$$_1q_2 = 418.9 - 84.0 = 334.9 \text{ kJ/kg}$$

The total energy added to the working fluid during the cycle is composed of the energy to heat the water to the boiling point, process 1-2, and the energy needed to vaporize the water, process 2-3. The total energy added is

$$_1q_3 = {_1q_2} + {_2q_3} = 334.9 + 2257.3 = 2592.2 \text{ kJ/kg}$$

Note that this heat transfer from state 1 to state 3 can also be evaluated by applying the first law to the boiler. This yields

$$_1q_3 = h_3 - h_1 = 2675.8 - 84.0 = 2591.8 \text{ kJ/kg}$$

The cycle efficiency is then

$$\eta_R = -\frac{_3w_4 + _5w_1}{|_1q_3|} = -\frac{-521.8 + 0.099}{2592.2} = 0.201 = 20.1\%$$

A similar analysis can be carried through for the refrigerant 12 cycle; a cycle efficiency of 15.9 percent results.

The various work, heat, and quality terms found in the calculation are gathered in this table:

	Steam	**Refrigerant 12**
$_3w_4$, kJ/kg	-521.8	-26.90
$_1q_2$, kJ/kg	334.9	94.00
$_2q_3$, kJ/kg	2257.3	61.84
$_5w_1$, kJ/kg	0.099	2.09
x_4	0.8432	0.9301
η_R, %	20.1	15.9

A Carnot cycle operating between the same two temperature limits would have an efficiency of

$$\eta_C = 1 - \frac{293}{373} = 0.214 = 21.4\%$$

Comments

In determining the efficiency of the Rankine cycle, the pump work is included in the net work done by the cycle. The pump may not be driven by energy derived from the cycle itself, as was the case, for example, with the compressor in the Brayton cycle. Refer to the table above. The pump work for the steam cycle is very small in comparison with the energy required to heat and vaporize the working fluid. In steam plants in particular, then, pump work can usually be neglected in determining cycle efficiency. For other working fluids, pump work can become significant; and for the refrigerant 12 cycle, it is 7.8 percent of the work done by the turbine.

A further observation is that the simple Rankine cycle efficiency for steam of 20.1 percent approaches the Carnot efficiency of 21.4 percent more closely than does the refrigerant 12 efficiency of 15.9 percent, even though both Rankine cycles work between the same two temperatures. Obviously the choice of working fluid is important. Again, referring to the table of energies, we see that 87.1 percent of the energy added in the steam cycle is added during the constant-temperature vaporization process. For refrigerant 12, only 39.7 percent of the energy is added during vaporization. Thus, for the refrigerant 12 cycle much of the energy is added at temperatures below the maximum. To most closely approach the Carnot efficiency, then, a working fluid should be chosen for which most of the heat transfer is added during vaporization and least is added during the liquid heating process. This requires a fluid with a large enthalpy of vaporization and a small liquid specific heat.

A final observation is that the quality of the working fluid leaving the turbine, x_4, is higher for the refrigerant 12 cycle than for the steam cycle. This is an important turbine design consideration. Normally, exit vapor quality should be greater than 90 percent to avoid excess liquid droplet formation.

6.3.1 Inefficiencies of Real Cycles

The idealized Rankine cycles examined in Example Problem 6.6 approach the Carnot efficiency rather closely. This is not true for actual engines that operate on the Rankine cycle. For real engines, many irreversibilities are present, chiefly in the turbine and pump. Also note that we are studying the *cycle* efficiency here; the efficiency of a power plant is usually defined in terms of the work (or electric energy) output of the plant compared with the energy carried by the fuel entering the *boiler,* not the energy added to the working fluid. This overall plant efficiency is called the *heat rate* of the plant, and it is often expressed in the mixed units of Btu of energy input per kilowatthour of electric output. Since boiler efficiencies may be as low as 60 percent, overall plant efficiencies will be considerably lower than cycle efficiencies.

Treating Turbine and Pump Inefficiencies

The effect on the cycle of turbine and pump irreversibilities can be included through the use of pump and turbine efficiencies, as defined in Sec. 5.10. The method of treatment is most easily shown by an example.

Example Problem 6.7

A Rankine cycle operates with steam under the same conditions as in Example Problem 6.6. The turbine has an efficiency of 70 percent, and the pump efficiency is 60 percent. Find the cycle efficiency, and sketch the *T-s* and *h-s* diagrams.

Solution

The entropy at states 4 and 1 must be modified from the values in Example Problem 6.6, since the pump and turbine can no longer be assumed isentropic. The table is now modified to include the properties for isentropic conditions (noted, as before, with subscript *s*) and for actual conditions (denoted by subscript *a*).

Cycle Diagrams

State	P, kPa	T, °C	v, m³/kg	h, kJ/kg	s, kJ/(kg · K)	x
1s	101.32	—	—	—	s_5	—
1a	101.32	—	—	84.1	—	—
2	(101.32)	100	(0.001043)	(418.9)	(1.3062)	0.0
3	(101.32)	100	(1.6699)	(2675.8)	(7.3554)	1.0
4s	2.34	(20)	—	—	s_3	—
4a	(2.34)	20	—	2310.5	—	(0.9070)
5	(2.34)	20	(0.001002)	(83.9)	(0.2965)	0.0

The actual state points at 1 and 4 are now found by using the definitions of the pump and turbine efficiencies. First, let us examine the turbine. If the turbine were isentropic, the work output would be as found in Example Problem 6.6, or $_3w_{4,s} = -521.8$ kJ/kg. By using the turbine efficiency of 70 percent, the actual work done by the turbine is

$$_3w_{4,a} = -(0.7)(521.8) = -365.3 \text{ kJ/kg}$$

The first law can be used to find $h_{4,a}$ from

$$h_{4,a} = h_3 + {_3w_{4,a}} = 2675.8 - 365.3 = 2310.5 \text{ kJ/kg}$$

and the other properties in state 4 can now be determined. Solving for $x_{4,a}$, for example, gives

$$x_{4,a} = \frac{h_{4,a} - h_{l,4}}{h_{lg,4}} = \frac{2310.5 - 83.9}{2455.0} = 0.9070$$

For the pump, the isentropic work was found previously to be 0.099 kJ/kg, so the actual work is

$$_5w_{1,a} = \frac{0.099}{0.6} = 0.165 \text{ kJ/kg}$$

which results in

$$h_{1,a} = h_5 + {_5w_{1,a}} = 83.9 + 0.165 = 84.1 \text{ kJ/kg}$$

The value of $_1q_{2,\,a}$ is

$$_1q_{2,\,a} = h_2 - h_{1,\,a} = 418.9 - 84.1 = 334.8 \text{ kJ/kg}$$

compared with 334.9 kJ/kg for the ideal cycle. The difference cannot be plotted for point 1 on the cycle diagrams. The cycle efficiency can be determined from

$$\eta_R = -\frac{_3w_{4,\,a} + {}_5w_{1,\,a}}{_1q_2 + {}_2q_3}$$

$$= \frac{365.1}{334.8 + 2257.3} = 0.141 = 14.1\%$$

compared with 20.1 percent for the ideal Rankine cycle and 21.4 percent for the Carnot efficiency between the same two temperatures.

Comments

The cycle efficiency has been reduced, but the change is due to only one factor; the work output of the turbine has been reduced to 70 percent of its ideal value.

Although more work is required to drive the real pump, this additional work increases the enthalpy of the water leaving the pump, which in turn reduces the amount of heat transfer needed to bring the water to state 2. The result is that the net energy *added* to the cycle is actually greater for the ideal than for the real cycle. However, if the cost of energy in the form of work to the pump is higher than that of heat transfer in the boiler, the real cycle energy input will cost more.

Line 3-4a on the diagrams for the real turbine expansions are really not defined, since this is not a quasi-equilibrium process. It is shown on the diagram to indicate the series of states within the turbine, but actually only the final state 4a is defined. The final state is found by application of the first and second laws to the turbine, and knowledge of the intermediate states is not required.

In problems of this type, complete specification of state 2 is not really necessary because only $_1q_3 = h_3 - h_1$ is needed to find the heat transfer to the cycle. Inclusion of state 2 does point out the relative amounts of energy needed to increase the boiler feed temperature to saturation and then to evaporate the saturated liquid.

Note finally that the exit quality for the real cycle is above 90 percent, because the less efficient real turbine did not reduce the exit enthalpy to as low a value as was predicted for the isentropic turbine. As noted earlier, the higher quality is of great practical importance because the presence of liquid droplets at lower qualities causes rapid turbine blade erosion.

6.3.2   Increasing the Rankine Cycle Efficiency

Increased Operating Pressure

From a look at the P-v diagram for the Rankine cycle, it might be inferred that the cycle efficiency could be increased by increasing the boiler pressure, since the area under the *P-v* diagram will be increased. This is generally the case.

A number of interrelated factors, however, determine overall cycle efficiency change with increasing pressure. Let us try an example, building on the data from Example Problem 6.7.

Example Problem 6.8

Determine the cycle efficiency of the Rankine engine in Example Problem 6.7 when the boiler pressure is raised to 4 MPa. Assume that the pump and turbine efficiencies are unchanged. Sketch the *P-v*, *T-s*, *P-h*, and *h-s* diagrams.

Cycle Diagrams

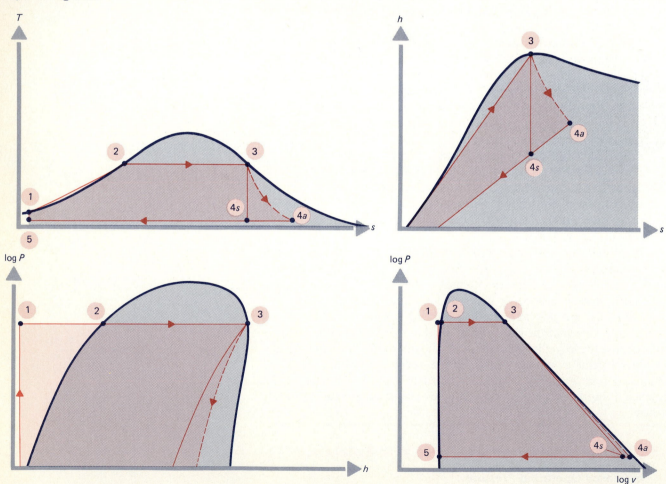

Solution

The solution follows exactly the method of Example Problem 6.7. This table of values results:

State	P, kPa	T, °C	v, m³/kg	h, kJ/kg	s, kJ/(kg · K)	x
$1s$	4000	—	—	—	s_5	—
$1a$	4000	(250.3)	—	90.6	—	—
2	4000	(250.3)	(0.001251)	(1087.2)	(2.7947)	0.0
3	4000	(250.3)	(0.0499)	(2801.1)	(6.0686)	1.0
$4s$	2.34	(20)	—	—	s_3	—
$4a$	(2.34)	20	—	2083.9	—	—
5	(2.34)	20	(0.001002)	(83.9)	(0.2965)	0.0

Some of the entries are not calculated for the table, since they are not required for solution of the cycle efficiency.

The actual energy values for this operating boiler pressure are

$$_3w_4 = -717.2 \text{ kJ/kg}$$

$$_1q_2 = 996.6 \text{ kJ/kg}$$

$$_2q_3 = 1713.9 \text{ kJ/kg}$$

$$_5w_1 = 6.7 \text{ kJ/kg}$$

$$x_4 = 0.8147$$

The cycle efficiency is then

$$\eta_R = -\frac{-717.2 + 6.7}{996.6 + 1713.9} = 0.262 = 26.2\%$$

compared with the efficiency of 14.1 percent for the 1-atm operating boiler pressure.

Comments

The Carnot efficiency for operation between the boiler and condenser temperatures of this cycle would be

$$\eta_C = 1 - \frac{20 + 273.15}{250.3 + 273.15} = 0.439 = 43.9\%$$

The Rankine cycle for the 4-MPa operating pressure thus reaches 59.7 percent of the Carnot efficiency. For the 1-atm steam engine of Example Problem 6.7, the cycle reached $(0.141/0.214)100 = 65.9$ percent of the Carnot efficiency. Thus, increasing the boiler steam pressure has increased the Rankine cycle efficiency from 14.1 to 26.2 percent, but the Rankine cycle has become relatively less efficient when compared with the Carnot efficiency between the same temperatures. This decrease in relative efficiency occurs because at higher pressures the Rankine cycle has a larger fraction of its energy addition in the subcooled heating region, $_1q_2$, rather than in the constant-temperature region, $_2q_3$.

The exit quality of the steam leaving the turbine is also quite low at 81.5 percent, compared with the 90.7 percent quality leaving the turbine of the 1-atm Rankine engine. Such a low quality is not acceptable in practice. Some other method of increasing cycle efficiency must be found aside from simply raising the boiler pressure.

. . . the auld fleet engineer
That started as a boiler-whelp—when steam and he were
   low. I mind the time we used to serve a broken pipe wi' tow!
Ten pound was all the pressure then—Eh! Eh!—a man
   wad drive;
An' here, our working' gauges give one hunder sixty-five!
Kipling, M'Andrew's Hymn

Superheating

Another way to improve Rankine cycle efficiency is to superheat the vapor leaving the boiler (Fig. 6.25). This is usually done by passing the steam through tubes exposed to hot combustion gases or some other energy source at a temperature above the boiler saturation temperature.

Figure 6.25 Rankine engine with super-heater.

The steam entering the turbine thus is at higher enthalpy than was the case for the simple Rankine cycle of Fig. 6.24.

Superheating the steam leaving the boiler has the effect of allowing the quality leaving the turbine to be higher than for the cycle using saturated steam, as illustrated in Example Problem 6.9.

Example Problem 6.9

Determine the cycle efficiency and turbine exit quality of the Rankine engine in Example Problem 6.8 when the boiler exit steam is superheated to 400°C. Sketch the *P-v*, *T-s*, *P-h*, and *h-s* diagrams.

Cycle Diagrams

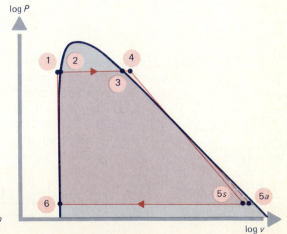

Solution

The superheater is assumed to be a constant-pressure heat transfer device. (This is not true in practice, since a considerable pressure drop owing to friction occurs as the steam passes through the superheater tubes.) The cycle diagrams for the cycle are shown, and these are the state point values:

State	P, kPa	T, °C	v, m³/kg	h, kJ/kg	s, kJ/(kg · K)	x
$1s$	4000	—	—	—	s_6	—
$1a$	4000	(250.3)	—	90.6	—	—
2	4000	(250.3)	(0.001251)	(1087.2)	(2.7947)	0.0
3	4000	(250.3)	(0.0499)	(2801.1)	(6.0686)	1.0
4	4000	400	(0.073377)	(3214.1)	(6.7699)	—
$5s$	2.34	(20)	—	—	s_4	—
$5a$	(2.34)	20	—	2351.7	—	(0.9237)
6	(2.34)	20	(0.001002)	(83.9)	(0.2965)	0.0

The energy values become

$$_4w_5 = -862.4 \text{ kJ/kg} \qquad _3q_4 = 413.0 \text{ kJ/kg}$$

$$_1q_2 = 996.6 \text{ kJ/kg} \qquad _6w_1 = 6.7 \text{ kJ/kg}$$

$$_2q_3 = 1713.9 \text{ kJ/kg} \qquad x_{5,a} = 0.9237$$

The cycle efficiency is then

$$\eta_R = -\frac{-862.4 + 6.7}{996.6 + 1713.9 + 413.0} = 0.274 = 27.4\%$$

compared with the engine of Example Problem 6.8, which used saturated steam from the boiler and no superheat to attain an efficiency of 26.2 percent.

Comments

Superheating usually has a marginal effect on cycle efficiency, but it does allow an increase in turbine output per unit mass of steam flow because the enthalpy of the steam entering the turbine has been increased.

Much of the energy added to the working fluid in the cycle with superheating occurs over a range of temperatures, and it is difficult to decide how to compare with the efficiency of a Carnot engine. If the maximum temperature in the cycle, 400°C, is used to find the Carnot efficiency,

$$\eta_C = 1 - \frac{20 + 273}{400 + 273} = 0.565 = 56.5\%$$

Figure 6.26   Ideal reheat in Rankine cycle.

and we have reached $(27.4/56.5)100 = 48.5$ percent of the Carnot efficiency. The quality of the steam leaving the turbine has reached 92.37 percent, providing better conditions for practical design than the 81.5 percent value for the system of Example Problem 6.8.

Reheating

To further increase Rankine cycle efficiency, the steam passing through the turbine can be reheated. In the simplest form, this process can be carried out as indicated in the system diagram of Fig. 6.26. Here, a two-stage turbine is used. As the steam leaves the high-pressure stage, it is piped back to the superheater and reheated before returning to the second (low-pressure) stage of the turbine for further expansion. A major benefit of reheating is that the quality of steam leaving each stage of the turbine can be maintained at a high value, and much higher boiler pressures can be used.

Example Problem 6.10

Examine the effect on cycle efficiency of adding reheat to the cycle using the basic system described in Example Problem 6.9. Assume that the outlet pressure from the high-pressure turbine is 1000 kPa and that the steam is reheated to the original superheat temperature of 400°C. So that we can compare the results to the cycle without reheat, assume also that each stage of the turbine has an efficiency of 70 percent. Sketch the cycle on P-v, T-s, P-h, and h-s diagrams on the top of the next page.

Solution

The state points for the reheat cycle in this problem are as follows:

State	P, kPa	T, °C	v, m³/kg	h, kJ/kg	s_a, kJ/(kg · K)	x
$1s$	4000	—	—	—	s_6	—
$1a$	4000	(250.3)	—	90.6	—	—
2	4000	(250.3)	(0.001251)	(1087.2)	(2.7947)	0.0
3	4000	(250.3)	(0.0499)	(2801.1)	(6.0686)	1.0
4	4000	400	(0.073377)	(3214.1)	(6.7699)	—
$5s$	1000	—	—	(2865.6)	s_4	—
$5a$	1000	—	—	2970.1	(6.9750)	—
$5'$	1000	400	—	(3262.7)	(7.4633)	—
$5''s$	2.34	(20)	—	—	$s_{5'}$	—
$5''a$	(2.34)	20	—	2508.5	—	(0.985)
6	(2.34)	20	(0.001002)	(83.9)	(0.2965)	0.0

The conditions at state 5 are known for the isentropic case, and the turbine efficiency can be used to find the actual outlet conditions of the high-pressure turbine. The result of these calculations for $P_5 = 1000$ kPa and $s_{5,s} = 6.7699$ kJ/(kg · K) is $h_{5,s} = 2865.6$ kJ/kg and $_4w_{5,s} = -348.5$ kJ/kg. Using the turbine efficiency of 70 percent then gives the actual work output of the turbine as -244.0 kJ/kg. The actual outlet conditions from the high-pressure turbine are then found to be $h_{5,a} = 2970.1$ kJ/kg and $s_{5,a} = 6.9750$ kJ/kg. This is indicated in the above table.

The heat transfer required to raise the temperature to 400°C along line 5a-5' is

$$_5q_{5'} = h_{5'} - h_{5,a} = 3262.7 - 2970.1 = 292.6 \text{ kJ/kg}$$

The usual calculation for the ideal turbine is now applied to the low-pressure turbine, and the actual turbine outlet conditions at state 5″ are found by again using the turbine efficiency. The actual work output of the low-pressure turbine is found to be $_{5'}w_{5''} = -754.0$ kJ/kg. Thus the energy values for the reheat cycle are

$$_4w_5 = -244.0 \text{ kJ/kg} \qquad _3q_4 = 413.0 \text{ kJ/kg}$$

$$_{5'}w_{5''} = -754.0 \text{ kJ/kg} \qquad _5q_{5'} = 292.6 \text{ kJ/kg}$$

$$_1q_2 = 996.6 \text{ kJ/kg} \qquad _6w_1 = 6.7 \text{ kJ/kg}$$

$$_2q_3 = 1713.9 \text{ kJ/kg} \qquad x_{5'',a} = 0.9877$$

Cycle Diagrams

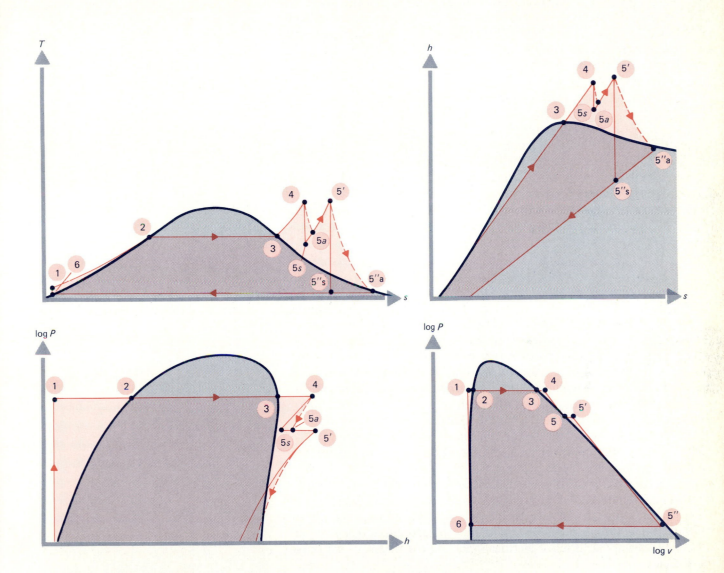

The work output of the combined turbine is 998.0 kJ/kg. The cycle efficiency is

$$\eta_R = -\frac{_4w_5 + {_5}'w_{5''} + {_6}w_1}{_1q_2 + {_2}q_3 + {_3}q_4 + {_5}q_{5'}}$$

$$= \frac{244.0 + 754.0 - 6.7}{996.6 + 1713.9 + 413.0 + 292.6} = 0.290 = 29.0\%$$

Comments

Adding reheat to the cycle has increased the efficiency of the Rankine cycle with superheat from 27.4 to 29.0 percent, or to 51.3 percent of the efficiency of a Carnot engine operating between the maximum and minimum temperatures of the steam cycle. Such an increase does not always occur; certain extraction pressures may actually cause a decrease in cycle efficiency.

Reheating has several advantages aside from the possible increase in cycle efficiency. A major advantage is that the turbine stages in the system each work over a smaller range of pressure. It is easier to design these individual turbines to operate at high turbine efficiency than to design a single high-efficiency turbine that must handle the entire expansion process. In large power plants, multiple turbine stages with reheat between each stage are used to increase turbine and cycle efficiency.

The outlet quality from the turbines exceeds the exit quality of 92.4 percent found at the outlet of the turbine in Example Problem 6.9, where no reheat was included.

Regeneration

As noted earlier in this chapter, the Rankine cycle efficiency is lower than the efficiency of a corresponding Carnot engine for a number of reasons, but one of the major deviations results because energy must be added to the liquid phase in order to bring the cool working fluid leaving the pump to the saturation temperature. Further isothermal heat transfer can then occur in the boiler at saturation conditions. The heat transfer to the liquid leaving the condenser and pump occurs over a range of temperatures. Thus a portion of the cycle heat addition is always at nonisothermal conditions, and so the Carnot efficiency cannot be reached.

It is possible, however, to use some of the energy carried by the working fluid after it leaves the boiler to heat the liquid phase. Then, a larger *fraction* of the heat transfer added to the working fluid from *external* sources will be added during vaporization in the boiler, which takes place at isothermal conditions. The various methods of using energy from one portion of the cycle to add to another portion are called *regeneration*. We have seen applications of this idea in the Stirling, Ericsson, and Brayton cycles.

In the Rankine cycle, the only energy available at temperatures that will allow a heat transfer to the liquid in or leaving the condenser must come from vapor passing through the turbine. Let us examine the ideal case in which exactly enough energy is extracted from the turbine vapor to completely preheat the working fluid from the condenser exit conditions to saturated liquid at the boiler pressure. In such a case, all heat transfer to the cycle (if there is no superheating) will take place isothermally; and if all components in the system could be made reversible, the Rankine cycle would have the same efficiency as the Carnot cycle.

Figure 6.27 Rankine cycle with ideal regeneration.

One possible method is to take working fluid leaving the condenser and pump it through cooling channels in the turbine in such a way that reversible heat transfer occurs (Fig. 6.27). The *T-s* diagram for such a process is shown in Fig. 6.28.

The temperature and entropy changes for the vapor and the regenerated liquid passing through the turbine will yield the same reversible heat transfer process. It follows that the heat transferred from the vapor in the turbine (area under 4-5) is exactly equal to the energy gained by the liquid (area under 1-2). The heat transfer from the boiler to the working fluid takes place isothermally at T_{sat} in the boiler except in the superheat region, since all preheating of the fluid has been done regeneratively.

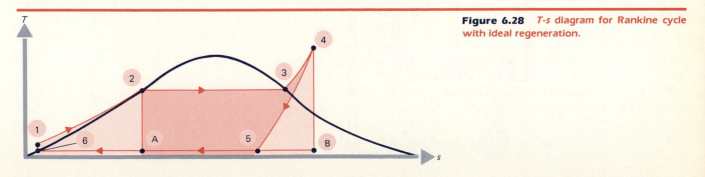

Figure 6.28 *T-s* diagram for Rankine cycle with ideal regeneration.

Thus, since heat rejection in the condenser is also an isothermal process, the cycle efficiency approaches the Carnot efficiency. If no superheating is present, the efficiency for this idealized cycle is equal to the Carnot efficiency.

Nature is not so obliging as to make such a system feasible in actual applications. It is not possible to design a turbine so that the heat transfer process is truly reversible; in practice, the heat transfer to the boiler feed is done externally to the turbine. Usually, only some portion of the turbine vapor is bled from one or more locations at different pressures from various turbine stages and is passed through one or more *feedwater heaters,* as shown in Fig. 6.29. The quality of the vapor-liquid mixture leaving the turbine would be considerably reduced if all the vapor were used for regeneration, an undesirable effect, as noted earlier.

Types of Feedwater Heaters

The type of feedwater heater shown in Fig. 6.29 is called a *closed* feedwater heater in which the bleed steam is not in direct contact with the fluid leaving the pump. Rather, the steam is condensed on the outer

Figure 6.29 Rankine cycle with regeneration by use of a feedwater heater.

surface of tubes carrying the water leaving the pump, and the bleed condensate is simply pumped back to the condenser entrance and mixed with the turbine exit steam, which slightly reduces the required condenser heat rejection.

In some feedwater heaters, the bleed steam is simply mixed directly with the water leaving the pump and enters the boiler feed stream at that point. Such a system is called an *open feedwater heater.* These heaters offer the advantage that air or other dissolved gases are usually driven out of the feed stream by the heating process and can be removed by an appropriate pump. Removal of inert gases aids in condenser performance and reduces corrosion in the system.

Example Problem 6.11

The Rankine cycle with superheat of Example Problem 6.9 is modified by the addition of a single closed feedwater heater. For 1 kg of mass flowing into the turbine, 0.1 kg is bled from the turbine at 1000 kPa and sent to a closed feedwater heater. Assume that the turbine has a 70 percent efficiency at every pressure. Find the cycle efficiency, and compare it with the efficiency of the cycle without regeneration.

Cycle Diagram

Solution

For the system shown in Fig. 6.29, the bleed flow is shown by the dashed line $5'a\text{-}1'$ on the $T\text{-}s$ diagram. The line $1\text{-}2\text{-}3\text{-}4\text{-}5'a\text{-}5a\text{-}6\text{-}1$ shows the main flow through the turbine and the remainder of the Rankine cycle. The flows combine along path $1'\text{-}2\text{-}3\text{-}4\text{-}5'$. The state table is on the next page.

The work output of the turbine between states $5'$ and 5 *per unit mass of fluid passing through the boiler* $(\dot{m}_5 = 0.9\dot{m}_4)$ is

$$_5w_{5'} = 0.9(h_{5a} - h_{5'a}) = 0.9(2351.7 - 2970.1) = -556.6 \text{ kJ/kg}$$

State	P, kPa	T, °C	v, m³/kg	h, kJ/kg	s, kJ/(kg · K)	x
1	4000	—	—	90.6	—	—
1′	4000	—	—	—	—	—
2	4000	(250.3)	(0.001251)	(1087.2)	(2.7947)	0.0
3	4000	(250.3)	(0.0499)	(2801.1)	(6.0686)	1.0
4	4000	400	(0.073377)	(3214.1)	(6.7699)	—
5′s	1000	—	—	(2865.6)	s_4	—
5′a	1000	—	—	2970.1	(6.9750)	—
5s	2.34	(20)	—	—	s_4	—
5a	2.34	(20)	—	2351.7	—	(0.9237)
6	(2.34)	20	(0.001002)	(83.9)	(0.2965)	0.0

It is now necessary to find the enthalpy of the feedwater at state 1′, so that the heat transfer still necessary to bring the feedwater to saturation, $_1q_2$, can be determined. This is done by writing the first law for the feedwater heater and assuming the outlet water to the boiler and the condensed bleed steam reach the same temperature, which is an idealization. For an adiabatic feedwater heater, this gives

$$\dot{m}_{5'}(h_{5'a} - h_{1'}) = \dot{m}_1(h_{1'} - h_1)$$

or

$$h_{1'} = \frac{0.1 h_{5'a} + h_1}{1.1} = \frac{(0.1)2970.1 + 90.6}{1.1} = 352.4 \text{ kJ/kg}$$

and

$$_1q_2 = h_2 - h_{1'} = 1087.2 - 352.4 = 734.8 \text{ kJ/kg}$$

The energy values are now (per unit of mass flow through the boiler)

$$_4w_{5'} = -244.0 \text{ kJ/kg} \qquad _3q_4 = 413.0 \text{ kJ/kg}$$
$$_{5'}w_5 = -556.6 \text{ kJ/kg} \qquad _6w_1 = 6.7 \text{ kJ/kg}$$
$$_1q_2 = 734.8 \text{ kJ/kg} \qquad x_{5,a} = 0.9237$$
$$_2q_3 = 1713.9 \text{ kJ/kg}$$

The cycle efficiency is then

$$\eta_R = -\frac{_4w_{5'} + _{5'}w_5 + _6w_1}{_1q_2 + _2q_3 + _3q_4}$$

$$= \frac{244.0 + 556.6 - 6.7}{734.8 + 1713.9 + 413.0} = 0.277 = 27.7\%$$

This compares with 27.4 percent for the cycle with superheat but no regeneration.

Comments
Is 10 percent of the flow stream a reasonable value to bleed off for the

regeneration stream? The amount depends on the other conditions in the cycle such as boiler pressure and condenser temperature; the bleed flow is usually found by specifying the desired feedwater heater outlet condition. Note that the outlet quality from the turbine is quite low; it is generally advantageous to combine reheat with regeneration to avoid low turbine exit quality.

Other Rankine Cycle Inefficiencies

The Rankine cycle usually does not attain the efficiencies predicted by the example problems presented here. Although the examples include the turbine and pump efficiencies and thus account for many real losses in the Rankine cycle, other important factors have been ignored. For example, it is difficult to design a condenser that will produce saturated liquid at the outlet. Real condensers are designed to subcool the liquid; this allows the system to operate in off-design conditions without passing vapor through the condenser and avoids having nearly saturated liquid enter the pump. If the latter is allowed, the pump can be destroyed by *cavitation,* an effect created when boiling and recondensation at low-pressure points behind the pump impeller rapidly erode the impeller.

Further inefficiency is caused by the frictional losses in pumping the working fluid through the cycle, heat losses from the high-temperature components in the system to the environment, reduced condenser performance because of air leakage into the subatmospheric condenser, and energy requirements to operate auxiliary equipment such as combustion air fans and other blowers.

Other Modifications to the Rankine Cycle

Modern Rankine cycle power plants operate with both reheat and regeneration and at as high a pressure as present mechanical design and materials properties allow. Some present steam cycles are designed to operate at pressures above the critical point. A diagram of a typical power plant cycle using reheat and regeneration is shown in Fig. 6.30.

Attention is also being paid to combined cycles of various types. For example, the rejected energy from a Brayton engine (gas turbine) can be used as the energy input to a Rankine engine, so that some fraction of the energy rejected by the gas turbine is recovered as useful work. The energy rejected by a well-engineered Rankine cycle using steam for power generation is usually at such a low temperature that its usefulness in driving another cycle is limited. However, by increasing the heat rejection temperature and lowering the power output, the waste heat can be used for space heating, plant steam, or other low-temperature application. Such combined power-heat cycles are called *cogeneration* systems. The economics of such systems depend on the particular use and value of the waste heat.

Figure 6.30 Rankine fossil-fuel power cycle with single reheat and eight-stage regeneration, $P = 24.2$ MPa (3515 psia), $T = 538°C$ (1000°F) steam. (From *Steam, Its Generation and Use,* 39th ed., Babcock and Wilcox Company, 1978, used with permission.)

6.4 Internal Combustion Cycles

We now return to consideration of air-standard versions of cycles, and we turn our attention to cycles that gain their energy from the combustion of fuel within the engine. Such engines are treated in their air-standard version as if the energy were introduced from an external source. Then the process of heat addition is approximated as one of the standard types treated previously, i.e., via constant temperature, constant volume, etc.

Intake   Exhaust

Fuel

Regenerator

Combustor

Turbine

−w

Compressor

Figure 6.31   Internal combustion version of Brayton cycle.

6.4.1   Brayton Cycle (Internal Combustion)

If energy is added to the working fluid of the Brayton cycle by the combustion of a fuel rather than by heat transfer in a heat exchanger, then the air-standard cycle analysis is effectively unchanged from that shown in Figs. 6.13 through 6.15. However, the cycle must now be treated as an open cycle. The working fluid is replaced during each cycle, so the exhaust gases are purged, and new oxidant is introduced for further combustion to occur. In addition, the hardware used to carry through the cycle will be much different than for the closed cycle. An example of the possible hardware arrangement is shown in Fig. 6.31.

The open Brayton cycle with internal combustion is the cycle used to analyze gas-turbine engines. In most stationary Brayton engines, the turbine work output is designed to be a maximum and thus exceed the required compressor work input. The excess shaft work is then used to drive, for example, an electric generator or pipeline compressor. The total energy of the gases leaving the turbine is minimized so that the turbine work output is maximized.

Another possible design is to *maximize* the energy of the exhaust gases, taking only enough energy that the turbine work is just sufficient to drive the compressor. The remaining energy of the exhaust gases is then used for propulsion, usually by ejecting the gases through a reaction nozzle. A compromise between the design for the jet engine and the stationary plant turbine is the turboprop engine. This engine uses much of the energy of the hot gases from the combustor for turbine work, to turn a propeller as well as to power the compressor, but also uses the remaining energy of the exhaust gases to provide additional thrust.

It is possible to use regeneration, reheat, and intercooling on stationary internal combustion Brayton engines, as discussed in Sec. 6.2.4. These efficiency-improving techniques can also be employed in propulsion turbines, but the increase in complexity and weight must be carefully analyzed to assess the net benefit.

6.4.2   Air-Standard Otto Cycle

The Otto cycle is an idealization of the cycle used in gasoline engines, although its original use was in engines using natural gas or other gaseous fuels. The air-standard Otto cycle deviates substantially from the cycle realized in practice, but it is still instructive to examine the idealized cycle.

In the so-called two-stroke Otto cycle used in many small gasoline engines, the cycle is as shown in Fig. 6.32 for the *P-v* diagram and in Figs. 6.33 and 6.34 for the *T-s* and *P-h* diagrams. In these diagrams, at state point 1, the piston begins the compression stroke, assumed to be an isentropic compression process, and compression is continued until the piston reaches the top of its stroke at point 2. The combustion process is then assumed to occur instantaneously and is modeled as raising the pressure by heat addition at constant volume to state 3. The piston begins the power stroke, again isentropically, and the expansion continues to point 4, where the piston reaches the limit of its travel. The exhaust valve opens, reducing the pressure instantaneously to state 1, while the cylinder volume remains constant. This process is modeled as a constant-volume heat rejection. The fuel-air charge for the next cycle is then assumed to enter the cylinder at state 1, and the cycle is repeated.

In the four-stroke air-standard cycle, illustrated in Fig. 6.35, the piston compresses the working fluid from state 1 until it reaches top dead center, state 2, and then combustion occurs instantaneously. The pressure thus rises at constant volume to point 3. The piston then descends in an isentropic expansion on the power stroke to point 4, as in the two-stroke engine. Now, the exhaust valve opens, and the pressure drops at constant volume to state 1. The piston begins its upward travel, purging the cylinder of combustion products. This exhaust, or "scavenging," stroke occurs at essentially constant pressure between states 1 and 1'. At top dead center, the exhaust valve closes, and the intake valve opens. The piston now descends, again at effectively constant pressure, and a new charge of unburned fuel and air is drawn into the cylinder on the intake stroke between states 1' and 1. Note that, as indicated on the *P-v* diagram, no net work is done on the combination of the exhaust and intake strokes. The intake valve closes at state 1, and the piston begins the upward compression stroke to state 2, continuing the cycle.

The efficiency of the ideal air-standard Otto cycle, either two- or four-stroke, is easily found by

Figure 6.32   *P-v diagram for two-stroke Otto cycle.*

Figure 6.33   *T-s diagram for two-stroke Otto cycle.*

$$\eta_O = 1 - \frac{|Q_L|}{|Q_H|}$$

$$= 1 - \frac{mc_v(T_4 - T_1)}{mc_v(T_3 - T_2)}$$

For the isentropic processes,

$$\frac{T_4}{T_3} = \left(\frac{v_3}{v_4}\right)^{k-1} = \left(\frac{v_2}{v_1}\right)^{k-1} = \frac{T_1}{T_2}$$

and substituting the resulting ratio of temperatures into the efficiency relation gives

$$\eta_O = 1 - \frac{T_4}{T_3} = 1 - \left(\frac{v_3}{v_4}\right)^{k-1} = 1 - \frac{1}{r_v^{k-1}} \tag{6.16}$$

where r_v is the compression ratio v_4/v_3.

The efficiency of the ideal air-standard Otto cycle (either two- or four-stroke, since the additional two strokes in the four-stroke cycle do not change the thermodynamics of the ideal air-standard cycle) is thus dependent on only the compression ratio of the engine.

Figure 6.34   *P-h diagram for two-stroke Otto cycle.*

6.4.3   Air-Standard Diesel Cycle

The diesel cycle is an attempt to approach the Carnot efficiency as closely as possible in an internal combustion engine. Again, the actual cycle differs considerably from the behavior predicted by the air-standard version.

The P-v diagram for the ideal air-standard diesel cycle is shown in Fig. 6.36. At state 1, the piston begins an isentropic compression of the combustion air and continues to state 2 where fuel is injected into the air. If a sufficiently high compression ratio is used, the temperature of the compressed air is high enough that the fuel will ignite without an external ignition source. The ignition process is assumed to occur at such a rate that the energy addition to the cylinder is just sufficient to hold the pressure in the cylinder constant while the piston moves, until state 3 is reached. The combustion process is assumed to end at this point, and the expansion continues as an isentropic process to point 4. The exhaust valve then opens, and the pressure drops to the initial condition at state 1. If the four-stroke cycle is used, then scavenge or intake strokes occur between states 1 and 1', at essentially constant pressure.

The diesel cycle efficiency is

$$\eta_D = 1 - \frac{|Q_L|}{|Q_H|} = 1 - \frac{c_v(T_4 - T_1)}{c_P(T_3 - T_2)} \tag{6.17}$$

Figure 6.35   *P-v diagram for four-stroke Otto cycle.*

Noting that $k = c_P/c_v$ and rearranging, we find

$$\eta_D = 1 - \frac{T_1}{kT_2}\frac{T_4/T_1 - 1}{T_3/T_2 - 1} \qquad (6.18)$$

If the compression ratio v_1/v_2 is denoted r_1 and the expansion ratio v_4/v_3 is denoted r_2, then the diesel ideal air-standard efficiency can be written as

$$\eta_D = 1 - \left(\frac{1}{r_1}\right)^{k-1}\frac{1}{k}\frac{(r_1/r_2)^k - 1}{r_1/r_2 - 1} \qquad (6.19)$$

The compression ratio of the diesel cycle is always greater than the expansion ratio (see Fig. 6.36). So for a given compression ratio r_1, the diesel engine always has lower efficiency than an Otto engine operating at the same compression ratio. The closer r_2 is made to r_1, the closer the diesel efficiency will approach that of the Otto engine.

Figure 6.36 *P-v diagram for diesel cycle.*

Example Problem 6.12

Compare the efficiencies of an Otto and a diesel engine, each operating at a compression ratio of 15. Take the state at the beginning of each compression stroke to be at 1 atm and 20°C, and assume the diesel engine has an expansion ratio of 9.

Solution
For the Otto cycle, the efficiency is simply

$$\eta_O = 1 - \frac{1}{(15)^{k-1}} = 1 - \frac{1}{15^{0.4}} = 0.661 = 66.1\%$$

from Eq. (6.16). For the diesel cycle, the efficiency is given by Eq. (6.19) as

$$\eta_D = 1 - \left(\frac{1}{15}\right)^{0.4}\left(\frac{1}{1.4}\right)\frac{(\frac{15}{9})^{1.4} - 1}{\frac{15}{9} - 1}$$

$$= 0.621 = 62.1\%$$

Comments
If the diesel engine is supposed to be more efficient in practice than other internal combustion engines, why is its efficiency less than that of an Otto engine at these conditions? Part of the answer is that the Otto engine cannot be operated at high compression ratios, since the temperature rise during compression of the fuel-air mixture will cause it to ignite before the piston reaches the top of its travel. This preignition (pinging) will damage the engine. In the diesel engine, this problem is avoided, since the air alone is compressed and the fuel is injected after compression occurs. Thus the diesel engine can be operated at compression ratios where its efficiency is better than Otto engines, which are limited in the compression ratio that can be practically used.

Cycle Diagrams

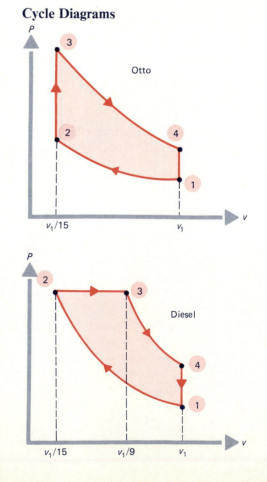

6.5 Refrigeration, Air Conditioning, and Heat Pump Cycles

It is possible to have a heat transfer from a lower- to a higher-temperature reservoir by doing work on a system. Common domestic air conditioners and heat pumps work by implementing this statement of the second law in a practical device using a *vapor-compression cycle*. By combining a heat engine with a vapor-compression cycle, a heat transfer can be used as the driving energy to the heat engine, which in turn drives the vapor-compression cycle. Thus, heat transfer rather than work can be used to provide the energy to drive the cycle. Other methods of using energy to provide cooling are also possible. A few are addressed in this section.

Systems that cool and dehumidify air for commercial or domestic structures are called *air conditioners;* those that provide cold water or other liquid are usually referred to as *chillers.* Systems that provide space heating by a heat transfer from a low-temperature reservoir and using a cycle that provides a heat transfer at the desired temperature are called *heat pumps.*

6.5.1 Coefficient of Performance for Air Conditioners and Chillers

The measure of performance for air-conditioning and chilling systems is the *coefficient of performance* (COP). COP is defined in Sec. 5.11 as the ratio of energy removed from the environment in the form of a heat transfer (the cooling provided) to the energy input. The energy input, E_{in}, to the system may be in the form of work or heat transfer and should include the energy necessary to drive all auxiliary equipment such as blowers, pumps, backup heaters, etc. The definition is then

$$\text{COP} = \frac{|q_L|}{|E_{in}|} \tag{6.20}$$

In the United States, a related figure of merit is the *energy efficiency rating* (EER). This is simply COP expressed in mixed units of Btu per hour of cooling per watt of energy input, resulting in the relation EER = $3.41 \times$ COP. The higher the COP or EER of a given system, the better its efficiency.

The cooling capacity of air-conditioning systems in the United States is often given in *tons.* One ton is the cooling rate necessary to freeze 1 ton (2000 lbm) of water in 1 day, which is 12,000 Btu/h or 3.517 kW.

For a reversible refrigeration or air-conditioning cycle that removes energy from a low-temperature reservoir at T_L and rejects the energy to a reservoir at a higher temperature T_H,

$$(\text{COP})_{rev} = \frac{|q_L|}{|E_{in}|} = \frac{|q_L|}{|q_H| - |q_L|} = \frac{1}{T_H/T_L - 1} \tag{6.21}$$

which is the ideal COP for a reverse Carnot cycle. All real systems operating between constant-temperature reservoirs must have a lower COP than predicted by Eq. (6.21).

Heat Pumps

Another use of the cycles described above is to take energy from a low-temperature source and make it available at a higher temperature by providing work to the cycle. In many cases, the work input is a small fraction of the energy made available at the high temperature; the remaining energy is drawn from the low-temperature reservoir. The figure of merit for such systems, called *heat pumps,* is again called the COP_{HP}. However, the definition is now changed to be the ratio of energy rejected in the form of a heat transfer at the high temperature to the energy input to the cycle in the form of work, or

$$COP_{HP} = \frac{|q_H|}{|w|} \qquad (6.22)$$

For a reversible heat pump cycle as in Sec. 5.11,

$$COP_{HP, rev} = \frac{|q_H|}{|q_H| - |q_L|} = \frac{1}{1 - T_L/T_H} \qquad (6.23)$$

which is the upper limit for real heat pump systems.

6.5.2 Vapor-Compression Systems

The most common method of providing air conditioning and chilling as well as heat pumping is probably by the vapor-compression cycle. First, the use of the cycle in air conditioning and chilling is described.

The cycle hardware is shown in Fig. 6.37, and the *P-h* and *T-s* diagrams are seen in Figs. 6.38 and 6.39. In this cycle, the working fluid is initially a saturated or slightly superheated vapor at a relatively low pres-

Figure 6.37 Vapor-compression air-conditioning cycle.

sure (state 1), and it is compressed to a high pressure (state 2), where it is condensed to a saturated or slightly subcooled liquid (state 3). Condensation may occur either by heat loss to the surroundings if the saturation temperature of the condensing vapor is sufficiently greater than that of the surroundings or by the use of a cooling tower to provide a lower-temperature heat sink. The condensed liquid is then expanded adiabatically through a throttling valve to a low pressure (state 4), where it becomes a mixture of saturated liquid and vapor of low quality. Its saturation temperature at this pressure is low, and a heat transfer from the surroundings can be used to evaporate the liquid fraction, thus providing the cooling effect and leaving the working fluid in the initial state 1. The resulting saturated or slightly superheated vapor is then fed to the compressor, completing the cycle.

The working fluid for this cycle is chosen so that the heat rejection in the condenser can take place at pressures obtainable from economical compressors, and the evaporation temperature in the evaporator can be easily compatible with the inlet pressure to the compressor while allowing efficient heat transfer from the surroundings. For residential and commercial air conditioners, various fluorinated hydrocarbons have the necessary properties; where lower temperatures are required, such as in commercial freezing equipment, ammonia is often used.

The vapor-compression heat pump cycle is exactly as shown in Figs. 6.37 through 6.39; only the objective of the cycle is changed. Now, the evaporator is placed outdoors where there is a heat transfer to the working fluid at environmental temperatures (i.e., heat transfer from the environment). The evaporator may be placed in conjunction with any energy source that is at a lower temperature than desired from the cycle. In domestic heating systems, the heat transfer can be obtained from the outdoor air (an *air-source* heat pump) or from a convenient body of water such as a lake, river, or well (a *water-source* heat pump). The condenser is placed where the higher-temperature heat transfer being rejected from the cycle is to be used. This can be in the residence for heating or in a heat exchanger in an industrial process.

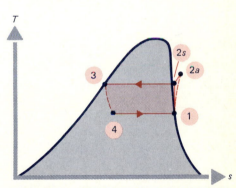

Figure 6.38   *P-h diagram for the vapor-compression cycle.*

Figure 6.39   *T-s diagram for vapor-compression cycle.*

Example Problem 6.13

Find the COP for a vapor-compression cycle that uses refrigerant 12 as the working fluid and is designed to operate at an evaporator temperature of 10°C and a condenser temperature of 40°C. The compressor efficiency is 65 percent.

Solution

Assume that the vapor entering the compressor is saturated. Two properties are known for each state. Now, by using the compressor efficiency, all properties in state 2 can be calculated. For the adiabatic expan-

Cycle Diagrams

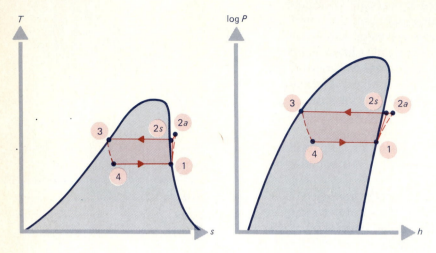

sion between states 3 and 4, the first law indicates that the enthalpy remains constant; this fact can be used along with the known final temperature to calculate all properties in state 4. These calculations result in this state table:

State	P, kPa	T, °C	v, m³/kg	h, kJ/kg	s, kJ/(kg · K)	x
1	(423.1)	10	(0.040934)	355.73	(1.5504)	1.0
2s	960.3	—	—	—	s_1	—
2a	960.3	(53.5)	—	377.74	(1.5743)	—
3	960.3	40	(0.00080)	(238.71)	(1.1311)	0.0
4	423.1	10	(0.0088)	238.71	(1.1372)	(0.201)

Now the COP can be computed:

$$\text{COP} = \frac{|q_L|}{|w|} = \frac{h_1 - h_4}{h_{2,a} - h_1}$$

$$= \frac{355.73 - 238.71}{377.74 - 355.73} = 5.317$$

So for every unit of work done on the compressor, this cycle can remove 5.32 units of heat transfer from the evaporator. The ideal COP [Eq. (6.21)] is

$$\text{COP}_{\text{rev}} = \frac{1}{T_H/T_L - 1} = \frac{1}{(273.15 + 53.5)/(273.15 + 10) - 1} = 6.509$$

where T_H has been taken as the maximum cycle temperature, not the condenser temperature.

Comments

The vapor compression system for cooling is relatively efficient in the sense that each unit of work input can provide more than one unit of cooling. Because of the various heat transfers in the system, frictional losses in the working fluid, electric motor efficiencies, etc., the usual COP values of commerically marketed systems range up to about 3.

Example Problem 6.14

Find the COP of the system described in Example Problem 6.13 when it is operated as a heat pump.

Solution

The final state table of Example Problem 6.13 applies, and the COP is given by

$$\mathrm{COP}_{\mathrm{HP}} = \frac{|q_H|}{|w|} = \frac{h_{2,a} - h_3}{h_{2,a} - h_1}$$

$$= \frac{377.74 - 238.71}{377.74 - 355.73} = 6.317$$

And now 6.32 units of heating at the condenser can be obtained at the expense of each unit of work to the compressor.

The ideal $\mathrm{COP}_{\mathrm{HP}}$ is [Eq. (6.23)]

$$\mathrm{COP}_{\mathrm{HP, rev}} = \frac{1}{1 - T_L/T_H} = \frac{1}{1 - 283.15/326.65} = 7.509$$

Comments

Note that this system is also efficient as a heat pump. Because of the definitions of COP for cooling and heating modes, it is also true in all cases that [see Eq. (5.175)]

$$\mathrm{COP}_{\mathrm{HP}} - \mathrm{COP} = 1 \tag{6.24}$$

which is seen to be true for this example.

As for the cooling mode, the heat pump described here will have significant deviations from the ideal calculated behavior.

When a vapor-compression cycle is used both as a heat pump and for cooling, as is common in residential and commercial heating, ventilating, and air-conditioning (HVAC) systems, the performance is usually somewhat below the cycle performance that could be obtained by two separate vapor-compression units. This occurs because the evaporator and condenser coils of the combined system are usually sized for best performance in the air-conditioning mode (which is often the primary use) and are thus somewhat off-design for the heat pumping mode of operation.

Figure 6.40 Open Brayton cooling cycle.

6.5.3 Other Cooling Cycles Driven by Work Input

All the external energy source engine cycles discussed earlier in this chapter can be envisioned as operating with each process in the cycle proceeding in the reverse direction. For example, the vapor-compression cycle is closely related to the Rankine cycle operated with each of its processes reversed. However, the boiler of the Rankine cycle has been replaced by the throttling valve in the vapor-compression cycle. The Stirling, Ericsson, and Brayton cycles can each be operated "backward" to act as cooling machines or heat pumps. The cycle diagrams will not be identical to the parent heat engine cycles when real effects such as turbine and compressor efficiencies are considered, since irreversibilities cause entropy increases that change the shapes of various process lines in the cooling cycles from those found for the heat engines.

Brayton Cooling Cycle

Perhaps the most widely used of the reverse cycles is the Brayton cooling cycle. It is used to provide cooling for aircraft cabins and avionics, and serves as a good example of the use of the reverse cycles.

The cycle operates by using work input to the turbine compressor shaft, using air as the working fluid on an open cycle. The hardware arrangement is shown in Fig. 6.40, and the corresponding cycle diagrams are in Fig. 6.41.

Figure 6.41 *T-s* and *P-h* diagrams for open Brayton cooling cycle.

The COP of the ideal cooling cycle (isentropic compressor and turbine) is given by

$$COP_B = \frac{|q_L|}{|w|} = \frac{|q_L|}{|q_H| - |q_L|}$$

$$= \frac{1}{[(T_{4,s} - T_1)/(T_3 - T_{2,s})] - 1}$$

$$= \frac{1}{(T_{4,s}/T_3)[(1 - T_1/T_{4,s})/(1 - T_{2,s}/T_3)] - 1} \quad (6.25)$$

For an isentropic compressor and turbine,

$$\frac{T_1}{T_{2,s}} = \frac{T_{4,s}}{T_3} = \left(\frac{P_H}{P_L}\right)^{[(k-1)/k]} \quad (6.26)$$

or

$$\frac{T_1}{T_{4,s}} = \frac{T_{2,s}}{T_3} \quad (6.27)$$

Substituting into Eq. (6.25) gives

$$COP_B = \frac{1}{T_{4,s}/T_3 - 1}$$

$$= \frac{1}{(P_H/P_L)^{[(k-1)/k]} - 1} \quad (6.28)$$

6.5.4   Cooling Cycles Driven by Heat Transfer

Heat-Engine-Driven Cycles

In some cases, heat transfer must be used to drive the cooling cycle. A convenient way to carry out the conversion of a high-temperature heat transfer into a cooling effect is to couple a heat engine with a work-driven refrigeration cycle. To determine the maximum performance of such a combined device, we can consider the case of a Carnot engine operating between the heat addition temperature of T_H and the environment temperature T_E. The Carnot engine drives a reverse Carnot cooling cycle that operates by a heat transfer addition at T_L and a subsequent rejection at the higher temperature T_E (Fig. 6.42).

For the heat engine, the work output is

$$|w| = |q_H|\left(1 - \frac{T_E}{T_H}\right) \quad (6.29)$$

The work input to drive the cooling machine must be [Eq. (5.172)]

$$\frac{|q_L|}{|w|} = \frac{T_L}{T_E - T_L} \quad (6.30)$$

Figure 6.42   Carnot cycle cooling machine.

Now, since the work output from the Carnot heat engine is equal to the work used to drive the Carnot cooling machine, $|w|$ from Eq. (6.29) may be substituted into (6.30) to give

$$\text{COP} = \frac{|q_L|}{|q_H|} = \frac{T_L}{T_H} \frac{T_H - T_E}{T_E - T_L} \qquad (6.31)$$

where, as before, the COP for the overall cooling cycle is defined as the cooling effect over the energy input to drive the cycle. Note that the COP depends not only on the heat transfer temperature T_H and the low temperature at which energy is removed, T_L, but also on the environment temperature T_E to which energy is rejected by both the heat engine and the cooling engine. Equation (6.31) predicts the upper limit of the COP for a cooling device driven by heat transfer rather than work.

Example Problem 6.15

A combined cycle Carnot refrigeration system is receiving heat transfer from a solar collector at $T_H = 200°C$. The environment is at $20°C$, and the system is being used to cool a residence. The cooling coil is to operate at $T_L = 5°C$. What is the maximum COP of the system?

Solution
Using Eq. (6.31), we find

$$\text{COP} = \frac{278}{473} \frac{473 - 293}{293 - 278} = 7.05$$

Comments
This result indicates the maximum possible COP of a heat-engine-driven cooling machine, because both machines operate on reversible cycles between fixed heat addition and rejection temperatures.

It is useful to compare other proposed heat-transfer-driven cycles with the COP of the combined Carnot cycle, since the Carnot device provides the upper limit of performance. The COP of the Carnot combined cycle cooling device is shown in Fig. 6.43. Note that

$$\frac{T_E}{T_H} > \frac{T_L}{T_H}$$

for any real system.

Other Cycles Driven by Heat Transfer
For some applications, it is useful to have a cycle that uses energy in the form of thermal energy rather than work to drive the cooling cycle,

Figure 6.43 Coefficient of performance of a combined Carnot cycle cooling machine.

and to not go through a conventional heat engine in the process. For example, if thermal energy from a solar collector, a geothermal source, or a source of waste heat is available, it may be quite inefficient to first drive a heat engine with the waste heat and then use the work output from the heat engine to drive the cooling cycle, as indicated by the results shown in Fig. 6.43 for low values of T_H. A cycle that efficiently uses the heat transfer directly to provide a cooling effect would be more practical.

Absorption Cycles

The most common heat-driven cooling cycle is the *absorption* cycle. The hardware used in the cycle is shown in Fig. 6.44. In this cycle, two working fluids are used, the *refrigerant* and the *absorbent*. Common pairs are ammonia and water and water and lithium bromide. The choice of the refrigerant-absorbent pair depends mainly on the ability of the absorbent to absorb the refrigerant at a suitable temperature and on the absorption process being reversible, so the refrigerant can be driven from the absorbent at a higher temperature.

The process works as follows. A solution of absorbent that is rich in refrigerant enters the *generator,* where a heat transfer drives off the refrigerant. This process occurs at a relatively high pressure (which is nevertheless subatmospheric for some refrigerant-absorbent pairs). The refrigerant leaving the generator at state 1 is then condensed to a liquid at state 2 and is expanded to a saturated mixture at low relative pressure through an expansion valve (state 3). The refrigerant then enters an evaporator, where heat transfer from the surroundings is used to evaporate the liquid fraction of the saturated refrigerant, producing the cooling effect and a saturated refrigerant vapor of 100 percent quality (state 4). The refrigerant vapor then enters an *absorber,* where it is absorbed in the rich absorbent

Figure 6.44 *Schematic diagram of absorption cooling cycle.*

which has been transferred from the generator after the refrigerant was driven off. The refrigerant-absorbent mixture in the absorber is then returned to the higher pressure of the generator by a pump. The heat exchanger (recuperator) between the absorber and the generator improves the system efficiency (COP) by cooling the hot absorbent before it enters the absorber (thus improving its absorption capability, which is strongly temperature-dependent). In addition, the mixture returning from the absorber to the generator is preheated by the recuperator, thus requiring less heat transfer from an external source to the generator.

The absorption system is identical in operation to the vapor-compression cycle except that the generator-recuperator-absorber system of the absorption cycle replaces the compressor in the vapor-compression cycle. The remaining components are the same. The advantage of the absorption cycle is that the only mechanical work needed is that necessary to drive the small liquid-return pump between the absorber and the generator (and even this can be eliminated by using an inert gas in the system which can circulate the working fluid by using a heat transfer to expand and contract it in a "thermosyphon"). The work to drive this pump is very small compared with the work needed to drive the compressor in the vapor-compression system. However, in terms of actual COP, the absorption systems are generally poor in comparison with vapor-compression systems; for example, the water–lithium bromide system has a COP of about 0.6. This can be improved by various cycle modifications, but at an increase in system capital cost.

The methods of treating the generation and absorption processes must await the development of two-component vapor-liquid equilibrium relations in Chap. 11.

Desiccant Cycles

Another class of cooling cycles that uses heat transfer as the driving energy operates by removing almost all water vapor from the air being conditioned. Cooling is then obtained by evaporating some water to re-humidify the air while evaporatively cooling. These processes are examined further in Chap. 9 in the sections on psychrometrics. The drying is carried out by contacting the humid air with a *desiccant,* i.e., a solid material with a strong affinity for absorbing water vapor on its surface (silica gel) or a liquid with the ability to absorb water vapor into solution (lithium chloride, lithium bromide, triethylene glycol, and others). After the desiccant becomes saturated with water, it is dried by a heat transfer to increase the desiccant temperature; and the water vapor is driven off, so the desiccant can be reused. These systems can operate with regeneration temperatures as low as 330 K, so low-grade energy sources can be used to drive them.

6.6   Concluding Remarks

Cycles and devices that can be used to harness energy transfers for useful purposes have been described in this chapter, and the methods of analyzing the efficiency of these devices have been outlined. Note that the cycles and their analysis as presented in this chapter are the result of some hundreds of years of study, experimentation, and development by engineers, scientists, inventors, and entrepreneurs. The excitement and romance of this rich history are difficult to impart in a text meant for an introduction of the principles of the subject. But this is a good opportunity for the reader to read App. A on the historical development of classical thermodynamics and then to visit a power plant or climb over the remains of a steam locomotive and reflect on what the human mind has brought us from this history.

> It was a fine, handsome old engine, much older than Jake Holman himself. He looked at it, massive, dully gleaming brass and steel in columns and rods and links arching above drive rods from twinned eccentrics, great crossheads hung midway, and above them valve spindles and piston rods disappearing into the cylinder block. He knew them all, each part in its place in the whole, and his eye followed the pattern, three times repeated from forward to aft, each one-third of the circle out of phase, and it was all poised and balanced there like three chunks of frozen music.
> Richard McKenna, The Sand Pebbles, *Harper and Row, New York, 1962 (used with permission).*

Problems

Problems marked with a star are lengthy; thus, computerized steam tables or an accurate Mollier chart should be used if these problems are to be completely worked.

6.1 Find P, v, and T at point 4 in Example Problem 6.1.

6.2 Plot the cycle diagram for the air-standard Carnot cycle operating with 1 kg of working fluid between 20 and 200°C if the maximum volume reached in the cycle is 0.1 m³ and the maximum pressure in the cycle is 14 MPa. Plot the P-v, T-s, P-h, and h-s diagrams. Assume that the specific heat of air can be taken as temperature-independent.

6.3 Find the power output of the Carnot engine of Prob. 6.2 if it operates at 10 cycles/min.

6.4 Assuming temperature-independent specific heats:
(a) Draw on the same graph the P-v diagrams for Carnot cycles using working fluids of air and argon. The cycles both have these state points (corresponding to Fig. 6.1):

State	P, kPa	v, m³/kg
1	3500	0.02
2	—	01
3	—	0.2

(b) Find the efficiency of each cycle.

6.5 Points 1 and 3 on air-standard Stirling, Ericsson, Brayton, and Carnot cycle P-v diagrams all lie at the same values: $P_1 = 10$ atm, $v_1 = 0.05$ m³/kg; $P_3 = 1$ atm, $v_3 = 0.40$ m³/kg. What is the fraction of the ideal Carnot efficiency at which the other cycles will operate? Assume that specific heat values are temperature-independent.

6.6 Air-standard Stirling, Ericsson, and simple Brayton external combustion engines all operate between minimum and maximum temperatures of 20 and 450°C. If each cycle has a maximum pressure of 4 MPa and a maximum specific volume of 0.06 m³/kg, plot the P-v, T-s, and P-h diagrams for the three cycles. (Plot the cycles for the three engines on the same P-v, T-s, and P-h graphs for comparison.) Assume that specific heat values are temperature-independent.

6.7 A Stirling engine operates between 25 and 450°C. The maximum pressure reached in the engine is limited to 50 atm, and the minimum operating pressure is 1 atm. Plot the T-s diagram for the cycle, and calculate the cycle efficiency and the work output per revolution (*a*) when the working fluid is air and (*b*) when the working fluid is hydrogen. Assume in each case that the specific heat of the working fluid is temperature-independent and can be evaluated at the arithmetic average of the cycle temperature limits.

6.8 A Stirling engine operates between 25 and 450°C and has a maximum operating pressure of 50 atm and a minimum operating pressure of 1 atm. The engine uses air as the working fluid. Determine the work output per cycle of the engine for two cases:

(*a*) The specific heat of air is temperature-independent.

(*b*) The properties of air are taken to be temperature-dependent.

6.9 An ideal air-standard gas-turbine cycle operates on a closed cycle between the pressures of 1 atm and P_2. Intake air to the compressor is at 27°C. The turbine is limited to a maximum temperature of 1227°C by materials limitations. Heat transfer occurs *to* the cycle at a value of $q_H = 200$ kJ/kg.

(*a*) What is the value of the allowable inlet pressure to the gas turbine?

(*b*) What is the cycle efficiency?

(*c*) What is the outlet temperature from the turbine?

6.10 An ideal air-standard closed Brayton cycle operates between the pressures of 1 atm and P_2. Intake air to the compressor is at 27°C and 1 atm. The turbine is limited to a maximum temperature of 1227°C. Heat rejection from the cycle is $q_L = 100$ kJ/kg.

(*a*) What is the value of the inlet pressure to the turbine?

(*b*) What is the cycle efficiency?

(*c*) What is the compressor outlet temperature?

6.11 A closed Brayton cycle has a two-stage turbine. The high-pressure (first) stage has inlet $T = 1100$ K at 10 atm and an efficiency of 90 percent. Argon is the working fluid. After leaving the first stage at 5 atm, the argon is reheated to 1100 K and enters the second stage, which also has an efficiency of 90 percent. The exhaust from the turbine is at 1 atm. The compressor in the cycle has efficiency of 95 percent, and the compressor inlet temperature is 27°C.

(*a*) Draw a *T-s* diagram for the cycle.

(*b*) Find the cycle efficiency (percentage).

(*c*) Find the cycle efficiency without reheat (percentage).

6.12 A closed Brayton cycle has a two-stage compressor, with each stage having an efficiency (compared to an isentropic compressor) of 95 percent. The working fluid is argon.

The compressor inlet is at 1 atm and 27°C. The compressor outlet is at 10 atm. Between the compressor stages, the argon is cooled at $P = 5$ atm to 100°C and then enters the second compressor stage. The single-stage turbine has an efficiency of 90 percent and an inlet temperature of 1100 K.

(a) Draw a T-s diagram for the cycle.

(b) Find the cycle efficiency (percentage).

(c) Find the cycle efficiency without intercooling (percentage).

6.13 A closed Brayton cycle uses argon as the working fluid. The engine has a two-stage compressor with intercooling, and each stage has an isentropic efficiency of 95 percent. The turbine is a two-stage type with reheat, and each stage has an efficiency of 90 percent. Reheat and intercooling occur at 5 atm, and the cycle operates between 1 and 10 atm. The turbine inlet temperature is 1100 K, and the compressor inlet is at 27°C. Reheating brings the second-stage inlet temperature to 1100 K, and intercooling brings the second-stage compressor inlet temperature to 100°C.

(a) Draw the T-s diagram for the cycle.

(b) Find the cycle efficiency (percentage).

(c) Find the cycle efficiency without intercooling (percentage).

(d) Find the cycle efficiency without reheat (percentage).

(e) Find the cycle efficiency with *neither* reheat *nor* intercooling (percentage).

6.14 An air-standard closed cycle Brayton engine takes in air at 27°C and 1 atm, and the compressor pressure ratio is 10. The compressor efficiency is 80 percent. The inlet temperature to the turbine is 900°C. What turbine efficiency is necessary to provide zero work output from the engine?

6.15 For the three cases of Example Problem 6.5, determine the fraction of the turbine work output and the fraction of the cycle work that go to drive the compressor in each case. At what value of the turbine efficiency will the compressor work equal the entire turbine work output (assume an ideal compressor)? For an ideal turbine, at what value of the compressor efficiency will the required compressor work equal the entire turbine work output?

6.16 Plot the cycle efficiency versus regenerator effectiveness for isentropic pressure ratios of 6, 8, and 10 for an air-standard Brayton cycle with ideal regeneration operating between $T_2 = 900$ K and $T_4 = 300$ K. Discuss how you would choose an optimum isentropic pressure ratio for a given regenerator.

6.17 A Carnot engine is connected to a bank of solar collectors in such a way that the outlet fluid at temperature T_H from the solar collectors is used to provide energy to the Carnot engine. Energy is rejected from the Carnot engine at the environment temperature T_L.

The efficiency of a solar collector is defined by $\eta_{sc} = Q_u/Q_s$, where Q_u is the useful energy used to raise the enthalpy of the fluid passing through the collector and Q_s is the amount of solar energy incident on the collector. For the collector used here, the efficiency is defined at a given time by

$$\eta_{sc} = 0.8 - 0.003(T_H - T_L)$$

If T_L is 300 K, what should the outlet temperature from the collectors T_H be to maximize the ratio of work output from the engine to the available solar energy? (Note that the efficiency of the Carnot engine increases with T_H, while the solar collector efficiency decreases.)

6.18 Thomas Newcomen's engines operated by spraying cold water into a cylinder filled with saturated steam near 1-atm pressure. The condensing steam then reduced the volume of the resulting liquid-vapor mixture, and atmospheric pressure pushed a piston into the cylinder (there is a diagram of the original Newcomen engine in App. A). A Newcomen engine with a piston diameter of 1 m and a stroke of 2 m reaches a minimum volume in the cylinder at the end of the stroke of 0.2 m³.

(a) Find the amount of water at 20°C necessary to inject into the cylinder to just complete the stroke.

(b) Find the work output per stroke from the engine.

(c) No work was obtained from the Newcomen engine on the return stroke, which was carried out by allowing a system of weights to return the piston to its original position. If the engine could operate at 1 cycle/min, and the piston is assumed to be of negligible weight, find the power output of the engine.

6.19 The P-v diagram for steam over the ranges commonly used in Rankine cycle analysis must be plotted on multicycle log-log paper for clarity. By using either the steam tables or computerized tables and plotting routines, plot the saturation curve for steam on a P-v plot for specific volume between 0.001 and 1000 m³/kg. On this graph, show the isentropes that pass through the points (P, v) of (4000, 0.02), (4000, 0.05), and (4000, 0.5), where P is in kilopascals and v is in cubic meters per kilogram. Discuss the linearity of the isentropes in terms of Eqs. (5.47) and (5.48).

*6.20 A simple Rankine cycle (no superheat, reheat, or regeneration) with an ideal pump and turbine is being planned with water as the working fluid. The condenser is to operate at 30°C. What boiler pressure will produce a cycle efficiency of 25 percent?

*6.21 A simple Rankine cycle (no superheat, reheat, or regeneration) has a turbine with 80 percent efficiency. Water is the working fluid, and the condenser pressure is 2 kPa. What boiler pressure will provide a cycle efficiency of 25 percent?

*6.22 Plot the simple ideal Rankine cycle efficiency versus boiler pressure for a condenser temperature of 27°C. Also plot the turbine exit quality versus boiler pressure on the same graph. Use a boiler pressure range of 0.7 to 7 MPa.

6.23 A simple Rankine cycle operates between a condenser pressure of 8 kPa and a boiler pressure of 6 MPa. Compare the cycle efficiency of this cycle with the efficiency of a cycle that operates between the same pressures but superheats the vapor leaving the boiler to 400°C. For both systems, the turbine efficiency is 88 percent. Compare and discuss the turbine exit quality for the two cases.

*6.24 Plot the turbine exit quality versus degrees superheat for a Rankine cycle operating at a boiler pressure of 4 MPa and a condenser pressure of 8 kPa for turbine efficiencies of 60, 70, and 80 percent. (*Degree of superheat* is defined as the temperature difference between the actual superheated steam temperature and the saturation temperature, both evaluated at the boiler pressure.) Use a degrees superheat range of 0 to 150°C.

6.25 Saturated steam leaves a boiler at $P = 200$ bars. Heat transfer occurs in a superheater at constant P until the temperature reaches 600°C. The steam then enters a steam turbine, where it does work on the turbine, and leaves the turbine at a pressure of $P = 7.0$ bars. The turbine has an efficiency of 75 percent.
(*a*) How much work is done on the turbine?
(*b*) How much heat transfer occurs in the superheater?
(*c*) If liquid water enters the boiler at 165°C and 200 bars, what is the Rankine cycle efficiency? (Neglect pump work.)

6.26 A Rankine cycle with superheat has a condenser pressure of 8 kPa, a boiler pressure of 6 MPa, and a turbine efficiency of 88 percent and superheats the steam leaving the boiler to 400°C. Compare the efficiency of this cycle to a reheat cycle operating with the same boiler, superheater, and condenser conditions in which steam leaves the high-pressure stage of a two-stage turbine at 300 kPa, is reheated to 400°C, and is then fed to the low-pressure turbine stage. Assume that both turbine stages have efficiencies of 88 percent.

6.27 A steam power plant operates with a boiler pressure of 1 MPa, and the steam leaving the boiler is superheated to 250°C. The turbine used in the plant is a two-stage type. The first stage operates at an efficiency of 70 percent. Steam leaving the first stage exits at 600 kPa, is reheated to 250°C, and is then returned to the second (low-pressure) stage, which has an efficiency of 77 percent. The low-pressure stage exhausts to the condenser at a pressure of 8 kPa.

(a) Sketch the equipment layout, and label the positions of states 1 through 7 on the layout to correspond with the state table.

(b) Sketch the cycle on a T-s diagram, being as accurate as you can. Label the states to correspond to the state table given here.

(c) Fill in the state table (using the steam tables).

(d) Fill in the energy table.

State Table

State	Location	T, °C	P, kPa	h, kJ/kg	s, kJ/(kg · K)	x
1	Boiler inlet	45	1000			—
2	Superheater inlet		1000			—
3	Superheater outlet	250	1000			—
4	High-pressure turbine outlet		600			—
5	Low-pressure turbine inlet	250	600			—
6	Low-pressure turbine outlet		8			
7	Pump inlet		8			0.0

Energy Table

$_1q_2$, kJ/kg =

$_2q_3$, kJ/kg =

$_4q_5$, kJ/kg =

$_3w_4$, kJ/kg =

$_5w_6$, kJ/kg =

$_7w_1$, kJ/kg =

Cycle efficiency, % =

6.28 A Rankine cycle with superheat has a condenser pressure of 8 kPa, a boiler pressure of 6 MPa, and a turbine efficiency of 88 percent and superheats the steam leaving the boiler to 400°C. Compare this cycle efficiency to that of a regeneration cycle using a closed feedwater heater in which 10 percent of the entering turbine mass flow is bled off at 200 kPa to the feedwater heater.

6.29 Repeat Prob. 6.28 for a regeneration cycle using an open feedwater heater.

6.30 Repeat Prob. 6.28 for bleed flows of 5, 10, 15, 20, and 25 percent of the entering turbine mass flow. Discuss why the cycle efficiency might reach a maximum at a particular bleed flow.

6.31 A Rankine cycle with superheat has a condenser pressure of 8 kPa and a boiler pressure of 6 MPa, and it superheats the steam leaving the boiler to 400°C. Steam leaves the high-pressure stage of a two-stage turbine at 300 kPa, is reheated to 400°C, and then is fed to the low-pressure turbine stage. Assume that both turbine stages have efficiencies of 88 percent. Of the entering turbine mass flow 10 percent is bled from the turbine at 200 kPa and is sent to a closed feedwater heater. Compare the efficiency of this superheat, reheat, regeneration cycle with the results of Probs. 6.23, 6.26, and 6.28.

*6.32 For a simple Rankine cycle with boiler pressure of 6 MPa and a turbine efficiency of 88 percent, plot the cycle efficiency and condenser heat rejection versus condenser pressure in the range $1 \le P_{cond} \le 100$ kPa.

*6.33 For a simple Rankine cycle with condenser pressure of 8 kPa and a turbine efficiency of 88 percent, plot the cycle efficiency, turbine exit quality, and condenser heat rejection versus boiler pressure over the range $0.1 \le P_{boiler} \le 6$ MPa.

6.34 An improved Rankine cycle has been suggested that operates as diagramed in Fig. P6.34. Some of the turbine work is used to drive an isentropic compressor that takes a saturated mixture of quality x_5 from the condenser and compresses it to saturated liquid at state 1. The turbine is assumed isentropic, the boiler operates at 6 MPa and the condenser at 8 kPa, and the boiler steam is superheated to 400°C.

(a) Find the efficiency of the proposed cycle.

(b) Find the efficiency of a usual Rankine cycle operating under the same conditions (i.e., use a pump in place of the compressor, with $x_5 = 0$).

(c) What are the ratios of the proposed and usual Rankine cycle efficiencies to the appropriate Carnot efficiency?

Figure P6.34

(d) Calculate the efficiency of the proposed cycle with no superheat.

6.35 A Rankine cycle operates with an open feedwater heater. The boiler pressure is 6 MPa, and the superheater provides a turbine inlet temperature of 400°C. Bleed steam is provided from the turbine at 500 kPa.
(a) What fraction of the boiler mass flow rate must be used in the bleed stream to provide a boiler feedwater inlet temperature of 150°C?
(b) What is the cycle efficiency?
(c) What is the cycle efficiency without regeneration?

6.36 An internal combustion Brayton turbine engine is being designed for jet propulsion. The engine has no regeneration. The turbine and compressor have efficiencies of 80 percent, the maximum allowable temperature on the turbine blades is 750°C, and air entering the compressor may be assumed to be at 1.0 atm and 0°C. What is the kinetic energy (per kilogram) of the gases leaving the engine if they exit at 70°C? Assume that constant air specific heat and a constant mass flow rate can be used for the gases passing through the engine and that the gases leave the turbine at 1 atm.

6.37 Repeat Prob. 6.36, using temperature-dependent properties for air.

6.38 A two-stroke motorcycle engine displaces 250 cm³ and operates at 4000 rpm. For a compression ratio of 8:1, what is the air-standard cycle efficiency, and how many kilowatts will the air-standard engine develop? (Assume that at the end of the power stroke, the cylinder pressure has dropped to 350 kPa.)

6.39 A hot-rod mechanic has an engine from a 1966 British sports car. The engine displaces 4.2 liters, has an original compression ratio of 10.2:1, and has six cylinders. Assume that the pressure at the end of the power stroke is 300 kPa just before the exhaust valve opens. The pistons in the original engine are 12 cm in diameter.

(a) If the mechanic bores the cylinders and replaces the pistons with new pistons that are 0.5 cm larger in diameter than the originals, by what percentage will the engine efficiency change (assuming the 300-kPa pressure is unchanged)?

(b) If the mechanic replaces the engine crankshaft and piston connecting rods, so that the original pistons have their stroke increased by 0.5 cm, by what percentage will the engine efficiency change?

(c) If the cylinder head is milled so that the original clearance is reduced to 0.4 cm, by what percentage will the engine efficiency change?

Assume in all cases that the ideal air-standard efficiency is desired.

6.40 A diesel engine displaces 6 liters and operates at 1600 rpm. The compression ratio of the engine is 12. The pressure in the cylinder reaches 2 atm at the end of the power stroke just as the exhaust valve opens. Graph the power developed by the air-standard engine for expansion ratios in the range 2 to 12.

6.41 A diesel engine has a compression ratio of $r_1 = 12$. Plot the diesel engine efficiency versus the expansion ratio r_2 for $1 \leq r_2 \leq 12$. Pay special attention to the case of $r_2 = 1$, and derive a relation for the cycle efficiency for that special case. Plot the cycle for $r_2 = 1$ on a P-v diagram.

6.42 Using the definition of efficiency for a control mass derived in Chap. 5, derive expressions for the cycle efficiency of air-standard Otto and diesel cycles with nonideal compression and expansion.

6.43 The COP of a reversible refrigeration cycle is 4.0. The cycle is able to remove 5 MJ/h from the refrigerator.

(a) What is the rate of work necessary to run the cycle?

(b) What temperature is maintained in the refrigerator if the cycle rejects a heat transfer to a room at $27°C$?

6.44 A 10-ton vapor-compression refrigeration device uses ammonia as the working fluid. The saturated liquid entering the throttling valve has $T = 30°C$, and the saturated vapor entering the compressor is at $-25°C$. The compressor efficiency is 65 percent.

(a) What is the ratio of the COP of this cycle to the COP of a Carnot cycle operating between the same temperatures?

(*b*) How much ammonia must be circulated through the device?

6.45 A vapor-compression refrigeration cycle using refrigerant 12 operates with a condenser temperature of 40°C and an evaporator temperature of 5°C. The compressor has an efficiency of 65 percent. If there is no subcooling of the liquid leaving the condenser, plot the cycle COP versus the degree of superheating of the refrigerant 12 leaving the evaporator for 0 to 20°C above the saturation temperature.

6.46 A vapor-compression heat pump system using refrigerant 12 operates with an evaporator temperature of 10°C and a condenser temperature of 40°C. There is no superheating of the vapor leaving the evaporator, and the compressor has an efficiency of 65 percent. Plot the COP_{HP} for the system versus the degrees of subcooling of the liquid leaving the condenser for subcooling of 0 to 15°C below the saturation temperature.

6.47 Aircraft commonly employ open Brayton reverse cycle systems for providing cabin and avionics cooling. In these systems, outside air is passed through a turbine to a low pressure, where heat transfer from the cabin air is provided by an air-to-air heat exchanger. The working fluid (air) is then compressed to the turbine inlet pressure and is exhausted. See Fig. P6.47.

If the outside air entering the turbine is at 70 kPa and 20°C and at $\mathbf{V} = 250$ m/s, and if the turbine and compressor have isentropic efficiencies of 70 percent, the required turbine outlet temperature is 5°C, and the compressor inlet temperature is 15°C, then:
(*a*) Find the required turbine outlet pressure.
(*b*) Find the net work necessary to drive the cycle.
(*c*) Find the COP.
(*d*) Find the heat transfer from the cabin air.

Figure P6.47

6.48 Derive the COP of a reverse Ericsson refrigeration cycle with a compressor isothermal efficiency of η_c and a turbine isothermal efficiency of η_T, where the isothermal efficiency is defined by $\eta_T = \dfrac{w_a}{w_{isothermal,\,rev}}$.

6.49 Derive an expression for the COP of a reverse Brayton cooling cycle in terms of the turbine efficiency η_t, the compressor efficiency η_c, the pressure ratio, and the ratio of turbine to compressor inlet temperatures. From the result, find expressions for the minimum allowable compressor efficiency for a given turbine efficiency and the minimum allowable turbine efficiency for a given compressor efficiency that will provide a given COP.

6.50 Find relations for the COP of cooling systems made of (*a*) a Stirling heat engine driving a reverse Stirling cooling cycle and (*b*) a Brayton heat engine driving a reverse Brayton cooling cycle. In both cases, assume all components of the cycle are reversible and that both cycles receive energy at a maximum temperature of T_H and reject heat to the surroundings at T_E, while providing cooling at T_L. The Brayton engine and Brayton cooling cycle have no regenerators and operate at atmospheric pressure when they are at T_E. Assume temperature-independent specific heats for the working fluids in each cycle.

6.51 An inventor plans a system that operates as follows: A heat engine receives energy from a source at 540°C, gives a work output of 0.50 kJ/kg for each 1 kJ/kg of heat transferred to the working fluid at that temperature, and rejects the remaining energy to the environment at 25°C. Part of the work output from the heat engine drives a heat pump with a COP_{HP} of 4. The heat pump is used to take energy from the environment at 25°C and pump it to 540°C, where it becomes the energy source for the heat engine. Since COP_{HP} is 4, only 0.25 kJ/kg of work from the heat engine is required to drive the heat pump for each 1 kJ/kg delivered to the heat engine. Thus, no outside energy source is required to drive the heat engine, and 0.25 kJ/kg is left of the engine work output to do useful work.

　　　Is there a flaw in this device? Discuss your answer in terms of (*a*) the second law of thermodynamics and (*b*) a detailed analysis of the possible efficiency of the heat engine and the possible COP_{HP}.

6.52 A Humphrey pump operates by using the liquid in the pump as a piston. The cycle of operation is shown in Fig. P6.52. Use assumptions similar to those used for the Otto cycle.
(*a*) Draw the *P-v* and *T-s* diagrams for the air-standard Humphrey pump cycle.
(*b*) Derive an expression for the thermodynamic efficiency of the air-standard Humphrey pump cycle.

1 Ignition, followed by power stroke.
2 Momentum of water column allows full expansion of combustion gases.
3 As chamber pressure drops below atmospheric, exhaust valve opens.

Intake valve closed

Exhaust valve closed

10 Falling water level draws a fresh combustible charge into chamber.
11 Inlet stroke continues until water height in standpipe causes reversal of column motion.

Intake valve open

Exhaust valve closed

4 Chamber level continues to drop and makeup water enters through water valves.
5 Pumping action continues until column velocity ceases.
6 Potential energy stored in standpipe height causes column motion to reverse.

Intake valve closed

Exhaust valve open

Water valve open

12 Potential energy stored in standpipe height accelerates column, which then compresses the charge.
13 As maximum chamber level is reached, spark ignites mixture and cycle repeats.

Intake valve closed

Exhaust valve closed

7 Exhaust gases are expelled as water rises in chamber.
8 Column accelerates until water slams exhaust valve shut.
9 Air trapped above exhaust valve acts as a gas spring and reverses column motion.

Intake valve closed

Exhaust valve closed

Figure P6.52

6.53 A *heat pipe* operates by evaporating a liquid by heat transfer in the evaporation section of a closed conduit. The resulting vapor moves to the condenser end, where heat transfer from the heat pipe causes the vapor to condense. The liquid is then returned to the evaporator section by capillary action in a wick or by gravity. The vapor pressure is nearly constant between the evaporator and condenser, so the heat pipe is a nearly constant-temperature device for heat transfer. See Fig. P6.53. Draw the *P-v* and *T-s* diagrams for a heat pipe process using water at $P = 50$ kPa.

Figure P6.53

6.54 An irrigation pump is to be powered by a solar-driven Rankine cycle. The energy striking the solar collector during daylight hours ($0 \leq t \leq 10$) is given by

$$\dot{Q}_{solar} = 900A \sin \frac{\pi t}{10} \quad \text{W}$$

where A is the collector area in square meters and t is the time after sunrise in hours. See Fig. P6.54.

The solar collector converts 40 percent of $\dot{Q}_{solar}$ to enthalpy change of the water entering the collector. The flow rate of water $\dot{m}(t)$ is varied in such a way that the collector outlet condition is always saturated vapor at 150°C. The turbine is 70 percent efficient and exhausts to 8 kPa. The pump must lift water a distance of 8 m, and the friction in the piping system is negligible. Both pumps have an efficiency of 80 percent.

(a) Find the collector area required to provide 500 kg/min at solar noon ($t = 5$, maximum pumping capacity of the system).

(b) Plot the pumping rate, $\dot{m}_p(t)$ versus time over the 10-h daylight period.

(c) Find the total amount of water pumped during the entire day.

Figure P6.54

7

Analysis Using the Second Law of Thermodynamics

A good many times I have been present at gatherings of people who, by the standards of traditional culture, are thought highly educated and who have with considerable gusto been expressing their incredulity at the illiteracy of scientists. Once or twice I have been provoked and have asked the company how many of them could describe the Second Law of Thermodynamics. The response was cold; it was also negative.

C. P. Snow, *The Two Cultures*

Vapor-compression chiller system, showing cutaway of centrifugal compressor, condenser, and evaporator. *(The Trane Company.)*

7.1　Introduction

The second law of thermodynamics is unique in its specification of the ideal limits on system processes and on system cycles. The second law, like the first law and conservation of mass, relates a system property to boundary transfers; unlike the first law and conservation of mass, the relation provides only a limit on the system performance. The connection between the system property (entropy) and the boundary transfer (heat transfer divided by absolute temperature) is an equality only in the limiting case of a reversible process. Otherwise, there is an entropy production or generation which is nonzero. The second law given in Table 5.1 is

$$\dot{S}_{\text{gen}} = \frac{\partial}{\partial t}\left(\int_{\text{CV}} s\rho \, dV\right) + \sum_{\text{out}} \int_A s \, d\dot{m}$$

$$- \sum_{\text{in}} \int_A s \, d\dot{m} - \sum_i \left(\frac{\dot{Q}_i}{T_i}\right)_{\text{CV}} \geq 0 \quad (7.1)$$

or, in differential form, from Eq. (5.127) as

$$\delta S_{\text{gen}} = dS_{\text{CV}} + \sum_{\text{out}} s \, dm - \sum_{\text{in}} s \, dm - \sum_i \left(\frac{\delta Q_i}{T_i}\right)_{\text{CV}} \geq 0 \qquad (7.2)$$

Entropy generation is the term which quantifies the irreversibility of the process. The identification of the sources of irreversibilities and their reduction are the desired goal of second law analysis, and this is specified by the reduction of entropy generation.

Comparisons of system performance are made in Chaps. 4 through 6. The first comparison is the *thermal efficiency,* defined as the desired output divided by the required input. This is often termed a *first law efficiency* since it is based entirely on first law quantities. The second law puts an upper limit on the first law efficiency, but does not directly enter into the definition. However, the *component* efficiencies in Sec. 5.10 are directly related to the second law. These efficiencies are ratios of the actual performance to the ideal performance. The ideal performance was usually (but not always) the performance of a device operated at the reversible adiabatic (isentropic) limit. Therefore, the second law specified the ideal or "upper limit" process for the system. Processes are compared in this chapter by using the entropy generation limitation given by the second law.

Second law analysis is closely tied to entropy generation and, therefore, to entropy transfer. Equations (7.1) and (7.2) reveal the relationship between the entropy generation and the boundary transfer. Boundary entropy transfer, which thermally connects the system to the heat transfer reservoir, sets the reference state for much of the subsequent analysis. The particular choice for the reference state depends on the problem, but many engineering considerations involve interactions with the earth's atmosphere. This reservoir is assumed to be an infinite heat transfer reservoir at temperature T_0. This reference state is often assumed in the analysis, and the temperature is 25°C unless specified otherwise. The atmosphere is also referred to as the *dead state* because it has the lowest naturally occurring temperature that can be used as a practical reservoir for a rejected heat transfer. Thus the energy in processes ending at this state cannot be used subsequently to obtain work.

In the next section we discuss the limits on the maximum possible work obtainable for an arbitrary process. This limiting expression for maximum, or *reversible*, work is developed by simultaneously considering the constraints imposed by the second law and the relation between the system energy and energy transfer prescribed by the first law. Particular definitions of thermodynamic properties are developed from the reversible work expression to help describe the available work that can be realized. The reversible, or maximum, work possible for an arbitrary process can then be compared to the actual work obtained. The possible processes are infinite and depend on the particular application, so in Sec. 7.6 we discuss process comparisons. In all the analyses in this chapter, we employ the fundamental laws previously presented. We present very important definitions and analytical techniques which are based on the previous laws, particularly the concept of entropy generation.

7.2   Reversible Work

The ability of the second law to predict the potential maximum work for a process is the focus of this section. The work output of a process is the desired quantity; yet, the statement of the second law does not directly include the work term (since energy transfer in the form of work causes no entropy generation). A relation is developed to include the entropy generation directly in the work prediction. This is accomplished by considering the first law and second law together.

A general control volume (CV) which considers an arbitrary number of inlets and outlets is shown schematically in Fig. 7.1. Also there are various heat transfers with the surroundings. The term for index $i = 0$ represents the heat transfer with the lowest naturally occurring reservoir (usually the earth's atmosphere). The energy transfer as work is grouped into a single work term. The first law for a general process is given from Eq. (5.126) as

$$dE_{\mathrm{CV}} + \sum_{\mathrm{out}} (e + Pv)\, dm - \sum_{\mathrm{in}} (e + Pv)\, dm - \sum_{i} \delta Q_i - \delta W = 0 \qquad (7.3)$$

Figure 7.1   *General control volume schematic.*

The second law is given by rewriting Eq. (7.2) as

$$dS_{CV} + \sum_{\text{out}} s \, dm - \sum_{\text{in}} s \, dm - \sum_i \frac{\delta Q_i}{T_i} - \delta S_{\text{gen}} = 0 \tag{7.4}$$

where the entropy generation term quantifies the irreversibilities of the process.

The first and second laws are now combined by multiplying the second law by temperature and subtracting the result from the first law. The temperature chosen is denoted by the constant value T_0, and the usual convention is to use the temperature of the surroundings or earth's atmosphere. This is the boundary temperature corresponding to the heat transfer, δQ_0. This choice is discussed further in Sec. 7.6. Multiplying Eq. (7.4) by T_0 and subtracting the result from Eq. (7.3) yield

$$d(E - T_0 S)_{CV} + \sum_{\text{out}} (e + Pv - T_0 s) \, dm - \sum_{\text{in}} (e + Pv - T_0 s) \, dm$$

$$- \sum_i \delta Q_i \left(1 - \frac{T_0}{T_i}\right) + T_0 \, \delta S_{\text{gen}} - \delta W = 0 \tag{7.5}$$

Solving for the work obtained for a process yields

$$\delta W = d(E - T_0 S)_{CV} + \sum_{\text{out}} (e + Pv - T_0 s) \, dm - \sum_{\text{in}} (e + Pv - T_0 s) \, dm$$

$$- \sum_i \delta Q_i \left(1 - \frac{T_0}{T_i}\right) + T_0 \, \delta S_{\text{gen}} \tag{7.6}$$

This is a very important relation in second law analysis.

Equation (7.6) appears to yield a somewhat arbitrary value for work since it depends on the specific choice of T_0. The mathematical manipulation of adding zero times T_0 [Eq. (7.4) times T_0] does not alter the work from the process. Alternatively, this mathematical manipulation can be viewed as an elimination of the heat transfer from the environment δQ_0 from the first law by expressing the heat transfer in terms of second law quantities. The work performed is not changed by the development and is not dependent on this mathematical manipulation.

The change in $E - T_0 S$ within the control volume [first term on the right-hand side of Eq. (7.6)] describes the work contributions resulting from change within the control volume. The $(e + Pv - T_0 s) \, dm$ terms in Eq. (7.6) represent the work contributions from the difference between the outlet and inlet flows. These terms have similar counterparts in the first law equation, but the entropy terms now incorporate the eliminated heat transfer. The last two terms in Eq. (7.6) are somewhat different. The heat transfer term does not include the contribution from δQ_0 since it was eliminated and incorporated in the entropy terms. The term $1 - T_0/T_i$ can be viewed as the efficiency of a *reversible*, two heat transfer reservoir engine operating between a low reservoir temperature of T_0 and a higher reservoir temperature of T_i (developed in Sec. 5.11). With a heat transfer input δQ_i into the engine then $\delta Q_i (1 - T_0/T_i)$ represents the reversible work output by this engine. This work output contributes to the work performed by the process. The $T_0 \, \delta S_{gen}$ term in Eq. (7.6) represents all the irreversibility contributions to the work performance. This term also includes the irreversibilities that result from the heat transfers, since the engine-type term was viewed as reversible.

Our sign convention on work makes the largest negative value the largest work output. The entropy generation is a positive quantity by definition [Eq. (7.2)], so this term reduces the work output of the process (or, the work is less negative). If the process work is positive work input, then the entropy generation term indicates that the irreversibilities require a larger work input as δS_{gen} increases.

The maximum work possible for a system undergoing a process is found in the reversible case. The reversible process has zero entropy generation, so Eq. (7.6) yields

$$\delta W_{rev} = d(E - T_0 S)_{CV} + \sum_{out} (e + Pv - T_0 s) \, dm$$

$$- \sum_{in} (e + Pv - T_0 s) \, dm - \sum_i \delta Q_i \left(1 - \frac{T_0}{T_i}\right)$$

$$= d(U - T_0 S + \tfrac{1}{2} m \mathbf{V}^2 + mgZ)_{CV}$$

$$+ \sum_{out} (h - T_0 s + \tfrac{1}{2} \mathbf{V}^2 + gZ) \, dm$$

$$- \sum_{in} (h - T_0 s + \tfrac{1}{2} \mathbf{V}^2 + gZ) \, dm - \sum_i \delta Q_i \left(1 - \frac{T_0}{T_i}\right) \qquad (7.7)$$

where the subscript *rev* indicates reversible. This maximum work is the largest negative value (work output) possible, and is the ideal, or maximum, work possible for the specified conditions. This expression sets limits on system processes. The heat transfer contributions, as discussed in the previous paragraph, are viewed as the work output from reversible engines operating between reservoirs at T_0 and T_i.

Example Problem 7.1

Air at 3000°C and 13 MPa is contained within an expandable piston cylinder. The cylinder is cooled by a heat transfer with the surroundings at $T_0 = 25$°C to a final temperature of 1500°C. The polytropic process is given by $Pv^{1.5} = $ constant. Evaluate the reversible work per unit mass for this process.

Diagrams

Solution

The properties are given by the ideal gas relations since the compressibility chart indicates that air is ideal for the stated conditions. The properties are given in Table 5.4. The initial state is given, and the final state is prescribed by 1500 K and

$$P_1 v_1^{1.5} = P_2 v_2^{1.5}$$

or

$$\frac{P_1}{P_2} = \left(\frac{v_2}{v_1}\right)^{1.5} = \left(\frac{T_1}{T_2}\right)^{1.5/(1.5-1)}$$

Thus $P_2 = 2.066$ MPa.

The reversible work is given by Eq. (7.7) for a control mass as

$$\delta W_{rev} = d(E - T_0 S)$$

since there are no inlets or outlets and $1 - T_0/T_0 = 0$. Integrating this expression from the initial to the final state yields

$$_1 W_{rev,\,2} = (E - T_0 S)_2 - (E - T_0 S)_1$$

or

$$_1 w_{rev,\,2} = (e - T_0 s)_2 - (e - T_0 s)_1$$

Neglecting the kinetic and potential energy changes yields

$$_1 w_{rev,\,2} = u_2 - u_1 - T_0(s_2 - s_1)$$

$$s_2 - s_1 = s_0(T_2) - s_0(T_1) - R \ln \frac{P_2}{P_1}$$

The change in entropy is evaluated with Eq. (5.60) to yield

$$_1 w_{rev,\,2} = u_2 - u_1 + T_0[s_0(T_1) - s_0(T_2)] + T_0 R \ln \frac{P_2}{P_1}$$

and, with values from Table D.2, this is

$$_1 w_{rev,\,2} = -1404 \text{ kJ/kg}$$

This is the maximum work possible between the given states. This maximum possible work is independent of the stated path (polytropic) and depends on only the end states.

Comments

The work for a quasi-equilibrium polytropic process in a control mass is given in Table 5.6 as

$$_1 w_2 = \frac{P_1 v_1 - P_2 v_2}{1 - n} = \frac{R(T_1 - T_2)}{1 - n} = -861 \text{ kJ/kg}$$

which is a work output that is ~ 39 percent less than the maximum reversible output. Also note that the heat transfer for this process is

$$_1 q_2 = u_2 - u_1 - {}_1 w_2 = -617.6 \text{ kJ/kg}$$

which is almost three-quarters of the work obtained. The entropy generation for the actual process is

$$_1s_{gen, 2} = s_2 - s_1 - \frac{_1q_2}{T_0} = 1.822 \text{ kJ/(kg} \cdot \text{K)}$$

which indicates, as expected, that the actual work should be less than the reversible case.

Example Problem 7.2

Determine the maximum work that can be produced by a steam turbine which has an inlet state of 3 MPa and 450°C and an outlet state as a saturated vapor at 0.1 MPa. The port properties are uniform, and there is a heat transfer with the surroundings at 25°C.

Solution
The reversible work expression for steady state conditions with a single inlet and single outlet is

$$\delta W_{rev} = (e + Pv - T_0 s)_{out} \, dm_{out} - (e + Pv - T_0 s)_{in} \, dm_{in}$$

Neglecting the changes in kinetic and potential energy and expressing the reversible work on a rate basis yield

$$\dot{W}_{rev} = (h - T_0 s)_{out} \dot{m}_{out} - (h - T_0 s)_{in} \dot{m}_{in}$$

or

$$w_{rev} = (h - T_0 s)_{out} - (h - T_0 s)_{in}$$
$$= (h_{out} - h_{in}) - T_0(s_{out} - s_{in})$$

With Tables D.9 and D.10, the properties are obtained. The reversible work is

$$w_{rev} = 2675.3 - 3343.1 - T_0(7.3598 - 7.0822) = -750.6 \text{ kJ/kg}$$

Comments
This is the maximum work possible between the specified states (i.e., saturated vapor out, $s_{out} \neq s_{in}$, and with the temperature of the surroundings at $T_0 = 25°C$). Remember that the inlet and outlet states are fixed. There are other processes that start at the inlet state and produce more work. However, they must terminate at a different outlet state. For example, the adiabatic reversible turbine starting at the inlet state and exhausting to the given outlet pressure produces

$$w = (h_{out} - h_{in})_s = -771.3 \text{ kJ/kg}$$

but the outlet state is at $P = 0.1$ MPa and $x = 0.954$. This is the comparison process for the turbine efficiency, and it should not be confused with the maximum work. Many different values for work are possible if the states are changed, but the maximum work is the *maximum* between the specified states.

7.3   **Availability**

An important consideration in engineering analysis is the maximum work available from a particular substance in a specified state. If steam is available in a particular state, what is the maximum work that can be obtained by the use of this fluid? A compressed-air storage system contains air at a high-pressure state. What is the maximum work that can be obtained from air in this state? There are numerous questions of this type, and they are generalized as: What is the maximum available energy to perform work that is contained by a substance in a specified state?

The maximum energy available in a particular state can be used to perform work as long as the state is not at standard environmental conditions. The state in equilibrium with the environment can no longer be used to obtain work. Therefore, the environmental conditions of $T_0 = 25°C$ and $P_0 = 0.101$ MPa are often referred to as the *dead state*. A substance has a maximum available energy at any state that is not in thermodynamic equilibrium with this dead state.

This maximum available energy is determined by consideration of Eq. (7.7) for the reversible work. The maximum work is obtained if all the heat transfers occur with the atmosphere. Thus $T_i = T_0$, and Eq. (7.7) reduces to

$$\delta W_{rev} = d(E - T_0 S)_{CV} + \sum_{out} (e + Pv - T_0 s)\, dm$$

$$- \sum_{in} (e + Pv - T_0 s)\, dm \quad (7.8)$$

Although this expression does not contain the symbol for heat transfer, the heat transfer is indirectly included. There is a reversible heat transfer with the environment at T_0. The second law has eliminated the symbol δQ, and Eq. (7.8) includes the entropy terms. We are also interested in the available work that is useful, so the work performed as the displacement of the atmosphere must be eliminated. The expansion of the control volume or control mass against the atmosphere (which is at T_0 and P_0) is not available for other useful purposes. Thus the useful reversible work is given as

$$\delta W_{rev,\, use} = \delta W_{rev} + P_0\, dV \quad (7.9)$$

and substituting Eq. (7.8) into (7.9) yields

$$\delta W_{rev,\, use} = d(E + P_0 V - T_0 S)_{CV} + \sum_{out} (e + Pv - T_0 s)\, dm$$

$$- \sum_{in} (e + Pv - T_0 s)\, dm$$

$$= d(U + P_0 V - T_0 S + \tfrac{1}{2} m\mathbf{V}^2 + mgZ)_{CV}$$

$$+ \sum_{out} (h - T_0 s + \tfrac{1}{2}\mathbf{V}^2 + gZ)\, dm$$

$$- \sum_{in} (h - T_0 s + \tfrac{1}{2}\mathbf{V}^2 + gZ)\, dm \quad (7.10)$$

This equation explicitly indicates the properties that prescribe the available energy in particular states. The combination of properties at the inlet and outlet has the form $h - T_0 s$, and the combination of properties within the control volume or for a control mass is $u + P_0 v - T_0 s$. The difference lies in the pressure term.

The combinations of properties that appear in Eq. (7.10) prescribe the available energy of a state to perform work. These property combinations are state properties in relation to the environment or dead state. They are defined as the *nonflow availability per unit mass* ϕ and the *flow availability per unit mass* ψ. The nonflow availability per unit mass is

$$\phi = (u + P_0 v - T_0 s) - (u_0 + P_0 v_0 - T_0 s_0)$$
$$= (u - u_0) + P_0(v - v_0) - T_0(s - s_0) \qquad (7.11)$$

and the flow availability is

$$\psi = (h - T_0 s) - (h_0 - T_0 s_0)$$
$$= (h - h_0) - T_0(s - s_0) \qquad (7.12)$$

These definitions do not include the kinetic and potential energies relative to a dead state reference plane; these terms are easily included if germane to a particular problem. The availabilities on an extensive basis are given the symbols Φ and Ψ, respectively. The quantities depend on the state of the particular substance and the environment. Therefore, they are properties relative to the defined environment. In all cases, unless specified otherwise, the environmental conditions are $P_0 = 0.101$ MPa and $T_0 = 25°C$.

Equation (7.11) shows that ϕ depends on three terms for available energy: $u - u_0$, the internal energy change relative to the dead state; $P_0(v - v_0)$, the work done against the environment; and $-T_0(s - s_0)$, the reversible heat transfer to the environment in coming to the dead state. Similarly, ψ in Eq. (7.12) shows that the flow availability comes from the enthalpy change between the initial and dead states less the reversible heat transfer between the control volume and the dead state.

These availability functions are quite similar. The difference lies in the pressure term. The nonflow availability includes a constant atmospheric pressure, while the flow availability includes a variable pressure within the enthalpy change term. It is important to note the physical process that each term is describing. The terms that arise from the inlet and outlet flow contributions are expressed through ψ, and the changes within the control volume are given by ϕ. Also, a control mass formulation includes only the nonflow availability; so ϕ is often referred to as the control mass availability, and ψ is the control volume availability. These availability functions are special cases of the more general definitions of the Helmholtz function, or Helmholtz free energy, $a = u - Ts$, and the Gibbs function, or the Gibbs free energy, $g = h - Ts$. These general functions are discussed in greater detail in Sec. 7.5.

The maximum useful work given in Eq. (7.10) is rewritten in terms of the availability functions as

$$\delta W_{rev,\ use} = d[(\phi + \tfrac{1}{2}\mathbf{V}^2 + gZ)m]_{CV} + \sum_{out} (\psi + \tfrac{1}{2}\mathbf{V}^2 + gZ)\ dm$$
$$- \sum_{in} (\psi + \tfrac{1}{2}\mathbf{V}^2 + gZ)\ dm \quad (7.13)$$

Conservation of mass has been used to eliminate the terms that include the state of the environment or dead state. Equation (7.13) clearly shows the importance and meaning of the availability functions. Consider steady state flow through a control volume with a single inlet and outlet with each port having uniform properties. The useful reversible work, or maximum available work, between the specified inlet state and an outlet state that exhausts to the environment is

$$\dot{W}_{rev,\ use} = -\dot{m}\psi_{in} \qquad\qquad (7.14)$$

where the changes in kinetic and potential energy are neglected. Thus the flow availability per unit mass at the inlet state quantifies the maximum useful work that can be obtained. For a control mass, the useful reversible work between a specified initial state 1 and the final state 2 in equilibrium with the environment is

$$W_{rev,\ use} = -\Phi_1 = -(m\phi)_1 \qquad\qquad (7.15)$$

Again, the availability indicates the maximum useful work available from the specified state.

Another application of interest is the availability from a high-temperature reservoir at T_H. The availability does not directly follow from the state functions [Eqs. (7.11) and (7.12)] defined above. The availability of a high-temperature reservoir is determined from the concept of the maximum available energy to perform work. A reservoir at T_H is available to perform work by means of some arbitrary system. The system receives a heat transfer from this reservoir. By starting from the expression for reversible, or maximum, work in Eq. (7.7), the maximum work for a single reservoir is

$$\delta W_{rev} = -\delta Q_H \left(1 - \frac{T_0}{T_H}\right) \qquad\qquad (7.16)$$

This is the maximum available energy to perform work for a reservoir at T_H, and it is viewed as the availability of the reservoir.

The above availability definitions indicate the close connection between the availability and the maximum useful work.

Example Problem 7.3

Determine the availability per unit mass for combustion products (treat as air) in an engine cylinder at 3000°C and 13 MPa.

Solution
Availability is a property, so it requires only state information. Also

consideration is directed to the combustion products contained in a cylinder, so the nonflow availability is desired. By taking the environment at $P_0 = 0.101$ MPa and $T_0 = 25°C$, Eq. (7.11) is

$$\phi = u - u_0 + P_0(v - v_0) - T_0(s - s_0)$$

Air is ideal in these states, but the specific heats are functions of temperature. So with Table D.2, $v = RT/P$, and Eq. (5.60),

$$\phi = u - u_0 + P_0 R\left(\frac{T}{P} - \frac{T_0}{P_0}\right) - T_0\left[s_0(T) - s_0(T_0) - R \ln \frac{P}{P_0}\right]$$

$$= 2940.6 - 213.40 + (101)(0.2870)\left(\frac{3273}{13,000} - \frac{298}{101}\right)$$

$$- 298\left(9.4328 - 6.6999 - 0.2870 \ln \frac{13,000}{101}\right)$$

$$= 2727.2 - 78.2 - 400.0 = 2250 \text{ kJ/kg}$$

Comments

As indicated in Eq. (7.15), this value represents the negative of the maximum useful work that can be obtained for air in this state. This can be compared to the result in Example Problem 7.1, which indicates the maximum work obtained between two specified states. The results of Example Problem 7.1 and this example are indicated graphically to the right. The available energy is a function of the specified state relative to the dead state, and the maximum work is evaluated between two process states. The actual work is less because of the irreversibilities and the unused heat transfer.

| Available energy (2250 kJ/kg) | Maximum work (1404 kJ/kg) | Actual work (861 kJ/kg) |

Example Problem 7.4

A geothermal energy source is the underground high-temperature water or steam which is heated by the earth's interior. Evaluate the steady flow availability per unit mass for water at 150°C and 0.6 MPa.

Solution

The availability per unit mass is given in Eq. (7.12) as

$$\psi = h - h_0 - T_0(s - s_0)$$

Table 5.5 gives the properties of an incompressible fluid as

$$\psi = c(T - T_0) + v(P - P_0) - T_0 c \ln \frac{T}{T_0}$$

For liquid water at 273 K (Table C.3), $c = 4.213$ kJ/(kg · K) and $v = 0.001$ m³/kg, so

$$\psi = 87.3 \text{ kJ/kg}$$

Comments
    If the temperature were 200°C, then the water would be superheated at 0.6 MPa. Tables D.8 and D.10 yield

$$\psi = 2850.2 - 104.9 - (298.15)(6.9669 - 0.3672)$$
$$= 777.6 \text{ kJ/kg}$$

which is much larger than the liquid case. The temperature increase needed to obtain this superheated vapor was 50°C.

7.4   Irreversibility

The second law of thermodynamics puts restrictions on the *potential* of possible processes. This restriction indicates the ideal limits that are obtainable. Therefore, comparisons of actual and ideal processes are possible. Some comparisons are made in Chap. 6 for both cycles and processes. The cycle efficiency compares the desired output to the required input. The component efficiencies usually compare the actual to the isentropic processes. This section presents a comparison between the actual process and the reversible process where the inlet and initial states of the actual and reversible processes are the same and the outlet and final states of the actual and reversible processes are the same. All heat transfers occur with the environment, so they are equivalent to a single heat transfer with a reservoir at T_0. This comparison is diagramed in Fig. 7.2.

Figure 7.2   **Irreversibility comparison.**

Inlets$_a$ = Inlets$_{rev}$
Outlets$_a$ = Outlets$_{rev}$
All heat transfers with T_0

Equation (7.6) gives the actual work obtained. For a single heat transfer, this equation yields

$$\delta W_a = d(E - T_0 S)_{CV,\,a} + \sum_{out} (e + Pv - T_0 s)_a \, dm$$

$$- \sum_{in} (e + Pv - T_0 s)_a \, dm + T_0 \, \delta S_{gen} \quad (7.17)$$

where the subscript a indicates that this is the actual case. The reversible work is given for this single heat transfer from Eq. (7.7) as

$$\delta W_{rev} = d(E - T_0 S)_{CV,\,rev} + \sum_{out} (e + Pv - T_0 s)_{rev} \, dm$$

$$- \sum_{in} (e + Pv - T_0 s)_{rev} \, dm \quad (7.18)$$

where the subscript *rev* indicates the reversible case. *Irreversibility I is defined as the difference between the reversible work output and the actual work output with both processes originating from the same state and both terminating at the same sta*te. Thus, subtracting Eq. (7.18) from Eq. (7.17) yields

$$\delta I = -(\delta W_{rev} - \delta W_a) = \delta W_a - \delta W_{rev} = T_0 \, \delta S_{gen} \quad (7.19)$$

The minus sign results from the sign convention on work which indicates that minus work is work output. On a rate basis, this is

$$\dot{I} = \dot{W}_a - \dot{W}_{rev} = T_0 \dot{S}_{gen} \quad (7.20)$$

Equations (7.1) and (7.2) can be substituted into (7.20) and (7.19), respectively, to yield

$$\delta I = T_0 \left(dS_{CV} + \sum_{out} s \, dm - \sum_{in} s \, dm - \frac{\delta Q_0}{T_0} \right) \quad (7.21)$$

and

$$\dot{I} = T_0 \left[\frac{\partial}{\partial t} \left(\int_{CV} \rho s \, dV \right) + \sum_{out} \int_A s \, d\dot{m} - \sum_{in} \int_A s \, d\dot{m} - \frac{\dot{Q}_0}{T_0} \right] \quad (7.22)$$

where all the heat transfers are with the reservoir at T_0. Irreversibility has units of work and, by its definition, measures the deviations from the ideal or reversible case. Therefore, the magnitude of δI or $\dot{I}$ is directly compared to the work performed and indicates a loss. A reduction in the entropy generation reduces the irreversibility and improves the work output of the actual process.

The irreversibility for steady state flow with uniform properties is given from Eq. (7.22) as

$$\dot{I} = T_0 \left(\sum_{out} s\dot{m} - \sum_{in} s\dot{m} - \frac{\dot{Q}_0}{T_0} \right) \quad (7.23)$$

For a single inlet and outlet, this reduces to

$$\dot{I} = T_0(s_{out}\dot{m}_{out} - s_{in}\dot{m}_{in}) - \dot{Q}_0 \tag{7.24}$$

or, on a unit-mass basis,

$$i = T_0(s_{out} - s_{in}) - q_0 \tag{7.25}$$

The irreversibility for a control mass proceeding from an initial state 1 to a final state 2 is obtained from Eq. (7.21) as

$$\delta I = T_0\left(dS_{CV} - \frac{\delta Q_0}{T_0}\right) \tag{7.26}$$

Integrating from the initial state to the final state yields

$$I = T_0(S_2 - S_1) - {}_1Q_{0,2} \tag{7.27}$$

On an intensive basis, this is

$$i = T_0(s_2 - s_1) - {}_1q_{0,2} \tag{7.28}$$

Example Problem 7.5

Evaluate the irreversibility per unit mass for a steady state steam turbine operating between 3 MPa, 450°C and 0.1 MPa, $x = 1.0$. The work produced by the turbine is 650 kJ/kg.

Solution

The irreversibility per unit mass is given as

$$i = T_0(s_{out} - s_{in}) - q_0$$

The specified states yield the entropy difference. The first law yields

$$q_0 = (h_{out} - h_{in}) - w_a$$

so

$$i = T_0(s_{out} - s_{in}) - (h_{out} - h_{in}) + w_a$$

Therefore, the tables give

$$i = T_0(7.3598 - 7.0822) - (2675.3 - 3343.1) - 650$$
$$= 100.6 \text{ kJ/kg}$$

Comments

This could alternatively be evaluated directly from

$$i = w_a - w_{rev}$$

where w_{rev} is given in Example Problem 7.2. This yields exactly the same result.

7.5   Exergy, Helmholtz Function, and Gibbs Function

The definitions for the nonflow availability and the flow availability are given in Sec. 7.3. They represent the ability to obtain useful work from a particular state. The intensive forms for nonflow and flow availability are given in Eqs. (7.11) and (7.12) as

$$\phi = u - u_0 + P_0(v - v_0) - T_0(s - s_0)$$

and

$$\psi = h - h_0 - T_0(s - s_0)$$

These definitions are also termed *nonflow exergy* and *flow exergy*, respectively [1,2]. There are other related definitions, and this section presents this nomenclature.

The availabilities given above express the difference between properties at the specified state and the dead state. For the nonflow case, the property is $u + P_0 v - T_0 s$, and for the steady flow case, the property is $h - T_0 s$. These definitions are also termed the *availability functions* [2,3]. The difference in the availability function at the specified state and the dead state yields the nonflow and flow availability used in this text. Essentially, the availability used here is directly referenced to the dead state, so that their values are zero at this dead state.

The *Helmholtz function*, or *Helmholtz free energy*, $a = u - Ts$ is a state function that depends on only the state of the substance (it is not tied to the dead state). The nonflow availability is related to the Helmholtz function for a special process. Consider a control mass that undergoes a constant-volume and constant-temperature process. The process occurs in thermal equilibrium with the atmosphere (i.e., at $T = T_0$). The maximum useful work for a process from state 1 to state 2 is obtained from Eq. (7.13) as

$$_1W_{rev, use, 2} = \phi_2 - \phi_1 \qquad (7.29)$$

where the changes in kinetic and potential energy are neglected. With the nonflow availability, the result for a constant-volume and constant-temperature ($T_1 = T_2 = T_0$) process is

$$_1W_{rev, use, 2} = (u_1 - T_1 s_1) - (u_2 - T_2 s_2) = a_1 - a_2 \qquad (7.30)$$

Thus, the nonflow availability is the Helmholtz function for a constant-volume and constant-temperature ($T_1 = T_2 = T_0$) process, and the difference gives the maximum useful work. Be sure to note that the above equation is *trivial* for a pure simple compressible substance, since the state postulate requires only two independent properties to determine the state. Therefore, with $V_1 = V_2$ and $T_1 = T_2$, the states are determined and $a_1 = a_2$. Equation (7.30) becomes useful for chemically reacting systems, which require more properties to specify the state.

The *Gibbs function, or Gibbs free energy, $g = h - Ts$ is also a state function and is related to the flow availability.* The maximum useful work for a control volume undergoing a constant-pressure and constant-temperature process in equilibrium with the atmosphere ($P_1 = P_2 = P_0$ and $T_1 = T_2 = T_0$) is given from Eq. (7.13) as

$$
\begin{aligned}
{}_1 W_{\text{rev, use, 2}} &= \psi_{\text{out}} - \psi_{\text{in}} \\
&= (h - Ts)_{\text{out}} - (h - Ts)_{\text{in}} \\
&= g_{\text{out}} - g_{\text{in}}
\end{aligned}
\tag{7.31}
$$

where the changes in kinetic and potential energy are neglected and there is a single inlet and outlet. The flow availability and Gibbs function are the same for the constant-pressure and constant-temperature process indicated above. The same final result is obtained for a control mass undergoing a constant-pressure and constant-temperature process. Again note that this difference is zero for a pure simple compressible substance, and Eq. (7.31) is useful for chemically reacting systems.

7.6 General Process Comparisons

This section develops a general comparison between processes and extends the previous analysis. Consider a general control volume with an arbitrary number of inlets, outlets, heat transfers, and work mechanisms. The system is depicted in Fig. 7.3. The actual process is indicated by the subscript *a*. The comparison is with a totally reversible process. This process is also diagramed in Fig. 7.3, where the subscript *rev* indicates reversible. At this point in the development, the inlets, outlets, initial states, and final states may be different for the two processes. The mass flow rates are assumed to be the same at all points in each process.

The first and second laws can be written for each process and then combined. The work for the *actual* process is given in Eq. (7.6) as

$$
\delta W_a = d(E - T_0 S)_{\text{CV}, a} + \sum_{\text{out}} (e + Pv - T_0 s)_a \, dm
$$
$$
- \sum_{\text{in}} (e + Pv - T_0 s)_a \, dm - \sum_i \delta Q_{i, a} \left(1 - \frac{T_0}{T_i}\right) + T_0 \, \delta S_{\text{gen}} \tag{7.32}
$$

where δS_{gen} is finite, since this is the actual process and is generally irreversible. The work for the *reversible* process is [same as Eq. (7.7)]

$$
\delta W_{\text{rev}} = d(E - T_0 S)_{\text{CV, rev}} + \sum_{\text{out}} (e + Pv - T_0 s)_{\text{rev}} \, dm
$$
$$
- \sum_{\text{in}} (e + Pv - T_0 s)_{\text{rev}} \, dm - \sum_i \delta Q_{i, \text{rev}} \left(1 - \frac{T_0}{T_i}\right) \tag{7.33}
$$

There is zero entropy generation for the reversible process.

The processes are compared on the basis of the work output, δW_a and δW_{rev}. The optimum that is expected from a process is the totally reversible process, and the actual process has irreversibilities or finite entropy generation contributions. The goal of the comparison is to determine the effect of the irreversibilities on the work output of a process. Equations (7.32) and (7.33) give the expressions for the work obtained by the actual and reversible processes. If a comparison of the work of an actual process and a reversible process is to be made, not all the terms in the equations describing them can be identical. The actual process has a finite entropy generation, so one of the other terms in Eq. (7.32) must be different from the corresponding term in Eq. (7.33). The specific term that differs in the two processes accounts for the different physical processes occurring—one is actual, and the other is reversible. This physical effect or term in the governing equation accommodates the differences between the actual and reversible processes.

The term that varies between the actual and reversible processes is completely arbitrary, yet conventions have been adopted. A standard process comparison considers the states and, therefore, properties within the control volume and the inlet and outlet states to be the same for both the reversible and the actual process. Therefore, the heat transfer is used to accommodate the irreversibility, or entropy generation, of the actual process. The heat transfer "floats" to account for the entropy generation. This comparison is presented in Sec. 7.4 and defined as irreversibility. The choice of the heat transfer as the term to account for the irreversibility is not universal. Thus, the general process comparison is left arbitrary here.

The comparison of the actual process with the reversible process is made by subtracting the actual work obtained from the reversible work. This yields [subtract Eq. (7.33) from Eq. (7.32)]

$$\delta W_a - \delta W_{rev} = T_0\, \delta S_{gen} + d[(E - T_0 S)_{CV,\,a} - (E - T_0 S)_{CV,\,rev}]$$
$$+ \sum_{out} [(e + Pv - T_0 s)_a - (e + Pv - T_0 s)_{rev}]\, dm$$
$$- \sum_{in} [(e + Pv - T_0 s)_a - (e + Pv - T_0 s)_{rev}]\, dm$$
$$- \sum_i (\delta Q_{i,\,a} - \delta Q_{i,\,rev})\left(1 - \frac{T_0}{T_i}\right) \quad (7.34)$$

This equation relates the maximum work that is possible for a chosen reference process to the actual work obtained by the real process under examination. The expression is in terms of a difference and can be interpreted as *lost available work* [2]. The resulting value is a work-type term which would be zero if the actual process were the reversible process. The first term on the right-hand side represents the irreversibilities of the actual process and is the reason for the comparison. The next three terms are differences in actual and reversible quantities. They are differences in properties within the control volume and differences at the inlets and outlets. The last term represents the difference in heat transfer for the actual and reversible reference processes.

Inlets$_a \neq$ Inlets$_{rev}$
Outlets$_a \neq$ Outlets$_{rev}$
$\delta Q_{k,a} \neq \delta Q_{k,rev}$
All heat transfers with same T_k's

Figure 7.3   General process comparison.

The comparison of the two processes in Fig. 7.3 and Eq. (7.34) might seem like a comparison of apples and oranges, since the processes compared appear to have no connection with each other. This is the case to this point in the development. To understand the generality of this expression, let us consider a few applications presented previously, recalling that Eq. (7.34) compares the actual and reference ideal processes.

Comparison 1

Consider the case of steady state conditions within a control volume with a single inlet and single outlet. There is only one heat transfer to the control volume from the surroundings at T_0. The sum over all the heat transfers in Eq. (7.34) includes only one term, and the temperature is T_0. Thus the heat transfer term does not appear directly in the equation, but it permits the process comparison. The reversible comparison considers the same inlet and outlet states as the actual process. Therefore, the inlet and outlet contributions in Eq. (7.34) are zero. The change within the control volume is zero since the process is steady state. Equation (7.34) reduces to

$$\delta W_a - \delta W_{rev} = T_0\, \delta S_{gen} \tag{7.35}$$

For this steady state process comparison, Eq. (7.35) on a rate basis is

$$\dot{W}_a - \dot{W}_{rev} = T_0 \dot{S}_{gen} \tag{7.36}$$

The difference between the reversible and actual case is the entropy generation of the actual process. These are the same as Eqs. (7.19) and (7.20) for irreversibility.

Comparison 2

Consider a control mass process with a heat transfer from the surroundings at T_0 with the actual and reversible processes that originate at a common state and terminate at another common state. A control mass does not have inlet or outlet ports, so Eq. (7.34) reduces to

$$\delta W_a - \delta W_{rev} = T_0 \, \delta S_{gen} \qquad (7.37)$$

This is the same result obtained for the irreversibility in Sec. 7.4. The common feature in both of the comparisons presented is the heat transfer with the surroundings at T_0. The states of the system are the same, and the process comparison is accommodated by the heat transfer term.

Comparison 3

A final comparison considers an actual and a reversible engine operating in a cycle between two heat transfer reservoirs at T_0 and T_1. The higher-temperature source is at T_1, and the low-temperature reservoir is at T_0. The second law for the cycle is [Eq. (7.2)]

$$S_{gen} = \frac{|Q_0|}{T_0} - \frac{|Q_1|}{T_1} \geq 0 \qquad (7.38)$$

Since the temperatures of the reservoirs are fixed, the heat transfers will differ for the actual and the reversible case. The integrated cyclic form of Eq. (7.34) is

$$W_a - W_{rev} = T_0 S_{gen} - (Q_{1,a} - Q_{1,rev})\left(1 - \frac{T_0}{T_1}\right) \qquad (7.39)$$

The comparison is arbitrary at this point since the actual and reversible heat transfer inputs are different. If the comparison is made with the basis being equal heat transfer inputs, then $Q_{1,a} = Q_{1,rev}$, so

$$W_a - W_{rev} = T_0 S_{gen} \qquad (7.40)$$

Note that Q_0 floats in the comparison.

Clearly numerous comparisons can be presented. The comparisons are based on the reversible work in Eq. (7.7) and the actual work. The difference is given in Eq. (7.34).

Example Problem 7.6

An idealized Rankine cycle operates with steam between a condenser temperature of 20°C and a boiler temperature of 100°C. The turbine inlet is a saturated vapor, and the pump inlet is a saturated liquid. The adiabatic turbine efficiency is 70 percent, and the adiabatic pump efficiency is 100 percent. Evaluate $w_a - w_{rev}$ for each process of the cycle as well as for the overall cycle. The ambient temperature is 15°C, and the energy source is at 150°C.

Diagram

Solution

The state information is as follows:

State	P, kPa	T, °C	h, kJ/kg	s, kJ/(kg · K)	x
1	101.32†	—	(84.0)	0.2965	—
2	(101.32)	100	—	(1.3062)	0.0
3	(101.32)	100	(2675.8)	(7.3554)	1.0
4a	(2.34)	20	2310.5	(7.8894)	(0.9070)
4s	2.34	(20)	(2153.9)	7.3554	(0.8432)
5	(2.34)	20	(83.9)	(0.2965)	0.0

† Entries not in parentheses determine the state.

The actual state at the exit of the turbine is found with the component efficiency. For the turbine,

$$\eta_t = \frac{h_{4,a} - h_3}{h_{4,s} - h_3} = 0.70$$

so

$$h_{4,a} = h_3 + (0.70)(h_{4,s} - h_3) = 2310.5 \text{ kJ/kg}$$

The pump is isentropic since the efficiency is 100 percent. Thus,

$$w_p = v_5(P_1 - P_5) = h_1 - h_5 = 0.099 \text{ kJ/kg}$$

and

$$h_1 = 84.0 \text{ kJ/kg}$$

All the states are determined, and the properties are obtained from Table D.8.

The first law analysis of this cycle yields the heat transfers and the work. Since some of this information is required in the lost available work analysis, it is presented here. The work for the pump and the work for the turbine are

$$_5w_1 = w_p = v_5(P_1 - P_5) = 0.099 \text{ kJ/kg}$$

and

$$_3w_{4,a} = w_t = h_{4,a} - h_3 = -365.3 \text{ kJ/kg}$$

so

$$w_{net} = w_t + w_p = -365.2 \text{ kJ/kg}$$

The heat transfers in the boiler and condenser are

$$_1q_3 = q_b = h_3 - h_1 = 2591.8 \text{ kJ/kg}$$

$$_{4,a}q_5 = q_c = h_5 - h_{4,a} = -2226.6 \text{ kJ/kg}$$

Equation (7.34) is the governing expression for the lost available work, and the entropy generation is given in Eq. (7.2). The steady state form eliminates the changes within the control volume. In the comparison between the actual and reversible component processes, the inlet and outlet states are the same, and the heat transfers are taken as equal. Thus, with Eq. (7.2) substituted into (7.34), the governing equation is

$$w_a - w_{rev} = T_0(s_{out} - s_{in}) - T_0 \sum_i \frac{q_i}{T_i}$$

The values of lost available work for each process, using $T_0 = 273.15 + 15 = 288.15$ K are

$$(w_a - w_{rev})_p = T_0(s_1 - s_5) = 0$$

$$(w_a - w_{rev})_t = T_0(s_{4,a} - s_3) = 153.9 \text{ kJ/kg}$$

$$(w_a - w_{rev})_b = T_0(s_3 - s_1) - T_0 \frac{q_b}{273.15 + 150} = 269.1 \text{ kJ/kg}$$

$$(w_a - w_{rev})_c = T_0(s_5 - s_{4,a}) - T_0 \frac{q_c}{T_0} = 38.7 \text{ kJ/kg}$$

The lost available work in the pump is zero since the process is reversible and adiabatic or isentropic.

The lost available work for the overall cycle is found from the above expression by noting that the entropy difference for a cycle is zero. Therefore,

$$(w_a - w_{rev})_{cycle} = -q_b \frac{T_0}{423.15} - q_c \frac{T_0}{T_0}$$

or

$$(w_a - w_{rev})_{cycle} = |q_c| - |q_b| \frac{288.15}{423.15} = 461.7 \text{ kJ/kg}$$

This value is the same as that obtained by summing the lost available work values for the turbine, boiler, and condenser.

Comments

Note that the lost available work expressions in this example are very similar to the irreversibility definition in Sec. 7.4. The difference here is that heat transfers are permitted with more than one surrounding.

This example, repeated for various turbine efficiencies, yields the following results:

η_t, %	$(w_a - w_{rev})_t$, kJ/kg	$(w_a - w_{rev})_b$, kJ/kg	$(w_a - w_{rev})_c$, kJ/kg	$(w_a - w_{rev})_{cycle}$, kJ/kg
70	153.9	269.1	38.7	461.7
80	102.6	269.1	37.8	409.5
90	51.3	269.1	36.9	357.3
100	0.0	269.1	36.0	305.1

The sum of the turbine, boiler, and condenser lost available work values equal the cycle value, as shown. The pump value is always zero since it is isentropic.

One final and important point. The lost available work for the adiabatic turbine given above is exactly the irreversibility in Sec. 7.4, as noted in comparison 1 of Sec. 7.5. It is equally important to compare the actual turbine work to that of the reversible isentropic turbine. This is a comparison which has different outlet states. The lost available work in Eq. (7.34) is

$$(w_a - w_{rev})_t = T_0(s_{4,a} - s_3) + (h - T_0 s)_{4,a} - (h - T_0 s)_{4,s}$$
$$= h_{4,a} - h_{4,s} + T_0(s_{4,s} - s_3)$$

But $s_{4,s} = s_3$, so

$$(w_a - w_{rev})_t = h_{4,a} - h_{4,s}$$

The values for this lost available work comparison are as follows:

η_t	$(w_a - w_{rev})_t$, kJ/kg
70	156.6
80	104.4
90	52.2
100	0.0

The values in this comparison are larger than those in the previous comparison. The value of w_{rev} in this last case is fixed at the isentropic value (turbine efficiency = 100 percent) and is the largest negative value possible. The previous comparison was between the *same* inlet and *same* outlet states for the actual and reversible work at a given turbine efficiency, so the comparison process changed with the turbine efficiency and had a smaller negative value than for the isentropic turbine.

7.7　Summary

The theme of this chapter is entropy generation, and the reduction of the entropy generation is the desired goal of second law analysis. Very important definitions and relations are developed in this chapter to analyze thermodynamic systems. A substance at other than the dead state contains available energy which is useful to meet our needs. Engineers try to use this available energy to meet the needs of their particular devices. Methods to calculate this available energy are presented in Sec. 7.3. Availability is directly related to the maximum useful work that can be obtained from a specified state (in relation to the dead state).

Other constraints may be imposed on thermodynamic devices that do not permit the use of all the available energy. Constraints imposed by other system components, materials, economics, etc., indicate whether a process is permitted between an initial and final state (or inlet and outlet state). Therefore, the maximum useful work between two specified states is important. In this case, the reversible work [Eq. (7.7)] and reversible useful work [Eq. (7.10)] are the relations for the maximum work that can be obtained. The measure of the inability of a real process to reach the ideal case between two specified states is the irreversibility [Eqs. (7.21) and (7.22)]. The irreversibility is directly related to entropy generation, and it can be directly compared to the difference between the actual and reversible work.

A general comparison equation for real and ideal processes is Eq. (7.34). This relation compares two arbitrary processes. The examples given in Sec. 7.6 indicate that irreversibility is a specific case of this general comparison. Specification of an actual process and a reversible comparison process enables the process comparison to be made.

Problems

7.1　The cooling water from a power plant leaves the condenser at 33°C and atmospheric pressure. The water is cooled and discharged into a reservoir at 25°C and atmospheric pressure. Determine the heat transfer required per unit mass and the entropy generation. Evaluate the reversible work that could be obtained from the specified states.

$Q = m(\Delta h) = m c_p \Delta T$

$S_2 = mc \ln \frac{T_2}{T_1}$

$W_{rev} = (h_{ex} - h_{in}) - T(s_{ex} - s_{in})$

7.2 The concept of cogeneration is that a Rankine cycle can be used with a turbine outlet condition that is above ambient conditions. The steam enthalpy at the turbine exit is then used for a second purpose such as process steam, operating a heating or cooling system, etc.

(a) Plot the availability of turbine exit steam versus exit pressure $(101.3 < P < 1000$ kPa) with quality $(0.5 < x < 1)$ as a parameter.

(b) If you were assigned Prob. 6.32, also note how the Rankine cycle efficiency in that case varies with turbine exit pressure. Discuss any limitations that you see on cogeneration.

7.3 Ocean thermal energy conversion (OTEC) systems seek to use the temperature gradient within the ocean as an energy source. What is the flow availability per unit mass for surface water at 25°C if the dead state is at 10°C?

7.4 An ocean thermal energy conversion system operates as a heat engine taking surface water at 28°C as the high-temperature reservoir and deep water at 14°C as the low-temperature reservoir. What is w_{rev} for this system?

7.5 Two sources of refrigerant 12 are available for use within a closed cylinder: 4 MPa, 150°C or 80 kPa, 200°C. Which source has the larger available energy? (Use $P_0 = 100$ kPa.)

7.6 A geothermal well produces steam at the wellhead at 220°C and 1 MPa. What are the flow availability and reversible work for this source?

7.7 A solar pond 3 m deep produces energy at the pond bottom that can be withdrawn at up to the saturation temperature at the pond bottom. Assume, for simplicity, that the specific volume of the pond water can be taken as an average of 8.83×10^{-4} m³/kg. What are the flow availability and reversible work for this source?

7.8 An air bottle contains air at 10 atm and 500 K. The air is cooled by heat transfer with the surroundings until it reaches 300 K.

(a) What is the availability ϕ of the air in the initial state?

(b) What is the reversible work in the cooling process?

(c) What is the irreversibility of the process?

7.9 Evaluate the Gibbs function g and the Helmholtz function a at the inlet to a turbine receiving steam at 350°C and 2 MPa.

7.10 An ideal gas is compressed isothermally at temperature T from volume V_1 to V_2. What is the ratio of the reversible to the actual work done? The surroundings are at T_0.

7.11 An ideal gas is compressed isobarically from volume V_1 to volume V_2. What is the ratio of the reversible to the actual work done? The surroundings are at T_0.

7.12 Starting with Eq. (7.34), derive a relation for the efficiency of a steam turbine, $\eta_t' = w_a/w_{rev}$, operating between T and P at the inlet and P_{out} at the exhaust. Use the same assumptions made in Sec. 5.10.1 in determining the turbine efficiency. Discuss the result, and compare it with the result normally used for turbine efficiency.

7.13 Calculate the irreversibility per unit mass for an adiabatic turbine with an inlet at 4 MPa and 350°C and an outlet pressure of 0.1 MPa. The turbine efficiency is 0.80, and the working fluid is water.

7.14 The tank shown in Fig. P7.14 initially contains 0.3 kg of an ideal gas, N_2, at 530 K and an adiabatic weightless and frictionless piston. The 0.12-m³ tank is connected to an infinite source of air at 530 K and 0.7 MPa. Air enters the tank until the temperature of the air is 560 K, and the final mass of air is 0.15 kg. Calculate the irreversibility of the process, assuming there is 8.6 kJ of heat transfer *from* the N_2 to the surroundings at 25°C.

7.15 A steam turbine receives superheated steam at 350°C and 2 MPa and exhausts to a pressure of 8 kPa. The isentropic turbine efficiency is 75 percent.
   (a) What fraction of the availability is produced as work by the turbine?
   (b) What fraction of the maximum work is produced by the turbine?

7.16 What is the irreversibility of the turbine in Prob. 7.15?

7.17 A solar tower generates steam at 515°C and 1 MPa and transfers energy to a mineral oil for storage at 304°C. Calculate the availability of the generated steam and of the stored hot mineral oil. Discuss whether it is more desirable to generate power directly from the steam or from stored energy via a steam generator.

7.18 Ammonia vapor at −10°C and 100 kPa is to enter an adiabatic compressor and be compressed to a final pressure of 250 kPa. The compressor has an isentropic efficiency of 70 percent.
   (a) Find the actual compressor work.
   (b) Calculate the reversible work for the prescribed process.
   (c) What is the flow availability of the ammonia entering the compressor?
   (d) Find the flow availability of the ammonia leaving the compressor.

Figure P7.14

7.19 Ammonia at $-10°C$ and 100 kPa is to enter an isothermal compressor and be compressed to a final pressure of 250 kPa. The compressor has an isothermal efficiency of 70 percent. The heat transfer in the ideal (isothermal) compressor is -150 kJ/kg.

(a) Find the actual compressor work.

(b) What is the reversible work for the prescribed process?

(c) Calculate the flow availability of the ammonia entering the compressor.

(d) Find the flow availability of the ammonia leaving the compressor.

7.20 A process heater is to increase the temperature of a working fluid by heat transfer at a rate $\dot{Q}$ from a reservoir at temperature T_R. The working fluid receives the heat transfer in a constant-pressure process. Derive an expression for the second law efficiency for this process.

7.21 The efficiency of a given heat engine as defined in Sec. 5.11 is a first law efficiency; i.e., it describes the work output relative to the heat transfer input. Discuss how you would describe the second law efficiency of a heat engine, and derive an expression for such an efficiency.

7.22 Derive an expression for the second law efficiency of a simple closed Brayton cycle operating between maximum and minimum temperatures of T_2 and T_4.

7.23 One form of the Clausius statement of the second law is: "It is not possible to transfer heat from a lower- to a higher-temperature reservoir without doing work on the system." Given this statement, consider an absorption cooling machine (see Sec. 6.5.4 for a description).

(a) Given that only heat transfer—no work—is necessary to operate the cycle, does it violate the Clausius statement of the second law?

(b) How would you characterize the second law efficiency of an absorption cooling system?

7.24 Find the second law efficiency of a Rankine steam cycle operating at a boiler pressure of 6 MPa with a superheat temperature of 400°C, a condenser pressure of 8 kPa, and an isentropic turbine.

7.25 Derive an expression for the second law efficiency of a general refrigeration cycle. Using this expression, find the second law efficiency of a vapor-compression refrigeration cycle using refrigerant 12 with an evaporator temperature of 5°C and a condenser temperature of 40°C. The cycle has no superheat at the evaporator exit and no subcooling at the condenser exit. The compressor has an efficiency of 65 percent.

References

1. J. E. Ahern, *The Exergy Method of Energy Systems Analysis,* Wiley, New York, 1980.
2. A. Bejan, *Entropy Generation through Heat and Fluid Flow,* Wiley-Interscience, New York, 1982.
3. J. H. Keenan, *Thermodynamics,* Wiley, New York, 1941.

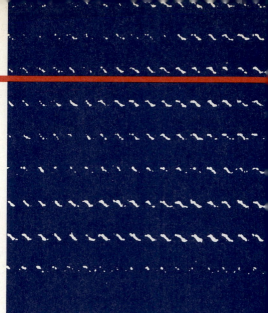

8

General Property Relations and Equations of State

When a greybearded scientist says something is possible, believe him; when he says that it's impossible, he's very likely wrong.
Arthur C. Clarke

Automotive heat exchanger (radiator) using fin-tube surfaces to increase area for heat transfer. *(Harrison Radiator Division, General Motors.)*

8.1 Introduction

There are many reasons for finding relationships among thermodynamic properties. First, it is, of course, desirable to minimize the amount of laboratory work, since obtaining reliable and accurate thermodynamic property data requires meticulous measurements. If a minimum set of property data is obtained and relationships are available for calculation of other properties from these measured values, much tedious work can be avoided. A second reason for developing relations among properties is that the engineer or scientist may require data on the thermodynamic properties of a particular substance for which little or no data are available in the literature. In such a case, it is certainly much easier to use the few existing data to generate other property values than to measure the required values.

In this chapter, we develop the general relations among thermodynamic properties, and we apply these relations to some common situations. In so doing, we see how various tables, such as those for ideal gases and for steam, have been generated from limited experimental property measurements.

8.2 Relations among Properties

8.2.1 Fundamental Equations and Maxwell's Relations

We see in Chap. 2 that for a pure simple compressible substance, the state of a system is defined by specifying any two independent thermodynamic properties. It follows that all other properties are thus fixed whenever two properties are specified. In other words, any two independent properties can be chosen, and all other dependent properties are then fixed.

Let us, for example, fix the two independent properties s and v. Then the internal energy at this state is fixed at the value $u(s, v)$. Let us see what this implies. The total differential of u as a function of its partial derivatives is then [Eq. (1.15)]

$$du = \left(\frac{\partial u}{\partial s}\right)_v ds + \left(\frac{\partial u}{\partial v}\right)_s dv \tag{8.1}$$

Now, recall Eq. (5.43):

$$du = T\, ds - P\, dv \tag{8.2}$$

Note that neither equation contains any assumptions about reversible processes, adiabatic systems, etc. Both are completely general, Eq. (8.1) being a mathematical relationship between total and partial derivatives and Eq. (8.2) being a general result of the first law providing a relation among thermodynamic properties of a simple system containing a pure substance.

Comparing the two equations, we find the identities

$$T = \left(\frac{\partial u}{\partial s}\right)_v \tag{8.3}$$

and

$$P = -\left(\frac{\partial u}{\partial v}\right)_s \tag{8.4}$$

These were the defining relations for T and P, stated in Chap. 5 in Eqs. (5.9) and (5.23). Thus, if data were available for internal energy u as a function of s and v, then *all* the thermodynamic properties at a given state could be calculated. When s and v are given, u is computed from the data; the corresponding T and P are found from Eqs. (8.3) and (8.4); and the enthalpy h, the Helmholtz function a, and the Gibbs function g are found from their definitions, $h = u + Pv$, $a = u - Ts$, and $g = h - Ts$.

It appears that the function $u(s, v)$ contains the necessary information to find all other properties at any given state. It is thus called a *fundamental equation,* or *characteristic thermodynamic function,* for the system. Is this true of *any* dependent property in terms of two independent properties? It is not. The reason is that if properties other than s and v are chosen as the independent variables, then we find it impossible to develop a set of equations equivalent to Eqs. (8.3) and (8.4), since we cannot match the partial derivatives with the properties in the equation of state. We will have introduced some degrees of freedom into the equations; that is, we will have fewer equations than unknown properties.

Example Problem 8.1

Develop the relations among properties starting with a known function $u(T, v)$ and Eq. (8.2).

Solution
The total derivative of the given function is

$$du = \left(\frac{\partial u}{\partial T}\right)_v dT + \left(\frac{\partial u}{\partial v}\right)_T dv$$

We also have, from Eq. (8.2),

$$du = T\,ds - P\,dv$$

Comparison of these two equations provides no way to relate property T

or P to the derivatives of u which are available to us. Thus $u(T, v)$ is *not* a fundamental equation.

How do we determine what sets of independent and dependent variables *are* fundamental equations? We must choose sets having dependent and independent variables that can be matched with a set of variables in a general equation of state such as Eq. (8.2). Two such equations are developed in Sec. 5.1:

$$du = T\,ds - P\,dv \tag{8.5a}$$

$$dh = T\,ds + v\,dP \tag{8.5b}$$

With the definitions of the Helmholtz and Gibbs functions in Sec. 7.5 and Eqs. (8.5a) and (8.5b), two other equations are

$$da = du - T\,ds - s\,dT = -P\,dv - s\,dT \tag{8.5c}$$

$$dg = dh - T\,ds - s\,dT = v\,dP - s\,dT \tag{8.5d}$$

Thus, it appears that the fundamental equations must be $u(s, v)$, $h(s, P)$, $a(v, T)$, and $g(P, T)$.

Let us repeat the development used in Eqs. (8.1) through (8.4), but we start with $h(s, P)$ and Eq. (8.5b). The total derivative dh is then

$$dh = \left(\frac{\partial h}{\partial s}\right)_P ds + \left(\frac{\partial h}{\partial P}\right)_s dP \tag{8.6}$$

However, without restriction, Eq. (8.5b) gives for a pure substance

$$dh = T\,ds + v\,dP \tag{8.7}$$

and, by inspection, we find

$$T = \left(\frac{\partial h}{\partial s}\right)_P \tag{8.8}$$

and

$$v = \left(\frac{\partial h}{\partial P}\right)_s \tag{8.9}$$

Now, if data are available for $h(s, P)$, we could, for a given s and P, find the corresponding T, v, $u = h - Pv$, $a = u - Ts$, and $g = h - Ts$.

Of course, laboratory data are usually taken in terms of the variables P, v, and T, usually as either $P(v, T)$ or $v(P, T)$. Seldom are thermodynamic property data available primarily in the form $u(s, v)$ or $h(s, P)$. We develop relations later in this chapter for finding u, h, s, g, and a in terms of data in P, v, and T.

Note that some insight into property behavior can be gleaned from the relations derived above. For example, because v is always a positive quantity, Eq. (8.9) shows that enthalpy h will always increase with pressure in an isentropic process. Similarly, Eq. (8.4) shows that internal energy u will always decrease with increasing volume in an isentropic process.

One additional piece of information has been developed. Note that Eqs. (8.3) and (8.8) both provide relations for temperature. Equating the right-hand sides of these equations results in the identity

$$\left(\frac{\partial u}{\partial s}\right)_v = \left(\frac{\partial h}{\partial s}\right)_P \tag{8.10}$$

Equations of the form of Eq. (8.10) occur because properties are exact differential quantities, and relations between the properties thus follow the relations developed for exact differentials in Sec. 1.7.4. For example, Eq. (1.31a) can be invoked to develop a variety of relations among the properties. Now, applying Eq. (1.31a) to Eqs. (8.5a) to (8.5d) in turn results in

$$\left(\frac{\partial T}{\partial v}\right)_s = -\left(\frac{\partial P}{\partial s}\right)_v \tag{8.11a}$$

$$\left(\frac{\partial T}{\partial P}\right)_s = \left(\frac{\partial v}{\partial s}\right)_P \tag{8.11b}$$

$$\left(\frac{\partial P}{\partial T}\right)_v = \left(\frac{\partial s}{\partial v}\right)_T \tag{8.11c}$$

$$\left(\frac{\partial v}{\partial T}\right)_P = -\left(\frac{\partial s}{\partial P}\right)_T \tag{8.11d}$$

These equations are called *Maxwell's relations.* They hold for the properties of a simple compressible substance. There are many other such relations when other work modes are considered.

Example Problem 8.2

Using Maxwell's relations, find an expression for the change in entropy with volume for an isothermal process involving an ideal gas.

Solution
Applying Eq. (8.11c), we find

$$\left(\frac{\partial s}{\partial v}\right)_T = \left(\frac{\partial P}{\partial T}\right)_v = \frac{R}{v}$$

so that the entropy changes inversely with the volume in such a process.

8.2.2 Clapeyron Equation

As a useful application of the Maxwell relations, consider Eq. (8.11c), repeated here:

$$\left(\frac{\partial P}{\partial T}\right)_v = \left(\frac{\partial s}{\partial v}\right)_T$$

If we consider an isothermal phase change process, say for the evaporation of a saturated liquid, then the right-hand side is simply

$$\frac{s_g - s_l}{v_g - v_l} = \frac{s_{lg}}{v_{lg}} \tag{8.12}$$

Also, in a phase change, pressure and temperature are dependent only on each other, and the derivative of P with respect to T becomes an ordinary rather than a partial derivative. Equation (8.11c) can now be written as

$$\left(\frac{dP}{dT}\right)_{sat} = \frac{s_{lg}}{v_{lg}} \tag{8.13}$$

For an isothermal phase change, Eq. (8.5b) gives $h_{lg} = T_{sat} s_{lg}$, and Eq. (8.13) finally becomes

$$\left(\frac{dP}{dT}\right)_{sat} = \frac{h_{lg}}{T_{sat} v_{lg}} \tag{8.14}$$

This equation, called the *Clapeyron equation,* provides a useful check on P-v-T data for saturated substances when h_{lg} is also available; alternatively, h_{lg} can be computed from easily measurable P-v-T data for saturated liquid and vapor. In addition, consider the P-T curve for a given substance, such as presented for water in Fig. 3.8. Equation (8.14) can be used to construct the P-T saturation curve by integrating between any two saturation states to obtain

$$(P_2 - P_1)_{sat} = \int_{T_1, \, sat}^{T_2, \, sat} \frac{h_{lg}}{T v_{lg}} \, dT \tag{8.15}$$

Because all the properties in the integrand are temperature-dependent, the integral must be carried out numerically.

Example Problem 8.3

Discuss the shape of the solid-liquid equilibrium curve on a *P-T* diagram
for a substance that contracts on freezing in comparison with the curve for
a substance that expands on freezing.

Solution

The Clapeyron equation, Eq. (8.14), can be rewritten for a solid-liquid phase change in the form

$$\left(\frac{dP}{dT}\right)_{\text{melt}} = \frac{h_l - h_s}{T_{\text{melt}}(v_l - v_s)} = \frac{h_{sl}}{T_{\text{melt}}v_{sl}}$$

This equation predicts that the slope of the *P-T* curve along the solid-liquid phase change curve for any material which contracts upon freezing
$(v_l - v_s = v_{sl} > 0)$ will be positive (since the heat of fusion h_{sl} is always
positive). For a substance that expands on freezing $(v_{sl} < 0)$, dP/dT will be
negative. As noted in Chap. 3, water expands as it freezes. The slope is
observed in Fig. 3.14.

8.2.3   Generation of Property Tables

Tables of properties and their use are discussed in Chap. 3. Here, we
briefly note how such tables are generated from available experimental
data.

Probably the most accurate and complete tables for the properties of
steam presently available are those of the National Bureau of Standards/
National Research Council [1]. In these tables, the Helmholtz function a
is used as the characteristic function. Values of a are readily obtained from
P-v-T data, which is the most easily measured set, as will be seen.

It is assumed that a particular equation of state (an expanded form
of the Ursell-Mayer equation of state for nearly hard ellipsoid molecules)
is valid for the value of a as a function of density ρ (rather than specific
volume) and temperature. This is a fundamental equation, as shown in
Sec. 8.2.1. The value of the Helmholtz function given by the equation of
state is called $a_{\text{basis}}(\rho, T)$. However, comparison of the Ursell-Mayer
equation of state with experimental data shows deviations, particularly at
high density or near the critical point. A correction term, called the *residual function* $a_{\text{resid}}(\rho, T)$ is thus obtained by fitting a function composed of
40 terms to the complete range of data. Finally, a term is added to account
for ideal gas behavior at very low densities where the other terms approach
zero. This term, $a_{\text{ideal}}(T)$, is a function of temperature only. The Helmholtz function is thus given by

$$a(\rho, T) = a_{\text{basis}}(\rho, T) + a_{\text{resid}}(\rho, T) + a_{\text{ideal}}(T) \tag{8.16}$$

Once this function is constructed, all other properties can be found by taking the appropriate derivatives, and complete property tables can be constructed. The results can be validated by comparison of computed property values (say c_v or c_P) with experimentally determined values. This is a severe test, because some of the computed properties depend on second derivatives of the Helmholtz function and can be considerably more inaccurate than the function itself. The comparisons presented in Ref. 1 show excellent agreement.

   All the above arguments presuppose that values of a can be found from P-v-T data so that the curve fits can be produced. This is done as follows. Values of u and s must be determined at each state, so that $a = u - Ts$ can be computed. Suppose that a graph of P versus T is constructed, with constant-volume lines superimposed from the measured P-v-T data. (Such data are obtained by placing a fixed mass into a container of known volume and measuring the pressure as the temperature of the fixed mass is varied. The experiment is then repeated for different fixed masses.) The graph will appear as in Fig. 8.1, which shows a portion of the superheat region for steam. From such graphical presentations, values of $P(T, v)$ are easily obtained, as are certain derivatives such as $(\partial P/\partial T)_v$.

   Let us now develop a method for computing the internal energy from P-v-T data. This can be done by letting u be a function of T and v, as $u(T, v)$. The total derivative is then

Figure 8.1 *P-T* diagram for superheated steam.

$$du = \left(\frac{\partial u}{\partial T}\right)_v dT + \left(\frac{\partial u}{\partial v}\right)_T dv \tag{8.17}$$

If we also assume $s(T, v)$, then

$$ds = \left(\frac{\partial s}{\partial T}\right)_v dT + \left(\frac{\partial s}{\partial v}\right)_T dv \tag{8.18}$$

Substituting Eq. (8.18) into Eq. (8.5a) gives

$$du = T\left(\frac{\partial s}{\partial T}\right)_v dT + \left[T\left(\frac{\partial s}{\partial v}\right)_T - P\right] dv \tag{8.19}$$

Comparing Eq. (8.19) with Eq. (8.17) shows that

$$\left(\frac{\partial u}{\partial T}\right)_v = T\left(\frac{\partial s}{\partial T}\right)_v \tag{8.20}$$

and

$$\left(\frac{\partial u}{\partial v}\right)_T = T\left(\frac{\partial s}{\partial v}\right)_T - P \tag{8.21}$$

The identity in Eq. (8.20) is not used here; however, we note that $(\partial u/\partial T)_v = c_v$. Also, note that the Maxwell relation, Eq. (8.11c), can be used in the second identity, so that the relation in Eq. (8.21) becomes

$$\left(\frac{\partial u}{\partial v}\right)_T = T\left(\frac{\partial P}{\partial T}\right)_v - P \tag{8.22}$$

And substituting Eq. (8.22) into Eq. (8.17) results in

$$du = c_v\, dT + \left[T\left(\frac{\partial P}{\partial T}\right)_v - P\right] dv \tag{8.23}$$

To find the change in u between any two states, say between a reference state and a final state, we can choose any path of integration between the states. The integrated form of Eq. (8.23) becomes

$$u - u_{\text{ref}} = \int_{T_{\text{ref}}}^{T} c_v\, dT + \int_{v_{\text{ref}}}^{v} \left[T\left(\frac{\partial P}{\partial T}\right)_v - P\right] dv \tag{8.24}$$

If we choose a reference state for which steam behaves as an ideal gas $(P \to 0)$ and integrate along a path on Fig. 8.2 at constant v (path A-B) until we intercept the desired constant-T line at point B, then the value of c_v along path A-B is a function of T only. Along isotherm B-C, the first term is zero. The first term in Eq. (8.24) can thus be evaluated by using only ideal gas properties for water vapor. The second term has a zero value along the constant-volume line and can be evaluated along B-C given only detailed $P(v, T)$ data for steam. We have reached our goal of finding u by using only P-v-T data. Numerical integration will probably be required, and the data must be quite precise because derivatives of the data appear in the relations we are using.

Figure 8.2 *P-T diagram.*

To determine the entropy at any state, a similar approach is used. The resulting relation for s is

$$s = s_{ref} + \left(\int_{T_{ref}}^{T} \frac{c_v}{T} \, dT \right)_{v_{ref}} + \left[\int_{v_{ref}}^{v} \left(\frac{\partial P}{\partial T} \right)_v dv \right]_T \qquad (8.25)$$

Because both Eqs. (8.24) and (8.25) require a reference state as $P \rightarrow 0$, they may not be completely useful in constructing a property table, which is usually based on some other reference condition such as the triple point. The equations are easily modified to find the difference in s or u between any two states (one of which can be a nonideal gas reference state) by simply subtracting the equation for state 2 from the equation for state 1. Then Eqs. (8.24) and (8.25) become

$$u_2 - u_1 = \left\{ \int_{v_1}^{v=const} \left[T \left(\frac{\partial P}{\partial T} \right)_v - P \right] dv \right\}_{T=T_1} + \left(\int_{T_1}^{T_2} c_v \, dT \right)_{v=const}$$
$$+ \left\{ \int_{v=const}^{v_2} \left[T \left(\frac{\partial P}{\partial T} \right)_v - P \right] dv \right\}_{T=T_2} \qquad (8.26)$$

and

$$s_2 - s_1 = \left(\int_{T_1}^{T_2} \frac{c_v}{T} \, dT \right)_{v=const} + \left[\int_{v_1}^{v=const} \left(\frac{\partial P}{\partial T} \right)_v dv \right]_{T=T_1}$$
$$+ \left[\int_{v=const}^{v_2} \left(\frac{\partial P}{\partial T} \right)_v dv \right]_{T=T_2} \qquad (8.27)$$

Now u and s relative to any reference point can be determined from data for $P(v, T)$, and the value of the Helmholtz function a can be found from $a = u - Ts$. The complete tables can now be constructed.

8.3   Principle of Corresponding States

In Chap. 3, the *principle of corresponding states* is expressed for a simple compressible substance as follows:

> All substances obey the same equation of state expressed in terms of the reduced properties.

This principle is an empirical statement based on some theoretical justification, but it is not a "law." That is, the results of its application are not, and should not be expected to be, exact. Indeed, we have seen by comparison of the compressibility factor at the critical point, which should have the same value for all substances if the principle is valid, that considerable deviations are exhibited by some substances (see Table C.4). Thus, although the principle provides a basis for examining the *behavior* of thermodynamic properties, we still must rely on measurements if we wish exact *values* for the properties.

8.3.1 Some Observations Based on van der Waals' Equation

van der Waals' equation of state [Eq. (3.26)] as discussed in Sec. 3.5 is

$$P = \frac{RT}{v-b} - \frac{a}{v^2} \tag{8.28}$$

To find the values of the constants a and b for a particular substance, we note that the critical isotherm has a horizontal inflection point as it passes through the critical point (see, for example, Fig. 3.5). Thus, the first and second partial derivatives of P with respect to v must be zero at the critical point for any universally valid equation of state (try the ideal gas law, which fails badly around the critical point). These derivatives, evaluated at the critical point by using van der Waals' equation, are

$$\left(\frac{\partial P}{\partial v}\right)_T = 0 = -\frac{RT_{cr}}{(v_{cr}-b)^2} + \frac{2a}{v_{cr}^3} \tag{8.29}$$

and

$$\left(\frac{\partial^2 P}{\partial v^2}\right)_T = 0 = \frac{2RT_{cr}}{(v_{cr}-b)^3} - \frac{6a}{v_{cr}^4} \tag{8.30}$$

Solving simultaneously and using Eq. (8.28) to eliminate v_{cr} result in

$$a = \frac{27R^2T_{cr}^2}{64P_{cr}} \quad \text{and} \quad b = \frac{RT_{cr}}{8P_{cr}} \tag{8.31}$$

Thus, van der Waals' equation follows the principle of corresponding states, because a single relation is provided for all substances, which is expressed in terms of the critical properties.

If van der Waals' equation is valid, it should work throughout the entire region of properties for which a given substance is a fluid, including the saturation region. It is not *universally* valid; that is, it is not valid, for example, for the solid phase, because it is based on a physical model for a fluid. Let us check that out. We can write Eq. (8.28) as

$$Pv^3 - (bP + RT)v^2 + av - ab = 0 \tag{8.32}$$

Note that this is a cubic equation in v, which can have three roots; that is, v can have three possible values at a given value of T and P. Let us plot the P-v curve for an isotherm that passes through the two-phase region for water ($a = 1.7044$ atm $\cdot$ m^6/kg^2 and $b = 1.690 \times 10^{-3}$ m^3/kg), say at $T = 180°C$. The resulting curve is shown in Fig. 8.3.

The curve has three values of v at $P = 1002$ kPa, the saturation pressure at $T = 180°C$. The smallest and largest values lie near the values for saturated liquid and saturated vapor, respectively. At the saturation pressure, van der Waals' equation predicts a specific volume for the satu-

Figure 8.3   *P-v diagram and van der Waals' equation for water.*

rated liquid, point A, as 0.00239 m³/kg, compared with a value from the steam tables of 0.001127 m³/kg. For the saturated vapor, the equation predicts (point D) that $v_g = 0.202$ m³/kg, compared with a table value of 0.1938. Between these values, the curve behaves much differently from the curve found experimentally. That is, between the minimum of the isotherm from van der Waals' equation (which is negative in value and so cannot be plotted on the log scale) and point C, the volume increases as the pressure increases, which is in contradiction to our observations. Keep in mind that the equation is valid only for either the vapor or the liquid, and fails in the two-phase mixture. We interpret the predicted P-v curve to mean that between the minimum of the isotherm and point C, the vapor in the two-phase mixture is mechanically unstable, and slight changes in pressure cause some of the vapor to change phase.

Although the numerical results of this exercise show considerable error in the prediction of saturation properties when compared with experimental data, study of van der Waals' equation does give some insight into how equations of state might be constructed and the characteristics we might expect them to have.

van der Waals' equation can have greater practical use by finding values of a and b that minimize the difference from experimental data, rather than using the values given by Eq. (8.31), which provide an exact fit at the critical point only.

8.3.2 Expanded Use of the Principle of Corresponding States

In Sec. 3.5, the compressibility factor $z(T_r, P_r)$ is introduced, and its use in determining the P-v-T behavior of any substance for which critical property data are available is discussed. The principle of corresponding states implies that *any* thermodynamic property for any substance should be describable in terms of its critical properties. So we should be able to provide charts for u, h, s, etc. that are in terms of reduced temperature and pressure, for example.

Two-Parameter Reduced Property Charts

The compressibility charts (Fig. C.1) relate P-v-T data for any substance in terms of the critical constants T_{cr} and P_{cr}. The latter two properties are used to find the *reduced* temperature and pressure $T_r = T/T_{cr}$ and $P_r = P/P_{cr}$, and the value of z is then presented as a function of those two parameters. The charts are constructed by averaging the P-v-T behavior of 30 gases, with a resulting error in the range of 4 to 6 percent except near the critical point. On the charts, the reduced specific volume v_r is crossplotted. However, the experimental determination of v_{cr} is quite difficult to achieve, so the reduced volume shown on the charts is a *calculated* value v_r' defined by the relation

$$v_r' = \frac{v}{RT_{cr}/P_{cr}} \tag{8.33}$$

The charts are thus usable if we know only the critical pressure and temperature.

Once the compressibility chart has been generated, it can be used to find the other thermodynamic properties in reduced form. Rather than presenting the property itself in terms of reduced pressure and temperature, it is convenient to find the deviation of the property from that of an ideal gas in the same state. The ideal gas property can be found from either the gas tables (such as Tables D.2 through D.7 for air, argon, nitrogen, oxygen, water vapor, and carbon dioxide) or by computations using ideal gas relations.

Let us see how the reduced chart for one of the properties, enthalpy, would be developed. Since we wish to present the data as a function of T and P, we choose as a starting point the total derivative of $h(T, P)$

$$dh = \left(\frac{\partial h}{\partial P}\right)_T dP + \left(\frac{\partial h}{\partial T}\right)_P dT$$

$$= \left(\frac{\partial h}{\partial P}\right)_T dP + c_P \, dT \tag{8.34}$$

Note that $dh = T\,ds + v\,dP$ [Eq. (8.5b)]. If we assume that entropy is a function of temperature and pressure, then the total derivative becomes

$$ds = \left(\frac{\partial s}{\partial T}\right)_P dT + \left(\frac{\partial s}{\partial P}\right)_T dP \tag{8.35}$$

Substituting Eq. (8.35) to eliminate ds in Eq. (8.5b) gives

$$dh = T\left[\left(\frac{\partial s}{\partial T}\right)_P dT + \left(\frac{\partial s}{\partial P}\right)_T dP\right] + v\, dP$$

$$= T\left(\frac{\partial s}{\partial T}\right)_P dT + \left[v + T\left(\frac{\partial s}{\partial P}\right)_T\right] dP \tag{8.36}$$

Comparison with Eq. (8.34) shows that

$$c_P = T\left(\frac{\partial s}{\partial T}\right)_P$$

$$\left(\frac{\partial h}{\partial P}\right)_T = v + T\left(\frac{\partial s}{\partial P}\right)_T \tag{8.37}$$

But from the Maxwell relation Eq. (8.11d), we know $(\partial s/\partial P)_T = -(\partial v/\partial T)_P$, so we can write Eq. (8.36) as

$$dh = c_P\, dT + \left[v - T\left(\frac{\partial v}{\partial T}\right)_P\right] dP \tag{8.38}$$

We have now generated an expression for the change in enthalpy as a function of P-v-T data and the specific heat at constant pressure c_P. Since our goal is to develop expressions for h as a function of reduced properties T_r and P_r, we can now use the equation of state in terms of the compressibility factor [Eq. (3.30)] and reduced properties as

$$v = \frac{zRT_r T_{cr}}{P_r P_{cr}} \tag{8.39}$$

and its partial derivative

$$\left(\frac{\partial v}{\partial T}\right)_P = \frac{R}{P_r P_{cr}}\left[z + T_r\left(\frac{\partial z}{\partial T_r}\right)_{P_r}\right] \tag{8.40}$$

Now, substituting Eqs. (8.40) and (8.39) into (8.38) results in

$$dh = c_P T_{cr}\, dT_r - \frac{RT_{cr} T_r^2}{P_r}\left(\frac{\partial z}{\partial T_r}\right)_{P_r} dP_r \tag{8.41}$$

or, by integrating between a reference state at $P_r = 0$ and any final state f,

$$\frac{h_f - h_{P_r=0}}{RT_{cr}} = \int_{T_{r,\text{ref}},\, P_r=0}^{T_{r,f}} \frac{c_P}{R}\, dT_r - \int_{T_{r,\text{ref}},\, P_r=0}^{P_{r,f}} \frac{T_r^2}{P_r}\left(\frac{\partial z}{\partial T_r}\right)_{P_r} dP_r \tag{8.42}$$

Suppose for a moment that we consider an ideal gas. For such a gas, $z = 1$ and $(\partial z/\partial T_r)_{P_r} = 0$, and Eq. (8.42) reduces to

$$\frac{h_f - h_{P_r=0}}{RT_{cr}} = \int_{T_{r,\mathrm{ref}}, P_r=0}^{T_{r,f}} \frac{c_P}{R}\, dT_r \tag{8.43}$$

At $P_r = 0$, all real gases behave as ideal gases, so $h_{P_r=0}$ is the same for the real and ideal gases. Thus, if we subtract Eq. (8.42) from (8.43), we find

$$\frac{(h_{\mathrm{ideal}} - h)_f}{RT_{cr}} = \int_{T_{r,\mathrm{ref}}, P_r=0}^{P_{r,f}} \frac{T_r^2}{P_r} \left(\frac{\partial z}{\partial T_r}\right)_{P_r} dP_r$$

$$= \int_{T_{r,\mathrm{ref}}, P_r=0}^{P_{r,f}} T_r^2 \left(\frac{\partial z}{\partial T_r}\right)_{P_r} d(\ln P_r) \tag{8.44}$$

and the left-hand side of Eq. (8.44) is called the *enthalpy departure function,* and is presented in Fig. C.2. It shows the deviation of the enthalpy of a real fluid (liquid or gas) from that of an ideal gas in the same state. Note, however, that choosing T and P as the independent variables in Eq. (8.34) precludes the use of this method for predicting the departure function in the saturation region, where P and T are *not* independent, although the saturation curve itself can be computed and drawn as shown on Fig. C.2.

The Clapeyron equation, Eq. (8.14), can be used to find the enthalpy change between the liquid and vapor states as a check on the saturation curve. This is done by first noting that

$$v_{lg} = v_g - v_l = \frac{RT_{\mathrm{sat}}}{P_{\mathrm{sat}}} (z_g - z_l) \tag{8.45}$$

and then using Eq. (8.14) to get

$$\left(\frac{dP}{dT}\right)_{\mathrm{sat}} = \frac{h_{lg}}{T_{\mathrm{sat}} v_{lg}} = \frac{P_{\mathrm{sat}} h_{lg}}{RT_{\mathrm{sat}}^2 (z_g - z_l)} \tag{8.46}$$

or

$$\frac{h_{lg}}{RT_{cr}} = \left(\frac{dP_r}{dT_r}\right)_{\mathrm{sat}} \frac{T_{r,\mathrm{sat}}^2}{P_r}(z_g - z_l) \tag{8.47}$$

The values of z_g and z_l are a function of $P_{r,\mathrm{sat}}$, and the corresponding values of P_r and $(dP_r/dT_r)_{\mathrm{sat}}$ are all available from the $z(P_r, T_r)$ charts. Note that

$$\left[\frac{h_{\mathrm{ideal}}(T) - h_g}{RT_{cr}}\right]_g - \left[\frac{h_{\mathrm{ideal}}(T) - h_l}{RT_{cr}}\right]_l = \frac{h_{lg}}{RT_{cr}} \tag{8.48}$$

and Eq. (8.47) can thus be used to check the h_{lg} values found on Fig. C.2 by using only $z(T_r, P_r)_{\mathrm{sat}}$ data.

Equation (8.44) can be evaluated for any unsaturated state by using data for $z(P_r, T_r)$ as presented, for example, in Fig. C.1. The resulting variation of the enthalpy departure function with T_r and P_r is shown in Fig. C.2. Of course, similar derivations can be used to find departure functions for the other common thermodynamic properties, and charts for entropy and for fugacity (defined in Sec. 8.4.5) are also presented in

Figs. C.3 and C.4, respectively. Separate charts are presented for low pressures in Fig. C.5. Note that the dimensionless ordinates used on these charts can be treated with each quantity on a molar basis if the universal gas constant is used or on a unit-mass basis if the gas constant specific to a particular gas is used.

We now can evaluate the thermodynamic properties v, h, s, and fugacity (and, of course, $u = h - Pv$) for any fluid in any state as long as the critical temperature and pressure are known for that fluid. The charts should be about as accurate as the compressibility factor chart, i.e., about 6 percent. Because of this accuracy limitation, the results are presented in graphical rather than tabular form, since the errors in reading the graphs introduce little additional uncertainty to the results.

Example Problem 8.4

Determine the change in the values of enthalpy and entropy of ethylene between an initial state at 1 atm and 273 K and a final state at 200 atm and 350 K, using the reduced property charts.

Solution

From Table C.4, for ethylene, $T_{cr} = 282.4$ K and $P_{cr} = 50.5$ atm. The molecular weight is 28.052. Thus, the values of reduced temperature and pressure at the initial state are

$$T_{r,1} = \frac{T_1}{T_{cr}} = \frac{273}{282.4} = 0.967$$

$$P_{r,1} = \frac{P_1}{P_{cr}} = \frac{1}{50.5} = 0.0198$$

and at the final state are

$$T_{r,2} = \frac{350}{282.4} = 1.24$$

$$P_{r,2} = \frac{200}{50.5} = 3.96$$

From Figs. C.2 and C.3, the pressure in the initial state is so low that the enthalpy and entropy departure functions are both zero, indicating that ethylene in the initial state behaves as an ideal gas. In the final state, $(h_{ideal} - h)/(RT_{cr}) = 2.90$ and $(s_{ideal} - s)/R = 1.70$. Using the particular gas constant $R = 8.31441/28.052 = 0.2964$ kJ/(kg · K) yields

$$h_2 = h_{ideal,\,2} - (2.90)(0.2964)(282.4)$$
$$= h_{ideal,\,2} - 243 \quad \text{kJ/kg}$$

$$s_2 = s_{ideal, 2} - (1.70)(0.2964)$$
$$= s_{ideal, 2} - 0.504 \quad kJ/(kg \cdot K)$$

To find the change in h and s, the values of h_1 and s_1 must be subtracted. Since in this case the enthalpy and entropy departure functions have zero value in the initial state, the equations for the changes in the properties become

$$h_2 - h_1 = (h_2 - h_1)_{ideal} - 243$$

$$s_2 - s_1 = (s_2 - s_1)_{ideal} - 0.504$$

The changes in h and s for the ideal gas can be found by using the values of specific heat at constant pressure from Table D.1:

$$(h_2 - h_1)_{ideal} = \int_{273}^{350} c_P \, dT$$

$$= \frac{1}{28.052} \left[3.95T + (15.64 \times 10^{-2}) \left(\frac{T^2}{2} \right) \right.$$
$$\left. - (8.344 \times 10^{-5}) \left(\frac{T^3}{3} \right) + (17.67 \times 10^{-9}) \left(\frac{T^4}{4} \right) \right] \Big|_{273}^{350}$$

$$= 350.65 - 226.9 = 123.7 \text{ kJ/kg}$$

$$(s_2 - s_1)_{ideal} = \int_{273}^{350} \frac{c_P}{T} \, dT - R \ln \frac{P_2}{P_1}$$

$$= \frac{1}{28.052} \left[3.95 \ln \frac{350}{273} + (15.64 \times 10^{-2})(350 - 273) \right.$$

$$- (8.344 \times 10^{-5}) \left(\frac{350^2 - 273^2}{2} \right)$$

$$\left. + (17.67 \times 10^{-9}) \left(\frac{350^3 - 273^3}{3} \right) \right] - 0.2964 \ln \frac{200}{1}$$

$$= -1.1727 \text{ kJ/(kg} \cdot K)$$

Thus, the changes in the properties are finally

$$h_2 - h_1 = 123.7 - 243 = -119 \text{ kJ/kg}$$

$$s_2 - s_1 = -1.1727 - 0.504 = -1.677 \text{ kJ/(kg} \cdot K)$$

Comments

Note that the entropy change is negative for this change of state even though the temperature increases because of the extremely large change in pressure. Ethylene is apparently far from an ideal gas in the final state, because the departure functions are quite large relative to the ideal gas terms in the final equations.

If tables of ideal gas properties are available for the substance under study, then the ideal gas terms can be evaluated from the tables. Care must be taken to include the pressure-dependent term in the entropy relations for the ideal gas, as is obvious in this problem. The specific entropy term in the tables does not include the pressure-dependent term, as discussed in Chap. 5.

Three-Parameter Corresponding-States Methods

At the beginning of the discussion of the principle of corresponding states, we noted that the principle is not exact but does provide guidance as to thermodynamic property behavior for many materials based on very limited data. Suppose an additional critical property were available, such as the critical volume. Could this additional piece of information be used to more accurately model the behavior of substances than is possible by using only P_{cr} and T_{cr}?

It is, indeed, possible. As stated earlier, the critical volume is a difficult quantity to measure accurately, so it is not an appropriate property to use in such a model; but other properties are available. Lee and Kesler [2] have presented a three-parameter corresponding-states model based on a proposal by K. S. Pitzer and coworkers that uses P_{cr}, T_{cr}, and ω as the required properties as well as a modified form of the Benedict-Webb-Rubin equation of state. The factor ω is called the *acentric factor,* and it is a measure of the deviation in the thermodynamic behavior of a real fluid from a simple fluid. A *simple fluid* is defined as a fluid with a spherical molecule, and the acentric factor indicates the degree of deviation of the molecular structure (acentricity) of the real molecule from that of the simple spherical model. For example, the compressibility factor z of a fluid with acentric factor ω is given by

$$z = z^{(0)} + \omega z^{(1)} \tag{8.49}$$

where $z^{(0)}$ is the compressibility factor for the spherical model of the molecule and $z^{(1)}$ incorporates the deviation due to the acentricity of the real molecule.

Departure functions can be evaluated by using the three-parameter method in a similar manner to that used in the two-parameter correlations. The resulting departure functions take the form

$$\frac{h_{ideal} - h}{RT_{cr}} = \left(\frac{h_{ideal} - h}{RT_{cr}}\right)^{(0)} + \omega \left(\frac{h_{ideal} - h}{RT_{cr}}\right)^{(1)} \tag{8.50}$$

and

$$\frac{s_{ideal} - s}{R} = \left(\frac{s_{ideal} - s}{R}\right)^{(0)} + \omega \left(\frac{s_{ideal} - s}{R}\right)^{(1)} \tag{8.51}$$

The terms on the right-hand sides of Eqs. (8.50) and (8.51) with superscripts (0) and (1) are termed *residual properties,* and they correspond to

$z^{(0)}$ and $z^{(1)}$ in Eq. (8.49). Values of the acentric factor are presented in Table C.4, and values of the three-parameter functions for compressibility, residual enthalpy, and residual entropy are given in Tables C.6 through C.8. Tables are used for the presentation of these data, because in most cases the accuracy of the results is greater than can be obtained from plots of the functions. The solid lines traversing the tables for $P_r < 1$ show the location of the vapor-liquid equilibrium (saturation) states.

Lee and Kesler also present relations for the fugacity coefficient and the constant-pressure and constant-volume specific heats. Comparison with experimental data shows that addition of the third parameter provides good accuracy. It is difficult to give a blanket number for accuracy, but Lee and Kesler show absolute average deviations in z of 1 to 2 percent over wide ranges of T_r and P_r for most hydrocarbons, with a few deviating by about 4 percent. Deviations are greatest for a given fluid near the critical point.

Example Problem 8.5

Repeat Example Problem 8.4, using the three-parameter reduced property tables, and compare your results with those of the previous example.

Solution

The additional information required to use the three-parameter method is the acentric factor, which for ethylene has the value 0.085 (Table C.4). Now the enthalpy departure in state 1 ($P_r = 0.0198$ and $T_r = 0.967$) is given from Eq. (8.50) and Table C.7 as

$$\frac{h_{\text{ideal}} - h}{RT_{\text{cr}}} \simeq 0$$

$$\frac{s_{\text{ideal}} - s}{R} \simeq 0$$

so that $h_1 = h_{\text{ideal}, 1}$ and $s_1 = s_{\text{ideal}, 1}$. For state 2 ($P_r = 3.96$ and $T_r = 1.24$), by using double interpolation on Tables C.7 and C.8,

$$\frac{h_{\text{ideal}} - h}{RT_{\text{cr}}} = 2.803 + (0.085)(1.088) = 2.895$$

$$\frac{s_{\text{ideal}} - s}{R} = 1.634 + (0.085)(1.138) = 1.731$$

As in Example Problem 8.4,

$$h_2 - h_1 = (h_2 - h_1)_{\text{ideal}} - 2.895RT_{\text{cr}}$$

$$s_2 - s_1 = (s_2 - s_1)_{\text{ideal}} - 1.731R$$

or, using the ideal gas property changes from Example Problem 8.4, we find

$$h_2 - h_1 = 123.7 - (2.895)(0.2964)(282.4)$$
$$= 123.7 - 242.3 = -118.6 \text{ kJ/kg}$$

$$s_2 - s_1 = -1.1727 - (1.731)(0.2964)$$
$$= -1.1727 - 0.5131 = -1.6857 \text{ kJ/(kg} \cdot \text{K)}$$

Comments

These results are in good agreement with those of Example Problem 8.4 [$h_2 - h_1 = -119$ kJ/kg and $s_2 - s_1 = -1.677$ kJ/(kg $\cdot$ K)]. We would expect the three-parameter calculations to be the more accurate. Note that it is possible to maintain accuracy to at least one more significant figure with the three-parameter results than is possible with the two-parameter graphical method.

Example Problem 8.6

Determine the value of internal energy for CO_2 at $P = 300$ atm and $T = 450$ K, using both the two-parameter and three-parameter reduced property methods.

Solution

The critical temperature, critical pressure, and acentric factor are obtained from Table C.4 as $T_{cr} = 304.2$ K, $P_{cr} = 72.9$ atm, and $\omega = 0.225$. The reduced temperature (1.479) and reduced pressure (4.115) are used with Fig. C.1b to obtain

$$z = 0.80$$

which indicates that this state cannot be accurately approximated as an ideal gas. The internal energy is calculated from the enthalpy as

$$u = h - Pv = h - zRT$$

The two-parameter method uses the z value above and h determined from Fig. C.2. The three-parameter method uses the Lee and Kesler values (Tables C.6 and C.7).

Two-Parameter Method

The critical temperature and pressure given above are used with Fig. C.2 to yield $(h_{ideal} - h)/(RT_{cr}) = 1.8$. The enthalpy determined from the ideal gas approximation is given in Table D.7 as $h_{ideal} = 352.19$ kJ/kg. Therefore,

$$h = h_{ideal} - 1.8RT_{cr}$$
$$= 352.19 - (1.8)(0.18892)(304.2)$$
$$= 304.2 \text{ kJ/kg}$$

and the internal energy is

$$u = h - zRT$$
$$= 249 - (0.80)(0.18892)(450)$$
$$= 181 \text{ kJ/kg}$$

Note that no more than two significant figures should be kept.

Three-Parameter Method

The compressibility factor is given by Eq. (8.49). Double interpolation in Table C.6 yields the values for $z^{(0)}$ and $z^{(1)}$:

$$z = z^{(0)} + \omega z^{(1)}$$
$$= 0.7949 + (0.225)(0.2294)$$
$$= 0.8465$$

The enthalpy is evaluated from Eq. (8.50) with Table C.7 as

$$\frac{h_{ideal} - h}{RT_{cr}} = \left(\frac{h_{ideal} - h}{RT_{cr}}\right)^{(0)} + \omega\left(\frac{h_{ideal} - h}{RT_{cr}}\right)^{(1)}$$
$$= 1.963 + (0.225)(0.098) = 1.985$$

With h_{ideal} given above as 352.19 kJ/kg, the enthalpy is

$$h = h_{ideal} - 1.985 RT_{cr}$$
$$= 352.19 - (1.985)(0.18892)(304.2)$$
$$= 238.1 \text{ kJ/kg}$$

and the corresponding internal energy is

$$u = h - zRT$$
$$= 238.1 - (0.8465)(0.18892)(450)$$
$$= 166.1 \text{ kJ/kg}$$

Comments

The improvement in the accuracy obtained by each level of approximation is noteworthy. A very inaccurate approximation is obtained if the state is assumed to be an ideal gas since $u = h - RT = 352.19 - (0.18892)(450) = 267.2$ kJ/kg. The number of significant figures is misleading, since the equation itself is invalid. This is expected since $z = 0.80$ was obtained from Fig. C.1b. The two-parameter method yields $u = 181$ kJ/kg while the most accurate, three-parameter method yields $u = 166.1$ kJ/kg.

8.4 Some Other Properties

Although the primary thermodynamic properties P, v, T, u, h, s, a, and g and their relationships form the basis of much of engineering thermodynamics, a number of other properties are of interest in certain applications. In particular, note that the derivatives of these properties with respect to other properties are also properties. We used this fact when the specific heats were defined as $c_P = (\partial h/\partial T)_P$ and $c_v = (\partial u/\partial T)_v$.

Other derivatives of the primary properties can also be defined as separate properties where this is useful. We briefly examine some of these derived properties.

8.4.1   Isothermal Compressibility

In some applications, it is useful to know the degree to which a substance can be compressed when the pressure is changed in an isothermal process. The fractional change in volume with pressure at constant temperature is called the *isothermal compressibility κ* and is defined as

$$\kappa = -\left(\frac{1}{v}\right)\left(\frac{\partial v}{\partial P}\right)_T \tag{8.52}$$

The minus sign is used because specific volume decreases with pressure, so the partial derivative carries a negative value. Values of κ can be tabulated from P-v-T data for any substance.

8.4.2   Coefficient of Thermal Expansion

To describe the change in volume with temperature of a substance at constant pressure, we can define the *coefficient of thermal expansion β* as

$$\beta = \left(\frac{1}{v}\right)\left(\frac{\partial v}{\partial T}\right)_P \tag{8.53}$$

This definition can be combined with the isothermal compressibility to yield the total derivative of volume. The total derivative of $V(T, P)$ divided by V is

$$\frac{dV}{V} = \frac{1}{V}\left(\frac{\partial V}{\partial T}\right)_P dT + \frac{1}{V}\left(\frac{\partial V}{\partial P}\right)_T dP \tag{8.54}$$

Substituting Eqs. (8.52) and (8.53) yields

$$\frac{dV}{V} = \beta\, dT - \kappa\, dP \tag{8.55}$$

Example Problem 8.7

Derive relations for the coefficient of thermal expansion and the isothermal compressibility of an ideal gas.

Solution
The derivative $(\partial v/\partial T)_P$ for an ideal gas is R/P, so Eq. (8.53) becomes

$$\beta = \left(\frac{1}{v}\right)\left(\frac{R}{P}\right) = \frac{1}{T}$$

So the coefficient of thermal expansion of an ideal gas is simply the inverse of the absolute temperature.

The derivative $(\partial v/\partial P)_T$ for an ideal gas is $-RT/P^2$, so Eq. (8.52) becomes

$$\kappa = \left(\frac{1}{v}\right)\left(\frac{RT}{P^2}\right) = \frac{1}{P}$$

and the isothermal compressibility of an ideal gas is the inverse of the absolute pressure.

8.4.3 Joule-Thomson Coefficient

In throttling processes, the enthalpy is constant across the throttling device. Of interest is whether the temperature of the throttled substance increases or decreases and the magnitude of the change. The property that tells us this is the *Joule-Thomson coefficient* μ, where

$$\mu = \left(\frac{\partial T}{\partial P}\right)_h \tag{8.56}$$

If the substance being throttled increases in temperature during passage through a throttling device, the coefficient is negative; if the temperature drops, the coefficient is positive.

Example Problem 8.8

Refrigerant 12 is being considered for use as a pressurant in a shaving cream container. The container has saturated refrigerant 12 vapor at 27°C, and the shaving cream exits the container nozzle to room conditions. What is the sign of the Joule-Thomson coefficient for this process? Will the shaving cream be warmer or colder than in the can?

Solution

Assuming the velocity of the shaving cream/refrigerant 12 leaving the can is small, we see this is a constant-enthalpy process. From Table D.11 or Fig. D.4 or computerized properties (used here), the initial state is $T = 27°C$, $P = 687.5$ kPa, and $h = 362.6$ kJ/kg. At the final state, the enthalpy is the same, and $P = 101.3$ kPa. The final temperature is found to be 11.5°C. The coefficient is thus roughly

$$\mu = \left(\frac{11.5 - 27}{101.3 - 687.5}\right)_h = 0.026 \text{ K/kPa}$$

The coefficient is positive, reflecting the fact that the pressurant temperature drops during the expansion. If we want automatically warm shaving cream, we must search for a different pressurant.

8.4.4   Specific Heats

The Maxwell relations have been used to provide two new ways of relating the specific heats to other properties. In Eq. (8.20) we found

$$c_v = \left(\frac{\partial u}{\partial T}\right)_v = T\left(\frac{\partial s}{\partial T}\right)_v \tag{8.57}$$

and in Eq. (8.37)

$$c_P = \left(\frac{\partial h}{\partial T}\right)_P = T\left(\frac{\partial s}{\partial T}\right)_P \tag{8.58}$$

It follows that

$$c_P - c_v = T\left[\left(\frac{\partial s}{\partial T}\right)_P - \left(\frac{\partial s}{\partial T}\right)_v\right] \tag{8.59}$$

Let us find a relation for $c_P - c_v$ in terms of P-v-T data. Using the substitution rule, Eq. (1.23), in the form

$$\left(\frac{\partial s}{\partial T}\right)_P = \left(\frac{\partial s}{\partial T}\right)_v + \left(\frac{\partial s}{\partial v}\right)_T\left(\frac{\partial v}{\partial T}\right)_P$$

results in

$$c_P - c_v = T\left[\left(\frac{\partial s}{\partial v}\right)_T\left(\frac{\partial v}{\partial T}\right)_P\right] \tag{8.60}$$

Now, the Maxwell relation Eq. (8.11c) is substituted to eliminate the derivative involving s, and we obtain

$$c_P - c_v = T\left[\left(\frac{\partial P}{\partial T}\right)_v\left(\frac{\partial v}{\partial T}\right)_P\right] \tag{8.61}$$

We can observe the following from this equation. Since $(\partial v/\partial T)_P$ is small for liquids and solids, the values of c_P and c_v should be nearly equal for all materials in those states. This is stated for incompressible substances in Sec. 3.5. Also, the values of $(\partial P/\partial T)_v$ and $(\partial v/\partial T)_P$ are easily found for an ideal gas, and for that case we find, as we have derived before, that $c_P - c_v = R$.

8.4.5   Fugacity

One other property is introduced here for use in later chapters, particularly when we deal with mixtures and equilibrium systems. The property is useful for certain single-component systems in which an isothermal process is carried out. This property is called the *fugacity,* and it is defined here by starting with Eq. (8.5d):

$$dg = v\, dP - s\, dT \tag{8.62}$$

For a constant-temperature process, this reduces to

$$dg = v \, dP \tag{8.63}$$

If we assume an ideal gas, then Eq. (8.63) becomes

$$dg = RT \frac{dP}{P} = RT \, d(\ln P) \tag{8.64}$$

Suppose we now introduce the fugacity f as the quantity that will produce an exact relation for a *real* gas in this equation ($T = $ constant):

$$dg = RT \, d(\ln f) \tag{8.65}$$

From such a defining relation for fugacity, we can infer various characteristics about the behavior of f. First, f will have units of pressure; second, at low pressure, f must approach P in value, since Eq. (8.64) is correct at low pressure. Or

$$\lim_{P \to 0} \frac{f}{P} = 1 \tag{8.66}$$

The fugacity thus acts as a "pseudo-pressure" which gives the correct behavior for dg when it is used in Eq. (8.65). Values for f can be found by writing Eq. (8.63) for a real gas in terms of the compressibility factors as

$$dg = v \, dP = zRT \, d(\ln P) = RT \, d(\ln f) \tag{8.67}$$

This can be rearranged to yield

$$(z - 1) \frac{dP}{P} = (z - 1) \frac{dP_r}{P_r} = (z - 1) \, d(\ln P_r) = \frac{d(f/P)}{f/P} = d\left[\ln \left(\frac{f}{P} \right) \right] \tag{8.68}$$

and it is clear that f/P (which is often referred to as the *fugacity coefficient*) can be tabulated as a function of reduced properties T_r and P_r. A graph of fugacity as a function of reduced properties is given in Figs. C.4 and C.5*c*.

Example Problem 8.9

A natural gas compressor station is to take methane at 1 atm and 25°C and compress it isothermally and reversibly to 60 atm. Find the compressor work per kilogram of methane.

Solution

The first law gives

$$q + w = h_{\text{out}} - h_{\text{in}}$$

and for a reversible isothermal compression, $q = T(s_{\text{out}} - s_{\text{in}})$, so

$$w = (h - Ts)_{\text{out}} - (h - Ts)_{\text{in}} = g_{\text{out}} - g_{\text{in}} = RT \ln \frac{f_{\text{out}}}{f_{\text{in}}}$$

The values of the reduced properties are

$$T_r = \frac{298}{191.1} = 1.559$$

$$P_{r,\text{in}} = \frac{1}{45.8} = 0.0218$$

$$P_{r,\text{out}} = \frac{60}{45.8} = 1.31$$

From Fig. C.5c, $(f/P)_{\text{in}} = 1.0$; from Fig. C.4, $(f/P)_{\text{out}} = 0.93$; and

$$w = (0.51835)(298) \ln \frac{(0.93)(60)}{(1)(1)} = 621 \text{ kJ/kg}$$

Comment

Clearly, for this type of problem, the use of fugacity is much simpler than going through the complete first and second law analysis.

8.5 Summary

In this chapter, we have shown that many useful relationships exist among the thermodynamic properties; that these relationships can be used to minimize the laboratory work necessary to determine property values and to construct tables and functions that describe all properties from data; that the principle of corresponding states allows the data for one substance to provide a good estimate of the property behavior of other substances; and that many properties are available and useful for specific problems and can be derived from the set we have already studied.

We have barely touched the wealth of literature available on property measurement, relationships, equations of state, etc. This is an ongoing area of research in thermodynamics. However, the material presented here should act as a guide for the engineer in finding and applying thermodynamic data.

Problems

Problems marked with a star are lengthy; thus, computerized steam tables or an accurate Mollier chart should be used if these problems are to be completely worked.

8.1 The virial equation of state is

$$\frac{P}{T} = \frac{R}{v}\left[1 + \frac{B(T)}{v} + \frac{C(T)}{v^2} + \cdots\right]$$

Use the fact that $(\partial P/\partial v)_T$ and $(\partial^2 P/\partial v^2)_T$ must be zero at the critical point [as was used in Eqs. (8.29) and (8.30)] to find the first two constants, $B(T)$ and $C(T)$. Neglect higher-order terms, and derive an equation of state.

8.2 Show whether the Benedict-Webb-Rubin equation [Eq. (3.29)] obeys the relations that $(\partial P/\partial v)_T$ and $(\partial^2 P/\partial v^2)_T$ equal zero at the critical point for methane.

*8.3 Plot the saturation curve for water on a P-v diagram, using van der Waals' equation in the range 10 kPa $< P < P_{cr}$. Do this by taking values of P_{sat} and T_{sat} from the steam tables, and using van der Waals' equation to find the corresponding values of v_1 and v_g. Plot the saturation curve from the steam tables on the same diagram.

8.4 Derive expressions for the isothermal compressibility $\kappa = -(1/v)\,(\partial v/\partial P)_T$ and the volumetric thermal expansion coefficient $\beta = (1/v)\,(\partial v/\partial T)_P$ for a van der Waals gas.

8.5 Show that, in general, $c_P - c_v = (Tv\beta^2)/\kappa$. And show that this relation reduces to $c_P - c_v = R$ for an ideal gas. Start with Eq. (8.61).

8.6 Using Eq. (8.55), show that for a constant-volume process

$$P_2 = P_1 + \frac{\beta}{\kappa}(T_2 - T_1)$$

8.7 Using the relation given in Prob. 8.5, show how c_P and c_v are related for an incompressible liquid.

8.8 Derive Eq. (8.25), following the procedure similar to that used for obtaining Eq. (8.24).

8.9 Starting with Eq. (8.38) and the fact that $h = u + Pv$, derive Eq. (8.23).

8.10 The acentric factor is defined by

$$\omega = [-\log P_{r,\,sat}]_{T_r=0.7} - 1$$

Find the saturation pressure of water and of refrigerant 12 at $T_r = 0.7$ from this relation. Compare the results with the values from the tables in App. D or the computerized property data.

8.11 Prove that the Joule-Thomson coefficient can be expressed by

$$\mu = \frac{1}{c_P}\left[T\left(\frac{\partial v}{\partial T}\right)_P - v\right]$$

8.12 Find an expression for the Joule-Thomson coefficient of an ideal gas.

8.13 Using the Clapeyron equation and P-v-T data for refrigerant 12 from the tables or the computerized property relations, find h_{1g} for refrigerant 12 at $P = 50$ and $P = 200$ kPa. Compare with the values from Table D.12.

8.14 Substitute the relations for a and b [Eq. (8.31)] into van der Waals' equation, and derive a value for the compressibility factor at the critical point for a van der Waals gas. Compare the result with the value from Fig. C.1.

8.15 n-Hexane is at $T = 507.9$ K and 17.94 atm. What is the difference between the actual enthalpy of n-hexane in this state and the enthalpy in the same state if n-hexane acts as an ideal gas? (Use the Lee-Kesler three-parameter method to obtain your result.)

8.16 Find an accurate value of g in kilojoules per kilogram for CO_2 at $P = 350$ atm and $T = 1065$ K.

8.17 Find h and s for nitrogen at 250 atm and 210 K, using (a) the departure charts and (b) the Lee-Kesler tables of departure functions.

8.18 You need the value of c_P for argon at $P = 30$ atm and $T = 600$ K. At this pressure the value may not be a function of T only. Use the Lee-Kesler tables to find the required value.

8.19 Find the Joule-Thompson coefficient for steam at $P = 25$ MPa and $T = 600°C$. Use the steam tables.

8.20 Ice increases in volume by about 11 percent when it is frozen from liquid water, and the heat of fusion of water is $h_{sl} = 345$ kJ/kg. How much pressure, in kilopascals, must be exerted on a block of ice to change its melting point by 2°C? Is the melting point increased or decreased when pressure is increased?

8.21 An extension of the Clapeyron equation, called the *Clausius-Clapeyron equation*, is

$$\frac{d(\ln P)}{dT} = \frac{h_{lg}}{RT^2}$$

It is based on the assumptions that $v_g \gg v_l$ and that the vapor phase acts as an ideal gas.
(a) Derive the Clausius-Clapeyron equation from the Clapeyron equation.
(b) Using steam table data, make a plot of $\ln P_{sat}$ versus $1/T_{sat}$ for steam in the range $10 < P_{sat} < 100$ kPa. From the plot and using the Clausius-Clapeyron equation, find h_{lg} for steam. Check your result against the steam tables.

8.22 From the two data points given, estimate the temperature of the triple point of water. Remember that the triple point occurs where the lines for vapor-liquid and liquid-solid equilibrium meet on a P-T diagram. (The Clausius-Clapeyron equation of Prob. 8.21 will be helpful for the vapor-liquid equilibrium line.)

Solid-liquid at $T = -10°C$, $P_{sat} = 119.$ MPa:

$\Delta v_{sl} = -1.1 \times 10^{-4}$ m³/kg

$h_{sl} = 345$ kJ/kg

Liquid-vapor at $P = 1$ kPa, $T = 7°C$:

$h_{lg} = 2393$ kJ/kg

8.23 Solar energy systems often store energy in the form of hot water in storage tanks. Because of the density difference between hot and cold water, the storage tank can often be "stratified"; i.e., hot water introduced at the top of the tank tends to remain there because of its high specific volume (low density) relative to the colder water below.

(a) Plot the volumetric expansion coefficient β for liquid water in the range $60°C \leq T \leq 100°C$, using data from the steam tables.

(b) Plot the volumetric expansion coefficient β for liquid water in the range $0° \leq T \leq 15°C$.

(c) In the light of the data generated above, discuss the feasibility of hot versus cold storage stratification for solar-powered cooling systems.

8.24 Plot the volumetric expansion coefficient for water in the range $0°C \leq T \leq 15°C$, using data from the steam tables. Using these data, discuss why lakes freeze first at the surface rather than at the bottom.

8.25 In a "solar pond," injection of salt at various concentrations induces a strong gradient in specific volume with depth in the pond. Such a pond is stable when the salt specific volume gradient equals or exceeds the gradient in specific volume due to the temperature gradient caused by the absorption of solar energy in the pond, i.e.,

$$\left[\frac{1}{v} \left(\frac{\partial v}{\partial T} \right)_P dT < \frac{1}{v} \left(\frac{\partial v}{\partial X} \right)_T dX \right]_z$$

where X is the salt concentration and z is the distance beneath the pond surface.

For a pond the temperature gradient in the solar-heated layer is given by

$T(z) = 20 + 30z \qquad °C$

where z is in meters. Find the required minimum concentration of salt if the specific volume of a saline solution is given by

$$v = v_1(X = 0) + [v_{sat} - v_1(X = 0)] \, X(z)$$

where v_{sat} is the specific volume of salt-saturated brine (0.00093 m³/kg) and $v_1(X = 0)$ is the specific volume of pure water at the local temperature. (Actually, the specific volume of the brine varies with T as well as X, which makes the problem more complicated.)

8.26 At what pressure does the fugacity of ammonia gas (NH_3) first deviate by more than 10 percent from the pressure when the ammonia temperature is 150°C?

References

1. Lester Haar, John S. Gallagher, and George S. Kell, *NBS/NRC Steam Tables: Thermodynamic and Transport Properties and Computer Programs for Vapor and Liquid States of Water in SI Units,* Hemisphere, Washington, 1984.
2. Byung Ik Lee and Michael G. Kesler, "A Generalized Thermodynamic Correlation Based on Three-Parameter Corresponding States," *American Institute of Chemical Engineering Journal,* vol. 21, no. 3, May 1975, pp. 510–527.

Multicomponent Systems without Chemical Reaction

The most important thing we can experience is the mysterious. It is the source of all true art and science.
Albert Einstein

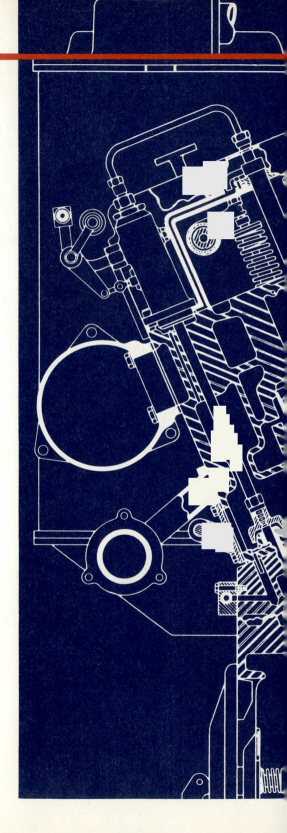

Cross-section of a marine diesel engine.
(General Electric.)

9.1 Introduction

A multicomponent fluid is composed of more than one chemical species. The most familiar example is air, which is a mixture N_2, O_2, Ar, and trace amounts of other chemical species. Another important example is moist air, which adds water vapor to the above list of chemical components. In this chapter we consider the thermodynamic properties of mixtures and chemical species, and analysis of systems that use multicomponent fluids.

Most of the thermodynamic analysis and systems considered to this point have addressed pure substances such as water or refrigerant 12. The thermodynamic properties for pure substances are considered in Chaps. 3 and 8, and the property tables are contained in the appendixes. There exists a large body of property data on pure substances. These data are available through numerical tables, equations, and programs for all phases — solid, liquid, and vapor — of numerous pure substances. We seek to use this wealth of information for multicomponent systems when possible.

One approach to obtaining thermodynamic properties for multicomponent fluids is to generate data for the specific mixture needed. This is exactly what has been done for air. As indicated, air is a mixture of various chemical species, and the property data have been obtained for air. This approach is useful since air, with fixed proportions of chemical species, is used in many engineering systems. The same approach is also required for other nondilute mixtures; however, a separate table is needed for every imaginable combination of chemical species. In some cases this is not required, and we consider these cases in this chapter.

The first goal of this chapter is to develop methods, through idealizations, to combine the properties of pure substances to obtain the properties for multicomponent fluids. Then we analyze systems that utilize multicomponent fluids. Last, we analyze nonideal mixtures.

9.2 Multicomponent Measures

A mixture is characterized in terms of the number of moles of each component i or in terms of the mass of each component i. The number of moles of component i is denoted n_i, and the total number of moles in the mixture is

$$n = \sum_i n_i \qquad (9.1)$$

The fraction of the total moles that is made up of component i is the *mole fraction*, denoted

$$y_i = \frac{n_i}{n} \tag{9.2}$$

The sum of all mole fractions is unity, so

$$\sum_i y_i = 1 \tag{9.3}$$

For a binary mixture, these relations are

$$n = n_A + n_B \tag{9.4}$$

and

$$y_A + y_B = 1 \tag{9.5}$$

A mixture is also characterized in terms of the mass of each component i, which is denoted m_i. The mass of the mixture is

$$m = \sum_i m_i \tag{9.6}$$

for nonreacting components. The *mass fraction* of component i is denoted

$$x_i = \frac{m_i}{m} \tag{9.7}$$

(The subscript denotes the component of a multicomponent fluid while x without a subscript is the quality, which specifically means m_g/m.) Therefore, the sum of all the mass fractions of the mixture yields

$$\sum_i x_i = 1 \tag{9.8}$$

For a binary mixture this is simply

$$m = m_A + m_B \tag{9.9}$$

and

$$x_A + x_B = 1 \tag{9.10}$$

The mass of component i is related to the number of moles of component i by the molecular weight of component i, M_i, as

$$m_i = n_i M_i \tag{9.11}$$

The total mass of the mixture is given by

$$m = \sum_i m_i = \sum_i n_i M_i \tag{9.12}$$

The *mean molecular weight* of the mixture, M, is defined so that $nM \equiv m$. Thus

$$M = \frac{m}{n} = \frac{1}{n} \sum_i n_i M_i = \sum_i y_i M_i \qquad (9.13)$$

For a binary mixture, this is $M = y_A M_A + y_B M_B$. Also, the relationship between the mass fraction and mole fraction is obtained from Eqs. (9.11) and (9.13) as

$$\frac{M_i}{M} y_i = x_i \qquad (9.14)$$

Figure 9.1   Mixture relations.

The above relations indicate the notation and definitions used for the mass or mole characterization of a mixture of various components. Since there are no chemical reactions, the number of moles of each component, or the mass of each component, sums to the total amount. The next step is to prescribe properties of a mixture in terms of the properties of the individual components. A simple sum of the individual component properties does not necessarily describe the mixture property. Consider a constant-pressure and constant-temperature mixing of two real gases. Gas A occupies an initial volume V_A, and gas B occupies an initial volume V_B. They are mixed. The final volume of the mixture is V_C, and *generally*

$$V_C \neq V_A + V_B \qquad (9.15)$$

This is diagramed in Fig. 9.1. Typical results for the complete range of mole fractions are shown in Fig. 9.2. Note, from a general viewpoint, that there is no reason to expect that the mixture volume is obtained from the addition of the original component volumes. Molecularly this is thought of as an interaction between the components that does not permit the volumes to add simply. The molecules of one gas can be imagined to "tuck in" between the molecules of the other. The above example is also valid for mixtures of liquids. This result for a general mixture holds for all extensive properties. The internal energy, enthalpy, entropy, and specific heats of the mixture of fluids at constant temperature and pressure are not the simple sum of pure component properties.

This complexity of mixture properties is simplified in the case of ideal gases.

9.3   Properties of a Multicomponent Ideal Gas

An ideal gas is a special case of a general mixture but is one with significant practical utility. Physically, a gas satisfies the ideal gas relation when the gas molecules are treated as independent of one another. A mixture of ideal gases also satisfies this molecular independence idealization. As is shown later, the mixture properties are simply mole or mass weighted averages of the individual pure-substance properties.

Reconsider the mixing of two gases presented in Sec. 9.2 with each gas satisfying the ideal gas relation. Ideal gas A is mixed with ideal gas B at

Figure 9.2   Mixture volume.

constant temperature and constant pressure. Since an ideal gas implies molecular independence, the final volume *is* the sum of the two volumes occupied by each gas independently. This is shown in Fig. 9.3 and is termed the *Amagat-Leduc law* of additive volumes. Repeating this experiment for various mole fractions yields a straight line on a *V-y* diagram (see Fig. 9.3).

This relation is demonstrated by the equation of state for a binary mixture of ideal gases. Each component is described by

$$P_A V_A = n_A \overline{R} T_A \tag{9.16a}$$

and

$$P_B V_B = n_B \overline{R} T_B \tag{9.16b}$$

and the mixture is given as

$$P_C V_C = n_C \overline{R} T_C \tag{9.16c}$$

For a constant-temperature and constant-pressure mixing, $T_A = T_B = T_C$ and $P_A = P_B = P_C$. Thus

$$V_A = \frac{n_A \overline{R} T_A}{P_A} = \frac{n_A \overline{R} T_C}{P_C} \tag{9.17a}$$

and

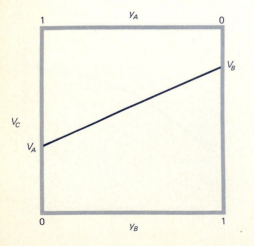

Figure 9.3   Ideal gas mixtures.

$$V_B = \frac{n_B \overline{R} T_B}{P_B} = \frac{n_B \overline{R} T_C}{P_C} \tag{9.17b}$$

The sum of the volumes yields

$$V_A + V_B = (n_A + n_B)\frac{\overline{R} T_C}{P_C} \tag{9.18}$$

but $n_A + n_B = n_C$ and $V_A + V_B = V_C$, so the mixture relation in Eq. (9.16c) is satisfied. Thus, the mixture volume is the sum of the individual gas volumes. This development could also be viewed as proof that the mixture is an ideal gas when the individual ideal gases are combined. Also, eliminating $\overline{R} T_C/P_C$ from Eqs. (9.17) and (9.16) yields the general relation

$$\frac{V_i}{V} = \frac{n_i}{n} = y_i \tag{9.19}$$

A multicomponent mixture of more than two ideal gases is given as

$$V = \sum_i V_i = \sum_i y_i V \tag{9.20}$$

where V_i is termed the *partial volume* of component i. The partial volume is the pure component volume that would result at the mixture temperature and pressure if that component were separated from the mixture.

An alternative viewpoint for a mixture of ideal gases considers the mixing of two ideal gases which are initially at different pressures but occupy equivalent volumes at the same temperature. Figure 9.4 shows this mixing. Physically, the pressure exerted on a plane within an ideal gas (force per unit area) results from the change in momentum of the independent molecules in random motion. Thus, the pressure contribution from mixing two ideal gases which have independent molecules should be the sum of the individual gas pressure contributions (with volumes and temperatures equal). This is termed *Dalton's law* of additive pressures.

The ideal gas equation of state for the components and mixtures is given in Eq. (9.16). For a constant-temperature and equal-volume mixing, $T_A = T_B = T_C$ and $V_A = V_B = V_C$. Thus,

$$P_A = \frac{n_A \overline{R} T_A}{V_A} = \frac{n_A \overline{R} T_C}{V_C} \tag{9.21a}$$

and

$$P_B = \frac{n_B \overline{R} T_B}{V_B} = \frac{n_B \overline{R} T_C}{V_C} \tag{9.21b}$$

Thus, the sum of the pressures is

$$P_A + P_B = (n_A + n_B)\frac{\overline{R} T_C}{V_C} \tag{9.22}$$

But $n_A + n_B = n_C$ and $P_A + P_B = P_C$, so the mixture relation in Eq.

Figure 9.4   Ideal gas mixture.

Ideal gas A $\quad$ Ideal gas B $\quad$ Ideal gas mixture C

(9.16c) is satisfied. Note that $\overline{R}T_C/V_C$ can be eliminated from Eq. (9.21) with Eq. (9.16) to yield the general relation

$$\frac{P_i}{P} = \frac{n_i}{n} = y_i \tag{9.23}$$

where P_i is termed the *partial pressure* of component i. The partial pressure of component i is the pressure that would be exerted by the component at the mixture temperature and volume if it were separated from the mixture. A general multicomponent mixture of ideal gases yields

$$P = \sum_i P_i = \sum_i y_i P \tag{9.24}$$

The above relations, which were stated as the Amagat-Leduc law and Dalton's law, yield identical information. Another way to state the equivalency is that the Amagat-Leduc law requires that Dalton's law be satisfied or Dalton's law requires the Amagat-Leduc law be satisfied (depending on the starting point). This is seen by comparing Eqs. (9.19) and (9.23). An important point is that these statements are founded on the molecular independence concept for ideal gases and are strictly valid only for ideal gases. Modifications of these statements are available for real gas properties, and are described in Sec. 9.8.

A mixture of ideal gases is an ideal gas. Therefore, it is often convenient to have an apparent ideal gas equation of state for this mixture. Equation (9.13) defines the mean molecular weight M of the ideal gas mixture, which can be viewed as the apparent molecular weight of the ideal gas mixture. On a mole basis, the mixture equation of state is [see Eq. (9.16c)]

$$PV = n\overline{R}T \tag{9.25}$$

where the properties are those of the mixture. On a mass basis, Eq. (9.13) is used to yield

$$PV = \frac{m}{M}\overline{R}T \tag{9.26}$$

An *apparent gas constant* for the mixture is defined as

$$R = \frac{\overline{R}}{M} \tag{9.27a}$$

and with Eq. (9.13), $M = m/n$. So

$$R = \frac{\overline{R}}{m}n = \frac{\overline{R}}{m}\sum_i \frac{m_i}{M_i} = \sum_i x_i R_i \tag{9.27b}$$

and the mixture equation of state is

$$PV = mRT \tag{9.28}$$

with the apparent gas constant given in Eq. (9.27).

The remaining ideal gas mixture properties are obtained from an extension of Dalton's law which is referred to as the *Gibbs-Dalton law.* This states that any extensive property of a multicomponent mixture is the sum of the individual pure-substance properties that would occupy the mixture volume at the mixture temperature. Therefore, the mixture properties are

$$U = \sum_i U_i(T, V) \tag{9.29a}$$

$$H = \sum_i H_i(T, V) \tag{9.29b}$$

$$S = \sum_i S_i(T, V) \tag{9.29c}$$

$$C_v = \sum_i C_{v,i}(T, V) \tag{9.29d}$$

$$C_P = \sum_i C_{P,i}(T, V) \tag{9.29e}$$

With the specific extensive property on a mole basis denoted with an overbar, Eqs. (9.29) become

$$U = n\bar{u} = n \sum_i y_i \bar{u}_i(T, V) \tag{9.30a}$$

$$H = n\bar{h} = n \sum_i y_i \bar{h}_i(T, V) \tag{9.30b}$$

$$S = n\bar{s} = n \sum_i y_i \bar{s}_i(T, V) \tag{9.30c}$$

$$C_v = n\bar{c}_v = n \sum_i y_i \bar{c}_{v,i}(T, V) \tag{9.30d}$$

$$C_P = n\bar{c}_P = n \sum_i y_i \bar{c}_{P,i}(T, V) \tag{9.30e}$$

On a mass basis, the mixture properties are

$$U = mu = m \sum_i x_i u_i(T, V) \tag{9.31a}$$

$$H = mh = m \sum_i x_i h_i(T, V) \tag{9.31b}$$

$$S = ms = m \sum_i x_i s_i(T, V) \tag{9.31c}$$

$$C_v = mc_v = m \sum_i x_i c_{v,i}(T, V) \tag{9.31d}$$

$$C_P = mc_P = m \sum_i x_i c_{P,i}(T, V) \tag{9.31e}$$

Note that all the pure-substance properties on the right-hand sides of the above equations are obtained at the volume and temperature of the mixture. This is explicitly indicated by the functional notation. From the discussion of Dalton's law of additive pressures, this is equivalent to evaluating the pure-substance properties at the mixture temperature and the component partial pressure. This is explicitly indicated in the following expressions:

$$U = n\bar{u} = n \sum_i y_i \bar{u}_i(T, P_i) = mu = m \sum_i x_i u_i(T, P_i) \qquad (9.32a)$$

$$H = n\bar{h} = n \sum_i y_i \bar{h}_i(T, P_i) = mh = m \sum_i x_i h_i(T, P_i) \qquad (9.32b)$$

$$S = n\bar{s} = n \sum_i y_i \bar{s}_i(T, P_i) = ms = m \sum_i x_i s_i(T, P_i) \qquad (9.32c)$$

$$C_v = n\bar{c}_v = n \sum_i y_i \bar{c}_{v,i}(T, P_i) = mc_v = m \sum_i x_i c_{v,i}(T, P_i) \qquad (9.32d)$$

$$C_P = n\bar{c}_P = n \sum_i y_i \bar{c}_{P,i}(T, P_i) = mc_P = m \sum_i x_i c_{P,i}(T, P_i) \qquad (9.32e)$$

For an ideal gas, the internal energy, enthalpy, and specific heats are functions of temperature only, so entropy is the only property requiring the use of the partial pressure.

Expanding Eq. (9.32c) to provide the specific entropy of an ideal gas mixture [noting $s_{ref} = s(T = 0 \text{ K}, P = 1 \text{ atm}) = 0$] results in

$$s = \sum_i x_i s_i(T, P_i) = \sum_i x_i \left[s_{0,i}(T) - R_i \ln \frac{P_i}{P_{ref}} \right] \qquad (9.33)$$

This is a general relation in terms of partial pressures. It can alternatively be expressed in terms of mole fractions. If we define

$$s_0(T) \equiv \sum_i x_i s_{0,i}(T) \qquad (9.34)$$

and use the properties of the natural logarithm to write

$$\ln \frac{P_i}{P_{ref}} = \ln \frac{P_i}{P_{tot}} + \ln \frac{P_{tot}}{P_{ref}}$$

then Eq. (9.33) becomes

$$s(T, P) = s_0(T) - \sum_i x_i R_i \ln \frac{P_{tot}}{P_{ref}} - \sum_i x_i R_i \ln \frac{P_i}{P_{tot}} \qquad (9.35)$$

However, $P_i/P_{tot} = y_i$, and Eq. (9.27b) shows that $\sum_i x_i R_i = R$. Further, from Eq. (9.14), $x_i R_i = y_i M_i R_i / M = y_i R$. Making these substitutions results in

$$s(T, P) = s_0(T) - R \ln \frac{P_{tot}}{P_{ref}} - R \sum_i y_i \ln y_i \qquad (9.36a)$$

or

$$\bar{s}(T, P) = \bar{s}_0(T) - \bar{R} \ln \frac{P_{tot}}{P_{ref}} - \bar{R} \sum_i y_i \ln y_i \qquad (9.36b)$$

For a mixture with *constant composition* undergoing a process between states 1 and 2,

$$s_2 - s_1 = s_0(T_2) - s_0(T_1) - R \ln \frac{P_{tot, 2}}{P_{tot, 1}}$$

and the final term in Eq. (9.36) drops out. In tabulations of properties for an ideal gas mixture, the $\sum_i y_i \ln y_i$ term is omitted from the tabulated s values.

Example Problem 9.1

Air is a mixture of nitrogen, oxygen, argon, and trace amounts of other components. Calculate u, h, s_0, and c_P at 1 atm and 300°C from Tables D.3 to D.5 (which all have the same reference state). Take air to be composed of 78.03 percent N_2, 20.99 percent O_2, and 0.98 percent argon on a volume basis.

Solution
Equation (9.19) indicates that the mole fractions are identically equal to the volume fractions. Thus the multicomponent measures of composition are as follows:

Component	Molecular Weight M	Volume Fraction V_i/V	Mole Fraction y_i	Mass Fraction x_i
N_2	28.016	0.7803	0.7803	0.7546
O_2	32.000	0.2099	0.2099	0.2319
Ar	39.944	0.0098	0.0098	0.0135
		1.0000	1.0000	1.0000

Here Eq. (9.13) yields

$$M = \sum_i y_i M_i = 0.7803(28.016) + 0.2099(32.000) + 0.0098(39.944)$$
$$= 28.969$$

and Eq. (9.14) gives

$$x_i = y_i \frac{M_i}{M}$$

which yields the values in the above table.
From Tables D.3, D.4, and D.5 and Eq. (9.32)

$$u = x_{N_2}u_{N_2} + x_{O_2}u_{O_2} + x_{Ar}u_{Ar}$$
$$= 0.7546(428.67) + 0.2319(384.95) + 0.0135(179.05)$$
$$= 415.2 \text{ kJ/kg}$$

$$h = x_{N_2}h_{N_2} + x_{O_2}h_{O_2} + x_{Ar}h_{Ar}$$
$$= 0.7546(598.89) + 0.2319(533.97) + 0.0135(298.42)$$
$$= 579.8 \text{ kJ/kg}$$

and

$$c_P = x_{N_2}c_{P,N_2} + x_{O_2}c_{P,O_2} + x_{Ar}c_{P,Ar}$$
$$= 0.7546(1.0699) + 0.2319(0.9954) + 0.0135(0.5207)$$
$$= 1.045 \text{ kJ/(kg} \cdot \text{K)}$$

The computation of entropy including variable specific heat requires use of Eq. (9.34) to yield

$$s_0 = (0.7546)(7.5261) + (0.2319)(7.0311) + (0.0135)(4.2163)$$
$$= 7.367 \text{ kJ/(kg} \cdot \text{K)}$$

These values are compared to the tabulated values in Table D.2. The maximum difference is approximately 0.01 percent, which is within the above specification for the volume fractions of air.

Comments

The above calculations are possible since all the values in the tables are referenced to the same state. The usual need for properties in thermodynamics involves the differences between values as calculated in this example. The above expressions can be used directly to obtain differences in properties.

9.4   Thermodynamic Analysis of Ideal Gas Mixtures

The properties of a mixture of ideal gases at any state can be obtained from the previous analysis. This section applies the first and second laws to multicomponent systems. Consideration is given to a system with fixed mole and mass fractions as well as to mixing problems. Since there are no chemical reactions, the number of moles or mass of each component can change only by the addition or deletion of a component. If a mixture of fixed composition is used in a process and there is no addition or deletion of specific components, then the mole and mass fractions are constant. This multicomponent system can be visualized as an ideal gas with fixed mixture properties. On the other hand, if a process is considered in which the mole and mass fractions change, a special approach is required. This last process is referred to as a *mixing process*.

The internal energy, enthalpy, and specific heats of an ideal gas are functions of temperature only. As indicated in Table 5.4, the changes in internal energy and enthalpy are

$$du_i = c_{v,i}\, dT \qquad (9.37a)$$

and

$$dh_i = c_{P,i}\, dT \qquad (9.37b)$$

on a mass basis. From Eq. (9.32)

$$dU = \sum_i n_i \bar{c}_{v,i}\, dT = \sum_i m_i c_{v,i}\, dT \qquad (9.38a)$$

$$dH = \sum_i n_i \bar{c}_{P,i}\, dT = \sum_i m_i c_{P,i}\, dT \qquad (9.38b)$$

With the mixture definitions for the specific heats in Eqs. (9.32d) and (9.32e),

$$dU = n\bar{c}_v\, dT = mc_v\, dT \qquad (9.39a)$$

$$dH = n\bar{c}_P\, dT = mc_P\, dT \qquad (9.39b)$$

Again the physical viewpoint for internal energy and enthalpy is one of independent gases simply being added, and the resulting mixture is an ideal gas with new averaged properties. This is not so simple for entropy.

The entropy of a mixture is given in Eq. (9.32c). Entropy for an ideal gas is a function of two independent variables. Recalling that the mixture is treated as a multicomponent system with pure-substance properties evaluated at the mixture temperature and component partial pressure, we see the entropy for each component is given from Table 5.4 as

$$d\bar{s}_i = \bar{c}_{P,i}\, \frac{dT}{T} - \bar{R}\, \frac{dP_i}{P_i} \qquad (9.40a)$$

on a mole basis, and

$$ds_i = c_{P,i}\, \frac{dT}{T} - R_i\, \frac{dP_i}{P_i} \qquad (9.40b)$$

on a mass basis. The change in mixture entropy is

$$dS = \sum_i n_i \bar{c}_{P,i}\, \frac{dT}{T} - \sum_i n_i \bar{R}\, \frac{dP_i}{P_i} \qquad (9.41a)$$

$$dS = \sum_i m_i c_{P,i}\, \frac{dT}{T} - \sum_i m_i R_i\, \frac{dP_i}{P_i} \qquad (9.41b)$$

The mixture specific heats are given in Eqs. (9.32d) and (9.32e), so

$$dS = n\bar{c}_P\, \frac{dT}{T} - \bar{R}\, \sum_i n_i\, \frac{dP_i}{P_i} \qquad (9.42a)$$

$$dS = mc_P \frac{dT}{T} - \sum_i m_i R_i \frac{dP_i}{P_i} \qquad (9.42b)$$

The change in entropy can be evaluated with constant specific heats or with variable specific heats as presented in Sec. 5.5.2. The partial pressures in Eq. (9.42) can also be expressed in terms of the mole or mass fractions [Eq. (9.23)].

First consider a mixture of ideal gases which proceeds from one state to another but does not change composition (the mole fractions and mass fractions are constant at all points in the process). The multicomponent measures are computed, and the mole and mass fractions are determined. Because the mole and mass fractions are constants, Eq. (9.23) indicates that the partial pressures vary with the total pressure. The mixture specific heats are evaluated by using Eqs. (9.32d) and (9.32e) as

$$dU = n\bar{c}_v \, dT = mc_v \, dT \qquad (9.43a)$$

$$dH = n\bar{c}_P \, dT = mc_P \, dT \qquad (9.43b)$$

$$dS = n\bar{c}_P \frac{dT}{T} - n\bar{R} \frac{dP}{P} \qquad (9.43c)$$

$$dS = mc_P \frac{dT}{T} - mR \frac{dP}{P} \qquad (9.43d)$$

where $P_i = y_i P$ and Eq. (9.27b) are used. These equations clearly indicate that the mixture is an ideal gas with averaged properties.

Next, consider the entropy change for a mixing process. The entropy change of a pure gas or gas mixture A when mixed with another pure gas or gas mixture B, where there are changes in the mole and mass fractions, is different from that presented above [Eq. (9.43)]. From Eq. (9.36), the change in entropy of gas A is

$$(\bar{s}_2 - \bar{s}_1)_A = \left\{ \bar{s}_0(T_2) - \bar{s}_0(\overline{T}_1) - \overline{R} \ln \frac{P_{tot,\,2}}{P_{tot,\,1}} \right.$$
$$\left. - \overline{R} \sum_i [(y_i \ln y_i)_2 - (y_i \ln y_i)_1] \right\}_A \qquad (9.44a)$$

or

$$(s_2 - s_1)_A = \left[s_0(T_2) - s_0(T_1) - R \ln \frac{P_{tot,\,2}}{P_{tot,\,1}} \right.$$
$$\left. - \left(R \sum_i y_i \ln y_i \right)_2 + \left(R \sum_i y_i \ln y_i \right)_1 \right]_A \qquad (9.44b)$$

For a *constant-pressure process* with $y_i = n_i/n_{tot}$, Eq. (9.44) reduces to

$$(S_2 - S_1)_A = n_A[\bar{s}_{0,A}(T_2) - \bar{s}_{0,A}(T_1)] - \overline{R} \sum_i \left(n_i \ln \frac{y_{i,2}}{y_{i,1}} \right)_A \qquad (9.45)$$

And if $\bar{c}_{P,A}$ is not a function of T, then with Eq. (5.58),

$$(S_2 - S_1)_A = n_A \bar{c}_{P,A} \ln \frac{T_2}{T_1} - \bar{R} \sum_i \left(n_i \ln \frac{y_{i,2}}{y_{i,1}} \right)_A \qquad (9.46)$$

where the last term indicates the change in entropy from mixing. Mixing of two gases results in $y_{i,2} < y_{i,1}$, so that the contribution to the entropy change is positive, reflecting the disorganization involved in mixing.

Example Problem 9.2

Nitrogen at 0.3 MPa and 100°C is to be adiabatically mixed with He at 0.15 MPa and 50°C. Determine the final temperature of the mixture and the entropy generation if the gases are mixed in the following two processes:

(a) Initially 0.5 kg of N_2 and 0.05 kg of He are separated by a partition, and they are mixed in a closed vessel.

(b) In a steady state flow device with an outlet pressure of 0.1 MPa, 0.5 kg/s of N_2 and 0.05 kg/s of He are mixed.

Diagrams

State 1   State 2

Case *a*

State 1   State 2

Case *b*

Solution

The first law is used to obtain state 2, and Eq. (9.41*a*) then gives the entropy change. The primary difference in the analysis of each of the mixing processes lies in the use of the internal energy for (*a*) and enthalpy for (*b*). Constant specific heats are assumed in both cases.

(a) The first law gives

$$U_2 - U_1 = m_2 u_2 - m_1 u_1 = 0$$

so

$$[m(u_2 - u_1)]_N + [m(u_2 - u_1)]_H = 0$$

where subscripts N and H represent nitrogen and helium, respectively. Thus

$$(mc_v)_N(T_2 - T_{N,1}) + (mc_v)_H(T_2 - T_{H,1}) = 0$$

since both components reach the same final equilibrium temperature. Thus

$$T_2 = \frac{(mc_v)_N T_{N,1} + (mc_v)_H T_{H,1}}{(mc_v)_N + (mc_v)_H}$$

Substituting the numerical values with Table C.2 yields

$$T_2 = 358.4 \text{ K} = 85.2°\text{C}$$

The evaluation of the entropy change requires the partial pressures at state 2. The partial pressures are determined from $P_i = m_i R_i T/V$ for the final mixture volume. Evaluating the initial volumes and summing yield

$$V_2 = V_{N,1} + V_{H,1} = \frac{m_N R_N T_{N,1}}{P_{N,1}} + \frac{m_H R_H T_{H,1}}{P_{H,1}} = 0.408 \text{ m}^3$$

Thus

$$P_{N,2} = \frac{(0.5)(0.2968)(358.38)}{0.408} = 130 \text{ kPa}$$

$$P_{H,2} = \frac{(0.05)(2.07703)(358.38)}{0.408} = 91.2 \text{ kPa}$$

The entropy generation is evaluated from Eq. (9.41a) as

$$_1 S_{\text{gen}, 2} = S_2 - S_1$$

$$= (mc_P)_N \ln \frac{T_2}{T_{N,1}} + (mc_P)_H \ln \frac{T_2}{T_{H,1}}$$

$$- (mR)_N \ln \frac{P_{N,2}}{P_{N,1}} - (mR)_H \ln \frac{P_{H,2}}{P_{H,1}}$$

$$= 0.1813 \text{ kJ/K}$$

(b) The first law for steady state adiabatic flow is

$$\dot{m}_1 h_1 = \dot{m}_2 h_2$$

$$\dot{m}_{N,1} h_{N,1} + \dot{m}_{H,1} h_{H,1} = \dot{m}_{N,2} h_{N,2} + \dot{m}_{H,2} h_{H,2}$$

Since $\dot{m}_{N,1} = \dot{m}_{N,2}$ and $\dot{m}_{H,1} = \dot{m}_{H,2}$, this yields

$$\dot{m}_N(h_2 - h_1)_N + \dot{m}_H(h_2 - h_1)_H = 0$$

$$(\dot{m}c_P)_N(T_2 - T_{N,1}) + (\dot{m}c_P)_H(T_2 - T_{H,1}) = 0$$

Solving for T_2 yields

$$T_2 = \frac{(\dot{m}c_P)_N T_{N,1} + (\dot{m}c_P)_H T_{H,1}}{(\dot{m}c_P)_N + (\dot{m}c_P)_H}$$

With the numerical values from Table C.3,

$$T_2 = 357 \text{ K} = 83°C$$

The entropy generation calculation requires the outlet mole fractions. They are evaluated from $\dot{n}_N = 0.5/28.016 = 0.0178$ kmol/s and $\dot{n}_H = 0.05/4.003 = 0.0125$ kmol/s. Then

$$y_{N,2} = \frac{0.0178}{0.0178 + 0.0125} = 0.588$$

$$y_{H,2} = \frac{0.0125}{0.0178 + 0.0125} = 0.413$$

The expression for the entropy generation for this steady flow process is

$$\dot{S}_{gen} = \dot{m}_2 s_2 - \dot{m}_1 s_1$$

With Eq. (9.44b) and $s_0(T_2) - s_0(T_1) = c_P \ln(T_2/T_1)$, this is expressed as

$$\dot{S}_{gen} = (\dot{m}c_P)_N \ln\frac{T_2}{T_{N,1}} + (\dot{m}c_P)_H \ln\frac{T_2}{T_{H,1}}$$
$$- (\dot{m}R)_N \ln\frac{P_2}{P_{N,1}} - (\dot{m}R)_H \ln\frac{P_2}{P_{H,1}}$$
$$- [(\dot{m}R)_N + (\dot{m}R)_H](0.588 \ln 0.588 + 0.413 \ln 0.413)$$
$$= 0.378 \text{ kW/K}$$

Comments

The difference in the analysis for each case is slight. The specific heats are changed in the first law analysis, and the partial pressure evaluation differed in the second law analysis. Note that the properties of He differ greatly from those of N_2. Entropy generation due to mixing is positive for both gases.

9.5 Multicomponent Analysis of Ideal Gas-Vapor Mixtures

The analysis in the previous sections is now applied to a mixture which includes a component that is near its saturation state. In Chap. 3 we indicated that a vapor and a gas are both gaseous fluids, but a vapor is at a temperature below the critical temperature (see Fig. 3.6). The important point is that the vapor in the mixture is close to the saturation region for that component, so it can condense out of the mixture to form another phase.

The most common example is the mixture of air and water vapor which occurs in the atmosphere. This mixture is viewed as one component in the gas phase, air, mixed with another component in the vapor phase, water. At atmospheric conditions of one atmosphere and 33°C, the meteorologist indicates that the relative humidity on that day is 90 percent. This is, indirectly, a specification of the mole and mass fractions of the mixture (this is explained subsequently). At the stated pressure and temperature, this corresponds to $x_w = 0.0285$ kg of water for each 1 kg of air–water vapor mixture. Thus the meteorologist has supplied the water concentration in the air-vapor mixture. Experience tells us that if this mixture were reduced in temperature while the total pressure was kept constant (atmospheric pressure), liquid droplets or dew would form, indicating the condensation process. This dew formation would occur if the ambient temperature were 31.2°C. This type of mixture of gas and vapor is considered in this section.

The analysis of gas-vapor mixtures is a direct application of the relations of Secs. 9.3 and 9.4. The gas and vapor are considered to obey the ideal gas relation, so the components are considered to be composed of independent molecules. The Gibbs-Dalton law states that the mixture properties are the sum of the pure-substance properties, each evaluated at the mixture temperature and the component partial pressure. Thus, the gas-vapor mixture, like any ideal gas mixture, can be treated as two independent components each at the *mixture temperature* and the component *partial pressure*. Also, the total pressure of the mixture is the sum of the partial pressures of the components. The partial pressure of the vapor component is called the *vapor pressure,* and the partial pressure of the gas component is called the *gas pressure.* The total pressure of the mixture is denoted P. The condensed phase is assumed to be pure (does not contain dissolved gas), and the saturation pressure of the condensed phase is not affected by the gas.

9.5.1   Measures and Properties

An ideal gas mixture of gas and vapor that undergoes any process is characterized by the relations in Secs. 9.3 and 9.4. The new aspect of the gas-vapor mixture is that the vapor phase can undergo a phase change. This fact requires an additional piece of information for a gas-vapor mixture which indicates the phase change process. The mixture temperature at which the vapor will undergo a phase change to liquid while at the partial pressure within the mixture is called the *dew point temperature T_d* of the gas-vapor mixture. Consider an air–water vapor mixture with the water phase in the vapor state shown in Fig. 9.5. At the partial pressure at state 1, the dew point temperature is the saturation temperature at the

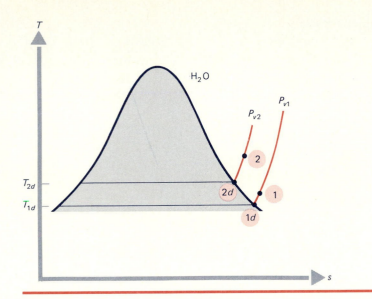

Figure 9.5 Dew point temperature.

water vapor partial pressure, or vapor pressure. This is indicated in Fig. 9.5 as $1d$. Also note that any mixture state that has the vapor pressure of P_{v1} has the same dew point temperature T_{1d}. For another state 2 with vapor pressure P_{v2}, the dew point temperature is T_{2d}.

A *saturated mixture* is a mixture with the vapor phase at its saturation pressure and temperature. Thus, an air–water vapor mixture with the water at state $1d$ or $2d$ in Fig. 9.5 is a saturated mixture. This mixture is also referred to as saturated air.

The concentrations of air-water mixtures are expressed in terms of two new, but related, quantities: the relative humidity ϕ and the humidity ratio ω. The *relative humidity* ϕ is defined as

$$\phi \equiv \frac{y_v(T, P)}{y_g(T, P)} \tag{9.47}$$

where $y_v(T, P)$ is the mole fraction of the vapor component (subscript v) and $y_g(T, P)$ is the mole fraction of the vapor at the saturated state, both corresponding to the mixture temperature and total pressure. There is no v subscript in the denominator since only the vapor component can undergo a phase change, which is indicated by the subscript g (as in Chap. 3). Since each component of the ideal gas mixture is treated as an ideal gas and the total pressures in the ratio in Eq. (9.47) are the same, the relative humidity is

$$\phi = \frac{P_v}{P_g} = \frac{P_v(T)}{P_{\text{sat}}(T)} \tag{9.48}$$

by Eq. (9.23). Figure 9.6 indicates the states that are compared in the definition of the relative humidity. This figure also shows that the relative

Figure 9.6   Relative humidity.

humidity is always less than or equal to unity since $P_v(T)$ is always less than or equal to $P_g(T)$. With the ideal gas relation for the vapor component, the relative humidity is also expressed as

$$\phi = \frac{P_v}{P_g} = \frac{RT/v_v}{RT/v_g} = \frac{v_g}{v_v} = \frac{\rho_v}{\rho_g} \tag{9.49}$$

Another measure of the concentration of the mixture is the *humidity ratio*, or *specific humidity*, defined as

$$\omega \equiv \frac{m_v}{m_a} \tag{9.50}$$

This is the mass of water vapor divided by the mass of dry air (without water vapor). The humidity ratio is related to the mass fraction of vapor $x_v = m_v/m$ by

$$\omega = \frac{1}{1/x_v - 1} \tag{9.51}$$

The humidity ratio is also expressed in terms of partial pressures by noting that the volumes occupied by air and water vapor are the same. Thus,

$$\omega = \frac{m_v}{m_a} = \frac{M_v}{M_a} \frac{n_v}{n_a} = \frac{M_v}{M_a} \frac{P_v V/(\bar{R}T)}{P_a V/(\bar{R}T)}$$

or

$$\omega = \frac{M_v}{M_a} \frac{P_v}{P_a} \tag{9.52}$$

With the molecular weights given in Table C.2,

$$\omega = 0.622 \, \frac{P_v}{P_a} = 0.622 \, \frac{P_v}{P - P_v} = 0.622 \, \frac{\phi P_g}{P - \phi P_g} \tag{9.53}$$

which is a function of the vapor pressure only. Another relation between the relative humidity and the humidity ratio is given by eliminating the vapor pressure from Eqs. (9.48) and (9.53):

$$\phi = \frac{\omega P_a}{0.622 P_g} \tag{9.54}$$

A final measure of the mixture concentration is the *degree of saturation* μ, defined as

$$\mu \equiv \frac{\omega(T, P)}{\omega_g(T, P)} \tag{9.55}$$

With Eqs. (9.53) and (9.48), this is

$$\mu = \frac{P_v(P - P_g)}{P_g(P - P_v)} = \phi \, \frac{P - P_g}{P - P_v} \tag{9.56}$$

The vapor pressure and the saturation pressure are generally much less than the total pressure (except at high temperatures), so the degree of saturation is approximately equal to the relative humidity.

The above measures of concentration are now used to obtain the mixture properties. The mixture properties are defined in Eq. (9.32). Many applications are based on a unit mass of dry air rather than a unit mass of mixture, so the humidity ratio is used in the property determination. In the case of a binary mixture of air and water vapor, the enthalpy per unit mass of dry air is given from Eq. (9.32b) as

$$\frac{H}{m_a} = \frac{m}{m_a}(x_a h_a + x_v h_v) = h_a + \frac{m_v}{m_a} \, h_v = h_a + \omega h_v \tag{9.57}$$

All the other properties given in Eq. (9.32) have a similar form when the enthalpy is replaced by the appropriate property.

Example Problem 9.3

Determine the humidity ratio, vapor mass fraction, degree of saturation, and dew point temperature for an air-water mixture at 0.10132 MPa, 33°C, and a relative humidity of 90 percent.

Solution
The relative humidity of 90 percent gives

$$\phi = \frac{P_v}{P_g} = 0.9$$

where P_g is the saturation pressure at 33°C. Table D.8 gives $P_g = 5.07$ kPa, so the vapor pressure is

$P_v = (0.9)(5.07) = 4.56$ kPa

Since the total pressure is 101.32 kPa, the partial pressure of the air is

$P_a = P - P_v = 96.76$ kPa

Equation (9.53) gives the humidity ratio as

$$\omega = 0.622 \frac{4.56}{96.76} = 0.0293 \frac{\text{kg H}_2\text{O}}{\text{kg dry air}}$$

The mass fraction is given from Eq. (9.51) as

$$x_v = \frac{m_v}{m} = \frac{m_v}{m_v + m_a}$$

$$= \frac{1}{1 + m_a/m_v} = \frac{1}{1 + 1/\omega} = 0.0285 \frac{\text{kg H}_2\text{O}}{\text{kg mixture}}$$

The degree of saturation is given from Eq. (9.56) as

$$\mu = 0.9 \frac{101.32 - 5.07}{101.32 - 4.56} = 0.895$$

Finally, the dew point temperature is obtained from Table D.9 at a pressure of 4.56 kPa: $T_d = 31.2$°C

Comments
The vapor pressure for water is generally small compared to the total pressure, as indicated. Also, the relative humidity and the degree of saturation are almost equal.

9.5.2 Thermodynamic Analysis

The first and second laws are now applied to air–water vapor mixtures. Many specific applications are considered in Sec. 9.7, but a few processes that exhibit many important features of gas-vapor mixtures are considered here.

Consider a control mass with a movable piston, so the air-water mixture is initially at state 1 and is cooled at constant pressure through the saturation state 2 to a temperature T_3 below the dew point temperature at state 1. Figure 9.7 diagrams the process. The mixture is initially in state 1, and as the mixture is cooled at constant pressure, the partial pressures remain constant until state 2. At this point, condensate forms at the bottom of the container, which is at state 2*l*. Subsequent lowering of the temperature forms liquid and saturated vapor at state 3. The vapor pressure from state 2 to 3 decreases, so the partial pressure of the gas phase increases to keep the total pressure constant. The mass of air is constant

Figure 9.7 Vapor condensation.

throughout the process, and so is the mass of the water. The first law for this process is

$$U_3 - U_1 = {}_1Q_3 + {}_1W_3 \tag{9.58}$$

which, for constant-pressure work, is

$$U_3 - U_1 = {}_1Q_3 - P(V_3 - V_1) \tag{9.59}$$

The heat transfer is given as

$${}_1Q_3 = H_3 - H_1 \tag{9.60}$$

Explicitly denoting the components and phases yields

$${}_1Q_3 = m_a(h_{a,3} - h_{a,1}) + m_{l,3}h_{l,3} + m_{g,3}h_{g,3} - m_{v,1}h_{v,1} \tag{9.61}$$

The mass of water is constant ($m_{v,1} = m_{l,3} + m_{g,3}$), so dividing this mass relation by the mass of dry air yields

$$\frac{m_{v,1}}{m_a} = \frac{m_{l,3} + m_{g,3}}{m_a} \qquad\qquad \omega = \frac{m_v}{m_a}$$

$$\omega_1 = \frac{m_{l,3}}{m_a} + \omega_3 \tag{9.62}$$

Thus the first law is

$$\frac{m_{l,3}}{m_a} = \omega_1 - \omega_3$$

$$\frac{{}_1Q_3}{m_a} = (h_{a,3} - h_{a,1}) + (\omega_1 - \omega_3)h_{l,3} + \omega_3 h_{g,3} - \omega_1 h_{v,1}$$

$$= h_{a,3} - h_{a,1} + \omega_1(h_{l,3} - h_{v,1}) + \omega_3 h_{lg,3} \tag{9.63}$$

This can be rearranged into the form given in Eq. (9.57) to yield

$$\frac{{}_1Q_3}{m_a} = h_{a,3} + \omega_3 h_{g,3} - (h_{a,1} + \omega_1 h_{v,1}) + (\omega_1 - \omega_3)h_{l,3} \qquad (9.64)$$

Example Problem 9.4

An air–water vapor mixture is initially at 101.32 kPa, 30°C, and $\phi = 90$ percent. The mixture is cooled at constant pressure to 10°C. Plot the process for both air and water on separate T-s diagrams. Also plot the humidity ratio and relative humidity as a function of process temperature. Evaluate the heat transfer per unit mass of dry air for the process.

Diagrams

Solution

The total pressure in this process is 101.32 kPa. The humidity ratio, vapor pressure, and partial pressure of air are evaluated from Tables D.8 and D.9, as in Example Problem 9.3. At 30°C, the saturation pressure is $P_g = 4.24$ kPa, so the vapor pressure is $P_v = \phi P_g = 3.82$ kPa. The dew point temperature at this vapor pressure is $T_d = 28.2$°C. The following table is obtained:

State	T, °C	ϕ	ω, $\dfrac{\text{kg H}_2\text{O}}{\text{kg dry air}}$	P_v, kPa	P_a, kPa
1	30.0	0.9	0.0244	3.82	97.50
2	28.2	1.0	0.0244	3.82	97.50
	25	1.0	0.0201	3.17	98.15
	20	1.0	0.0147	2.34	98.98
	15	1.0	0.0107	1.71	99.61
3	10	1.0	0.0076	1.23	100.09

The process plots are given, and the plot of (ϕ, ω) versus T is given here:

The heat transfer is evaluated from Eq. (9.63). Using $dh = c_P\, dT$ for air yields

$$\frac{{}_1Q_3}{m_a} = c_{P,a}(T_3 - T_1) + \omega_1(h_{l,3} - h_{v,1}) + \omega_3 h_{lg,3}$$

With numerical values given in Tables C.2 and D.8, this yields

$$\frac{{}_1Q_3}{m_a} = 1.0052(-20) + 0.0244(42.0 - 2556.8) + 0.0076(2478.4)$$

$$= -62.6 \text{ kJ/kg air}$$

The value for the enthalpy of water vapor at $P_v = 3.82$ kPa and $T = 30°C$ is taken as $h_{v1} = h_{g1}(30°C) = 2556.8$ kJ/kg, since water vapor in this state acts as an ideal gas so $h = h(T$ only) (see Fig. D.1 or the compressibility chart, Fig. C.1).

Comments

The humidity ratio does not change until condensation occurs. Condensation removes water vapor from the gas-vapor mixture, so the humidity ratio decreases. The relative humidity does not change after the mixture is saturated, because ϕ depends on only the water vapor state and not that of the air.

The heat transfer per unit mass of air to reduce the temperature from 30 to 20°C at state 1 is given as above,

$$\frac{{}_{30}Q_{20}}{m_a} = 1.0052(-10) + 0.0244(83.9 - 2556.8) + 0.0147(2455.0)$$

$$= -34.3 \text{ kJ/kg air}$$

The required heat transfer per unit mass of air to go from 20 to 10°C is

$$\frac{_{20}Q_{10}}{m_a} = \frac{_1Q_3}{m_a} - \frac{_{30}Q_{20}}{m_a} = -28.3 \text{ kJ/kg air}$$

This shows the required heat transfer for various portions of the process.

Another important problem is a constant-pressure steady flow process through the adiabatic system shown in Fig. 9.8. The air-water mixture enters at T_1 and P_1 with a humidity ratio of ω_1 and leaves the system at T_2, $P_2 = P_1$, and a humidity ratio ω_2. The air flows over the water within the system, and the water evaporates and is carried in the air as the air–water vapor mixture. Liquid water is supplied at T_2 to maintain the water level in the system.

The mass of air entering is equal to that which leaves, and the same is true for the water component. The first law, if we neglect changes in kinetic and potential energy, is

$$\dot{m}_a h_{a,1} + \dot{m}_{v,1} h_{v,1} + \dot{m}_l h_{l,2} = \dot{m}_a h_{a,2} + \dot{m}_{v,2} h_{v,2} \qquad (9.65)$$

Dividing by $\dot{m}_a$ yields

Figure 9.8   Adiabatic saturation process.

$$h_{a,1} + \omega_1 h_{v,1} + \frac{\dot{m}_l}{\dot{m}_a} h_{l,2} = h_{a,2} + \omega_2 h_{v,2} \qquad (9.66)$$

Conservation of mass for water is

$$\dot{m}_{v,1} + \dot{m}_l = \dot{m}_{v,2} \qquad (9.67)$$

or

$$\omega_1 + \frac{\dot{m}_l}{\dot{m}_a} = \omega_2 \qquad (9.68)$$

Therefore, the first law is

$$h_{a.1} + \omega_1 h_{v,1} + (\omega_2 - \omega_1)h_{l,2} = h_{a,2} + \omega_2 h_{v,2} \qquad (9.69)$$

If the system is sufficiently long, then the mixture leaving is saturated and the exit temperature is referred to as the *adiabatic saturation temperature.* Equation (9.69) is rearranged to give the entering humidity ratio as

$$\omega_1 = \frac{h_{a,1} - h_{a,2} - \omega_2(h_{g,2} - h_{l,2})}{h_{l,2} - h_{v,1}} \qquad (9.70)$$

Therefore, with the inlet temperature and adiabatic saturation temperature known for this constant-pressure saturation process, the inlet humidity ratio is obtained. This is one method to experimentally determine the humidity ratio for a mixture.

The adiabatic saturation temperature obtained at the outlet of the above system requires the outlet state to be saturated. This is difficult to obtain since the channel must be very long. The adiabatic saturation system is approximated by a thermometer that is covered with a wetted wick material. The mixture of unknown humidity passes over the wetted wick, and an equilibrium of the heat transfer and mass transfer is obtained. This measured temperature is called the *wet-bulb temperature,* and it approximates the adiabatic saturation temperature. Thus, the wet-bulb temperature is a measure of the water concentration of the mixture as given in Eq. (9.70). To distinguish between this wet-bulb temperature and the mixture temperature as measured by a standard uncovered thermometer, the mixture temperature is referred to as the *dry-bulb temperature.*

9.6   Psychrometric Chart

The previous equations give the measures of concentration and properties for air–water vapor mixtures. The expressions can be used with the tables in the appendixes for air and water to yield the necessary solutions for air–water vapor systems. Since these mixtures occur often, the quantities are presented in a chart form, which is termed the *psychrometric chart.* A portion of the psychrometric chart in Fig. D.5 is presented in Fig. 9.9.

Figure 9.9 in margin: **Figure 9.9** Psychrometric chart.

The state of the air–water vapor mixture is determined by pressure, temperature, and concentration. The total pressure for the chart is 1 atm or 101.325 kPa (corrections are available for other pressures, but they are not discussed in this text). Therefore, if the pressure is at a single value, the state is determined by the temperature and concentration. The measure of concentration is taken as the humidity ratio. Figure 9.9 shows the two axes as the dry-bulb temperature and the humidity ratio. Since Eq. (9.48) indicates that the humidity ratio is a function of the vapor pressure only, an equivalent concentration measure is the vapor pressure.

The air–water vapor mixture is bounded by the saturated mixture condition. This is given by $\phi = 1$ and has the basic trends of a $P\text{-}T$ diagram for pure water (see Chap. 3). The gas–water vapor mixture states are below this line. Above this line, the water must be in the liquid phase. Since $\phi = 1$ is the saturated state, the dry-bulb temperature at the saturated state is also the wet-bulb temperature. This temperature at $\phi = 1$ also corresponds to the dew point temperature for the vapor pressure. Enthalpy and the wet-bulb temperature for the mixture states are also indicated. Note that the enthalpy is based on a unit mass of dry air, that is, $(h_a + \omega h_v)_i$.

The psychrometric chart combines the ideal gas charts with the steam tables. The reference states are different for each component, and this practice of combining properties with different reference states is not recommended in general. In this case, this is acceptable since only differences in properties are required. This is indicated in Eqs. (9.60) and (9.65).

Example Problem 9.4 can now be solved by using the psychrometric chart.

9.7 Applications

This section considers some of the basic elements of systems which operate with multicomponent mixtures of gas and vapor. Combinations of these basic elements are used to construct various complex systems.

9.7.1 Heat Transfer at Constant ω

An air–water vapor mixture that undergoes a heat transfer while the component concentrations are kept constant is the same process as considered in Sec. 9.4. For a steady state flow at atmospheric pressure, this process is diagramed in Fig. 9.10. Since the process is at atmospheric pressure, the psychrometric chart is used. The process is a horizontal line on the psychrometric chart, and the direction of the process depends on the direction of the heat transfer. The process shown is a heating process, so the enthalpy increases and the relative humidity decreases. The vapor pressure and air partial pressure are also constants since ω is a constant, and these processes are also shown in Fig. 9.10. A cooling process goes in the opposite direction, so the enthalpy decreases and the relative humidity increases. The first law for the diagramed process is

$$h_{a,1} + \omega_1 h_{v,1} + \frac{\dot{Q}}{\dot{m}_a} = h_{a,2} + \omega_2 h_{v,2} \tag{9.71}$$

Figure 9.10 Heat transfer with constant ω.

9.7.2 Humidification

Many applications require an addition of water to air. This process is termed *humidification*. Figure 9.11 diagrams a steady humidification process where water is injected into a mixture of air and water vapor. The addition of water increases the humidity ratio. The psychrometric chart indicates the larger value of ω at the outlet state of the process, but the process direction (with respect to the dry-bulb temperature) depends on the enthalpy of the entering water. If liquid water enters at the outlet air temperature and the process ends at $\phi = 1$, then the process is the adiabatic saturation process described in Sec. 9.5.2. Equation (9.69) describes this process as

$$(h_a + \omega h_v)_1 + (\omega_2 - \omega_1)h_l = (h_a + \omega h_v)_2 \qquad (9.72)$$

Since $(\omega_2 - \omega_1)h_l$ is small relative to $(h_a + \omega h_v)$, this process is essentially a process with constant $h_a + \omega h_v$. This enthalpy is a function of the wet-bulb temperature only. Thus, the process follows approximately a constant wet-bulb temperature line. The lowest dry-bulb temperature possible is the saturation temperature. If water vapor is added rather than liquid water, the process can reach a higher dry-bulb temperature at the outlet, since h_l in Eq. (9.72) is now $h_{v,3}$ and its value can be large. This process is also shown in Fig. 9.11, ending at state 2'.

Figure 9.11   Humidification.

Figure 9.12 Dehumidification.

9.7.3 Dehumidification

Dehumidification is the process of taking water out of the mixture. A process used to achieve this goal is the cooling process at constant ω to the saturation state and subsequent condensation of the liquid from the mixture. The process is discussed in Sec. 9.5 for a control mass and is presented in Fig. 9.7. On a psychrometric chart, the mixture undergoes a process shown in Fig. 9.12.

9.7.4 Mixing of Air–Water Vapor Streams

A specific air–water vapor mixture can be obtained from separate streams by mixing the streams to obtain the desired outlet state. Figure 9.13 diagrams an adiabatic process with two entering streams and a single outlet stream. Conservation of mass yields

$$\dot{m}_{a,1} + \dot{m}_{a,2} = \dot{m}_{a,3} \tag{9.73}$$

and

$$\dot{m}_{v,1} + \dot{m}_{v,2} = \dot{m}_{v,3} \tag{9.74}$$

or

Figure 9.13 Mixing of streams.

$$\dot{m}_{a,1}\omega_1 + \dot{m}_{a,2}\omega_2 = \dot{m}_{a,3}\omega_3 \tag{9.75}$$

These conservation-of-mass statements yield

$$\frac{\dot{m}_{a,1}}{\dot{m}_{a,2}} = \frac{\omega_2 - \omega_3}{\omega_3 - \omega_1} \tag{9.76}$$

The first law is written as

$$(\dot{m}_a h_a + \dot{m}_v h_v)_1 + (\dot{m}_a h_a + \dot{m}_v h_v)_2 = (\dot{m}_a h_a + \dot{m}_v h_v)_3 \tag{9.77}$$

Rewriting the first law for a unit mass of dry air yields

$$\dot{m}_{a,1}(h_a + \omega h_v)_1 + \dot{m}_{a,2}(h_a + \omega h_v)_2 = \dot{m}_{a,3}(h_a + \omega h_v)_3 \tag{9.78}$$

Substituting Eq. (9.73) into Eq. (9.78) yields

$$\frac{\dot{m}_{a,1}}{\dot{m}_{a,2}} = \frac{(h_a + \omega h_v)_2 - (h_a + \omega h_v)_3}{(h_a + \omega h_v)_3 - (h_a + \omega h_v)_1} \tag{9.79}$$

Combining Eqs. (9.76) and (9.79) gives the following relation:

$$\frac{\dot{m}_{a,1}}{\dot{m}_{a,2}} = \frac{\omega_2 - \omega_3}{\omega_3 - \omega_1} = \frac{(h_a + \omega h_v)_2 - (h_a + \omega h_v)_3}{(h_a + \omega h_v)_3 - (h_a + \omega h_v)_1} \tag{9.80}$$

Thus, fixing the inlet states and mass flow rates of the inlets specifies the outlet state. The graphical interpretation is shown in Fig. 9.13.

Example Problem 9.5

An important application requiring the analysis of a multicomponent mixture of gas and vapor is a cooling tower. The cooling-water outlet of a power plant is often at too high a temperature to be rejected into a lake or river, so the energy is rejected into the atmosphere by evaporation of the water. Some makeup water is added to the cooling water before the water is returned to the power plant.

The water leaving the power plant condenser is at 35°C and is to be cooled to 25°C. The water mass flow rate leaving the plant into the cooling tower is 20,000 kg/s. The air–water vapor mixture enters the cooling tower at 15°C and a relative humidity of 50 percent, and leaves at 30°C as saturated air. Determine the mass flow rates of the dry air and the makeup water.

Solution

This process is similar to the humidification process discussed earlier except that the desired result is a reduction of temperature of the power plant cooling water rather than humidification of the air. Also, this particular case has $\phi = 1$ at the air outlet, so this is also similar to the adiabatic saturation process discussed in Sec. 9.5.2. The first law is

$$\dot{m}_{a,3}h_{a,3} + \dot{m}_{v,3}h_{v,3} + \dot{m}_{l,1}h_{l,1} = \dot{m}_{a,4}h_{a,4} + \dot{m}_{v,4}h_{v,4} + \dot{m}_{l,2}h_{l,2}$$

Diagram

Air + vapor out

4

H_2O in 1

H_2O out 2

Air + vapor in

3

Dividing by $\dot{m}_{a,3} = \dot{m}_{a,4} = \dot{m}_a$ gives

$$(h_a + \omega h_v)_3 + \frac{\dot{m}_{l,1}}{\dot{m}_a} h_{l,1} = (h_a + \omega h_v)_4 + \frac{\dot{m}_{l,2}}{\dot{m}_a} h_{l,2}$$

Conservation of mass for water yields

$$\frac{\dot{m}_{l,2}}{\dot{m}_a} + \omega_4 = \frac{\dot{m}_{l,1}}{\dot{m}_a} + \omega_3$$

This is substituted into the first law to yield

$$(h_a + \omega h_v)_3 + \frac{\dot{m}_{l,1}}{\dot{m}_a} h_{l,1} = (h_a + \omega h_v)_4 + \left(\frac{\dot{m}_{l,1}}{\dot{m}_a} + \omega_3 - \omega_4\right) h_{l,2}$$

or

$$\frac{\dot{m}_{l,1}}{\dot{m}_a} = \frac{(h_a + \omega h_v)_4 - (h_a + \omega h_v)_3 + (\omega_3 - \omega_4) h_{l,2}}{h_{l,1} - h_{l,2}}$$

The values are obtained from the psychrometric chart in Fig. D.5 and Table D.8. This yields

$$\frac{\dot{m}_{l,1}}{\dot{m}_a} = \frac{100.0 - 28.7 + (0.0052 - 0.0273)(104.8)}{146.5 - 104.8} = 1.65$$

Therefore, $\dot{m}_a = \dot{m}_{l,1}/1.65 = 20,000/1.65 = 12,100$ kg/s.

The makeup water is given by $\dot{m}_{l,1} - \dot{m}_{l,2}$. Conservation of mass for the water yields

$$\frac{\dot{m}_{l,1} - \dot{m}_{l,2}}{\dot{m}_a} = \omega_4 - \omega_3$$

or

$$\dot{m}_{l,1} - \dot{m}_{l,2} = \dot{m}_a(\omega_4 - \omega_3)$$

Thus, the required makeup water is

$$\dot{m}_{l,1} - \dot{m}_{l,2} = 12,100(0.0273 - 0.0052) = 267 \text{ kg/s}$$

Comments

The makeup water mass flow rate is $267/20,000 = 0.0134$, or 1.34 percent, of the inlet mass flow rate. This must be supplied to the power plant since it is lost to the surrounding environment.

9.8 Nonideal Mixtures

When the components of a mixture cannot be treated as ideal gases, then the relations governing the properties of mixtures become more complex.

9.8.1   Mixtures of Nonideal Gases

As a relatively simple way to treat gases that are at pressures and/or temperatures that make their behavior nonideal, we can proceed as follows. Take Eq. (5.44) on a molar basis,

$$d\bar{h} = T\,d\bar{s} + \bar{v}\,dP \tag{9.81}$$

where we understand each of the properties to apply to the mixture. Now, substitute the relations for the specific extensive properties of the mixture $\bar{h}$, $\bar{s}$, and $\bar{v}$ from Eqs. (9.30*b*), (9.30*c*), and (9.20), to give

$$d\left(\sum_i y_i\bar{h}_i\right) = T\,d\left(\sum_i y_i\bar{s}_i\right) + \left(\sum_i y_i\bar{v}_i\right)dP \tag{9.82}$$

This substitution makes the implicit assumption that the Gibbs-Dalton and Amagat-Leduc laws apply to a mixture of nonideal gases, which is, of course, an approximation.

Equation (9.82) can now be rearranged to give

$$\sum_i y_i(d\bar{h}_i - T\,d\bar{s}_i - \bar{v}_i\,dP) = 0 \tag{9.83}$$

Because y_i is independent of $\bar{h}_i$, $\bar{s}_i$, and $\bar{v}_i$ (i.e., the mixture can have any concentration of the components regardless of the specific extensive properties of the components), the only way that Eq. (9.83) can be satisfied is for the term in parentheses to be zero for every term in the series; i.e.,

$$d\bar{h}_i = T\,d\bar{s}_i + \bar{v}_i\,dP \tag{9.84}$$

Now, reduced property charts for each component in the mixture could be developed by using Eq. (9.84) and the method described in Sec. 8.3.2. Note, however, that temperature and pressure in Eq. (9.84) are the *total* pressure and the temperature of the *mixture,* not the component (partial) pressure. The result of these observations is that the existing reduced property charts can be used for determining the properties of the components of the mixture *provided* that the mixture temperature and mixture total pressure are used in finding the reduced properties; that is, $P_{r,\,i} = P_{tot}/P_{cr,\,i}$ and $T_{r,\,i} = T/T_{cr,\,i}$ for each component of the mixture. The mixture properties are computed by using the Gibbs-Dalton relations [Eqs. (9.29) to (9.31)] or their extensions, Eq. (9.32). In using Eq. (9.32), however, we are implicitly evaluating the component properties, using P_{tot} instead of P_i as implied by the notation.

This is not an exact method of determining the properties of a nonideal mixture, because it involves using properties of nonideal gases and then using ideal gas mixture relations to define the mixture properties. The results are, of course, more accurate than the simple ideal gas relations.

9.8.2 Kays' Rule

An alternative and quite simple method for determining nonideal mixture properties is to *first* define the reduced temperature and pressure of the mixture and *then* use the reduced property charts to find the mixture, rather than component, properties. W. B. Kays suggested that *pseudocritical* properties be defined for the mixture by the relations

$$T_{cr} = \sum_i y_i T_{cr,\, i}$$

$$P_{cr} = \sum_i y_i P_{cr,\, i}$$

(9.85)

The values of z and enthalpy and entropy deviation for the mixture are then found directly from the reduced property charts by using $P_r = P/P_{cr}$ and $T_r = P/P_{cr}$ as for any gas.

9.8.3 Equations of State

It would be quite useful to have an accurate equation of state for a mixture of gases. For example, the Benedict-Webb-Rubin equation [Eq. (3.29)] is available for a wide range of hydrocarbons. Yet, in many applications, a mixture of hydrocarbons is present, and then the coefficients for the equation are not available. Is there a way to combine the coefficients for the components of the mixture to provide averaged coefficients so that the equation will apply? Considerable effort is still devoted to methods for doing so, and no completely satisfactory method is at hand.

We have seen that the ideal gas equation of state can be modified for use with mixtures; and as long as we are in the region of ideal gas behavior, this works well. Some extension is possible with van der Waals' equation [Eq. (3.26)]. To use the general equation, constants a and b [Eqs. (3.27) and (3.28)] are computed for the mixture from

$$a = \left[\sum_i y_i (a_i)^{1/2} \right]^2$$

(9.86a)

$$b = \sum_i y_i b_i$$

(9.86b)

Use of the Benedict-Webb-Rubin equation for mixtures is much more complex and is discussed by Kyle [1].

9.9 General Mixture Relations

To this point, consideration has been limited to gases and to simplified relations for computing the properties of gas mixtures. The methods presented so far apply to a wide variety of engineering problems, including air-conditioning design and combustion-product analysis. However, in some cases, the relations developed so far are not applicable. In particular, the approximate relations for nonideal gases may not be accurate enough for the required calculations or, more generally, when one or more of the mixture components are a liquid or a solid. In this section, we develop relations for general mixtures and show the application to a few problems of engineering interest.

9.9.1 Partial Molal Properties

In general, any mixture extensive property Z can be expressed as a function of the two intensive properties T and P and the number of moles of the individual components, $n_A, n_B, n_C, \ldots$, as

$$Z = Z(T, P, n_A, n_B, n_C, \ldots) \tag{9.87}$$

Extending Eq. (1.15) to more than two independent variables gives

$$dZ = \left(\frac{\partial Z}{\partial T}\right)_{P, n_A, n_B, \ldots} dT + \left(\frac{\partial Z}{\partial P}\right)_{T, n_A, n_B, \ldots} dP$$
$$+ \left(\frac{\partial Z}{\partial n_A}\right)_{T, P, n_B, n_C, \ldots} dn_A + \left(\frac{\partial Z}{\partial n_B}\right)_{T, P, n_A, n_C, \ldots} dn_B + \cdots \tag{9.88}$$

Let us define the properties

$$\overline{Z}_A = \overline{Z}_{T, P, n_B, n_C, \ldots} = \left(\frac{\partial Z}{\partial n_A}\right)_{T, P, n_B, n_C, \ldots}$$

$$\overline{Z}_B = \overline{Z}_{T, P, n_A, n_C, \ldots} = \left(\frac{\partial Z}{\partial n_B}\right)_{T, P, n_A, n_C, \ldots} \tag{9.89}$$

$$\overline{Z}_C = \text{etc.}$$

as *partial molal properties*. Note the use of a capital letter with an overbar to denote a partial molal property. This is the only use of this combination of notation, and it is generally different from a lowercase letter with an overbar, which is a molar intensive quantity.

At constant temperature and pressure, Eq. (9.88) can be integrated to give

$$Z_{T, P} = \overline{Z}_{T, P, n_B, n_C, \ldots} n_A + \overline{Z}_{T, P, n_A, n_C, \ldots} n_B + \cdots$$
$$= \overline{Z}_A n_A + \overline{Z}_B n_B + \cdots \tag{9.90}$$

Let us relate Eq. (9.90) to what we know about ideal mixtures. For the case of the volume of a gas mixture composed of two gases, Eq. (9.90) becomes

$$V_{T, P} = \overline{V}_{T, P, n_B} n_A + \overline{V}_{T, P, n_A} n_B \tag{9.91}$$

For an ideal gas, $\overline{V}_{T, P} = (\partial V/\partial n)_{T, P} = \overline{v}$, and Eq. (9.91) becomes

$$V_{T, P} = \overline{v}_A n_A + \overline{v}_B n_B \tag{9.92}$$

as given by the Amagat-Leduc law.

   Now the general mixture curve for a general multicomponent mixture (Fig. 9.2) can be used to show how partial molal properties can be computed from experimental data. For a two-component mixture, Fig. 9.14 shows the volume per mole of mixture versus mole fraction relation that can be obtained experimentally. For such a curve, T and P are held constant.

   In this figure, the partial molal volumes of each component can be found graphically at any composition as follows: At the given mixture composition (point c), draw a tangent to the composition curve and extend it to the $y_B = 0$ and $y_B = 1$ axes. At the $y_B = 0$ intercept (point d), we read the value of $\overline{V}_A$, the partial molal volume of component A; and at the $y_B = 1$ axis (point e), we read the partial molal volume of component B, or $\overline{V}_B$. Clearly these endpoints to the tangent line are the partial molal volumes given the definitions of the partial molal properties.

   Note that

$$V_c = (n_A + n_B)\bar{v}_c \tag{9.93}$$

and from Eqs. (9.89) and (9.93), at the mixture composition at point c,

$$\overline{V}_A = \left(\frac{\partial V}{\partial n_A}\right)_{T,P,n_B} = \bar{v}_c + (n_A + n_B)\left(\frac{\partial \bar{v}}{\partial n_A}\right)_{T,P,n_B} \tag{9.94}$$

Since $y_A = n_A/(n_A + n_B)$, it follows that $dy_A = n_B\, dn_A/(n_A + n_B)^2$, or

$$\frac{dy_A}{1 - y_A} = \frac{dn_A}{n_A + n_B} \tag{9.95}$$

Equation (9.94) can then be written as

$$\overline{V}_A = \bar{v}_c + (1 - y_A)\left(\frac{\partial \bar{v}}{\partial y_A}\right)_{T,P,n_B}$$

$$= \bar{v}_c + (1 - y_A)\frac{\bar{v}_c - \bar{v}_d}{y_A - 1} = \bar{v}_d \tag{9.96}$$

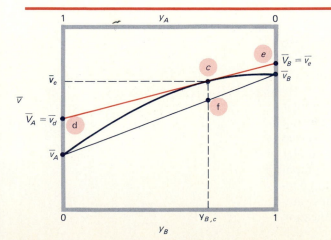

Figure 9.14   *Nonideal mixture volumes.*

and we see that the partial molal volume of gas A at point c is indeed equal to $\bar{v}_d$, the molar specific volume at the tangent intercept where $y_B = 0$.

All mixture relations can be expressed in terms of partial molal properties. The definition of mixture internal energy, for example, is

$$U = H - PV \qquad (9.97)$$

Holding constant T, P, and all molar amounts n_i except for n_A, we see the total derivative of Eq. (9.97) is simply

$$\left(\frac{\partial U}{\partial n_A}\right)_{T, P, n_i} = \left(\frac{\partial H}{\partial n_A}\right)_{T, P, n_i} - P\left(\frac{\partial V}{\partial n_A}\right)_{T, P, n_i} \qquad (9.98)$$

or

$$\bar{U}_A = \bar{H}_A - P\bar{V}_A \qquad (9.99)$$

Thus, all partial molal mixture properties can be developed from a minimal set of P-v-T data at a given concentration by using the methods in Chap. 8, which provide all information from a given fundamental equation.

9.9.2 Property Changes during Mixing

For a mixture of real materials, in general, the volume of the mixture is not equal to the sum of the volumes of the components. Let us restrict our attention to a two-component mixture that is mixed at constant temperature and pressure. The difference between the mixture volume and the volume of the components is

$$\Delta V = (n_A + n_B)\bar{v} - n_A\bar{v}_A - n_B\bar{v}_B = V - n_A\bar{v}_A - n_B\bar{v}_B \qquad (9.100)$$

By using Eq. (9.91), this becomes

$$\begin{aligned}
\Delta V &= \bar{V}_A n_A + \bar{V}_B n_B - n_A\bar{v}_A - n_B\bar{v}_B \\
&= n_A(\bar{V}_A - \bar{v}_A) + n_B(\bar{V}_B - \bar{v}_B)
\end{aligned} \qquad (9.101)$$

The change in volume on mixing is seen to be due to the difference between the partial molal volume of the components when mixed as compared with the molar volumes when unmixed. This difference is given by $\Delta\bar{v} = \bar{v}_c - \bar{v}_f$ in Fig. 9.14.

Changes due to mixing for other properties are found in a similar way. Note that, for an ideal gas, the partial molal volumes are given by Eq. (9.92), and substitution into Eq. (9.101) reveals that $\Delta V = 0$, as predicted by Amagat's law. We can define an *ideal solution* as one with the molar volume equal to the sum of the partial molal volumes for each component, so that the change in volume during a constant-temperature constant-pressure mixing process is zero. An ideal gas is obviously one example of an ideal solution. Ideal solutions offer considerable simplification in treating the properties of multicomponent systems with vapor-liquid equilibrium, which are examined in Chap. 11.

9.10 Summary

This chapter presents the thermodynamic analysis of multicomponent systems. General definitions and relations for systems composed of more than one component are given. Note that most of the analysis and properties are presented for ideal gases where the components are viewed as being composed of independent molecules. This independence characteristic of an ideal gas permits the combination of individual component properties to obtain mixture properties. The individual components are viewed as being independent at the mixture temperature and component partial pressure.

When one component exists in a state that is close to the saturation state for that component, a phase change is possible. Section 9.5 addresses mixtures of gas and vapor, and Sec. 9.6 presents the psychrometric chart which is applicable to mixtures of air and water vapor. The applications given in Sec. 9.7 involve air–water vapor mixtures and demonstrate the use of the psychrometric chart. Throughout Sec. 9.7, the processes for both individual components (air and water) are diagramed. The general relations in Sec. 9.5 for mixtures of ideal gas and vapor are also valid for mixtures other than air and water vapor. The gases leaving a power plant contain the products of combustion (hydrocarbons, CO_2, H_2O, N_2, etc.) and can be analyzed with the relations presented. The water vapor in this mixture can condense and present corrosion difficulties which must be avoided. The process for each component can be diagramed as in Sec. 9.7. (The psychrometric chart cannot be used since it is applicable to air–water vapor mixtures only.)

Nonideal mixtures can be approximated by ideal mixture relations with modified properties. Although the result is not accurate in all cases, the relations are quite useful. General mixture relations are presented that require the partial molal properties.

Problems

9.1 Methane and oxygen are mixed in a volume ratio of $1:5$. The total pressure of the mixture is 100 kPa. Find the following:

(a) The mole fraction of methane in the mixture.
(b) The mass fraction of methane in the mixture.
(c) The mean molecular weight of the mixture.
(d) The partial pressure of the methane. $P = \Sigma\, y_i\, P = 1.667 \times 10^{4}$
(e) The apparent gas constant for the mixture. $R = \bar{R}/m.w. = 8.314/29.341 = .28$
(f) The volume of 1 kmol of mixture at 300 K. $PV = nRT$

Gas	mole Fraction y_A	MW	$y_i M_i$	mass fraction $\dfrac{y_i M_i}{MW}$
CH_4	.1667	16.04	2.67	9.11×10^{-2}
O_2	.833	32	26.67	.90 885
			$\underline{29.341}$ $\leftarrow$ mw of mixt.	

9.2 For the mixture of Prob. 9.1 at 300 K, find (*a*) the specific internal energy *u* of the mixture in kJ/kg *and* kJ/kmol and (*b*) the specific entropy of the mixture in kJ/(kg · K) *and* kJ/(kmol · K). Assume $T_{ref} = 1$ K.

9.3 A 1.0-kg/h stream of argon at 27°C and 1 atm is mixed adiabatically with 0.5-kg/h CO_2 at 1 atm and 300°C.
(*a*) What is the final temperature of the mixture?
(*b*) Find the rate of entropy generation during the process.
(*c*) Calculate the entropy of the final mixture.

9.4 It is desired to adiabatically separate air (21 percent O_2 and 79 percent N_2 by volume) into pure oxygen and nitrogen, by carrying out the separation process at 1 atm and 27°C in a continuous reactor.
(*a*) What is the entropy change for the process?
(*b*) What is the reversible work required if the surroundings are at 27°C?

9.5 The gases leaving a natural-gas-fired boiler have the following composition:

Component	Volume Percentage
CO_2	8.7
H_2O	16.0
N_2	72.1
O_2	3.2
	100.0

At what temperature will condensation of the water vapor occur if the stack gases are at 1-atm total pressure?

9.6 What is the value of *h* for the gas mixture of Prob. 9.5 when the given mixture is saturated with water vapor?

9.7 If the stack gases in Prob. 9.5 are initially at 600°C, how much heat transfer can occur before condensation of the water vapor occurs?

9.8 A gas turbine provides combustion products at 540°C with the following composition (by weight): O_2, 17 percent; H_2O, 4 percent; N_2, 79 percent. These products are to be mixed with air (25°C, 1 atm), and the resulting gas mixture will be used as combustion air in a natural-gas-fired furnace. The combustion air must be at $T = 375$°C upon entering the furnace. See Fig. P9.8.
(*a*) How much air must be added (kilograms of air per kilogram of turbine exhaust)?
(*b*) What will be the value of *h* of the mixture entering the furnace?

Figure P9.8

(c) How much energy is saved by this system over an alternative system that uses fuel to preheat 100 percent ambient air to the furnace to 375°C? (Put the answer in terms of energy saved per kilogram of O_2 entering the furnace.)

9.9 Air at 30°C and 1 atm has a relative humidity of 60 percent. Find (a) the dew point of the wet air, (b) the humidity ratio of the wet air, and (c) the enthalpy of the wet air.

9.10 A stream of air at 1 atm, 20°C, and relative humidity of 80 percent has a mass flow rate of 0.05 kg/s. A second stream of air at 1 atm, 35°C, and relative humidity of 40 percent is mixed adiabatically with the first stream to give a mixed stream temperature of 30°C.
(a) What mass flow rate of the second stream is required?
(b) What is the relative humidity of the mixed stream?

9.11 Desert air at 33°C with relative humidity 10 percent is passed through a *swamp cooler*, a device that contacts the dry air with a wet surface. Water in the swamp cooler is evaporated until the desert air is saturated with water vapor. The swamp cooler is assumed to operate adiabatically, and the water in the swamp cooler is at the temperature of the air leaving the swamp cooler. What will be the temperature of the saturated air leaving the swamp cooler?

9.12 If 0.01 kg/s of desert air enters the swamp cooler of Prob. 9.11, how much makeup water must be added to the swamp cooler?

9.13 Air at 1 atm, 30°C, and 80 percent relative humidity is passed over the cold evaporator coil of an air conditioner. The air leaves the coil at 18°C. The wet air has a mass flow rate of 0.02 kg/s. How much water is condensed from the wet air?

9.14 What is the rate of heat transfer from the air in Prob. 9.13?

9.15 An air-conditioning system works as shown in Fig. P9.15. Find (a) T_B, (b) ϕ_B, (c) T_C, (d) ϕ_C, and (e) $T_{H_2O, \text{ out}}$

Figure P9.15

Handwritten solution:

9.15

State	Dry Bulb temp °C	wet °C	Dew A.	Humidity Ratio Kg H_2O/Kg dry	Rel. hum.
A	37	28.9	28.0	.02411	60
B	10.3	12.3	10.3	.0078	100
D	25.0	19.4	16.7	.011012	60

	υ m^3/Kg	h KJ/Kg	Sensible h KJ/Kg	Latent enthalpy KJ/Kg
A	.91295	116.595	55.042	61.503
B	.81344	47.898	28.247	19.051
D	.86109	73.169	42.984	30.185

Mass balance on water to mixing box:

$\dot{m}_B W_B + \dot{m}_C W_C = (\dot{m}_C + \dot{m}_B) W_D$

$W_C = \dfrac{\dot{m}_C + \dot{m}_B}{\dot{m}_C} W_D - \dfrac{\dot{m}_B}{\dot{m}_C} W_B$

$= \dfrac{.65}{.15} \times .011012 - \dfrac{1}{5} \times .0078 = .02016$

Energy balance on mixing box

$\dot{m}_B h_B + \dot{m}_C h_C = (\dot{m}_B + \dot{m}_C) h_D$

$h_C = \dfrac{.65}{.5} \times 73.17 - 2 \times 47.9 = 123.71$

From chart (extrapolated) or $h = \int c_p dt$

$T = 53.2 °C \quad Q_C = 21.9 \text{ KJ/kg}$

Energy balance on cooling coil:

$\dot{Q}_{air} = \dot{m}_a (h_B - h_A) + \dot{m}_A (W_A - W_B) T_B$

$= 1(47.9 - 116.6) + 1(.02411 - .0078) 43.2$

$= -68 \text{ KJ/s}$

$-\dot{Q}_{air} = \dot{Q}_{H_2O} = \dot{m}_{H_2O} C_{p H_2O} \Delta T_{H_2O}$

$T_{H_2O, \text{ out}} = 5 + \dfrac{68}{10 \cdot 4.8} = 6.4°C$

Figure labels (diagram):

Cooling coil — A, B
Mixing box — D
$\dot{m}_a = 1$ kg/s
$T = 37$ °C
$\phi = 60\%$
$\omega = 0.0078$
ϕ_B
T_B
Water (liquid)
$T_{in} = 5°C$
$\dot{m} = 10$ kg/s
Condensate
T_C
ϕ_C
C
$\dot{m}_a = 0.5$ kg/s
$T = 25$ °C
$\phi = 60\%$
$T_{H_2O, \text{ out}}$
Return air

9.16 Outside air at 35°C and $\phi = 65$ percent is cooled to 20°C and then mixed with recirculated air at 25°C, $\phi = 50$ percent. Recirculated air is mixed with outside air in the ratio of 1:3 on a dry air mass basis. What is the humidity ratio of the resulting mixed air? Sketch the process on a psychrometric chart.

9.17 An old and inefficient air-conditioning system used in commercial buildings when energy was very cheap worked as follows: Outside air (warm or cold, dry or humid) was passed over a cooling coil to condense water vapor and thus control humidity. The air was then heated to raise the temperature back to comfort conditions. The conditions are shown in Fig. P9.17.
(*a*) Evaluate the heat transfer in the heater.
(*b*) How much heat transfer occurred in the cooling coil?

9.18 Plot the density ρ ($\rho = 1/v$) for air versus relative humidity at $T = 30$°C. If aircraft lift is proportional to density at a given speed, what effect on required runway length would you expect humidity to have?

9.19 The temperature of the atmosphere on a given day follows roughly the relation

$$T(y) = T_0 - \frac{2y}{300}$$

where T_0 is the temperature at sea level and y is the height above sea level in meters. On this day, the temperature at sea level is 30°C, and the relative humidity at sea level is 70 percent. At what altitude would you predict clouds to form if sea-level air were being carried aloft by convective currents and cooled to the temperature of the altitude it reaches?

9.20 A passenger is on an aircraft that takes off from sea level on a day when the dry-bulb temperature is reported as 27°C and the relative humidity is 80 percent. She notes that fog appears to form in a sheet on the top of the wing as the plane takes off, and she assumes that this is due to condensation of water vapor.

(a) Explain this phenomenon in terms of what you know about air–water vapor mixtures and Bernoulli's equation relating velocity and pressure.

(b) What is the *maximum* pressure on the wing surface at which this phenomenon can occur at the given conditions?

9.21 Desiccant systems control air humidity by absorbing or adsorbing water vapor from the air into liquids or onto solids with a high affinity for water molecules. During the process, enthalpy of absorption is released, and the air-desiccant system increases in temperature.

Draw a schematic of the desiccation process described on a psychrometric chart, and show how to find the enthalpy change during the desiccant absorption process.

9.22 Using Kay's rule, find the fugacity of a mixture of methane and *n*-butane in the molar ratio of 2 : 1, respectively, when the mixture is at $T = 500°C$ and $P = 60$ atm.

9.23 Find h, s, and v for a mixture of 20 mol percent methane, 40 mol percent ethane, and 40 mol percent propane at $500°C$ and 100 atm by using (a) Kay's rule and (b) the nonideal mixture relations and reduced property data (Sec. 9.8). Assume $T_{ref} = 1$ K.

9.24 Using Kay's rule and the Lee-Kesler tables, plot the specific volume of a mixture of SO_2 and methane at $P = 300$ atm and $T = 400$ K versus the mole fraction of SO_2 present. Using this plot, determine the partial molal volume of SO_2 in the mixture at $y_{SO_2} = 0.5$.

9.25 The molar volume of a mixture of components A and B at T and P is given by

$$\bar{v}_{T, P} = y_A \bar{v}_A + (1 - y_A)^2 \bar{v}_B$$

What are the partial molal volumes of components A and B at $y_A = 0.75$?

Reference

1. B. G. Kyle, *Chemical and Process Thermodynamics*, Prentice-Hall, Englewood Cliffs, N.J., 1984.

Chemical Reactions and Combustion

When primitive man learned how to make fire, he had discovered controllable energy, which then became a "servant" destined to perform an endless series of "miracles." . . . This discovery may have been the single most vital factor which allowed mankind to develop modern civilization.

Herman Kahn, *The Next 200 Years*

V-16 diesel engine for marine service.
(General Electric.)

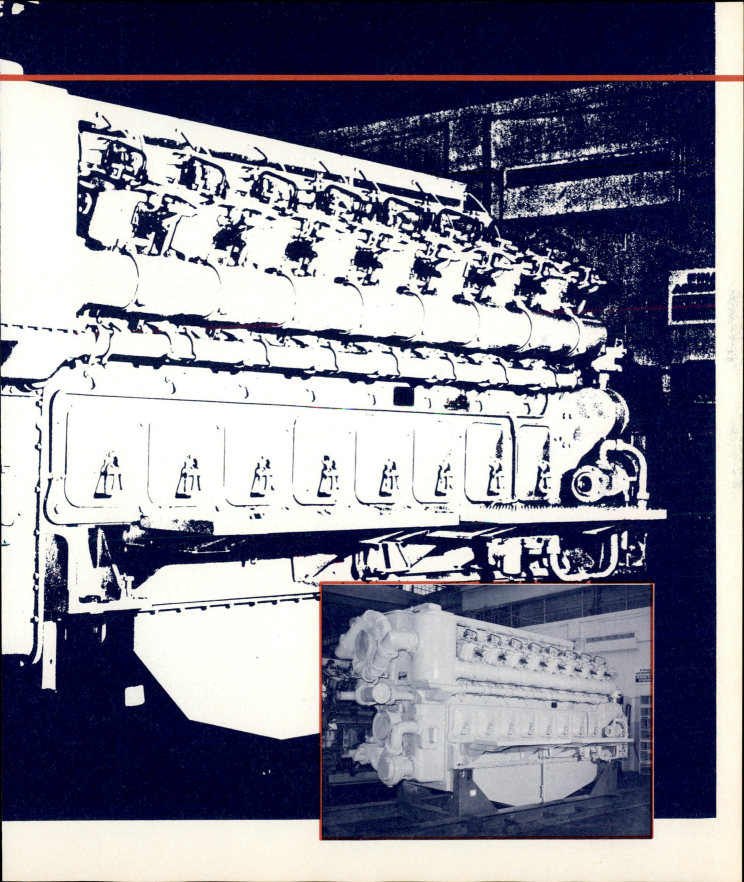

10.1　Introduction

Chemically reacting systems, and particularly systems undergoing combustion, are of interest to the engineer dealing with design in the chemical process industry, internal and external combustion engines of all types, fossil-fueled power plants, solid waste conversion systems, and studies of the initiation and spread of fires, among many other related subjects.

In this chapter, we develop the concepts and methods necessary to treat a broad variety of combustion problems of engineering interest, and we apply the methods to some problems.

10.2　Establishing a Common Basis for Combustion Processes

In treating pure simple substances as in Chap. 8, we noted that any state could be chosen as the reference state for property tabulations, since only *differences* between the properties in two states were of interest. When two or more substances undergo a chemical reaction, however, the establishment of a reference state that is common to all substances participating in the reaction becomes necessary. More than this, we must account for changes in the internal energy of a system that occur because of the chemical energy released or absorbed during the reaction of the chemical species. Thus, a reference value must be established that can be used to find the internal energy of the reactants *before* the reaction and of the products *after* the reaction. We establish the necessary reference values in this section.

10.2.1　Zero-Enthalpy Basis

A simple choice for a reference state would appear to be to set $h^o = 0$ kJ/kg at $T = 0$ K, so that all enthalpies take on positive values. (We denote property values at the chosen standard reference state by the superscript o, as in h^o.) However, as noted in Chap. 12 on examination of the microscopic behavior of materials, we find that some energy is possessed by most materials, even at absolute zero temperature. This energy is in the form of nuclear binding energy, molecular binding forces, and other forces that are not the same for all molecules. The situation is simpler for the *elements,* as opposed to more complex molecules. The elements can be assigned enthalpy values relative to one another at an arbitrary state. The effects of molecular energies can then be computed relative to the elements that make up the molecules. For convenience in comparison with laboratory data, the *standard reference state with $h^o = 0$ kJ/kg for all elements is assigned at $T = 25°C$ (77°F) and $P = 1$ atm.*

Thus for all elements $h^o = 0$ kJ/kg at the standard reference state. When the element exists in the chemically stable form of a diatomic gas at the reference state (N_2, O_2, H_2, etc.), the diatomic gas is assigned a reference value of zero at the standard reference state. The value of h for elements in any other state is found by applying the relation

$$h(T, P) - h_{\text{ref}} = h(T, P) - h^o = h(T, P) = \int_{\substack{T=25°C, \\ P=1 \text{ atm}}}^{T, P} dh \qquad (10.1)$$

The value of u at the standard reference state u^o is easily found from

$$u^o = h^o - P^o v^o = -P^o v^o \qquad (10.2)$$

and the value of v^o is found from P-v-T data at the reference state for the particular element.

10.2.2 Enthalpy of Formation

In Sec. 10.2.1 we established a common basis for the enthalpy of all elements. When compounds (molecules) are formed from the elements, chemical energy may be required to complete the reaction, or energy may be released. In either case, the energy, and therefore the enthalpy, of the molecule is different from the sum of the enthalpies of the atoms of which the molecule is composed. We can denote the enthalpy of the resulting molecule as

$$h = \Delta h_f + \sum_i N_i h_i \qquad (10.3)$$

where Δh_f is the enthalpy required to form the molecule from its constituent atoms at a given state, or the *enthalpy of formation.* The summation term accounts for the enthalpy added to the compound by the N_i atoms of each element i that make up the molecule.

If we use Eq. (10.3) to find the enthalpy of the compound at the standard reference state, then

$$h^o = \Delta h_f^o + \sum_i N_i h_i^o \qquad (10.4)$$

However $h_i^o = 0$ kJ/kg for all elements at the standard reference state, so *the enthalpy of a compound at the standard reference state is equal to the enthalpy of formation at the standard reference state.* Thus we can provide a reference value for all compounds by setting

$$h^o = \Delta h_f^o \qquad (10.5)$$

at the standard reference state of $T = 25°C$ and $P = 1$ atm. The values of the enthalpy of formation at the standard reference state are usually determined experimentally, although in some cases the methods of statistical thermodynamics can be used to predict values. A tabulation of Δh_f^o for various compounds is found in Table C.9. Note that the values presented in the table are actually $\Delta \bar{h}_f^o = M \Delta h_f^o$. That is, the values in the table are on a molar rather than a mass basis, and the molecular weight of the compound M is used to convert to the mass basis.

The enthalpy of a compound can now be computed in any state by using the enthalpy of the compound in the standard reference state as the reference value. The enthalpy in the standard reference state, given by Eq. (10.5), is called the *standardized reference enthalpy* of the compound. The enthalpy in another state is then

$$h(T, P) = h^o + \int_{\substack{T=25°C, \\ P=1 \text{ atm}}}^{T, P} dh = \Delta h_f^o + \int_{\substack{T=25°C, \\ P=1 \text{ atm}}}^{T, P} dh \qquad (10.6)$$

where $h(T, P)$ is the *standardized enthalpy* of the compound.

Example Problem 10.1

Find the standardized enthalpy of the following:
   (a) Argon gas at 100°C (assume it acts as an ideal gas)
   (b) Liquid water at 100°C
   (c) Water vapor at 100°C and 1 atm

Solution
   (a) The change in h for argon between 25 and 100°C can be found from Table D.3 as

$$\int_{25°C}^{100°C} dh = h(T = 100°C) - h(T = 25°C)$$
$$= 194.29 - 155.24 = 39.05 \text{ kJ/kg}$$

The ideal gas assumption allows us to consider h as independent of pressure. Because argon is an element, the value of the enthalpy at the standard reference state of 25°C and 1 atm is taken as zero. Thus, the value of the enthalpy at $T = 100°C$ is given by

$$h(T = 100°C) = h^o + \int_{25°C}^{100°C} dh = 0 + 39.05 = 39.05 \text{ kJ/kg}$$

   (b) For liquid water, assuming incompressibility, we use

$$h(T = 100°C) = h^o + \int_{25°C}^{100°C} dh = \Delta h_f^o + \int_{25°C}^{100°C} dh$$
$$= \Delta h_f^o + c_P(100 - 25)$$

The value of Δh_f^o is found from Table C.9 as $(-286.0 \times 1000)/18.02 = -15,870$ kJ/kg, and c_P for water is found in Table C.3 as 4.178 kJ/(kg $\cdot$ K) (by using the value at 323 K = 50°C). Substituting results in

$$h(T = 100°C) = -15,870 + 4.178(75) = -15,560 \text{ kJ/kg}$$

(c) For water vapor, we can take the value found in (b) for the liquid water standardized enthalpy and simply add h_{lg} from the steam tables. From Table D.8, $h_{lg} = 2257.0$ kJ/kg, so

$$h(T = 100°C) = -15,560 + 2257.0 = -13,300 \text{ kJ/kg}$$

Comments

Note the negative enthalpy value for liquid water as well as water vapor at the final state. This results from the negative enthalpy of formation at the standard reference state. Tabular data or equations of state that allow calculations of changes in enthalpy between states are found to be convenient aids in converting to standardized enthalpies.

10.2.3 Zero-Entropy Basis

In Sec. 10.2.2 we found a standard reference state for enthalpy and inferred reference values for internal energy at the same state. When we discuss second law considerations in combustion systems, we also need a standard set of entropy values. Here, a key observation from statistical thermodynamics is brought into play. As noted in Chap. 5, the property of entropy is related to our knowledge about the state of a particular system. As discussed more fully in Chap. 12, as the temperature of a pure substance is reduced, the possible range of energies that can be possessed by the atoms and molecules making up the system becomes smaller. Finally, as absolute zero is approached, the energy of all the constituent particles becomes small and indeed also approaches zero. Our knowledge about the state of the system becomes complete (the system is ordered), and our uncertainty about the state of the system (the entropy) becomes zero! This observation, originally formulated by Nernst, is embodied in the *third law of thermodynamics:*

> The entropy of each pure substance in thermodynamic equilibrium approaches zero as the temperature approaches absolute zero.

Thus, to find standard values of entropy in the standard reference state of 25°C and 1 atm, we integrate from absolute zero to the standardized state and tabulate the resulting values. We must account for all phase transitions between absolute zero and the standardized reference state. Entropies computed in this way are referred to as *absolute entropies.* Since the absolute entropies for all substances are referred to a common reference state, the entropies on this scale are appropriate for use in analyzing chemical reactions and combustion problems in particular. Some absolute entropies in the standard reference state are presented in Table C.10. To find absolute entropies in other states, the same methods are used as for the standardized enthalpy values carried out in Example Problem 10.1.

All the material presented above for standardized properties is general. For the particular case of combustion, it is usually valid to consider the products of the combustion reaction, the combustion air, and, quite often, the fuel to be ideal gases. When that is the case, many of the property data necessary for the treatment of combustion reactions are at hand in Tables D.2 through D.7, which list the temperature-dependent absolute entropy s_0 for various ideal gases.

10.3   Standards for the Comparison of Fuels

Generally, two standards are used in choosing a fuel. The first addresses the amount of energy released in the combustion process on a mole or mass basis, and the second addresses the temperature of the combustion process. To determine the appropriate standards, we can examine the properties of various fuels in idealized situations.

10.3.1   Enthalpy of Combustion

Consider the combustion of a general hydrocarbon C_nH_m in air. Assume that exactly enough oxygen is present to chemically react with the fuel that is present. We refer to such a situation as having the *stoichiometric* amounts of fuel and oxidant present. The combustion reaction is

$$C_nH_m + A_1(O_2) + A_2(N_2) \rightarrow A_3(CO_2) + A_4(H_2O) + A_5(N_2) \qquad (10.7)$$

where the A_i are the *stoichiometric coefficients* for this reaction. Balancing the atoms of each type on both sides of the equation results in $A_2 = A_5$, $A_1 = n + m/4$, $A_3 = n$, and $A_4 = m/2$. And the stoichiometric coefficient on the fuel has been arbitrarily set to unity.

In combustion with air, the nitrogen and oxygen in the air are in the volumetric ratio (and, therefore, if they act as ideal gases, in the molar ratio) of $A_2/A_1 = 79/21 = 3.76$ as found in the atmosphere. (It is common practice to lump all the other inert gases in the atmosphere with the nitrogen, since other gases commonly found in ambient air, such as argon, CO_2, etc., are present in small quantities. In any case, their properties of interest in combustion have values that are fairly close to those of nitrogen.) The value of $A_2 = A_5$ must thus be equal to $3.76A_1 = 3.76(n + m/4)$. The combustion equation then becomes

$$C_nH_m + \left(n + \frac{m}{4}\right)O_2 + 3.76\left(n + \frac{m}{4}\right)N_2 \rightarrow$$

$$nCO_2 + \frac{m}{2}H_2O + 3.76\left(n + \frac{m}{4}\right)N_2 \quad (10.8)$$

We have assumed in writing Eqs. (10.7) and (10.8) that *complete combustion* of the hydrocarbon occurs, and so the only combustion products are CO_2 and H_2O. Unless perfect mixing of the fuel and oxygen can be achieved, complete combustion will not occur, and carbon monoxide and oxides of nitrogen will be formed during combustion, as will a series of intermediate hydrocarbons, which are in turn subject to combustion. In practice, more oxygen is usually introduced to the combustion zone than is required according to the balanced chemical reaction equation shown above. This "excess air" will allow completion of the combustion reaction even if perfect mixing is not achieved; however, a penalty is paid, as we shall see.

When complete combustion occurs, the stoichiometric coefficients apply and show the relative number of moles of products and reactants that take part in the reaction. The mixtures of products and reactants in proportion to the stoichiometric coefficients are said to be in *stoichiometric ratio,* or simply to be *stoichiometric.*

Small amounts of CO and oxides of nitrogen are still formed. The amounts are small enough that if they are ignored, they do not affect the first law as written for a control volume. The CO and nitrogen oxides can, however, cause significant pollution problems, even in small relative amounts, if a large amount of a hydrocarbon is burned.

First, let us examine the case of an isothermal constant-flow reactor to which are fed the fuel at unit molar flow rate and air at a rate that gives the molar ratios to the fuel specified by Eq. (10.8). The reactor is shown in Fig. 10.1. In general, there is a heat transfer from the reactor in order for the control volume to remain isothermal. The first law for this control volume is

$$\dot{Q} = \sum_p \dot{m}_p h_p - \sum_r \dot{m}_r h_r \quad (10.9)$$

where the subscript r denotes reactants and the subscript p denotes products of the reaction. Let us carry out the reaction *under the particular conditions of the standard reference state,* that is, 1-atm pressure and with the temperature maintained at 298 K by heat transfer from the reactor. Then Eq. (10.9) on a molar basis becomes

$$\dot{Q}^o = \sum_p \dot{n}_p \bar{h}_p^o - \sum_r \dot{n}_r \bar{h}_r^o \quad (10.10)$$

Figure 10.1 *Isothermal constant-flow reactor.*

Combustion products

$\dot{n}_f = 1$ kmol/s
Fuel

$\dot{n}_{air}$ Air

$-\dot{Q}$, heat-transfer rate

where, as before, the overbar on a quantity indicates that it is on a molar basis. Now, divide through Eq. (10.10) by the molar flow rate of the fuel $\dot{n}_f$, and the equation becomes

$$\frac{\dot{Q}^o}{\dot{n}_f} = \bar{q}^o = \frac{1}{\dot{n}_f}\left(\sum_p \dot{n}_p \bar{h}_p^o - \sum_r \dot{n}_r \bar{h}_r^o\right) \qquad (10.11)$$

Equation (10.11) gives the amount of heat transfer required per mole of fuel from a reactor operating at the standard reference condition, which is equal to the net enthalpy difference between the products leaving the reactor and the enthalpy of the reactants entering. The quantity $\bar{q}^o$ is called the *enthalpy of combustion,* or sometimes the *heat of combustion.* Because we interpret the first law sign convention for Q as heat transfer to the control volume being positive, we expect $\bar{q}^o$ to generally carry a negative sign. The value of $\bar{q}^o$ as defined by Eq. (10.11) is fixed for a given fuel, and it is a measure of the chemical energy content per mole of the fuel when burned with oxygen (or air).

Example Problem 10.2

Determine the enthalpy of combustion of methane.

Solution
For methane (CH_4) the combustion equation, Eq. (10.8), becomes

$$CH_4 + (1+1)O_2 + 3.76(1+1)N_2 \rightarrow CO_2 + 2H_2O + 3.76(1+1)N_2$$

We must determine the values of $\bar{h}^o$ for each species in the reaction for use in Eq. (10.11). These are as follows:

Species	M	$\bar{h}^o$, kJ/kmol	
CH_4	16.042	-74.90×10^3	(Table C.9)
O_2	32.00	0	(element)
N_2	28.016	0	(element)
CO_2	44.01	-393.8×10^3	(Table C.9)
H_2O — liquid	18.016	-286.0×10^3	(Table C.9)
vapor		-242.0×10^3	(Table C.9)

Substituting into Eq. (10.11) and assuming that the water is produced in the form of vapor, we find

$$\bar{q}^o = \frac{1}{n_{CH_4}} [(n\bar{h}^o)_{CO_2} + (n\bar{h}^o)_{H_2O} + (n\bar{h}^o)_{N_2} - (n\bar{h}^o)_{CH_4} - (n\bar{h}^o)_{O_2} - (n\bar{h}^o)_{N_2}]$$

$$= \frac{1}{1} [-393.8 \times 10^3 + (2)(-242.0 \times 10^3) + 0 - (1)(-74.90 \times 10^3)$$

$$\quad - 0 - 0]$$

$$= -393,800 - 484,000 + 74,900 = -802,900 \text{ kJ/kmol}$$

Note that $\dot{n}/\dot{n}_f = n/n_f$ for a steady flow process. The enthalpy of combustion per mole of methane is thus $\bar{q}^o = -802,900$ kJ/kmol, or $q^o = -802,900/16.042 = -50,050$ kJ/kg.

Comments

The nitrogen in the air did not affect the result, so the enthalpy of combustion in the standard reference state is the same whether the methane is burned in stoichiometric air, pure oxygen, or excess air.

Note that if we had assumed that the water produced in the combustion reaction was in the form of liquid rather than vapor, then the calculated enthalpy of combustion would have been larger ($-890,900$ kJ/ kmol, or $-55,540$ kJ/kg of fuel). This occurs because more of the combustion energy is required to vaporize the water, and then less is available to remove from the combustion reactor in the form of heat transfer. The enthalpy of combustion when liquid water is produced is often called the *higher heating value* (HHV) of the fuel (usually the minus sign is dropped when this name is used). And the enthalpy of combustion is called the *lower heating value* (LHV) when water vapor is produced (again we drop the minus sign). In practical devices, the LHV is the only choice that we have, since combustion reactions in practice usually produce water vapor as a combustion product.

Values of the enthalpy of combustion (lower values) are tabulated for many substances, and some are given in Table C.11. A fuel with a higher enthalpy of combustion per unit mass is obviously more desirable than one with a lower enthalpy of combustion per unit mass, if their costs are the same.

10.3.2 Adiabatic Flame Temperature

One measure of the usefulness of a particular fuel is its enthalpy of combustion. However, in some instances, the engineer is more interested in producing a high temperature than a large energy release. This is the case in the design of process equipment that requires a given minimum temperature of operation, such as dryers and chemical process furnaces. In this section, we determine a method for comparing the temperatures available from the combustion of a given fuel.

Suppose that the combustion reactor of Fig. 10.1 is operated in a different mode from that used to determine the enthalpy of combustion. Rather than removing sufficient energy from the reactor by heat transfer, so that the reactor is maintained at isothermal conditions, let us heavily insulate the reactor so that *no* heat transfer occurs and the reactor operates adiabatically. The reactor then appears as in Fig. 10.2. In such a case, the first law for the reactor as given by Eq. (10.9) reduces to

$$\sum_p \dot{m}_p h_p = \sum_r \dot{m}_r h_r \tag{10.12}$$

Suppose that the *reactants enter the reactor at the standard reference state.* Equation (10.12) on a per-mole-of-fuel basis then becomes

$$\frac{1}{n_f}\sum_p n_p \bar{h}_p = \frac{1}{n_f}\sum_r n_r \bar{h}_r^o \tag{10.13}$$

If enthalpy-of-combustion data are available for the fuel being used, then it is convenient to substitute Eq. (10.11) into Eq. (10.13) to eliminate the reactant term:

$$\frac{1}{n_f}\sum_p n_p \bar{h}_p = \frac{1}{n_f}\sum_p n_p \bar{h}_p^o - \bar{q}^o \tag{10.14}$$

The enthalpy carried from the reactor by the products of combustion equals the enthalpy of the products at the standard reference state plus (since $\bar{q}^o$ carries a negative value) the energy released by the combustion reaction.

Solution of Eq. (10.13) for the temperature of the combustion products (assumed to be the same for all products because of mixing within the reactor) is iterative. A temperature must be assumed, and the values of $\bar{h}_p$ are found from tables or are computed. The $\bar{h}_p$ values are substituted into the product side of Eq. (10.13), and the result is checked against the reactant side of Eq. (10.13). The process is repeated until a temperature is found for which the equality is satisfied. For the case of no excess air, the temperature is called the *theoretical adiabatic flame temperature* for the fuel. The theoretical adiabatic flame temperature is the highest temperature that can be obtained from the fuel used in the combustion reaction (unless we preheat some of the reactants to temperatures above the standard reference state), because no heat transfer is allowed to occur to the surroundings, and no excess air is present to absorb any of the enthalpy of combustion (Fig. 10.3). Values of theoretical adiabatic flame temperature for some common fuels are shown in Table C.11. Note that the value of the theoretical adiabatic flame temperature is based on the assumed chemical reaction equation. The actual temperature is often lower because of dissociation or ionization of the combustion products or the presence of reactions which we have assumed to not occur. (In some cases, operation at stoichiometric conditions may actually produce a lower flame temperature than at slightly off-stoichiometric conditions. For ex-

Figure 10.2   Adiabatic reactor.

Figure 10.3   **Relation of enthalpy of combustion to adiabatic flame temperature.**

ample, additional reactions may occur at the temperatures in the combustor; the CO_2 may dissociate to form CO, or significant amounts of oxides of nitrogen may form. In that case, higher adiabatic flame temperatures may actually occur by running at conditions away from stoichiometric, where these reactions are suppressed.)

If tables or computer programs are available for $\bar{h}(T_p, P)$ for each product, then these can be used to determine the adiabatic flame temperature by iteratively choosing values of T_p until Eq. (10.14) is satisfied. Care must be taken that the sources of all the data use the standard reference state, or the user must convert the source data to the standard reference state. For such data, based on ideal gas properties, Refs. 1 to 3 are useful.

Another method of finding T_p is to rewrite Eq. (10.14) as

$$\sum_p n_p(\bar{h}_p - \bar{h}_p^o) = -n_f \bar{q}^o \tag{10.15}$$

and then to replace the enthalpy changes in terms of the specific heat for a constant-pressure combustor, or by assuming the enthalpy to be independent of pressure (the ideal gas assumption), so that Eq. (10.14) becomes

$$\sum_p n_p \int_{T^o}^{T_p} \bar{c}_{P,p}\, dT = -n_f \bar{q}^o \tag{10.16}$$

Now, the polynomial expressions of Table D.1 can be used in the left-hand side and iteration used until convergence. Note that for a fixed T^o, the individual integrals for each product species have the form

Figure 10.4   Mean specific heats of combustion gases.

$\int_{T^o}^{T_p} \bar{c}_P(T)\, dT$. We could define a temperature-averaged value of $\bar{c}_P$ by the relation

$$\bar{c}_P^*(T) = \frac{1}{T - T^o} \int_{T^o}^{T_p} \bar{c}_{P,p}(T)\, dT \tag{10.17}$$

and Eq. (10.16) reduces to

$$(T_p - T^o) \sum_p n_p \bar{c}_{P,p}^*(T_p) = -n_f \bar{q}^o \tag{10.18}$$

Now the iteration procedure is somewhat simplified. A T_p value is assumed, values of $\bar{c}_P^*(T)$ are found at that T_p for each species, and convergence is checked by testing the equality in Eq. (10.18). A graph of $\bar{c}_P^*(T)$ is presented in Fig. 10.4.

Example Problem 10.3

Calculate the theoretical adiabatic flame temperature for methane.

Solution

The enthalpy of combustion for methane (for the production of water vapor) is found in Example Problem 10.2 to be $-802,900$ kJ/kmol. The enthalpies at the standard reference state for the products are also given in that example. The chemical reaction is

$$CH_4 + 2O_2 + 7.52N_2 \rightarrow CO_2 + 2H_2O + 7.52N_2$$

Substituting the known values into Eq. (10.14) for 1 kmol of fuel results in

$$\sum_p n_p \bar{h}_p = \bar{h}^o_{CO_2} + 2\bar{h}^o_{H_2O} - \bar{q}^o$$

$$= -393.8 \times 10^3 + 2(-242 \times 10^3) - (-802,900)$$
$$= -74,900 \text{ kJ/kmol}$$

The right-hand side of this equation does not include N_2 since its standard reference state enthalpy is zero. The term $\sum_p n_p \bar{h}_p$ on the left-hand side (LHS) is

$$\sum_p n_p \bar{h}_p = \bar{h}_{CO_2} + 2\bar{h}_{H_2O} + 7.52\bar{h}_{N_2}$$

The enthalpy of the N_2 becomes important and must be included since this last expression is at the temperature of the products leaving the reactor.

Method 1

Let us first use the polynomial expressions for specific heats to determine the adiabatic flame temperature. The expressions for the LHS by this method are as follows: For CO_2,

$$\int_{T^o}^{T_p} \bar{c}_P(T)\, dT = 22.26(T_p - 298) + (5.981 \times 10^{-2})\left(\frac{T_p^2 - 298^2}{2}\right)$$

$$- (3.501 \times 10^{-5})\left(\frac{T_p^3 - 298^3}{3}\right) + (7.469 \times 10^{-9})\left(\frac{T_p^4 - 298^4}{4}\right)$$

For H_2O,

$$2\int_{T^o}^{T_p} \bar{c}_P(T)\, dT = 2\left[32.24(T_p - 298) + (0.1923 \times 10^{-2})\left(\frac{T_p^2 - 298^2}{2}\right) \right.$$

$$\left. + (1.055 \times 10^{-5})\left(\frac{T_p^3 - 298^3}{3}\right) - (3.595 \times 10^{-9})\left(\frac{T_p^4 - 298^4}{4}\right) \right]$$

and for N_2,

$$7.52\int_{T^o}^{T_p} \bar{c}_P(T)\, dT = 7.52\left[28.90(T_p - 298) \right.$$

$$- (0.1571 \times 10^{-2})\left(\frac{T_p^2 - 298^2}{2}\right)$$

$$\left. + (0.8081 \times 10^{-5})\left(\frac{T_p^3 - 298^3}{3}\right) - (2.873 \times 10^{-9})\left(\frac{T_p^4 - 298^4}{4}\right) \right]$$

We now choose values of T_p, evaluate the LHS of the first law relation by summing the three terms above, and compare to the right-hand side (RHS). Graphical presentation of the iterations is usually helpful in interpreting the results.

T_p (guessed), K	$n[\bar{h}(T_p) - \bar{h}^o]$, kJ			SUM, kJ	RHS, kJ
	CO$_2$	H$_2$O	N$_2$		
2300	110,700	174,200	499,200	784,100	802,900
2400	117,300	183,900	523,700	824,900	802,900
2346	113,700	178,700	510,500	802,900	802,900

The graph of $\text{SUM} = \sum_p n_p(\bar{h}_p - \bar{h}_p^o)$ versus T_p(guessed) is shown here.

Note that the third guess was chosen based on linear interpolation between the guesses at 2300 and 2400 K, and this yielded the correct result. This is often the case for combustion problems because $\bar{h}$ is very nearly linear with T over fairly large temperature ranges, and the iterative procedure converges very quickly.

Note also that the largest portion of the product enthalpy is carried by the nitrogen rather than the reaction products CO$_2$ and H$_2$O.

The computed value does not agree well with the tabulated value of 2285 K for the adiabatic flame temperature of methane shown in Table C.11. This discrepancy is partly due to the use of the $\bar{c}_P(T)$ relations outside the temperature range for which they were intended (the range is usually 273 to 1800 K).

Method 2

Using the definition of mean specific heat, we can write Eq. (10.18) for this case as

$$(T_p - T^o)\{[n\bar{c}_P^*(T_p)]_{CO_2} + [n\bar{c}_P^*(T_p)]_{H_2O} + [n\bar{c}_P^*(T_p)]_{N_2}\} = -\bar{q}^o$$

A simple procedure which converges quickly is to rewrite this equation as

$$T_p = T^o - \frac{\bar{q}^o}{[n\bar{c}_P^*(T_p)]_{CO_2} + [n\bar{c}_P^*(T_p)]_{H_2O} + [n\bar{c}_P^*(T_p)]_{N_2}}$$

Now, values of T_p are assumed, and $\bar{c}_P^*(T_p)$ are taken from Fig. 10.3 and substituted into the RHS of the last equation. A new value of T_p is then computed from the equation, and the procedure is repeated until convergence. For example,

	$\bar{c}_P^*(T)$, kJ/kmol-K			
T_p(guessed), K	CO$_2$	H$_2$O	N$_2$	RHS, K
2300	$6.56\bar{R}$	$5.31\bar{R}$	$4.05\bar{R}$	2325

The first guess in this case produced a new value for T_p that is within the accuracy of the graphs.

Method 3

For the final method, tabulated values of $\bar{h}(T_p) - \bar{h}^o$ from the ideal gas tables are used directly in the equation

$$(\bar{h} - \bar{h}^o)_{CO_2} + 2(\bar{h} - \bar{h}^o)_{H_2O} + 7.52(\bar{h} - \bar{h}^o)_{N_2} = 802{,}900$$

Again, values of T_p are assumed, and $h - h^o$ values are found directly from the tables. The process is continued until the equation is satisfied.

	$\bar{h} - \bar{h}^o = M(h - h^o)$, kJ/kmol			
T_p(guessed), K	CO$_2$	H$_2$O	N$_2$	LHS, kJ/kmol
2300	109,100	88,420	67,250	791,700
2350	112,300	91,070	68,860	812,300
2329	110,800	89,960	68,310	804,400

Linear interpolation between the first and third table entries gives $T = 2326$ K.

Comments

The three methods have produced theoretical flame temperature predictions for methane of 2346, 2325, and 2326 K, respectively, while Table C.11 gives 2285 K. Which should we believe? As noted, method 1 required use of polynomial expressions that were extended to temperatures outside their range of validity, so the results of that method are less accurate. In method 2, graphical values good to no more than three significant figures were used, so the results cannot be precise to more than ± 20 K at best. The final method makes use of quite precise tables with five significant figures; however, the tables are for ideal gases and do not account for nonideal effects or for dissociation of the gases. Dissociation does occur to a small degree at the temperatures in the flame, and chemical energy is needed to sever the bonds holding the molecules together. This lowers the flame temperature slightly, and the values in Table C.11 account for this effect.

10.4 Applications to Combustion Systems

Now that we have found methods for comparing fuels on the basis of their energy release (enthalpy of combustion) and relative combustion temperatures (adiabatic flame temperature), we can examine the application of thermodynamics to some engineering combustion problems. In particular, we examine systems in which combustion occurs, but in which we deviate from the idealized conditions used in Sec. 10.3, i.e., reactants all entering at the standard reference state, adiabatic or isothermal combustion, stoichiometric amounts of oxygen (air) provided for combustion, etc. We assume that the combustor operates at constant pressure.

10.4.1 Accounting for Excess Air

In practice, more air must be provided for the combustion process than is indicated by the balanced chemical reaction equation. Excess air is necessary because it is difficult, if not impossible, to achieve perfect mixing of the fuel with the oxygen carried by the air during the period of residence of these components in the combustion reactor.

If air in an amount above that necessary for stoichiometric combustion is provided, then we refer to the extra as *excess air*. And, for example, if 20 percent more air by volume (or mole) is provided than the stoichiometric amount, then the system is receiving 20 percent excess air. (Sometimes the term *theoretical air* is used. Thus 100 percent theoretical air means the stoichiometric amount of air, and 120 percent theoretical air is the same as 20 percent excess air.)

For a combustor receiving $X \times 100$ percent excess air, the chemical reaction equation for combustion of 1 mol of a general hydrocarbon C_nH_m as in Eq. (10.8) now becomes

$$C_nH_m + (1 + X)\left(n + \frac{m}{4}\right)O_2 + 3.76(1 + X)\left(n + \frac{m}{4}\right)N_2 \rightarrow$$

$$nCO_2 + \frac{m}{2}H_2O + 3.76(1 + X)\left(n + \frac{m}{4}\right)N_2$$

$$+ X\left(n + \frac{m}{4}\right)O_2 \quad (10.19)$$

Comparison with Eq. (10.8) shows that $3.76X(n + m/4)$ mol of N_2 and $X(n + m/4)$ mol of unreacted O_2 now appear on the right-hand side as products. Neither of these participates in the chemical reaction; they simply enter with the reactants and are carried through the combustor and out with the products.

The ratio of the number of moles of air to the number of moles of fuel is called the *air-fuel ratio* (AFR). Examination of Eq. (10.19) shows that for each mole of fuel there will be $(1 + X)(n + m/4)$ mol of oxygen and 3.76(1 + X)(n + m/4) mol of nitrogen. So in general the AFR for the burning of a hydrocarbon is given by AFR = 4.76(1 + X)(n + m/4). Knowing the AFR permits quick calculation of the amount of air that must be provided for a known combustion reaction in excess air, since for an ideal gas the AFR must be the volume ratio as well as the molar ratio of air to fuel.

The general first law for the combustor is given by Eq. (10.9). If we assume that the reactants enter at the standard reference state, then changing to a molar basis and subtracting Eq. (10.10) result in

$$\bar{q} - \frac{1}{n_f} \sum_p n_p[\bar{h}_p(T) - \bar{h}_p^o] = \bar{q}^o \tag{10.20}$$

For many large combustors, the energy loss to the combustor surroundings by heat transfer is small; however, heat transferred to a reacting chemical stream (in a chemical furnace) or to steam (as in a power plant boiler) must be included in $\bar{q}$. Let us, for a moment, restrict our attention to an adiabatic furnace, so that the effect of excess air becomes obvious. In that case, Eq. (10.20) reduces to

$$\frac{1}{n_f} \sum_p n_p[\bar{h}_p(T) - \bar{h}_p^o] = -\bar{q}^o \tag{10.21}$$

which is the same form as was used in determining the adiabatic flame temperature, Eq. (10.14). Now, however, the left-hand side of the first law equation contains more terms in the series. Now the excess unused oxygen must be included, and the number of moles of nitrogen is also larger because of the additional nitrogen carried into (and out of) the combustor as a result of the excess air. Both of these factors decrease the temperature reached by the combustion products in comparison with the theoretical adiabatic flame temperature.

Example Problem 10.4

Determine the AFR and the temperature of the combustion products of methane burning with 20 percent excess air. Assume the reactants enter at the standard reference state.

Solution
The chemical balance equation is

$$CH_4 + 2.4O_2 + 9.024N_2 \rightarrow CO_2 + 2H_2O + 9.024N_2 + 0.4O_2$$

Then

$$AFR = 4.76(1 + 0.2)(1 + 1) = 11.4$$

The first law, Eq. (10.21), becomes

$$[\bar{h}(T_p) - \bar{h}^o]_{CO_2} + 2[\bar{h}(T_p) - \bar{h}^o]_{H_2O} + 9.024[\bar{h}(T_p) - \bar{h}^o]_{N_2}$$
$$+ 0.4[\bar{h}(T_p) - \bar{h}^o]_{O_2} = -(-802,900) \text{ kJ/kmol}$$

An iterative solution using method 1 of Example Problem 10.3 yields $T_p = 2075$ K.

Comment

The presence of the excess air, while necessary for such practical reasons as preventing unburned fuel and the products of incomplete combustion from leaving the combustor, has also lowered the flame temperature by 271°C from that predicted for stoichiometric conditions with complete mixing.

10.4.2 Accounting for Air or Fuel Preheating

We have assumed so far that the reactants enter the combustor at the standard reference state. This is rarely the case, so in this section we include the effect of preheating the fuel and/or air entering the combustor. As we shall see, the enthalpy of the combustion products can be raised significantly by such preheating, especially preheating of the combustion air. Industrial furnaces in particular use such preheating whenever a convenient source of waste energy is available.

If the reactants are preheated, then the first law is given by Eq. (10.9), which, given an adiabatic combustor and a molar basis, becomes

$$\sum_p n_p \bar{h}_p = \sum_r n_r \bar{h}_r \qquad (10.22)$$

Subtracting Eq. (10.11) from Eq. (10.22) and assuming one mole of fuel result in

$$\sum_p n_p[\bar{h}_p(T_p) - \bar{h}_p^o] = \sum_r n_r[\bar{h}_r(T_r) - \bar{h}_r^o] - \bar{q}^o \qquad (10.23)$$

Now the effect of each of the reactants (fuel and air) entering the combustor at their respective temperatures T_r can be considered. Examination of Eq. (10.23) shows that the higher the temperature T_r of each reactant, and therefore the higher the $\bar{h}_r(T_r)$ of each reactant, the larger will be the right-hand side of Eq. (10.23). Thus, the larger must be the value of T_p on the left-hand side of the equation.

Equation (10.23) illustrates why evaluation of enthalpy for all substances based on a standard reference state is so convenient. If this were not done, the summations in that term could not be conveniently evaluated, since each substance would have enthalpy changes based on different reference states.

Example Problem 10.5

Determine the temperature of the combustion products leaving an adiabatic furnace in which methane is burned with 20 percent excess air. The combustion air is preheated to 200°C, and the methane is preheated to 80°C.

Solution

This problem is similar to Example Problem 10.4, except that the fuel and air are preheated. The chemical balance equation remains the same. The first law, Eq. (10.23), for this case is

$$[\bar{h}(T_p) - \bar{h}^o]_{CO_2} + 2[\bar{h}(T_p) - \bar{h}^o]_{H_2O} + 9.024[\bar{h}(T_p) - \bar{h}^o]_{N_2}$$
$$+ 0.4[\bar{h}(T_p) - \bar{h}^o]_{O_2} = [\bar{h}(T = 80°C) - \bar{h}^o]_{CH_4}$$
$$+ 2.4[\bar{h}(T = 200°C) - \bar{h}^o]_{O_2} + 9.024[\bar{h}(T = 200°C) - \bar{h}^o]_{N_2}$$
$$- (-802,900) \text{ kJ/kmol}$$

Substituting the polynomial expressions as in Example Problem 10.3 and iterating to find T_p yield a temperature of 2200 K. The values of terms on the right-hand side for the enthalpies of the reactants at their respective temperatures are found to be dominated by the nitrogen (46,300 kJ), with a lesser contribution by the oxygen (12,700 kJ) and a small contribution by the methane (2000 kJ).

Comments

Preheating the reactants increases the resulting enthalpy and thus the temperature of the products. When a preheater can be used to withdraw the available energy from stack gases or other waste heat source and increase the temperature of the reactants, particularly the air, it is often a good investment.

10.4.3 Applications

Now that we are able to handle combustion calculations under most conditions, there remains only the case of combustors in which significant heat transfer occurs from the combustor. For such a case, the heat transfer and shaft work terms are retained in the first law, and Eq. (10.23) becomes

$$\bar{w} + \bar{q} = \sum_p n_p[\bar{h}_p(T_p) - \bar{h}_p^o] - \sum_r n_r[\bar{h}_r(T_r) - \bar{h}_r^o] + \bar{q}^o \qquad (10.24)$$

Again, each term is based on one mole of fuel. The shaft work term $\bar{w}$ is usually zero for combustion systems.

Example Problem 10.6

A power plant boiler is being designed to transfer 500 MW of thermal power to the working fluid. The boiler is to be fired by natural gas (assume methane properties) with 20 percent excess air. The stack gases from the plant are to be maintained at or above a temperature that exceeds the dew point of the combustion products by at least 150°C (so that water vapor will not condense in the stack and cause corrosion problems). The methane fuel is preheated to 80°C, and the air is preheated to 200°C. What volume flow rates of methane and air will be required (evaluated at standard conditions of 25°C and 1 atm)?

Solution

First, the dew point of the combustion gases is determined. When that is known, the first law can be used to determine the heat transfer per 1 kmol of fuel, and that value plus the required power output sets the fuel flow rate.

The chemical reaction equation is the same as that given in Example Problem 10.4. The total moles of product gas per mole of fuel equal $n_{CO_2} + n_{H_2O} + n_{N_2} + n_{O_2} = 1 + 2 + 9.024 + 0.4 = 12.42$. The mole fraction of H_2O vapor is then $2/12.42 = 0.161$. Assuming that water vapor acts as an ideal gas mixture, we see the partial pressure of water vapor is then $P_{H_2O} = 0.161 \times 101.3 = 16.3$ kPa. The saturation temperature corresponding to this partial pressure is (Table D.9) $T_{sat} = 56°C$. At 150°C above the dew point of 56°C or 206°C = 479 K, the first law [Eq. (10.24)] is

$$\bar{q} = \sum_p n_p[\bar{h}_p(T_p) - \bar{h}_p^o] - \sum_r n_r[\bar{h}_r(T_r) - \bar{h}_r^o] + \bar{q}^o$$

$$= [\bar{h}(T_p = 206°C) - \bar{h}^o]_{CO_2} + 2[\bar{h}(T_p = 206°C) - \bar{h}^o]_{H_2O}$$
$$+ 9.024[\bar{h}(T_p = 206°C) - \bar{h}^o]_{N_2} + 0.4[\bar{h}(T_p = 206°C) - \bar{h}^o]_{O_2}$$
$$- [\bar{h}(T = 80°C) - \bar{h}^o]_{CH_4} - 2.4[\bar{h}(T = 200°C) - \bar{h}^o]_{O_2}$$
$$- 9.024[\bar{h}(T = 200°C) - \bar{h}^o]_{N_2} + (-802,900) \quad \text{kJ/kmol}$$

Note that iteration is not required here, since the temperatures are known for evaluation of every property. Using the polynomial expressions for $\bar{c}_P$ to find the enthalpy changes results in $\bar{q} = -794,100$ kJ/kmol of fuel. The required molar flow rate of methane is then

$$\dot{n}_{CH_4} = \frac{\dot{Q}_{plant}}{\bar{q}} = \frac{500 \times 10^3 \text{ kW}}{794.1 \times 10^3 \text{ kJ/kmol}} = 0.630 \text{ kmol/s}$$

From the ideal gas relation, the value of $\bar{v}$ for methane at $T = 298$ K and 1 atm is

$$\bar{v}_{CH_4} = \frac{\bar{R}T}{P} = \frac{0.082057 \text{ m}^3 \cdot \text{atm/(kmol} \cdot \text{K)} \times 298 \text{ K}}{1 \text{ atm}}$$

$$= 24.45 \text{ m}^3/\text{kmol}$$

The required volumetric flow rate of methane is then

$$\dot{V}_{CH_4}(T = 298 \text{ K}, 1 \text{ atm}) = \dot{n}\bar{v}_{CH_4} = (0.630 \text{ kmol/s})(24.45 \text{ m}^3/\text{kmol})$$
$$= 15.4 \text{ m}^3/\text{s}$$

From Example Problem 10.5, the AFR was found to be 11.4, so the volumetric flow rate of air required is

$$\dot{V}_{air}(T = 298 \text{ K}, 1 \text{ atm}) = (\text{AFR})(\dot{V}_{CH_4}) = 11.4 \times 15.4 \text{ m}^3/\text{s} = 176 \text{ m}^3/\text{s}$$

Comments

As the stack gas (products) temperature is allowed to rise, less and less energy is available from the combustion process to provide heat transfer to the steam in the boiler. The figure shows the variation in heat transfer with increasing stack gas temperature for this fuel, excess air, and preheating. Note that the available energy for heat transfer approaches zero as the temperature approaches that found for the adiabatic combustor of Example Problem 10.5.

10.5   Application of the Second Law to Combustion Processes

All the calculations of fuel properties (enthalpy of combustion, adiabatic flame temperature) and applications of these properties (fuel and air preheating, excess-air effects, furnace and boiler calculations) have depended on the application of only the first law for reacting systems. The second law also has important applications in combustion and chemical reactions. We discuss applications of the second law of two general types: the determination of whether a particular chemical reaction can occur under a given set of conditions and the analysis of the overall thermodynamic efficiency of a system in which a chemical reaction plays a part.

10.5.1   Determination of the Possibility of Reaction: Adiabatic Combustion

One consequence of the second law is that the entropy generation in any adiabatic process must be greater than or equal to zero. We have tacitly assumed so far that the chemical reactions used in this chapter in fact occur; that is, methane when mixed with air can undergo a combustion reaction that results in the formation of combustion products in the form of CO_2 and H_2O. Experience tells us that this can occur; however, unless special conditions are met (the presence of an ignition source, for example), the reaction may not occur. The second law can be used to tell us whether a given chemical reaction is possible, as we shall see. Whether the reaction *will*, in fact, occur and at what *rate* it can occur depend on other factors (the presence of an ignition source or a catalyst, for example). These other factors are studied under the heading of chemical kinetics and are outside the scope of this text.

Application of the second law to determine whether a given reaction can occur is straightforward. We check to see whether, in an adiabatic steady flow reactor, the entropy generation is positive. If so, the reaction can possibly occur. If, however, the calculated entropy generation is negative, then the proposed reaction is not possible. If the entropy generation is zero, then the chemical reaction is reversible. (This idea is further developed in Chap. 11 for the case of reactions that do not proceed to completion, where chemical equilibrium is reached.)

The second law (Table 5.3) requires, for an adiabatic steady flow reactor, that

$$\dot{S}_{\text{gen}} = \sum_p \dot{n}_p \bar{s}_p - \sum_r \dot{n}_r \bar{s}_r \geq 0 \qquad (10.25)$$

Example Problem 10.7

Methane is preheated to 80°C and is burned in 20 percent excess air that is preheated to 200°C. Is this reaction possible according to the second law?

Solution

The chemical reaction equation is the same as that used in Example Problems 10.4 and 10.5. The product temperature was found in Example Problem 10.5 to be 2200 K. Required data are taken from Table C.10:

Reactant	Moles	$\bar{s}^o$, kJ/(kmol · K)	Product	Moles	$\bar{s}^o$, kJ/(kmol · K)
CH_4	1	186.3	CO_2	1	213.7
O_2	2.4	205.03	O_2	0.4	205.03
N_2	9.024	191.5	N_2	9.024	191.5
	12.424		H_2O (vapor)	2	188.7
				12.424	

The values of entropy at the conditions present for the reactants and products must now be computed. Given that the combustion occurs at a total pressure of 1 atm and that the materials act as a mixture of ideal gases, the partial pressure of each constituent i of the reactants and the products is given by $P_i = (n_i/n)(101.2 \text{ kPa})$. The entropy of each constituent at the known conditions of temperature and partial pressure is given by modifying Eq. (9.40a) to give the standard entropy:

$$\bar{s} = \bar{s}^o + \int_{298}^{T} \frac{\bar{c}_P(T)}{T} dT - \bar{R} \ln \frac{P_i}{P^o}$$

$$= \bar{s}^o + \int_{298}^{T} \frac{\bar{c}_P(T)}{T} dT - \bar{R} \ln \frac{n_i}{n}$$

$$= \bar{s}^o + \int_{298}^{T} \frac{\bar{c}_P(T)}{T} dT - \bar{R} \ln P_i$$

Note that the final two forms of the equation only hold for a total pressure in the combustor of 1 atm. For all substances in this problem except methane, Tables D.2 through D.7 are convenient and give values of

$$\bar{s}^o + \int_{298}^{T} \frac{\bar{c}_P(T)}{T} dT$$

directly. (*Note:* The table values are per unit mass rather than per mole.) For methane the polynomial expressions for $\bar{c}_P(T)$ in Table D.1 may be used.

These calculations result in the following:

Reactant	T, K	$\dfrac{n_i}{n}$	$\bar{s}(T, P)$, kJ/(kmol · K)	Product	T, K	$\dfrac{n_i}{n}$	$\bar{s}(T, P)$, kJ/(kmol · K)
CH_4	353	0.0805	213.5	CO_2	2200	0.0805	336.1
O_2	473	0.1932	232.6	O_2	2200	0.0322	301.0
N_2	473	0.7263	207.8	N_2	2200	0.7263	258.3
				H_2O	2200	0.1610	284.8

Substituting into Eq. (10.21) for 1 kmol of fuel results in

$$\bar{s}_{\text{gen}} = (1 \times 336.10 + 0.4 \times 301.0 + 9.024 \times 258.3$$
$$+ 2 \times 284.8) - (1 \times 213.5 + 2.4 \times 232.6 + 9.024 \times 207.8)$$
$$= 3357.0 - 2646.9 = +710.1 \text{ kJ/(kmol · K)} > 0$$

so the reaction is possible, as we expected.

Comment

References 1 to 3 contain excellent data sources to aid in solution of this type of problem, which greatly simplifies the computations necessary for accurate results.

10.5.2 Determination of the Possibility of Reaction: General Combustion Problems

In a reactor that is adiabatic, it is possible to remove or add energy by means of heat transfer with the surroundings. The second law must be modified to include the entropy changes that occur because of the heat transfer. The entropy generation must remain positive for the reaction being considered to have the possibility of going to completion. In such a case, the entropy generation of a system composed of the combustor and the surroundings is

$$S_{gen} = \sum_p n_p \bar{s}_p - \sum_r n_r \bar{s}_r - \frac{Q}{T_0} \tag{10.26}$$

where T_0 is the absolute temperature of the surroundings to which there is a heat transfer. In this equation, the entropy of mixing (Chap. 9) has not been explicitly stated but is included in the $\sum n\bar{s}$ terms. If the reactants are not mixed before entering the combustor, then the entropy-of-mixing term should not be included; quite often, however, they are mixed. The products are invariably mixed, and the mixing term should be included. In most engineering problems, the effect of the entropy-of-mixing term on the results for S_{gen} is minor. Let us divide Eq. (10.26) by the number of moles of fuel and find the entropy generation per mole of fuel, or

$$\bar{s}_{gen} = \frac{1}{n_f}\left(\sum_p n_p \bar{s}_p - \sum_r n_r \bar{s}_r\right) - \frac{\bar{q}}{T_0} \tag{10.27}$$

Now the general first law from Eq. (10.24) with $\bar{w} = 0$ can be used to eliminate $\bar{q}$, resulting in (for one mole of fuel)

$$\bar{s}_{gen} = \sum_p n_p \bar{s}_p - \sum_r n_r \bar{s}_r - \frac{\sum_p n_p[\bar{h}_p(T_p) - \bar{h}_p^o] - \sum_r n_r[\bar{h}_r(T_r) - \bar{h}_r^o] + \bar{q}^o}{T_0}$$

$$\tag{10.28}$$

This last equation can also be obtained from Eq. (7.6) with modifications for chemical reactions and zero work. Rearranging Eq. (10.28) gives

$$\bar{s}_{gen} = \frac{1}{T_0}\left\{\sum_p n_p[T_0\bar{s}(T) - \bar{h}(T)]_p - \sum_r n_r[T_0\bar{s}(T) - \bar{h}(T)]_r\right\}$$
$$+ \frac{1}{T_0}\left(\sum_p n_p\bar{h}_p^o - \sum_r n_r\bar{h}_r^o - \bar{q}^o\right) \tag{10.29}$$

Noting that the expression in parentheses of the final term is zero [Eq. (10.11)] results in the form

$$\bar{s}_{gen} = \frac{1}{T_0}\left\{\sum_p n_p[T_0\bar{s}(T) - \bar{h}(T)]_p - \sum_r n_r[T_0\bar{s}(T) - \bar{h}(T)]_r\right\} \tag{10.30}$$

Some special cases can be examined. For an adiabatic combustor, the final term in Eq. (10.27) or (10.28) goes to zero, and the equation reduces to Eq. (10.25).

Standard Gibbs Function of Reaction

For an isothermal combustor operated at the temperature $T_0 = T_r = T_p$, Eq. (10.30) becomes, after the Gibbs function ($\bar{g} = \bar{h} - T\bar{s}$) is used, for one mole of fuel,

$$\bar{s}_{\text{gen}} = \frac{1}{T_0}\left[\sum_r n_r\bar{g}_r(T_0) - \sum_p n_p\bar{g}_p(T_0)\right] \tag{10.31}$$

Since $\bar{s}_{\text{gen}}$ must be greater than or equal to zero for the reaction to proceed, to make the reaction proceed isothermally and reversibly at the temperature of the surroundings requires that

$$\sum_p n_p\bar{g}_p \leq \sum_r n_r\bar{g}_r \tag{10.32}$$

The Gibbs function is obviously a very important property for determining whether a particular reaction will proceed.

Values of the Gibbs function at the standard reference state $\bar{g}^o$ are easily calculated from

$$\bar{g}^o = \bar{h}^o - 298\bar{s}^o \tag{10.33}$$

If the reaction in question is carried out at the standard reference state, then

$$\Delta\bar{g}_f^o = \sum_p n_p\bar{g}_p^o - \sum_r n_r\bar{g}_r^o \leq 0 \tag{10.34}$$

The value of $\Delta\bar{g}_f^o$ is called the *standard Gibbs function of reaction,* or the *standard free energy of reaction,* and it must be negative in order for the reaction to have the possibility of proceeding spontaneously, as shown by Eq. (10.32). Values of $\Delta\bar{g}_f^o$ are given in Table C.9. Applications and extensions of the second law relations developed here to other types of chemical reactions are discussed in Chap. 11.

Useful Work Output from a Combustor

To determine the useful reversible work output available from a combustor operating in steady flow, we can refer to Eq. (7.10) and modify that relation for the presence of the combusting mixture, to give

$$\bar{w}_{\text{rev}} = \sum_p n_p(\bar{h} - T_0\bar{s})_p - \sum_r n_r(\bar{h} - T_0\bar{s})_r \tag{10.35}$$

for one mole of fuel, where changes in kinetic and potential energy have been neglected. If the combustion is carried out isothermally at the temperature of the surroundings, then

$$\bar{w}_{\text{rev}} = \sum_p n_p\bar{g}(T_{\text{surr}})_p - \sum_r n_r\bar{g}(T_{\text{surr}})_r \tag{10.36}$$

and again we see the importance of the Gibbs function. If the temperature of the surroundings is taken to be the standard reference temperature of 298 K, then

$$\bar{w}_{rev} = \sum_p n_p \bar{g}_p^o - \sum_r n_r \bar{g}_r^o = \Delta \bar{g}_f^o \qquad (10.37)$$

and the reversible work from the combustion reaction carried out adiabatically at the standard reference temperature and pressure is seen to be simply equal to the standard Gibbs free energy of reaction. Noting that

$$\bar{w}_{rev} = \Delta \bar{g}_f^o = \Delta \bar{h}_f^o - 298 \, \Delta \bar{s}^o \qquad (10.38)$$

we can interpret the reversible work as coming from the standard enthalpy of reaction less the "unavailable work" $298 \, \Delta \bar{s}^o$.

Example Problem 10.8

The combustion system described in Example Problem 10.6 is used to drive a generator. Determine the maximum mechanical power that can be obtained from the combustor when the surroundings are at 25°C.

Solution

The values of the entropy of the products and reactants must be found for use in Eq. (10.35) to determine the reversible work done per mole of fuel. These values are available for the reactants from Example Problem 10.7. For the products, $\bar{s}$ must be calculated at the outlet temperature of 479 K found in Example Problem 10.6. The values are computed as in Example Problem 10.7:

Reactant	T, K	$\dfrac{n_i}{n}$	$\bar{s}(T, P)$, kJ/(kmol · K)	Product	T, K	$\dfrac{n_i}{n}$	$\bar{s}(T, P)$, kJ/(kmol · K)
CH_4	353	0.0805	213.5	CO_2	479	0.0805	254.0
O_2	473	0.1932	232.6	O_2	479	0.0322	248.0
N_2	473	0.7263	207.8	N_2	479	0.7263	208.2
				H_2O	479	0.1610	220.1

Substituting into Eq. (10.35) gives

$$\bar{w}_{rev} = \sum_p n_p(\bar{h} - T_0\bar{s})_p - \sum_r n_r(\bar{h} - T_0\bar{s})_r$$

$$= \sum_p (n\bar{h})_p - \sum_r (n\bar{h})_r - 298 \left[\sum_p (n\bar{s})_p - \sum_r (n\bar{s})_r \right]$$

Note that the value of $\sum_p (n\bar{h})_p - \sum_r (n\bar{h})_r$ is equal to $-794{,}100$ kJ/kmol

of fuel, since it is equal to $\bar{q}$ by the first law. Making this substitution and entering the entropies of the products and reactants result in

$$\bar{w}_{rev} = -794{,}100 - 298(2672.2 - 2646.9) = -810{,}600 \text{ kJ/kmol}$$

Using the molar flow rate of $\dot{n} = 0.630$ kmol/s found in Example Problem 10.6 results in a reversible power output for the combustor of

$$\dot{W}_{rev} = \dot{n}\bar{w}_{rev} = (0.630 \text{ kmol/s})(-810{,}600 \text{ kJ/kmol}) = -505 \text{ MW}$$

Comments

Note that the reversible work ($-810,600$ kJ/kmol) is close to the value of the energy removed by heat transfer from the combustor ($-794,100$ kJ/kmol), because the value of the $T_0\bar{s}$ terms is relatively small in comparison with the values of the enthalpy terms. This is usually the case for the combustion of hydrocarbons. Also, the reversible power is greater than the power provided by the furnace in Example Problem 10.6 (5 MW), because some power is carried from that furnace by the stack gases.

10.6 Applications to Real Devices: Efficiency of Combustion Devices

For an actual device in which work output is the objective, such as an internal combustion engine, the amount of energy used is less than that calculated as being available from the combustion process. We have seen that the reversible work available from the combustion reaction is less than the energy release calculated from the first law, so that the upper limit of the energy that can be obtained from any real device where combustion occurs is set by Eq. (10.36). It would seem reasonable, then, to define the efficiency of a real device in terms of the fraction of the reversible work attained. However, historically a "first law" efficiency has been used for combustion processes in which work is produced, as is the case for power cycles described in Chap. 6. Thus, an alternative efficiency for a heat engine is defined as the actual work output over the negative of the enthalpy of combustion of the fuel, or

$$\eta_{th} = \frac{w}{-q^o} \tag{10.39}$$

Remember that the value of q^o depends on whether the water in the combustion products is assumed to be liquid or vapor. The choice obviously changes the value of the thermal efficiency computed for a given engine.

For devices in which maximum heat transfer (HT) is the desired result rather than work, such as furnaces, water heaters, steam generators, etc., the *efficiency* is defined as the actual heat transferred in the device divided by the higher heating value of the fuel, or

$$\eta_{HT} = \frac{-q}{HHV} \tag{10.40}$$

Finally, as mentioned earlier, in some devices the desired result of combustion is to attain a high temperature. This is true in certain combustion processes in internal combustion engines with efficiencies that depend on the combustion temperature and in some process furnaces where a particular temperature must be reached. In such cases, attainment of the desired temperature often requires that additional fuel be injected over that necessary for stoichiometric combustion; in other words, a fuel-rich mixture is needed. Thus, the actual AFR is reduced below the stoichiometric value. The *combustion efficiency* is then defined as

$$\eta_{comb} = \frac{(AFR)_{actual}}{(AFR)_{ideal}} = \frac{(FAR)_{ideal}}{(FAR)_{actual}} \tag{10.41}$$

where FAR is the fuel-air ratio (the inverse of AFR).

Example Problem 10.9

For the combustor described in Example Problem 10.6, find the heat transfer efficiency.

Solution

The amount of heat transfer per mole of fuel was found to be $-794,100$ kJ/kmol of fuel. The HHV of methane was found in Example Problem 10.2 to be 890,900 kJ/kmol, so the heat transfer efficiency [Eq. (10.40)] is

$$\eta_{HT} = -\frac{-794,100}{890,900} = 0.89 = 89\%$$

Problems

10.1 Using the enthalpy of formation data of Table C.9, calculate the enthalpy of combustion for methane (CH_4), ethane (C_2H_6), and propane (C_3H_8). Based on the differences between succeeding members of this hydrocarbon family, predict the value of the enthalpy of combustion of butane (C_4H_{10}). Check your prediction against the result in Table C.11.

10.2 Determine the standard enthalpy of combustion of (*a*) H_2, (*b*) CO, and (*c*) C_2H_5OH.

10.3 Plot the standard enthalpy for methane versus temperature in the range $25 < T < 200°C$. From these data and data for $\Delta \bar{g}_f^o$ from Table C.9, plus the Gibbs-Helmholtz relation, estimate the Gibbs function of combustion at $200°C$.

10.4 Derive a relationship between the lower and higher

heating values of a general hydrocarbon fuel C_nH_m in terms of the mass fraction of hydrogen in the fuel and the heat of vaporization of water.

10.5 One mole of methane is reacted stoichiometrically with air in an adiabatic constant-volume vessel. If the reactants are initially at 25°C and 1 atm, what final temperature is reached in the vessel? Compare your result with the result of Example Problem 10.3. (The temperature calculated for a constant-volume process is sometimes called the *explosion temperature* of a gas.)

10.6 The gasohol sold by one service station is effectively 10 percent ethyl alcohol (C_2H_5OH) and 90 percent octane (C_8H_{18}) by volume. Calculate the enthalpy of combustion of this mixture, assuming that it is initially in gaseous form and that all combustion products are gaseous.

10.7 The materials often used for biomass conversion to methane have the effective chemical formulas and heating values listed here. For each, compute the amount of energy released by combustion of the resulting methane per kilogram of original biomass. Compare the result with the energy released per kilogram of each material by direct combustion. Assume all combustion processes occur at standard conditions of 1 atm and 25°C and with no excess air.

Substance	Formula	HHV, kJ/kg
Wood chips (oak)	$C_{10}H_{24}O_{11}$	14,600
Cattle manure	$CH_{2.4} \cdot 4H_2O$	13,900
Bagasse (sugar cane processing by-product)	$C_4H_6O_5 \cdot 6H_2O$	9,300
Typical municipal waste	$CH_3 \cdot 3H_2O$	20,000
Chemical plant waste	$CH_{1.6}$	32,600

10.8 Set up the stoichiometric coefficient relations for the combustion reaction with X amount of excess air in the form

$$C_nH_m + A_1\text{Air} \rightarrow A_2CO_2 + A_3H_2O + A_4\text{Air} + A_5N_2$$

and determine $A_1, A_2, \ldots, A_5$ in terms of m, n, and X.

10.9 Find the theoretical adiabatic flame temperature of butane.

10.10 Find the temperature of the combustion products of butane when it is burned in an adiabatic steady flow combustor at 1 atm with 40 percent excess air preheated to 200°C.

10.11 Plot the stack gas temperature leaving an adiabatic steady flow combustor versus excess air in the range 0 to 150 percent. The fuel is propane, and the fuel and air both enter the combustor at 1 atm and 25°C.

10.12 Plot the stack gas temperature versus air preheat temperature for an adiabatic steady flow combustor in which propane is burned with 20 percent excess air. The fuel and air both enter the combustor at 1 atm, and the fuel enters at 25°C. Air preheating temperatures range from 25 to 400°C.

10.13 A natural-gas-fired furnace is used to evaporate saturated water at 300 kPa. The natural gas has a volume flow rate of 10 m^3/s and is burned with 20 percent excess air. The air and gas enter the furnace at 25°C and 1 atm. If the combustion gases must leave the furnace at a temperature no lower than 300°C, what flow rate, in kilograms per second, of water can be evaporated in the furnace? (Assume that the natural gas has the same properties as methane.)

10.14 A water heater in a farmhouse uses butane to heat the water. The butane burns with 50 percent excess air, and the air and butane enter the burner of the heater at 25°C and 1 atm. The flue gases leave the heater at 130°C. If the maximum gas flow rate to the burner when the burner runs continuously is 0.005 m^3/s, how much water, in kilograms per second, can be heated from 20 to 75°C?

10.15 The farmer who owns the farm in Prob. 10.14 decides to change bottle-gas dealers. The new supplier provides propane instead of butane. All conditions remain the same.
(a) How much water can be heated, in kilograms per second, by using propane?
(b) What ratio of costs per kilogram should the farmer expect to pay for the two fuels in relation to their energy content?

10.16 Plot the entropy generation per mole of fuel versus temperature when carbon monoxide is burned adiabatically with the stoichiometric amount of air. Use temperatures from 600 to 2000°C. The reactants enter at 25°C and 1 atm. What do you conclude from this plot?

10.17 Calculate the reversible work per mole of fuel that is given by the combustion of carbon monoxide with stoichiometric air when the combustion products are mixed at 400°C and the reactants enter at 25°C and 1 atm. The surrounding temperature is 25°C.

10.18 Jet A fuel has a standard enthalpy of formation of $-22,570$ kJ/kmol and an equivalent chemical formula of $CH_{1.9}$. In the combustor of an aircraft turbine engine, Jet A is burned with 10 percent excess air. The air enters the combustion chamber at 1 atm and $-50°C$. The fuel enters at the standard reference state and is gaseous. What is the temperature of the gas leaving the combustor?

10.19 Coal can be gasified by heating and mixing with steam by the *water-gas shift* reaction

$$C + H_2O \Leftrightarrow CO + H_2$$

The resulting water gas ($CO + H_2$) was sold in many cities as illuminating gas. Assume that the water gas has equal volumes of CO and H_2.
(*a*) Find the higher heating value of water gas.
(*b*) Find the adiabatic flame temperature of water gas when it is burned with stoichiometric air.

10.20 An advanced supersonic transport (super SST, or SSST) has been proposed with a goal of engine efficiency $= 0.55$. The turbine pressure ratio will be 20, and the thrust-to-weight ratio is 8.0. The turbine inlet temperature of this design will be 1560°C. Assuming that Jet A fuel ($CH_{1.9}$, heat of formation $= -22,570$ kJ/kmol) is used, how much excess air must be used to obtain this temperature? Assume the inlet fuel is gaseous at the reference state and that inlet air is at $-30°C$ and 1 atm.

10.21 Hydrogen gas is to be burned with 20 percent excess air in an adiabatic combustor. Both enter at standard reference conditions. What is the dew point of the combustion products?

10.22 Propane is to be burned with 10 percent excess air in an adiabatic furnace. The combustion air is to be preheated in a heat exchanger (air preheater) by heat transfer from the combustion products. The temperature of the combustion products is reduced to T_{out} in the preheater. The air enters the preheater at 25°C. See Fig. P10.22.

What will be the temperature of the combustion products leaving the preheater? (The propane enters the furnace at the standard reference condition.) Discuss the effect of the presence of the preheater on your result. How would you find the temperature T of the air leaving the preheater?

Figure P10.22

10.23 Butane and 30 percent excess air enter the combustor of a simple Brayton engine at 5 atm. The air enters at the temperature at which it left the compressor (efficiency 70 percent). The butane enters at standard reference temperature. After the combustion process, the hot combustion products enter the turbine (75 percent efficiency) and leave at 1 atm. Air enters the compressor at 25°C and 1 atm. See Fig. P10.23.

What is the cycle efficiency? (Do not forget to do mass balances!)

10.24 For an adiabatic combustor operating with a given fuel and given amount of excess air, derive an expression for the second law efficiency of the furnace.

10.25 Methane is burned with 15 percent excess air, both of which enter at standard reference conditions. The combustion products are then mixed with 10 percent by volume of cold air at 25°C. See Fig. P10.25. What are the temperature and dew point of the resulting mixture?

Figure P10.23

Figure P10.25

References

1. Joseph H. Keenan, Jing Chao, and Joseph Kaye, *Gas Tables: International Version,* 2d ed., Wiley, New York, 1983.
2. Joseph H. Keenan, Jing Chao, and Joseph Kaye, *Gas Tables (English Units),* 2d ed., Wiley, New York, 1980.
3. GASPROPS, "A Computer Program for the Calculation of the Thermodynamic Properties of Gases and Combustion Products," Wiley, New York, 1984.
4. Henry C. Barnett and Robert R. Hibbard, "Basic Considerations in the Combustion of Hydrocarbon Fuels with Air," NACA Rept. 1300, 1959.

11

Phase and Chemical Equilibrium

Human subtlety . . . will never devise an invention more beautiful, more simple, or more direct than does nature, because in her inventions nothing is lacking and nothing is superfluous.
Leonardo da Vinci

The compressor section of a 9000-kW gas turbine, showing the turbine blades on the rotating center shaft and the stationary (stator) blades in the split housing. *(Solar Turbines, Inc.)*

11.1 The Gibbs-Duhem Relation

In Chap. 5, entropy is discussed as an extensive equilibrium property, and it is shown that for an isolated system, the entropy could only remain constant or increase. The focus was limited to a simple compressible substance. In this chapter, we expand the treatment of equilibrium systems to include the discussion of single-component and multicomponent systems in which more than one phase is present and to systems in which chemical reactions can occur. Much of the groundwork for this chapter has already been laid — the ideas of thermal and mechanical equilibrium in Chap. 5, the introduction of partial molal properties in Chap. 9, and the use of standard reference properties in Chap. 10.

The Gibbs function for simple systems $g = h - Ts$ has some limited usefulness in problems involving single components and for some problems involving mixtures of nonideal gases. However, the Gibbs function enjoys a central place in the discussion of systems involving phase and chemical equilibrium, as does the Helmholtz function $a = u - Ts$. The Gibbs function appears in formulations involving constant-pressure processes, while the Helmholtz function usually is found in descriptions of constant-volume processes. Because many practical engineering problems occur at constant pressure, we begin with a review of the properties of the Gibbs function.

Let us assume that the Gibbs function is dependent on the variables $T, P, n_A, n_B, n_C, \ldots$. We can then write the total differential as

$$dG = \left(\frac{\partial G}{\partial T}\right)_{P, n_A, n_B, n_C, \ldots} dT + \left(\frac{\partial G}{\partial P}\right)_{T, n_A, n_B, n_C, \ldots} dP$$
$$+ \left(\frac{\partial G}{\partial n_A}\right)_{T, P, n_B, n_C, \ldots} dn_A + \left(\frac{\partial G}{\partial n_B}\right)_{T, P, n_A, n_C, \ldots} dn_B + \cdots \quad (11.1)$$

Note that $(\partial G/\partial n_i)_{T, P, n_{j(j \neq i)}}$ is defined in Sec. 9.9 as the *partial molal Gibbs function* $\overline{G}_A$. For a process at constant pressure and temperature, Eq. (11.1) can be expressed as

$$dG_{T, P} = \overline{G}_A \, dn_A + \overline{G}_B \, dn_B + \overline{G}_C \, dn_C + \cdots \quad (11.2)$$

Commonly, the partial molal Gibbs function is called the *electrochemical potential,* or simply the *chemical potential,* and is given the symbol $\overline{\mu}$ and defined as

$$\left(\frac{\partial G}{\partial n_i}\right)_{T, P, n_A, \ldots, n_{i-1}, n_{i+1}, \ldots} \equiv \overline{\mu}_i \quad (11.3)$$

so that Eq. (11.2) becomes

$$dG_{T,P} = \bar{\mu}_A\, dn_A + \bar{\mu}_B\, dn_B + \bar{\mu}_C\, dn_C + \cdots \tag{11.4}$$

As we discuss in the next section, the chemical potential is the driving potential for the molar transfer of components. Equation (11.4) can be integrated (if we remember that T and P are held constant) by using a particular approach. First, observe that $\bar{\mu}_i$ is an *intensive* property that depends on the *relative* amount of each species that is present. If the dn_i are changed in such a way that the mole fractions y_i are unchanged (by adding or withdrawing mass from the phase under consideration), then $\bar{\mu}_i$ is held constant and Eq. (11.4) can be integrated to give

$$G_{T,P} = \bar{\mu}_A n_A + \bar{\mu}_B n_B + \bar{\mu}_C n_C + \cdots \tag{11.5}$$

This results for a single phase with multiple components at any constant temperature and pressure, and relates the Gibbs function of a solution or mixture to the chemical potentials and amounts of each component in the phase. Now the total differential of Eq. (11.5) for a system with varying composition and varying chemical potential at constant temperature and pressure is

$$dG_{T,P} = \bar{\mu}_A\, dn_A + \bar{\mu}_B\, dn_B + \bar{\mu}_C\, dn_C + n_A\, d\bar{\mu}_A + n_B\, d\bar{\mu}_B + n_C\, d\bar{\mu}_C + \cdots \tag{11.6}$$

Comparing Eq. (11.6) with Eq. (11.4), both of which are general relations for a constant-T and constant-P process, indicates that

$$n_A\, d\bar{\mu}_A + n_B\, d\bar{\mu}_B + n_C\, d\bar{\mu}_C + \cdots = 0 \tag{11.7}$$

This is an important relation, as we shall see. It is called the *Gibbs-Duhem equation,* and it applies to any given phase in a multicomponent, multiphase system.

11.2   Equilibrium in Nonreacting Systems

Let us examine a system in which two or more components exist in two or more phases. If we limit discussion to a control mass maintained at constant temperature and pressure with no changes in kinetic or potential energy, then the first law is

$$dU = \delta Q + \delta W \tag{11.8}$$

If the system exchanges energy with the surroundings only, then Eq. (5.74) can be substituted to give the inequality

$$T_0\, dS + \delta W - dU \geq \delta Q + \delta W - dU = 0 \tag{11.9}$$

where the inequality becomes an equality only when the control mass undergoes reversible heat transfer. Further, if the only work mode present is by boundary movement, Eq. (11.9) becomes

$$T_0\, dS - P\, dV - dU \geq 0 \tag{11.10}$$

For constant temperature and pressure, the differential of the Gibbs function $G = H - TS$ is

$$dG_{T,P} = dH - T\, dS = d(U + PV)_{T,P} - T\, dS = dU + P\, dV - T\, dS \tag{11.11}$$

and substituting Eq. (11.10), we find

$$dG_{T,P} + (T - T_0)\, dS \leq 0 \tag{11.12}$$

11.2.1   Isolated Systems

For an isolated system in thermodynamic equilibrium, $dS = 0$, so that Eq. (11.12) becomes, by using Eq. (11.4),

$$\begin{aligned}
dG_{T,P} &= \sum_j \left[\sum_i (\bar{\mu}_i\, dn_i) \right]_j \\
&= (\bar{\mu}_A\, dn_A + \bar{\mu}_B\, dn_B + \cdots)_a + (\bar{\mu}_A\, dn_A + \bar{\mu}_B\, dn_B + \cdots)_b \\
&\quad + (\cdots)_c + \cdots \leq 0
\end{aligned} \tag{11.13}$$

where the sum on i indicates the sum over each component in phase j and the sum on j is the sum over the phases. Again, in thermodynamic equilibrium, the equality holds.

An important identity for the chemical potential is also obtained from Eq. (11.5). A single component existing in a single phase eliminates all the terms in Eq. (11.5) except one, so

$$\bar{\mu} = \bar{g} \tag{11.14}$$

Thus, the chemical potential for a single component in one phase is equal to the molar Gibbs function.

11.2.2   Single-Component Phase Equilibrium

Let us now examine a two-phase system with a single component (the subscript i is therefore eliminated). For such a system, the total number of moles of material is constant. So if mass is transferred from phase a to phase b, it follows that

$$d(n_a + n_b) = 0 \tag{11.15}$$

or

$$dn_a = -dn_b \tag{11.16}$$

Equation (11.13) then gives, for thermodynamic equilibrium,

$$dG_{T,P} = \bar{\mu}_a\, dn_a - \bar{\mu}_b\, dn_a = 0 \tag{11.17}$$

or

$$\overline{G}_a = \overline{\mu}_a = \overline{\mu}_b = \overline{G}_b \qquad (11.18)$$

Thus, the chemical potential (partial molal Gibbs function) in both phases of a single-pure-component system in thermodynamic equilibrium must be the same. This argument can be extended to multiple components in multiple phases, with the conclusion that the chemical potential of a given component is the same in every phase.

An important physical interpretation of the chemical potential is obtained from Eq. (11.13). For a single-component two-phase system, Eq. (11.13) yields

$$(\overline{\mu}_a - \overline{\mu}_b)\, dn_a \leq 0 \qquad (11.19)$$

The inequality implies that if $\overline{\mu}_a > \overline{\mu}_b$, then $dn_a < 0$, or the mass in phase a tries to move away from a higher chemical potential. Alternatively, if $\overline{\mu}_b > \overline{\mu}_a$, then $dn_a > 0$, which again indicates the motion from a higher to lower chemical potential. These arguments presenting the chemical potential as the driving force for mass transfer should be compared to the presentation of temperature and pressure as the driving forces for heat transfer and boundary movement work, respectively, given in Sec. 5.4.

Now rewrite Eq. (11.17) with Eq. (11.14) as

$$dG_{T,P} = \sum_j dG_j = \overline{g}_a\, dn_a + \overline{g}_b\, dn_b$$
$$= (\overline{h} - T\overline{s})_a\, dn_a + (\overline{h} - T\overline{s})_b\, dn_b = 0 \qquad (11.20)$$

which, after substituting Eq. (11.16), results in

$$\overline{s}_a - \overline{s}_b = \frac{\overline{h}_a - \overline{h}_b}{T} \qquad (11.21)$$

If Eq. (8.11c) is substituted into Eq. (11.21) and we remember that all the development in this section is for an isothermal and isobaric process, the result is

$$\left(\frac{dP}{dT}\right)_{\text{phase change}} = \frac{\overline{h}_{ab}}{T\overline{v}_{ab}} = \frac{h_{ab}}{Tv_{ab}} \qquad (11.22)$$

which is the Clapeyron equation [Eq. (8.14)], derived in a different way in Chap. 8.

11.2.3   Ideal Solutions

Ideal solutions are introduced in Chap. 9, where they are described as mixtures of two or more components in which the total volume of the solution equals the sum of the volumes of the components of the solution; i.e., the partial molal volumes equal the specific molar volumes. We now extend the idea of an ideal solution to other properties, and we reach some useful conclusions about the equilibrium properties of ideal solutions.

Suppose that we consider a general liquid solution of two components that is in equilibrium with the vapors of the components, with the entire control mass held at constant temperature and pressure. Remember that the chemical potential of each component has the same value in the liquid and vapor phases. For this case, if there are $n_{A,\,l}$ and $n_{B,\,l}$ moles of the two components in the liquid phase, then the Gibbs-Duhem equation, Eq. (11.7), can be divided by the total number of moles in the liquid, $(n_A + n_B)_l$, and can then be written for the liquid phase (denoted by a subscript l) as

$$(y_A \, d\bar{\mu}_A + y_B \, d\bar{\mu}_B)_l = 0 \qquad (11.23)$$

The change in chemical potential of component i depends on only the amount of component i present (if T and P are constant), so that

$$d\bar{\mu}_l = \left(\frac{\partial \bar{\mu}_i}{\partial y_i}\right)_{T,\,P} dy_i \qquad (11.24)$$

Noting that $y_A + y_B = 1$ and substituting Eq. (11.24), we see that Eq. (11.23) becomes

$$\left\{\left[\frac{\partial \bar{\mu}_A}{\partial (\ln y_A)}\right]_{T,\,P} = \left[\frac{\partial \bar{\mu}_B}{\partial (\ln y_B)}\right]_{T,\,P}\right\}_l \qquad (11.25)$$

Let us now consider the question of how to provide numerical values for the chemical potential, since we have derived in Eq. (11.25) a means of relating the chemical potentials of two components in a liquid solution. In Chap. 8, we define the fugacity f of a gas or vapor by the relation [Eq. (8.65)]

$$d\bar{g}_T = \bar{R}T \, d(\ln f) \qquad (11.26)$$

However, from the definition of the chemical potential as the partial molal Gibbs function, it also follows that the value of $\bar{\mu}_i$ for any component of a vapor or gas mixture at constant temperature and pressure is

$$d\bar{\mu}_i = (d\bar{g}_i)_{T,\,P} = \bar{R}T \, d(\ln f_i) \qquad (11.27)$$

Since the chemical potential of component i is the same in both phases, it also follows that Eq. (11.27) can be used to determine the chemical potential of each component of the liquid. In addition, it follows that the fugacity of component i must be the same in every phase! Substituting Eq. (11.27) into Eq. (11.25) results in

$$\left\{\left[\frac{\partial (\ln f_A)}{\partial (\ln y_A)}\right]_{T,\,P} = \left[\frac{\partial (\ln f_B)}{\partial (\ln y_B)}\right]_{T,\,P}\right\}_l \qquad (11.28)$$

which is called the *Duhem-Margules equation*. We have made no assumptions as to ideality in its derivation, except that it applies for changes in composition that occur at constant temperature and pressure. If, however, the vapor phase is in a state where the vapor mixture can be treated as an ideal gas mixture, then, from Chap. 8, $f \to P$, and Eq. (11.28) can be revised to

$$\left\{ \left[\frac{\partial(\ln P_A)}{\partial(\ln y_A)} \right]_{T, P} = \left[\frac{\partial(\ln P_B)}{\partial(\ln y_B)} \right]_{T, P} \right\}_l \qquad (11.29)$$

where P_A and P_B are the vapor pressures of the two constituents that are in equilibrium with the liquid, which has mole fractions y_A and y_B.

We can now define an *ideal liquid solution* as one which obeys the relation

$$f_i^* \equiv y_{i, l} f_i \qquad (11.30a)$$

or, at low pressures,

$$P_i = y_{i, l} P_{\mathrm{sat}, i}(T) \qquad (11.30b)$$

where P_i and f_i^* are the vapor pressure and vapor-phase fugacity of component i in the solution, respectively; $y_{i, l}$ is the mole fraction of component i *in the solution,* and $P_{\mathrm{sat}, i}(T)$ is the saturation pressure of component i at the control mass temperature. Equation (11.30a) is called the *Lewis-Randall rule.* If the vapor acts as an ideal gas mixture, then Dalton's law can be used to give

$$P_i = y_{i, g} P_{\mathrm{tot}} = y_{i, l} P_{\mathrm{sat}, i}(T) \qquad (11.31)$$

This relation is called *Raoult's law.*

If Raoult's law holds for either component of a two-component ideal solution, then Eq. (11.29) shows that it must also hold for the second component. Although we have limited the discussion to two-component mixtures, the results are easily extended to multiple components. We can now prescribe all the thermodynamic properties of an ideal liquid solution and the vapor in equilibrium with the solution.

Example Problem 11.1

An ideal liquid solution is composed of 80 mol % ammonia and 20 mol % water. What are the mole percentages of the ammonia and water in the vapor that is in equilibrium with the liquid solution if the temperature of the control mass is $20°C$?

Solution

The mole fractions in the liquid are given. According to Raoult's law, the partial pressures of the vapor are given by

$$P_{\mathrm{H_2O}} = 0.2 P_{\mathrm{sat, \, H_2O}}(T = 20°C) = 0.2(2.34) = 0.468 \text{ kPa}$$

$$P_{\mathrm{NH_3}} = 0.8 P_{\mathrm{sat, \, NH_3}}(T = 20°C) = 0.8(857.6) = 686.1 \text{ kPa}$$

and the total pressure of the vapor is 686.6 kPa. The mole fractions in the vapor are then

$$y_{g,\,H_2O} = \frac{P_{H_2O}}{P_{tot}} = \frac{0.468}{686.6} = 6.82 \times 10^{-4}$$

$$y_{g,\,NH_3} = 1 - y_{g,\,H_2O} = 0.9993$$

Given the mole fractions in each phase, all the thermodynamic properties of the ideal mixtures can be computed by using the relations in Chap. 9.

Comments

Note how much larger the mole fraction of the volatile component (ammonia) is in the vapor phase than in the liquid phase. If the mole fraction of component A in the vapor is plotted against the mole fraction of A in the liquid, generally the relation is not linear even for an ideal solution. Note that the nonideality of a mixture can be treated more fully by assuming Raoult's law to be of the general form

$$f_i^* = y_{i,\,l} f_{sat,\,i}(T)$$

which allows use of real gas properties in the vapor phase. However, for this problem, examination of the reduced property charts shows that $f/P = 1$, so the use of the ideal gas form of Raoult's law is justified.

11.2.4   Phase Rule

We have limited discussion so far to one- and two-component two-phase mixtures. In the single-component two-phase case, we found it sufficient to specify only one intensive property (usually T or P) plus the amount of each phase to completely describe the state of the system. For a two-component system with two phases, it was necessary to know the values of two intensive variables, normally T and P, plus the concentration (usually mole fraction) of *one* of the components in each phase. The concentration of the second component is then known, since the mole fractions in each phase must sum to unity. How many independent intensive properties N_{pr} must be specified to determine that a system of N_{ph} phases made up of N_c components is in thermodynamic equilibrium? In general, the number that must be specified is given by

$$N_{pr} = N_c - N_{ph} + 2 \tag{11.32}$$

which is called the *Gibbs phase rule*. The number 2 results from the need to specify two independent properties for a single-phase simple compressible substance. There is a need to specify $N_c - 1$ mole fractions (there is 1 less than N_c since the sum of all mole fractions must equal 1) for each of the N_{ph} phases or $N_{ph}(N_c - 1)$ properties. However, there are $N_{ph} - 1$ equations for the chemical potential equalities for each component, or $N_c(N_{ph} - 1)$ equations. Thus, this gives $N_{ph}(N_c - 1) - N_c(N_{ph} - 1) = N_c - N_{ph}$ other variables required, as indicated in Eq. (11.32).

For a single-component, single-phase control mass ($N_c = 1$ and $N_{ph} = 1$), two independent intensive properties must be known to specify that the system will be in equilibrium. For a single-component, two-phase system, only one intensive variable needs to be specified. For a single-component, three-phase system, *no* independent intensive properties can be specified. These requirements are in accord with our observations on simple compressible substances. The three-phase system for water, for example, corresponds to the behavior at the triple point, where only one pressure and temperature correspond to the equilibrium three-phase mixture, and these cannot be independently specified.

For a two-component, two-phase system, we apparently must set two intensive properties to ensure equilibrium. We have done this by fixing the temperature and pressure of the system.

The phase rule tells us nothing about the relative amounts of each phase present; it tells us only that if we specify a certain number of intensive properties, the system will be in thermodynamic equilibrium. For steam in a saturated state, then, if either P or T is specified and two phases are present, the state will be an equilibrium state for *any* proportions of liquid and vapor; but those proportions are not fixed by the phase rule.

11.3 Equilibrium in Systems with Chemical Reaction

Let us now extend our considerations of equilibrium states to systems in which chemical reactions occur. Many chemical reactions proceed to completion; that is, the reactants undergo complete reaction and are consumed in the production of reaction products. This was the case for all the combustion reactions studied in Chap. 10. In other cases, however, the reaction may not proceed to completion because the products themselves react to re-form into the original reactants. An equilibrium mixture of products and reactants will result. In this section, we examine such situations, limiting our discussion to *homogeneous* reactions in which all the reaction takes place in the gas phase.

Examine a general chemical reaction in which elements or compounds A and B react to form elements or compounds C and D. (The discussion is easily extended to more or fewer reactants and products.) The general reaction equation is

$$v_A A + v_B B \leftrightarrow v_C C + v_D D \tag{11.33}$$

where the v's are the stoichiometric coefficients.

The rate at which the number of moles of A, n_A, is being depleted because of reaction with B at any time depends on the concentrations of A

and B; i.e., if either A or B is missing, then no reaction can occur. If we denote the concentration of A in arbitrary units by $[A]$, then the rate of depletion of A, mol/s, can be written as

$$\left(\frac{d[A]}{dt}\right)^{+} = -k_1[A]^{v_A}[B]^{v_B}$$

(11.34)

where k_1 is a proportionality constant, called the *rate constant,* that describes the rate of the particular reaction. The stoichiometric coefficients appear as exponents because the reaction rate must depend on how many molecules (moles) of B are required to react with v_A molecules (moles) of A. If, for example, $v_A = 1$ and $v_B = 2$, then the reaction rate must depend on $[B]^2$, because 2 molecules of B must be present for each A molecule. The reaction rate for this case is

$$\left(\frac{d[A]}{dt}\right)^{+} = -k_1[A][B][B] = -k_1[A][B]^2$$

(11.35)

The value of k_1 depends not only on the particular reaction but also on the units chosen for expressing concentration (e.g., partial pressure, moles per liter, etc.). The minus sign appears in Eq. (11.34) because the concentration of A must decrease as the reaction proceeds, so $(d[A]/dt)^{+}$ must be negative.

For many of the reactions described by Eq. (11.33), the products C and D will themselves react to form A and B. For such a case, the rate at which A is formed by this *reverse reaction* is given by reasoning similar to that used for Eq. (11.34) as

$$\left(\frac{d[A]}{dt}\right)^{-} = k_2[C]^{v_C}[D]^{v_D}$$

(11.36)

When equilibrium is reached, the rates of depletion $(d[A]/dt)^{+}$ and of formation $(d[A]/dt)^{-}$ of A must sum to zero. Summing Eqs. (11.34) and (11.36) results in

$$K = \frac{k_1}{k_2} = \frac{[C]^{v_C}[D]^{v_D}}{[A]^{v_A}[B]^{v_B}}$$

(11.37)

where K is a new constant, called the *equilibrium constant* for the reaction. We see that K is related to the equilibrium concentrations of the products, $[C]$ and $[D]$, and of the reactants, $[A]$ and $[B]$. The value of K also depends on the units chosen for concentration (unless $v_C + v_D$ happens to be equal to $v_A + v_B$, in which case K is dimensionless for any concentration units).

Because k_1 and k_2 are often not known for a given reaction, K must usually be found by other means than Eq. (11.37). We now use the information previously developed about equilibrium to find values for K for a given reaction.

If the reaction of Eq. (11.33) is not complete and equilibrium final concentrations are reached, then we can see what happens if the reaction proceeds to an extent α, where $\alpha = 0$ indicates only reactants are present. We are left with a mixture of reactants and products that has the following composition:

Component	Final Number of Moles
A	$v_A(1 - \alpha)$
B	$v_B(1 - \alpha)$
C	αv_C
D	αv_D

The *change* in the number of moles of each component is then $dn_A = -v_A \, d\alpha$, $dn_B = -v_B \, d\alpha$, $dn_C = v_C \, d\alpha$, and $dn_D = v_D \, d\alpha$, which shows that

$$d\alpha = \frac{-dn_A}{v_A} = \frac{-dn_B}{v_B} = \frac{dn_C}{v_C} = \frac{dn_D}{v_D} \tag{11.38}$$

Given the value of α, then, the final number of moles of each component in the equilibrium chemical reaction is known from the table. Our immediate problem is to find the value of α for a given chemical reaction at a given temperature and pressure.

At the condition of chemical equilibrium, we can substitute Eq. (11.38) into the relation between the Gibbs function and the chemical potential of each component, as given by Eq. (11.4), to find

$$dG_{T, P} = (-\bar{\mu}_A v_A - \bar{\mu}_B v_B + \bar{\mu}_C v_C + \bar{\mu}_D v_D) \, d\alpha \tag{11.39}$$

We now must find a relation for the chemical potential of a component of the system after the reaction comes to equilibrium at a given temperature and pressure. Equation (11.27) gives the differential of the chemical potential of a component of a mixture. To define a *standard* chemical potential, we must define a standard reference state. *For the equilibrium mixture, the standard reference state is taken to be an ideal solution with the same mole fraction of component i as in the equilibrium mixture.* The ideal reference solution is assumed to be *at 1-atm total pressure and at the equilibrium mixture temperature.* Let us integrate Eq. (11.27) at constant temperature from the ideal solution reference state to the equilibrium mixture state:

$$\int_{\bar{\mu}_i^o}^{\bar{\mu}_i} (d\bar{\mu}_i)_T = \bar{\mu}_i(T) - \bar{\mu}_i^o(T) = \int_{f_i^{*o}}^{f_i^*} \bar{R}T \, d(\ln f_i^*)_T$$

$$= \bar{R}T \ln \frac{f_i^*}{f_i^{*o}} \tag{11.40}$$

Or, since $f_i^{*o} = y_i f_i^o$ in an ideal solution, where f_i^o is the fugacity of the pure component at 1 atm and T, this becomes

$$\bar{\mu}_i(T) = \bar{\mu}_i^o(T) + \bar{R}T \ln \frac{f_i^*}{y_i f_i^o}$$

$$= \bar{\mu}_i^o(T) - \bar{R}T \ln y_i + \bar{R}T \ln \frac{f_i^*}{f_i^o} \tag{11.41}$$

Now, remembering from Chap. 9 [e.g., Eqs. (9.97) through (9.99)] that the general thermodynamic relations among properties can also be expressed for partial molal properties, we can write

$$\bar{\mu}_i^o(T) - \bar{R}T \ln y_i = (\bar{G}_{T,P}^o)_i - \bar{R}T \ln y_i$$

$$= (\bar{H}^o - T\bar{S}^o)_{T,P,i} - \bar{R}T \ln y_i \tag{11.42}$$

For an ideal solution, which was taken as the reference state, the partial molal properties are equal to the actual properties of the component, so $\bar{H}_i^o = \bar{h}_i^o$ and $\bar{S}_i^o + \bar{R} \ln y_i = \bar{s}_i^o$. Equation (11.41) then finally becomes

$$\bar{\mu}_i(T) = \bar{h}_i^o(T) - T\bar{s}_i^o(T) + \bar{R}T \ln \frac{f_i^*}{f_i^o}$$

$$= \bar{g}_i^o(T) + \bar{R}T \ln \frac{f_i^*}{f_i^o} \tag{11.43}$$

which is the equation for the standard chemical potential, with a reference state taken as an ideal solution at 1 atm and T.

In Chap. 10, Eq. (10.34) is given for the standard Gibbs function per mole of fuel for a combustion reaction that goes to completion while at standard conditions of 25°C and 1 atm. Here, it is convenient to define a slightly different although closely related function which we denote $\Delta \bar{G}^o(T)$. This is the change in Gibbs function that occurs for a given reaction *that goes to completion* at a given temperature (not necessarily the reference temperature), and from this definition,

$$\Delta \bar{G}^o(T) = (v_C \bar{g}_C^o + v_D \bar{g}_D^o - v_A \bar{g}_A^o - v_B \bar{g}_B^o)_T \tag{11.44}$$

where the $\bar{g}_i^o$ are the molar Gibbs functions for the components in their pure state at standard pressure and at the temperature of the reaction. This implies that $\Delta \bar{G}^o(T)$ applies to a reaction in which the reactants were initially separated but were brought together, the reaction between them went to completion, and the products were then separated.

Now, we can substitute Eqs. (11.43) and (11.44) into Eq. (11.39), which results in

$$dG_{T,P} = \left[\Delta \bar{G}^o(T) + \bar{R}T \ln \frac{a_C^{v_C} a_D^{v_D}}{a_A^{v_A} a_B^{v_B}} \right] d\alpha \tag{11.45}$$

where we have used the definition $a_i \equiv f_i^*/f_i^o$, and a_i is called the *activity* of component i. The activity should not be confused with the Helmholtz function, which uses the same symbol.

Because $dG_{T,P} = 0$ at equilibrium, this equation becomes

$$\ln \frac{a_C^{\nu_C} a_D^{\nu_D}}{a_A^{\nu_A} a_B^{\nu_B}} = \frac{-\Delta \overline{G}^o(T)}{\overline{R}T} = \ln K \qquad (11.46)$$

The K is the equilibrium constant found in Eq. (11.37) by using the concentrations in terms of activities, which are dimensionless. Now K is defined by

$$K = \frac{a_C^{\nu_C} a_D^{\nu_D}}{a_A^{\nu_A} a_B^{\nu_B}} \qquad (11.47)$$

Equation (11.46) is a complete definition of the equilibrium constant; however, it is difficult to find the activities (or fugacities) of each component in many cases. If we treat the equilibrium gas phase as an ideal gas mixture, then

$$a_i = \frac{f_i^*}{f_i^o} = \frac{f_i^*}{P^o} = y_i \frac{P}{P^o} \qquad (11.48)$$

and the equilibrium constant becomes

$$K = \frac{y_C^{\nu_C} y_D^{\nu_D}}{y_A^{\nu_A} y_B^{\nu_B}} \left(\frac{P}{P^o} \right)^{\nu_C + \nu_D - \nu_A - \nu_B} \qquad (11.49)$$

and the equilibrium constant is determined solely by the mole fractions of the components, the stoichiometric coefficients, and the pressure of the equilibrium mixture (although K itself is *not* a function of pressure, as we shall see). The mole fractions themselves can be put in the form of the degree of reaction α; and if K is known for a given reaction, the degree of reaction is then fixed. The equilibrium coefficient can be obtained through Eq. (11.46) as

$$K = \exp \left[\frac{-\Delta \overline{G}^o(T)}{\overline{R}T} \right] \qquad (11.50)$$

This evaluation of α was the goal of our derivation.

Example Problem 11.2

The reaction

$$CO_2 + H_2 \leftrightarrow CO + H_2O$$

takes place at 2000 K and 1 atm. What are the mole fractions of each component in the resulting equilibrium mixture?

Solution

Table C.12 gives the value of the equilibrium constant for various reactions but does not give this specific reaction. The equilibrium constant for this reaction is obtained by noting that the reaction equation is the sum of three basic reactions:

Reaction 1: $C + O_2 \leftrightarrow CO_2$ $K_1 = \dfrac{y_{CO_2}}{y_C y_{O_2}}$

Reaction 2: $C + \frac{1}{2}O_2 \leftrightarrow CO$ $K_2 = \dfrac{y_{CO}}{y_C y_{O_2}^{1/2}}$

Reaction 3: $H_2 + \frac{1}{2}O_2 \leftrightarrow H_2O$ $K_3 = \dfrac{y_{H_2O}}{y_{H_2} y_{O_2}^{1/2}}$

With Eq. (11.33) and the definition of K in Eq. (11.49), the sum of reactions 2 and 3 minus reaction 1 yields $H_2 \leftrightarrow -CO_2 + CO + H_2O$, or

$$H_2 + CO_2 \leftrightarrow CO + H_2O \qquad K = \frac{y_C y_{O_2} y_{CO} y_{H_2O}}{y_{CO_2} y_C y_{O_2}^{1/2} y_{H_2} y_{O_2}^{1/2}} = \frac{K_2 K_3}{K_1}$$

Table C.12 gives the values $K_1 = 2.25 \times 10^{10}$, $K_2 = 2.94 \times 10^7$ and $K_3 = 3470$ at 2000 K. Thus

$$K = \frac{(2.94 \times 10^7)(3470)}{2.25 \times 10^{10}} = 4.53$$

The final relative concentrations are as follows:

Component	Relative Equilibrium Concentration
CO_2	$1 - \alpha$
H_2	$1 - \alpha$
CO	α
H_2O	α

Equation (11.49) then gives

$$K = 4.53 = \frac{y_{CO} y_{H_2O}}{y_{CO} y_{H_2}} = \frac{\alpha^2}{(1 - \alpha)^2}$$

Solving the resulting quadratic equation for the value of α that lies in the range $0 < \alpha < 1$ gives $\alpha = 0.681$. The equilibrium mole fractions are then as follows:

Component	Relative Equilibrium Concentration	Relative Mole Fraction
CO_2	0.319	0.1595
H_2	0.319	0.1595
CO	0.681	0.3405
H_2O	0.681	0.3405
	2.000	1.0000

11.3.1 Temperature Dependence of K

The definition of the equilibrium constant is given through Eq. (11.46). For an ideal gas mixture, combining Eqs. (11.49) and (11.50) yields

$$K = \frac{y_C^{v_C} y_D^{v_D}}{y_A^{v_A} y_B^{v_B}} \left(\frac{P}{P^o}\right)^{v_C + v_D - v_A - v_B} = \exp\left[-\frac{\Delta \overline{G}^o(T)}{\overline{R}T}\right] \tag{11.51}$$

The right-hand side is a function of temperature only, so the equilibrium constant is also a function of temperature only. Differentiating Eq. (11.51) with respect to temperature yields

$$\frac{dK}{dT} = \frac{K}{\overline{R}}\left[\frac{\Delta\overline{G}^o(T)}{T^2} - \frac{1}{T}\frac{d\,\Delta\overline{G}^o(T)}{dT}\right] \tag{11.52}$$

or

$$\frac{1}{K}\frac{dK}{dT} = \frac{d(\ln K)}{dT} = \frac{1}{\overline{R}T^2}\left[\Delta\overline{G}^o(T) - T\frac{d\,\Delta\overline{G}^o(T)}{dT}\right] \tag{11.53}$$

The total differential of the Gibbs function given in Eq. (8.5d) shows that $(\partial g/\partial T)_{P,\,n_i} = -s$. This relation is used to replace the temperature derivative of the Gibbs function in Eq. (11.53) by

$$\frac{d\,\Delta\overline{G}^o(T)}{dT} = -\Delta\overline{S}^o \tag{11.54}$$

The definition of the Gibbs function gives

$$\Delta\overline{G}^o(T) = \Delta\overline{H}^o - T\,\Delta\overline{S}^o \tag{11.55}$$

Therefore, Eq. (11.53) yields

$$\frac{d(\ln K)}{dT} = \frac{\Delta\overline{H}^o}{\overline{R}T^2} \tag{11.56}$$

This equation is called the *van't Hoff equation*. This expression can be used to evaluate $\Delta\overline{H}^o$ from equilibrium mole fractions or, alternatively, to evaluate the equilibrium constant from $\Delta\overline{H}^o$ information. Specifically, if $\Delta\overline{H}^o$ is independent of temperature, then $\ln K$ is linear in $1/T$. The slope on a plot of $\ln K$ versus $1/T$ then yields $\Delta\overline{H}^o/\overline{R}$.

11.3.2 Pressure Dependence of Equilibrium Concentrations

In the last section we found that K itself is independent of pressure. Yet, pressure appears explicitly in Eq. (11.49). It follows that the equilibrium concentrations in Eq. (11.49) must change in such a way that K remains constant when P is varied. At constant T, Eq. (11.49) can be rewritten in the form

$$\frac{y_C^{v_C} y_D^{v_D}}{y_A^{v_A} y_B^{v_B}} = \frac{K}{(P/P^o)^{v_C + v_D - v_A - v_B}} \tag{11.57}$$

Examining the exponent of P/P^o shows that if $v_C + v_D = v_A + v_B$, then pressure has no effect on the equilibrium concentrations. However, if $v_C + v_D > v_A + v_B$, then the concentrations of the products will be decreased relative to the reactants as the pressure is increased; and if $v_C + v_D < v_A + v_B$, the concentrations of the reactants will be decreased relative to the products as pressure is increased. Thus, *increasing pressure tends to increase the concentrations of the constituents that are on the side of the reaction with the smallest sum of stoichiometric coefficients.* This observation is known as *LeChatelier's principle.*

Example Problem 11.3

Methane and water vapor react to form CO and H_2. To attain a high concentration of the products, should the reaction be carried out at high or low pressure?

Solution
The reaction is balanced to give

$$CH_4 + H_2O \leftrightarrow CO + 3H_2$$

The sum of the stoichiometric coefficients is greater on the right, indicating that higher pressure will hinder product formation. A low-pressure reaction is thus favored.

11.3.3   Phase Rule

The Gibbs phase rule for nonreacting equilibrium properties is given in Eq. (11.32). For reacting mixtures now additional N_{ch} chemical reaction equations must be satisfied. The number of independent intensive properties is therefore the same as for the previous case [Eq. (11.32)], minus N_{ch}. Thus,

$$N_{pr} = N_c - N_{ph} - N_{ch} + 2 \tag{11.58}$$

11.4   General Equilibrium

To generalize the concepts of phase and chemical equilibrium, take the general definition of the Gibbs function $G = H - TS$ and then take the total derivative to find

$$dG = dH - T\,dS - S\,dT \tag{11.59}$$

We can now substitute Eq. (5.42), which is a general relation among properties, to give

$$dG = V\,dP - S\,dT \tag{11.60}$$

Now, examine Eq. (11.1), and substitute Eq. (11.3) into that equation. The result is

$$dG = \left(\frac{\partial G}{\partial T}\right)_{P,\,n_A,\,n_B,\dots} dT + \left(\frac{\partial G}{\partial P}\right)_{T,\,n_A,\,n_B,\dots} dP + \bar{\mu}_A\,dn_A + \bar{\mu}_B\,dn_B + \cdots \tag{11.61}$$

For a system in chemical and phase equilibrium, we can substitute Eq. (11.13), which gives

$$dG|_{n_A,\,n_B,\dots} = \left(\frac{\partial G}{\partial T}\right)_{P,\,n_A,\,n_B,\dots} dT + \left(\frac{\partial G}{\partial P}\right)_{T,\,n_A,\,n_B,\dots} dP \tag{11.62}$$

Comparing Eqs. (11.62) and (11.60) gives the identities

$$\left(\frac{\partial G}{\partial T}\right)_{P,\,n_A,\,n_B,\dots} = -S \tag{11.63}$$

and

$$\left(\frac{\partial G}{\partial P}\right)_{T,\,n_A,\,n_B,\dots} = V \tag{11.64}$$

Now, Eq. (11.1) can be written in the completely general form

$$dG = -S\,dT + V\,dP + \bar{\mu}_A\,dn_A + \bar{\mu}_B\,dn_B + \cdots \tag{11.65}$$

We observed previously that thermal equilibrium requires $dT = 0$, mechanical equilibrium requires $dP = 0$, and chemical and phase equilibrium require that $\sum_j \sum_i (\bar{\mu}_i\,dn_i)_j = 0$. If all these requirements are met simultaneously, then Eq. (11.65) predicts that $dG = 0$. Thus, the Gibbs function incorporates a general description of the thermodynamic properties of any system [since $G(T, P, n_A, n_B, \dots)$ is a fundamental equation (Chap. 8)], but in addition we note that any thermodynamic system must have $dG = 0$ if the system is in thermodynamic equilibrium.

The Gibbs function is extremely useful in describing equilibrium in control volumes, where T and P and the component concentrations are the useful properties. For a control mass, the Helmholtz function $A = U - TS$ plays a similar role.

In this section, we used many of the most basic ideas that are introduced in various chapters. The use of a fundamental equation is introduced in Chap. 8, along with the identification of various properties with the corresponding partial derivatives. The concept of equilibrium requiring that the changes in certain properties be zero is used in Chap. 5 to relate entropy changes to other properties and in this chapter in discussing chemical and phase equilibrium. Here, we have gathered this information to provide a comprehensive prescription for describing a system, its properties, and its deviation from thermodynamic equilibrium.

11.5 Concluding Remarks

In this chapter, some general concepts of equilibrium are applied to problems of phase and chemical equilibrium. Much wider applications are found in the study of chemical thermodynamics and physical chemistry, but the fundamentals given here (mostly formulated originally by J. Willard Gibbs) form the basis for all these studies.

Note that this chapter deals entirely with thermodynamic systems in thermodynamic equilibrium. We have not discussed the question of how long it will take an initial mixture of reactants to reach the predicted equilibrium concentration of products plus reactants. Some reactions proceed very quickly (explosions) while others take years (rust, decay). Such questions are the province of the field of *chemical kinetics.*

It should be obvious from Sec. 11.4 that the ideas of mechanical, thermal, and chemical equilibrium form a coherent whole. Indeed, one approach to classical thermodynamics is to consider the requirements for general thermodynamic equilibrium and then to generate the necessary properties and relationships that describe changes in the general equilibrium state from that viewpoint.

Problems

11.1 For the following nonreacting mixtures, how many independent intensive properties must be specified for an equilibrium mixture?
(*a*) A three-phase two-component mixture
(*b*) A solid that consists of two crystalline forms and has three components
(*c*) A LiBr-H_2O mixture that has a liquid and a vapor phase

11.2 For saturated steam at 200°C, calculate the value of the Gibbs function for saturated liquid, saturated vapor, and a saturated mixture with quality $x = 40$ percent.

11.3 Without using any other data, find the value of h_l for saturated refrigerant-12 at $T = 20°C$, given $T_{sat} = 20°C$, $h_g = 359.81$ kJ/kg, and $s_{lg} = 0.48071$ kJ/(kg $\cdot$ K).

11.4 For the ammonia-water mixture at $T = 20°C$, construct a graph of ammonia vapor pressure versus mole fraction of ammonia in the liquid phase. On the same graph, plot ammonia vapor pressure versus mole fraction in the vapor phase. Assume ideal behavior.

11.5 An ammonia-water absorption chiller operates with an evaporator temperature of $+2°C$ and a condenser temperature of $50°C$ with no superheating of vapor in the evaporator and no subcooling in the condenser. The NH_3 constitutes 95 mol percent of the vapor in both the absorbor and the generator. What is the pressure in (*a*) the absorber and (*b*) the generator? What is the mole fraction of ammonia in the liquid solution in (*c*) the absorber and (*d*) the generator? Assume ideal solution behavior.

11.6 Plot the vapor mole fraction versus liquid mole fraction for ammonia-H_2O ideal mixtures at $T = +2°C$ and at $T = +50°C$ on the same graph. Comment on the meaning of such a graph for an absorption chilling device (Sec. 6.5.4) using ammonia and H_2O as the refrigerant-absorbent pair.

11.7 A LiCl-H_2O absorption air-conditioning system is to be designed for no vapor superheating in the evaporator and no liquid subcooling in the condenser. The LiCl liquid concentration in the generator is 45 weight percent, and in the absorbor it is 20 weight percent. Using the data given and assuming ideal solution behavior where necessary, find for an absorber temperature of $5°C$ and a generator temperature of $30°C$:

(*a*) Absorber pressure
(*b*) Generator pressure
(*c*) Mole fraction of LiCl in the generator

The following is the LiCl vapor pressure expression:

$$\log P = A - \frac{B}{T + 230}$$

$$A = 7.5713 + (3.7964 \times 10^{-2})(X)$$
$$- (9.0133 \times 10^{-4})(X^2)$$

$$B = 1.5821 \times 10^3 + 10.364X$$
$$- (1.4462 \times 10^{-1})(X^2)$$

where P = vapor pressure, mmHg
T = temperature, °C
X = concentration, wt % LiCl

11.8 One mole each of nonreacting refrigerant-12 and water is in a closed vessel at 1 atm and $25°C$. Find the concentration (mole fraction) of refrigerant-12 in each phase, liquid and vapor, assuming ideal behavior in both phases.

11.9 For a system of I nonreacting components in a single phase, derive the Gibbs-Duhem equation, Eq. (11.7), starting with Eq. (11.13).

11.10 Show that the Gibbs-Duhem equation [Eq. (11.7)] applies for the cases of any component i among J phases and for the I components in any phase j. [Start with Eq. (11.13).]

11.11 For the case of one component in J phases, show that the chemical potential $\bar{\mu}_j$ is the same in all phases. [Start with Eq. (11.13).]

11.12 For the case of I components in J phases, prove that the chemical potential of any component i, $[\bar{\mu}_i]_j$, is the same in all phases. [Start with Eq. (11.13).]

11.13 For the reaction

$$2CO + O_2 \Leftrightarrow 2CO_2$$

show the relationship of K for this reaction to the value of K: (a) when N_2 is present during the reaction and (b) for the reaction $2CO_2 \Leftrightarrow 2CO + O_2$.

11.14 An inventor claims that she has a method for producing hydrogen and oxygen from water vapor with a new catalyst by the reaction

$$2H_2O \Leftrightarrow 2H_2 + O_2$$

The reaction is carried out at 4000 K and 0.1 atm.
(a) What are the equilibrium mole fractions of H_2 and O_2 at this condition?
(b) In a continuous-flow reactor with inlet at $25\,^{\circ}C$, how much heat transfer is necessary per kilogram of H_2 produced?
(c) How much of the heat transfer could be recovered by cooling the products to $40\,^{\circ}C$?

11.15 At 3000 K and $P = 1$ atm, the reversible reaction

$$CO + \tfrac{1}{2}O_2 \Leftrightarrow CO_2$$

takes place.
(a) Find the value of y_{CO_2} in the mixture.
(b) Find the value of y_{CO_2} if the pressure is 10 atm.

11.16 The reaction

$$N_2 + O_2 \Leftrightarrow 2NO$$

occurs in automobile engines. Calculate the equilibrium concentration of NO at $T = 327\,^{\circ}C$ and $P = 1$ atm (assumed conditions at the tailpipe exit). Discuss how the NO concentration in the real case, which is much higher than found by this calculation, could occur.

11.17 Ammonia is synthesized for fertilizer production by reacting nitrogen and hydrogen at high pressure and temperature through the reaction

$$N_2 + 3H_2 \Leftrightarrow 2NH_3$$

    (*a*) At $T = 1000$ K and $P = 40$ atm, what will be the mole fraction of NH_3 in the mixture of gases? (Assume ideal behavior.)

    (*b*) Would increasing the pressure of the synthesis aid or hurt ammonia production?

11.18 After nitrogen and hydrogen react to form ammonia at any given P and T, the nitrogen and hydrogen in the equilibrium mixture are in the mole ratio $1/r$ when the mole fraction of ammonia formed is x. Show that $r = 3$ for the maximum value of x that is possible.

11.19 Carbon dioxide dissociates to form carbon monoxide and oxygen under certain conditions. For pure CO_2 at 3000 K, plot the mole fraction of CO in the equilibrium mixture versus total pressure in the range $0.01 < P_{tot} < 5$ atm. Use semilog paper for the plot.

11.20 Carbon dioxide at 1 atm may dissociate to form CO and O_2. Plot the mole fraction of CO versus T in the range $600 < T < 3000$ K on semilog paper.

11.21 Following the development of Eqs. (11.8) through (11.12), derive an expression for chemical equilibrium in a constant-volume constant-temperature control mass, using the Helmholtz function.

11.22 Find $\Delta \bar{G}^o(T)$ for the reaction

$$CO + \tfrac{1}{2}O_2 \Leftrightarrow CO_2$$

at $T = 1000$ K, and use the result to calculate $K(T)$ for this reaction. Compare the result with $K(T)$ from Table C.12.

11.23 A certain reaction has an equilibrium constant of $K(T = 500$ K$)$ with $\Delta \bar{h}^\circ = 8812$ kJ/kmol. Estimate the value of $K(T = 700$ K$)$ in terms of $K(T = 500$ K$)$.

11.24 Near $T = 1000$ K, a plot of the equilibrium constant for a particular reaction is curve-fitted and yields the relation

$$\ln K = -\frac{4200}{T}$$

Estimate the value of $\Delta \bar{H}^o(T = 1000$ K$)$ for this reaction.

11.25 For the reaction described in Prob. 11.24, estimate the value of K at $T = 1200$ K.

11.26 For a control mass, it is often convenient to work with the Helmholtz function $A = U - TS$ rather than the Gibbs function. Following the development of Eqs. (11.59) through (11.65), derive a general expression for

dA in terms of the appropriate control mass properties. Discuss the implied conditions for the various types of equilibrium (thermal, mechanical, and chemical) that result.

11.27 Prove that the chemical potential $\bar{\mu}$ that appears in Eq. (11.65) is the same as the chemical potential that appears when the equivalent expression to Eq. (11.65) is derived for the Helmholtz function A.

12

Introduction to Microscopic Thermodynamics

We have no right to assume that any physical laws exist, or if they have existed up to now, that they will continue to exist in a similar manner.
Max Planck

Hydrogen cooler used for heat transfer from large generating equipment. Hydrogen is used inside the casing of the generator because of its good thermal properties and its low frictional resistance to the rotating parts. *(Westinghouse.)*

12.1 Introduction

Reference has been made to this chapter at earlier points, generally with respect to the ability to use a microscopic approach to thermodynamics that provides insight into the behavior of thermodynamic systems which is unavailable from the classical approach. To do so, we present here an overview of the microscopic approach to thermodynamics, with emphasis on providing that insight rather than on a rigorous introduction to statistical thermodynamics. Nevertheless, on completion of this chapter, the reader should have the tools to do some things that are not available from the classical approach. In particular, we see how properties can be computed for materials that can be modeled simply on a microscopic level; how entropy is related to our knowledge about the state of a system, and what this implies about irreversibility; and how we might develop equations of state that are based on more than empirical observation of macroscopic behavior.

12.2 Defining a Microscopic System

What do we need to know to completely describe a microscopic system in thermodynamic equilibrium? This question has led to detailed and interesting investigations, and the answer depends to some extent on how we intend to apply the information. The description of a plasma composed of free electrons and ions may require us to provide different sorts of information than is necessary for describing an ideal gas. For our purposes, we might agree on the following: We must specify as a minimum the number of individual particles that compose the system under study, the position and velocity of each particle, and the internal energy of each particle. With all this information, it is possible to calculate the quantities that we have found to be of interest in our study of classical thermodynamics.

How do we obtain such information for a given system? Generally, we can find the number of particles by use of the standard methods of chemistry and physics that allow us to relate the number of particles N_p to the total mass of the system, m_{sys}, the molecular weight M, and Avogadro's number $N_a = 6.023 \times 10^{26}$ particles per kilomole by

$$N_p = \left(\frac{m_{sys}}{M}\right) N_a \tag{12.1}$$

How many of these particles have a given microscopic property value, and what are their locations? Generally, we cannot answer these questions, because the particle energies and positions change with time. This is obvious for the case of a gas, where the individual gas molecules are in constant motion and the molecules exchange energy with one another through collisions. Even if we somehow were able to know the positions and energies of all particles at a given time, a short time later the values would have changed, and our information would no longer be valid. (At one period in the history of physics, it was believed that given the required information at one particular time, Newton's laws would allow us to trace the history of all particles through time so that our required information would be available at any later time. As we shall see, this hope of a "deterministic" solution is no longer a possibility, even if we had a computer with enough speed and memory to carry out such a computation.)

If we cannot expect to obtain the required information, how can we proceed with microscopic calculations? A different approach is obviously required, and we must work with the data that we can obtain. Such information is usually in the form of statistical distributions and their averages rather than the details of the properties of each individual particle that makes up the averages. Much information is lost to us in this way. Consider knowing the average age of a group of 150 people at a cocktail party. Such information tells us little about the actual age of any individual. We might have some other information, such as that the minimum age of any individual is, say, 16, while the upper bound on ages might be on the order of 85. We might further speculate on the statistical distribution of ages within this range, perhaps based on the distribution in the national population. But we can never predict with certainty the exact age of individual A or B at the party from the given information. This is the sort of dilemma that we face in the microscopic approach.

In this section, we address two questions: First, if we have statistical information about how many particles have a given microscopic property value (kinetic energy, for example) in the sense of a known statistical distribution of the property per particle, how are these distributions related to properties that we can observe? Second, how can the required distributions be developed?

12.2.1 General Properties

We are familiar with properties that describe the macroscopic state of a system in thermodynamic equilibrium, such as P, T, v, u, h, s, and others. For our discussion of microscopic thermodynamics to be useful, we should be able to relate the microscopic system properties to the macroscopic properties.

Suppose we examine particle i from the N_p particles that make up a control mass or that are instantaneously in a control volume under study. Further, let us make the same assumptions on a microscopic level that we do on a macroscopic level, i.e., that the system is homogeneous, so that if we look at any subvolume dV within the system, the average properties within that subvolume are the same as in any other subvolume. On a microscopic level, this means that particles may move in and out of dV, but the net number of particles within any dV at any time is, on the average, the same. We can now define some properties in terms of averages. The simplest example is the system mass, which must be made up of the masses of the individual particles, or

$$m_{sys} = \sum_{i=0}^{N_p} m_i \tag{12.2}$$

An alternative way of writing Eq. (12.2) is to realize that there may be J groups of particles, and each group will have N_j particles of the same mass m_j, so that the total mass is

$$m_{sys} = \sum_{j=0}^{J} N_j m_j \tag{12.3}$$

or, if all particles have identical mass,

$$m_{sys} = mN_p = N_p \frac{M}{N_a} \tag{12.4}$$

The specific volume of a system of N_p identical particles is then

$$v = \left(\frac{V}{m}\right)_{sys} = \frac{VN_a}{N_p M} \tag{12.5}$$

For particle i, there is some associated energy ε_i. This energy can be in various forms; if the particle is a gas molecule, the energy may be made up of translational kinetic energy, vibrational or rotational energy, etc. At the moment, the form of the energy is immaterial. The total energy of the N_p particles in the system is the sum of the energies of all the particles. However, more than one particle may have the same energy, so we can take the number of particles N_j with energy ε_j and sum over the resulting J groups, to give

$$U_{sys} = \sum_{j=0}^{J} N_j \varepsilon_j \tag{12.6}$$

The specific energy of the material in the system is then

$$u_{sys} = \left(\frac{U}{m}\right)_{sys} = \frac{1}{m_{sys}} \sum_{j=0}^{J} N_j \varepsilon_j \tag{12.7}$$

Any specific extensive property ϕ, such as system energy, can be written in a similar form as

$$\phi_{sys} = \left(\frac{\Phi}{m}\right)_{sys} = \frac{1}{m_{sys}} \sum_{j=0}^{J} N_j \phi_j \tag{12.8}$$

Thus, if we know how many particles have individual property ϕ_j, then the macroscopic specific extensive property ϕ_{sys} can be computed. We are left, then, with the problem of determining how many particles N_j have various values of the property ϕ_j.

12.2.2 Allowable Microstates

Answering the question of how many particles N_j have a certain value of the property ϕ_j at a given state would lead us through much of the development of modern physics, which is not our purpose here. However, we can observe the results of the reasoning that went into answering the question and some of the important discoveries that ensued.

Maxwell-Boltzmann Energy Distribution

Classical physics allows us to proceed a certain distance toward answering the question of the distribution of a given property ϕ_j among the N_p particles being studied. James Clerk Maxwell derived the distribution of velocities that exists among the molecules of a monatomic gas, and Ludwig Boltzmann extended Maxwell's result to general forms of energy, assuming that a particle could have any value of energy. This assumption will prove to be only an approximation and leads to significant errors in certain cases.

The derivation by Maxwell and Boltzmann shows that N_p particles in a system of fixed mass have energy distributed among the particles so that $dN_p(\varepsilon)$ particles have energy in the range $d\varepsilon$ around ε according to the relation

$$\frac{dN_p(\varepsilon)}{N_p} = 2(kT)^{-3/2}\left(\frac{\varepsilon}{\pi}\right)^{1/2} \exp\left(\frac{-\varepsilon}{kT}\right) d\varepsilon \tag{12.9}$$

In this equation, $k = \overline{R}/N_a$ is the *Boltzmann constant,* which has magnitude in SI units of 1.3806×10^{-26} kJ/K and can be interpreted as the gas constant per particle. Note that if we plot $(1/N_p) \, dN_p(\varepsilon)/d\varepsilon$ versus ε, the only parameter is the absolute temperature, or more commonly, the product kT. Such a plot is shown in Fig. 12.1.

The peak of the curve, or most probable energy, occurs at $\varepsilon_{mp} = kT/2$.

From Eq. (12.9), the average energy per particle can be found:

$$\bar{\varepsilon} = \frac{1}{N_p} \sum_{j=0}^{J} N_j \varepsilon_j$$

$$= 2(k\dot{T})^{-3/2}\left(\frac{1}{\pi}\right)^{1/2} \int_0^\infty \varepsilon^{3/2} \exp\left(\frac{-\varepsilon}{kT}\right) d\varepsilon \tag{12.10}$$

In Eq. (12.10), the summation has been replaced by an integration because of the assumption that *any* energy can be possessed by the particles, and so the energy levels form a continuous distribution, as plotted in Fig. 12.1. The integration in Eq. (12.10) can be done by a change of variables,

Figure 12.1   Maxwell-Boltzmann energy distribution.

by letting $X^2 = \varepsilon/(kT)$, which results in

$$\bar{\varepsilon} = \frac{4kT}{\pi^{1/2}} \int_0^\infty X^4 \exp(-X^2)\, dX = \frac{4kT}{\pi^{1/2}} \frac{3(\pi^{1/2})}{8}$$

$$= \tfrac{3}{2}kT \tag{12.11}$$

Some observations are in order here. As temperature increases, both $\bar{\varepsilon}$ and ε_{mp} increase in proportion. Also, the distributions of particles with respect to energy in Fig. 12.1 become more spread out as temperature increases. This is in accordance with the arguments made in Chap. 5 concerning the relation of entropy change and heat transfer, where we observed that the entropy, or "uncertainty," decreased with a heat transfer from a control mass. We see from Fig. 12.1 that a heat transfer from a control mass reduces the temperature of the control mass, but the energy distribution among the particles becomes compacted; i.e., we are more sure of the energy of any given particle, since the range of values is not so spread out at lower temperatures.

Some Predictions Using the Maxwell-Boltzmann Distribution

Let us see whether the Maxwell-Boltzmann distribution can lead us even further. Suppose the particles under study are assumed to be point masses within an adiabatic container, that the particles undergo perfectly elastic collisions with one another and the surface of their container, and that the particles have no other energy but their kinetic energy. This model has most of the characteristics of a monatomic gas.

For such an idealized gas, the pressure exerted on the container surface can be found from the rate of change of momentum, and thus force, exerted on the wall as the particles strike and rebound in the perfectly elastic collisions from the surface. For each particle, the change in momentum $\Delta \mathbf{p}$ on striking the wall is (Fig. 12.2)

$$\Delta \mathbf{p} = m\mathbf{V}_z - m(-\mathbf{V}_z) = 2m\mathbf{V}_z \tag{12.12}$$

Figure 12.2   Particle-wall interaction.

The number of particles at speed $\mathbf{V}_z$ striking the wall per unit time and area is found by examining Fig. 12.3. Any particle with velocity component $\mathbf{V}_z$ that is within a distance $\Delta L = \mathbf{V}_z \, \Delta t$ of the wall will strike the wall within time interval Δt. If there are $dN(\mathbf{V}_z)/V$ particles per unit volume with speed $\mathbf{V}_z$, then $dN(\mathbf{V}_z)/\Delta t = A \, \Delta L \, dN(\mathbf{V}_z)/(V \, \Delta t) = A\mathbf{V}_z \, dN(\mathbf{V}_z)/V$ particles with speed $\mathbf{V}_z$ will strike the wall per unit time.

The total rate of change of momentum at the wall is due to the number of particles per unit time striking the wall times the momentum change per particle, integrated over all particle speeds, or

$$F = \frac{dp}{dt} = \frac{m}{dt} \int_0^\infty \mathbf{V}_z \, dN(\mathbf{V}_z) = \frac{Am}{V} \int_0^\infty \mathbf{V}_z^2 \, dN(\mathbf{V}_z) \tag{12.13}$$

However, the integral on the right is simply the mean square velocity times the number of particles per unit volume, i.e.,

$$\int_{\mathbf{V}_z=0}^\infty \mathbf{V}_z^2 \, dN(\mathbf{V}_z) = N\overline{\mathbf{V}_z^2} \tag{12.14}$$

so that Eq. (12.13) can be written as

$$V\left(\frac{F}{A}\right) = PV = mN\overline{\mathbf{V}_z^2} \tag{12.15}$$

The average kinetic energy of a gas particle is given by summing the three components of velocity, to give

$$\bar{\varepsilon} = \left(\frac{m}{2}\right)(\overline{\mathbf{V}_x^2} + \overline{\mathbf{V}_y^2} + \overline{\mathbf{V}_z^2}) \tag{12.16}$$

If we assume that the three components of mean square velocity are equal (otherwise, all the particles would end up in one portion of the container), then the average energy per particle is

$$\bar{\varepsilon} = \left(\frac{3m}{2}\right)\overline{\mathbf{V}_z^2} \tag{12.17}$$

Figure 12.3   Rate of particles striking wall.

According to the Maxwell-Boltzmann distribution, the average energy per particle [Eq. (12.11)] is simply $\frac{3}{2}kT$, so

$$\mathbf{V}_z^2 = \left(\frac{2}{3m}\right)\bar{\varepsilon} = \left(\frac{2}{3m}\right)\frac{3kT}{2} = \frac{kT}{m} \tag{12.18}$$

Substituting into Eq. (12.15) gives

$$PV = NkT = \left(\frac{N}{N_a}\right)(kN_a)T \tag{12.19}$$

or, since $kN_a = \overline{R}$ and $N/N_a = n$, the number of moles of gas, Eq. (12.19), becomes

$$PV = n\overline{R}T \tag{12.20}$$

Our model of a gas follows the ideal gas law! In itself, this is perhaps not so surprising; however, we have developed some powerful insight into just what an ideal gas is on the microscopic level. In addition, because we know something about the energy contained in such a gas and how it is related to temperature, we may be able to go further with our analysis.

Let us examine the energy contained by the gas. Since we have assumed that only translational kinetic energy is present (in order to invoke the Maxwell-Boltzmann distribution), the energy contained by 1 mol of gas is

$$\overline{u} = N_a\overline{\varepsilon} = N_a(\tfrac{3}{2}kT) = \tfrac{3}{2}\overline{R}T \tag{12.21}$$

and the internal energy of the gas depends on the temperature only, which we know to be correct for an ideal gas. Taking the derivative of $\overline{u}$ with respect to temperature gives

$$\frac{d\overline{u}}{dT} = \overline{c}_v = \tfrac{3}{2}\overline{R} \tag{12.22}$$

We have thus predicted the value of a property $\overline{c}_v$ for an ideal gas. Indeed, we predict $\overline{c}_v$ for all gases to be the same! This is obviously not true; however, if we examine molar values of $\overline{c}_v$ for *monatomic* gases, we find the prediction to be quite accurate over broad ranges of temperature. For more complex molecules, the assumption that only translational energy makes up the internal energy $\overline{u}$ is not valid, and we must extend the model.

In the classical approach, we were unable to predict property values; we could develop relations among properties, but we were always forced to make a certain minimal set of experimental measurements to develop the remaining properties.

Extensions to Diatomic Molecules
Is it possible to extend the results of the simple model to diatomic or even more complex molecules? Many attempts at such an extension have been made. One simple and useful one is to note that the translational kinetic energy is made up of the three components of velocity according to Eq. (12.16). Further, this translational energy was set equal to the average energy found from the Maxwell-Boltzmann distribution, $\tfrac{3}{2}kT$. Thus it might be possible to assign to each component of kinetic energy the value

$\frac{1}{2}kT$; in other words, we can *partition* the total kinetic energy of a particle into its components, each with value $\frac{1}{2}kT$.

Suppose we now examine a diatomic molecule (Fig. 12.4) and assume that it acts as two masses attached by a rigid rod. Such a model is called a *rigid rotator.* Let an axis pass through the two masses, and two other mutually perpendicular axes originate at the center of mass of the molecule. If we assume that the moment of inertia of the molecule is negligible for rotation around the axis through the masses, there are two remaining rotational axes for storing kinetic energy of rotation in the rigid rotator. If each of these axes can store rotational kinetic energy in the amount $\frac{1}{2}kT$, then the total internal energy of a diatomic molecule is $\frac{5}{2}kT$, that is, $\frac{3}{2}kT$ in translational kinetic energy and $\frac{2}{2}kT$ in rotational energy. The internal energy of 1 mol of a diatomic gas is then

$$\bar{u} = \tfrac{5}{2}N_a kT \qquad (12.23)$$

resulting in a prediction of

$$\bar{c}_v = \tfrac{5}{2}\bar{R} \qquad (12.24)$$

This value for the molar specific heat of a diatomic ideal gas is also quite accurate at low temperatures; however, as the absolute temperature of the gas is increased, the measured specific heat increases. For nitrogen, for example, the $\bar{c}_v$ versus T data from Table D.4 are plotted in Fig. 12.5. At higher temperatures, the value of c_v appears to be approaching $\frac{7}{2}\bar{R}$. It appears, then, that if our partitioning method is correct, two more ways of storing internal energy in the diatomic molecule are occurring at higher temperature. The usual model invoked is that the two atoms making up the molecule are no longer rigidly attached; rather, the rigid rod should instead be modeled as a spring, which allows both potential and kinetic energy to be stored as the spring vibrates more and more intensely at increasing temperature. The value of $\bar{c}_v$ for a vibrating diatomic molecule should then approach $\frac{7}{2}\bar{R}$ at high temperature.

These results should raise a flag as to the correctness of the model. Based on our model, the vibrating diatomic molecule should have 7 energy storage modes, usually called *degrees of freedom,* while the rigid rotator should have 5. However, either the degrees of freedom exist for the molecule, or they do not. They should not slowly become active as a function of temperature. Why does the observed slow transition from 5 to 7 degrees of freedom occur as the temperature is raised? Additionally, it seems artificial to assign a separate degree of freedom to both the potential and the kinetic energy of vibration. After all, the kinetic and potential energies are simply traded back and forth during the vibration process such that their sum remains constant. Is this really two *modes* of energy storage? Yet we must find two such modes to make the model agree with experimental data. Our model must be still incomplete or at least not completely understood.

Figure 12.4   **Rigid rotator diatomic molecule.**

Figure 12.5   **Specific heat versus temperature for nitrogen.**

As more complex (polyatomic) molecules must be modeled, the kinetic theory approach we have been using becomes less fruitful. However, it does provide a simple model for the behavior of a gas, and, along with the principle of the equipartition of energy (which assigns $\frac{1}{2}kT$ to each possible energy mode), is historically important in that together they give some useful insight into the macroscopic observations available in classical thermodynamics. We have also found some useful equations for the internal energy and specific heat of ideal gases that agree with observation and indeed allow accurate predictions for these properties when no experimental data are available.

With this success, how do we proceed to the prediction of other properties and to a better understanding of macroscopic thermodynamics? When we try to take further steps in kinetic theory, we find that some earlier assumptions prove to be invalid in important cases. It is worthwhile to examine the consequences of our earlier assumptions: in particular, what happens if all possible energies are not available to the particles which we are studying, as we assumed for Eq. (12.9)? And do all particles then follow the Maxwell-Boltzmann distribution, or must some other distribution of energies be used?

It proves quite difficult to build models of the detailed behavior of atoms and molecules based on their assumed microscopic structure. A more fruitful approach is to look at the statistical average behavior of a group of molecules whose exact structure is unknown but whose average behavior can be inferred. Such an approach is called *statistical mechanics,* as opposed to the kinetic theory approach we have used so far, which is closely tied to the assumed structural properties of atoms and molecules.

The mathematical foundations of statistical mechanics, indeed almost all the detail that is required, were laid out in the early twentieth century by U.S. thermodynamicist Josiah Willard Gibbs. At that time, however, the statistical description of molecular behavior was limited to the Maxwell-Boltzmann distribution, which, as we have implied, has some shortcomings. Thus, although the *method* proposed by Gibbs is complete and correct, the information available at the time he proposed the method was incomplete and led to predictions which were in error. To avoid that pitfall, we take a detour through a most interesting area, quantum mechanics, and then return to our goal of surveying statistical approaches to thermodynamics with more complete information than was available to Gibbs.

12.3 Influence of Quantum Effects

So far, all our modeling (including the Maxwell-Boltzmann distribution) has been based on the assumption that newtonian mechanics provides a valid description of particle interactions at the microscopic level. There is much evidence to indicate that this is not the case, particularly in the observations of the behavior of the properties of solids at low temperatures, the absorption of radiant energy by gases, and even in our observations of the predicted versus observed specific heats of gases.

The inability of classical mechanics to provide satisfactory explanations of observed phenomena led physicists in the early twentieth century to try a different approach. Max Planck, in trying to predict the observed distribution of radiant energy with respect to wavelength from an ideal radiating surface, was forced to model the possible values possessed by the radiated energy as being discrete; that is, not all energies were possible, only discrete multiples of a basic value. Only by making such an assumption could he reconcile predictions with experiment, where predictions based on classical physics had all failed. Neils Bohr used this idea of discrete energy states in constructing a model of the hydrogen atom, which was successful in exactly predicting the radiation absorption characteristics of hydrogen gas, again after all attempts based on classical physics had failed. These successes, along with others by Albert Einstein, led to further development and final acceptance of Planck's original idea.

A further major advancement in the development of quantum theory was triggered by Louis de Broglie's suggestion that energy and matter were closely related and that both should therefore possess wave-like characteristics. His suggestion was almost immediately given striking confirmation by the results of experimental work by Davisson and Weeks on the patterns resulting from the scattering of electrons from solid surfaces, which could be explained only by ascribing a wavelength to the particles used in their experiments. This work led to a mathematical description of the wavelike properties of matter by Erwin Schrödinger (1887–1961). His equation describes the probability that matter will exist in a given location at a given time, and it gave impetus to the idea that our knowledge about the behavior of microscopic systems could, at best, be on a statistical or probabilistic basis, rather than on the deterministic basis prescribed by newtonian mechanics. We need not go into detailed solutions of the Schrödinger equation, except to note its time-independent form in cartesian coordinates:

$$\frac{\partial^2 \psi}{\partial x^2} + \frac{\partial^2 \psi}{\partial y^2} + \frac{\partial^2 \psi}{\partial z^2} + \left(\frac{8\pi^2 m}{h^2}\right)(\varepsilon - \text{PE})\psi = 0 \qquad (12.25)$$

where ψ is the *wave function,* which describes the amplitude or "presence" of a "matter wave" in space, PE is the potential energy, and h is Planck's constant, 6.6262×10^{-34} J · s.

12.3.1 An Example of Quantization

Let us examine a single particle which moves only along the x coordinate between two infinite parallel planes separated by a distance L. There is no potential energy for such a case, and Schrödinger's equation reduces to

$$\frac{d^2\psi_x}{dx^2} + \left(\frac{8\pi^2 m}{h^2}\right)\varepsilon\psi_x = 0 \tag{12.26}$$

This is a second-order linear ordinary differential equation, and we must prescribe two boundary conditions to obtain a complete solution. If the particle cannot escape from the enclosure and ψ_x is related to the probability that the particle exists at some position, then we can require $\psi_x = 0$ at $x \le 0$ and at $x \ge L$. The solution of Eq. (12.26) is

$$\psi_x = A\sin\left(\frac{8\pi^2 m\varepsilon}{h^2}\right)^{1/2} x + B\cos\left(\frac{8\pi^2 m\varepsilon}{h^2}\right)^{1/2} x \tag{12.27}$$

Applying the boundary conditions $\psi_x = 0$ at $x = 0$ gives $B = 0$, and applying the condition $\psi_x = 0$ at $x = L$ results in

$$0 = A\sin\left(\frac{8\pi^2 m\varepsilon}{h^2}\right)^{1/2} L \tag{12.28}$$

For Eq. (12.28) to hold without the trivial solution $A = 0$ requires that the argument of the sine term be a multiple of π, or

$$n'\pi = \left(\frac{8\pi^2 m\varepsilon}{h^2}\right)^{1/2} L \qquad n' = 0, 1, 2, 3, \ldots \tag{12.29}$$

Thus the energy of the particle must be

$$\varepsilon = \frac{n'^2 h^2}{8mL^2} \qquad n' = 0, 1, 2, 3, \ldots \tag{12.30}$$

If Schrödinger's equation is indeed correct (and there is ample evidence from other predictions that it is), then this is a profound result. The energy of the particle as given by Eq. (12.30) can have only the values $h^2/(8mL^2)$, $h^2/(2mL^2)$, $9h^2/(8mL^2)$, ... This *quantization* of the energy into a certain discrete progression of values is seen to be forced on the system by the form of the equation and the boundary conditions; yet, the prediction by Planck that such discrete energy values must exist for massless radiative energy is borne out here for a single particle of mass m.

If we bat a perfectly elastic tennis ball and allow it to bounce continuously between two walls, this result should apply for the energy values that can be possessed by the ball. We have, after all, placed no restriction on the mass of the particle or the size of the enclosure in our analysis. Yet we do not observe tennis balls to have only discrete energies (velocities). Is our analysis flawed? Let us see how large the difference is between allowed energy states j and k. Equation (12.30) predicts

$$\varepsilon_j - \varepsilon_k = (j^2 - k^2)\left(\frac{h^2}{8mL^2}\right) \tag{12.31}$$

For a 0.1-kg ball and walls 1 m apart, the energy difference between the two states is

$$\varepsilon_j - \varepsilon_k = \frac{(j^2 - k^2)(6.6262 \times 10^{-34} \text{ J} \cdot \text{s})^2}{8 \times 0.1 \text{ kg} \times 1^2 \text{ m}^2}$$

$$= (j^2 - k^2)(5.49 \times 10^{-67}) \qquad \text{J} \tag{12.32}$$

For any pair of integers that differ by unity, the difference between the quantized energy values is so small that we could not observe it. Thus our intuition is not violated. For very low-mass particles in small "containers," however, we can see that the difference in the energies between the quantized states could become appreciable.

The values of the integer n' that appear in the solution are called the *quantum numbers* for this system. If more coordinates than simply x are present, then additional quantum numbers appear in the mathematical solution of the resulting partial differential equation.

We could continue the solution to express the final form and properties of the wave function and further examine the probability of the particle being at a given point (which is proportional to ψ^2). However, our purpose here is simply to show that quantized energy states are a *result* of the probabilistic view of the behavior of matter. They are not imposed on the system by the analyst; they result from a model that has been proved capable of explaining observed behavior to a very high degree of accuracy.

If you are uncomfortable with the thought that you are composed of matter waves that interact in such a way that mass appears with the probability required in the locations necessary to compose your particular shape, then you are in good company. Einstein is widely quoted as saying, "God does not play dice," and Schrödinger said, "I don't like it, and I'm sorry I ever had anything to do with it." Nevertheless, the quantum theory has proved remarkably successful, and the understanding of the behavior of matter that it provides has led, among other things, to the laser, the transistor and integrated circuits, nuclear energy, the development of molecular biology, and our understanding of DNA, and seems on the brink of major breakthroughs in understanding the behavior of subatomic particles.

12.3.2 Uncertainty Principle

One consequence of the Schrödinger equation (and other approaches to quantum theory) is that the product of our uncertainty about the momentum (and thus velocity) of a particle $m \Delta \mathbf{V}$ and the uncertainty in the particle position Δx is greater than or equal to a constant. More exactly,

$$m \Delta \mathbf{V} \Delta x \geq \frac{h}{4\pi} \tag{12.33}$$

This relation was first observed and commented on by Werner Heisenberg in 1926. Equation (12.33) indicates that it is impossible to measure both the momentum and the position of a particle to an arbitrary degree of precision, since the product of the uncertainties in the two quantities must always exceed $h/(4\pi)$. The relation comes out of the interpretation of the wave equation, which implies that a particle has only a probability, not a certainty, of being at a given place at a given time. *Heisenberg's uncertainty principle* gives the limits on our possible knowledge about the location and momentum of a particle. This inherent uncertainty is what finally ends the dream of classical physics of a deterministic universe in which we could track the paths of all particles into the future if we could only know their present states. The more accurately we determine the position of a particle, the less accurately we can know its momentum, and vice versa.

The uncertainty principle applies not only to the product of the uncertainties in momentum and position, but also to any pair of *conjugate variables,* i.e., variables that multiply to give the units possessed by h, J · s, such as the obvious pair, energy × time.

The size of h (6.6262×10^{-34} J · s) is assurance that the uncertainty principle will not be a factor in our everyday laboratory experiments; it does become important when we deal with particles such as electrons.

12.3.3 Bose-Einstein Statistics

Satyendra Bose developed and Albert Einstein extended the study of the statistics that govern the possible energy states in a collection of particles that are indistinguishable. In this context, *indistinguishable* means that if we exchange two particles in the set under study, we cannot distinguish any difference in the set. Photons and single-component ideal gases fall in this category. Particles of this type are called *bosons.*

Bose and Einstein showed that the number of ways N_p bosons could be arranged among all the possible energy levels, when a given energy level can have g_j different quantum states (i.e., sets of quantum numbers) with the same energy, is given by

$$w_{\text{BE}} = \prod_{j=0}^{\infty} \frac{(g_j + N_j - 1)!}{(g_j - 1)!N_j!} \tag{12.34}$$

where w represents the number of possible particle arrangements among the possible energy states. Here, the symbol Π means the product of all the terms from $j = 0$ to infinity. The quantity g_j is called the *degeneracy* of level j, and it reflects the fact that various sets of quantum numbers can predict a state with the same energy. It is convenient to group all states with the same energy, rather than to sum them separately.

Using Eq. (12.34) for the number of possible ways of arranging the particles among the energy levels plus general Eq. (12.3), which requires the number of particles in all levels to add to the total number of particles, and Eq. (12.6), which requires the energy of the particles in all levels to sum to the total energy in the system, we find the Bose-Einstein (BE) energy distribution for bosons to be

$$\frac{N_j}{N_p} = \frac{g_j}{Q_{BE}\{\exp\left[\varepsilon_j/(kT)\right] - 1\}} \tag{12.35}$$

where

$$Q_{BE} = \sum_{j=0}^{\infty} \frac{g_j}{\exp\left[\varepsilon_j/(kT)\right] - 1}$$

Equation (12.35) describes the distribution of particles among the possible energy levels for a system in thermodynamic equilibrium.

12.3.4 Fermi-Dirac Statistics

Wolfgang Pauli in 1925 observed that no more than one particle of certain types could have the same set of quantum numbers. In such a case, more than one indistinguishable particle could not occupy the same quantum state. Since g_j degenerate quantum states might have the same energy, it is still possible for indistinguishable particles to be arranged in many ways among the degenerate states, as long as no more than one particle is in any given one of the degenerate states in level j.

For such a case, the statistics are slightly different from the Bose-Einstein case. The problem was worked out independently by Enrico Fermi and Paul Dirac (hence the FD subscript) in 1925 and 1926. The results are

$$w_{FD} = \prod_{j=0}^{\infty} \frac{g_j!}{N_j!(g_j - N_j)!} \tag{12.36}$$

and

$$\frac{N_j}{N_p} = \frac{g_j}{Q_{FD}\{\exp\left[\varepsilon_j/(kT)\right] + 1\}} \tag{12.37}$$

where

$$Q_{FD} = \sum_{j=0}^{\infty} \frac{g_j}{\exp\left[\varepsilon_j/(kT)\right] + 1}$$

As before, the distribution of particles among the possible energy levels [Eq. (12.37)] is for a system in thermodynamic equilibrium. Particles that obey these statistics include electrons and protons and are usually called *fermions.*

Let us make an important observation applicable to particles that make up the substances studied in this text. Examination of Eqs. (12.35) and (12.37) indicates that they are much different in form than the Maxwell-Boltzmann distribution, Eq. (12.9).

12.3.5   Maxwell-Boltzmann Statistics

For *dilute gases,* the number of degenerate states g_j at any energy level is very large compared with the number of particles at that level N_j. Note that in the limit $g_j \gg N_j$, for Bose-Einstein statistics,

$$\frac{(g_j + N_j - 1)!}{(g_j - 1)!} = (g_j + N_j - 1)(g_j + N_j - 2) \cdots (g_j + 1)(g_j) \frac{(g_j - 1)!}{(g_j - 1)!}$$

$$= (g_j + N_j - 1) \cdots (g_j + 1)g_j \approx g_j^{N_j} \qquad (12.38)$$

Similarly, for Fermi-Dirac statistics

$$\frac{g_j!}{(g_j - N_j)!} = \frac{g_j(g_j - 1)(g_j - 2) \cdots (g_j - N_j + 1)(g_j - N_j)!}{(g_j - N_j)!}$$

$$= g_j(g_j - 1) \cdots (g_j - N_j + 1) \approx g_j^{N_j} \qquad (12.39)$$

Substituting Eqs. (12.38) and (12.39) into (12.34) and (12.36), respectively, gives

$$w_{\text{BE}} = w_{\text{FD}} = \prod_{j=1}^{\infty} \frac{g_j^{N_j}}{N_j!} \qquad (12.40a)$$

in the limit of $g_j \gg N_j$. Thus, in the limit of a dilute gas with many more available energy states than particles, the bosons and fermions both obey the same statistics. If the particles are *distinguishable,* that is, interchanging any two particles gives a definable new state, then the possible number of arrangements is given by

$$w_{\text{MB}} = N_p! \prod_{j=0}^{\infty} \frac{g_j^{N_j}}{N_j!} \qquad (12.40b)$$

which is the result for Maxwell-Boltzmann statistics. Using Eq. (12.40a) along with Eqs. (12.3) and (12.6) to develop an energy distribution for a dilute gas results in

$$\frac{N_j}{N_p} = \frac{g_j}{Q \exp[\epsilon_j/(kT)]} \qquad (12.41)$$

where $Q = \sum_j g_j / \exp[\varepsilon_j/(kT)]$. This particular form of Q is usually given the symbol Z (from the German word *Zustandssumme,* or sum of states) and is called the *partition function.*

Equation (12.40b) can also be derived directly from the statistics governing *distinguishable* particles, that is, particles in systems where we can tell the effect of interchanging any two particles. Note from Eq. (12.40b) that the number of possible states w for such a system is much greater than that for systems described by Bose-Einstein (BE) or Fermi-Dirac (FD) statistics, since every interchange of any two particles in such a system results in another state. These additional states do not occur if the particles are indistinguishable.

Particles belonging to systems described by Eqs. (12.40a) and (12.41) are called *boltzons,* because these equations result for a quantized system that otherwise is described by the classical or continuous Maxwell-Boltzmann distribution. So Eq. (12.41) is also called the Maxwell-Boltzmann distribution. Note that a plot of Eq. (12.41) is not possible until the system and the particles in the system are fully described through quantum mechanics. When that is done, the values of the energy levels ε_j and the degeneracy of each level g_j can be found, and Z for that system can be evaluated.

12.4 Application of Microsystem Information: Entropy and Other Properties

Properties

Entropy is described in Chap. 5 as a measure of randomness, disorder, or uncertainty. We can now examine the meaning of such concepts. Note that the quantity w is the number of ways that a system of particles can be arranged among the possible energy states available. As w increases, we are less and less certain of the particular distribution of particle energies that does exist, since more possibilities exist.

For boltzons, which are representative of the particles in an ideal gas, the number of possible arrangements of particles is given by Eq. (12.40a). If entropy is a measure of uncertainty about the actual state of the system, we would expect S to be directly related to w. However, w has a very large value for any system composed of N_p particles, so we might try an equation of the form

$$S = C \ln w \tag{12.42}$$

where C is a constant of proportionality. There are more fundamental reasons for the choice of this form, but here we simply observe that it works well. It also has the advantage that for a system which has only one possible state ($w = 1$), the value of S is zero, reflecting the fact that we are certain of the state of such a system!

We can now proceed with this equation for entropy, apply it to a dilute gas, and see what results are obtained. Substituting Eq. (12.40a) to eliminate w results in

$$S = C \ln \prod_{j=0}^{\infty} \frac{g_j^{N_j}}{N_j!} = C \sum_{j=0}^{\infty} \ln \frac{g_j^{N_j}}{N_j!} = C \sum_{j=0}^{\infty} (\ln g_j^{N_j} - \ln N_j!)$$

$$= C \sum_{j=0}^{\infty} (N_j \ln g_j - \ln N_j!) \tag{12.43}$$

The final term can be evaluated by the use of *Stirling's approximation,* which states that

$$\ln N_j! \approx N_j \ln N_j - N_j \tag{12.44}$$

Substituting Eq. (12.44) into (12.43) results in

$$S = C \sum_{j=0}^{\infty} N_j \left(\ln \frac{g_j}{N_j} + 1 \right) \tag{12.45}$$

Now, Eq. (12.41) for the equilibrium distribution of particles among the energy levels can be solved for g_j/N_j and substituted into Eq. (12.45) to give

$$S = C \sum_{j=0}^{\infty} N_j \left\{ \ln \frac{Z \exp [\varepsilon_j/(kT)]}{N_p} + 1 \right\}$$

$$= C \sum_{j=0}^{\infty} N_j \left(\ln \frac{Z}{N_p} + \frac{\varepsilon_j}{kT} + 1 \right) \tag{12.46}$$

Finally, we can substitute Eqs. (12.2) and (12.6) to obtain

$$S = C \left(N_p \ln \frac{Z}{N_p} + \frac{U}{kT} + N_p \right) \tag{12.47}$$

Can we identify the constant of proportionality C? We can take the partial derivative of S with respect to U at constant V (in which case Z is constant) and obtain

$$\left(\frac{\partial S}{\partial U} \right)_V = \frac{C}{kT} \tag{12.48}$$

However, Eq. (5.9) gives $(\partial U/\partial S)_V = T$, so the constant must be $C = k$, the Boltzmann constant, if our microscopic and macroscopic viewpoints are to agree. Thus, for a dilute gas of boltzons, the entropy is given by

$$S = k \ln w = k \left(N_p \ln \frac{Z}{N_p} + \frac{U}{kT} + N_p \right) \tag{12.49}$$

Note that this is the absolute entropy of the system. We have succeeded in relating the entropy directly to the quantum behavior of the system through the partition function Z.

We can use Eq. (12.6) to find U by substituting Eq. (12.41) to eliminate N_j:

$$U = \sum_{j=0}^{\infty} N_j \varepsilon_j = \frac{N_p}{Z} \sum_{j=0}^{\infty} g_j \varepsilon_j \exp \left(\frac{-\varepsilon_j}{kT} \right) \tag{12.50}$$

Note that

$$\frac{dZ}{dT} = \frac{1}{kT^2} \sum_{j=0}^{\infty} g_j \varepsilon_j \exp \left(\frac{-\varepsilon_j}{kT} \right) \tag{12.51}$$

so that

$$U = \frac{N_p kT^2}{Z}\frac{dZ}{dT} = N_p kT^2 \frac{d(\ln Z)}{dT} \qquad (12.52)$$

This result can be substituted into the final relation for S [Eq. (12.49)] so that S depends on only N_p, T, and Z.

Finally, the Helmholtz function A is given by

$$A = U - TS = -N_p kT\left(\ln\frac{Z}{N_p} + 1\right) \qquad (12.53)$$

Recall that the Helmholtz function as used here is a fundamental equation; now that we have a relation for it, given in Eq. (12.53), we can find all other thermodynamic properties for a system of boltzons at thermodynamic equilibrium. The only missing piece of information is the partition function; once it is known, all the properties of the gas can be derived.

Partition Function for a Monatomic Gas

Our purpose is to show only the methodology and power of statistical thermodynamics, not the details, but we can take one case in some detail. Suppose we examine an ideal monatomic gas which has only translational energy. (Actually, at higher temperatures, the energy stored in electronic energy levels can become important, but we ignore that effect here.)

For such a case, the quantized translational kinetic energy is $\varepsilon_j = m\overline{\mathbf{V}}_j^2/2$. The partition function is then

$$Z = \sum_{j=0}^{\infty} g_j \exp\left(\frac{-\varepsilon_j}{kT}\right) = \sum_{j=0}^{\infty} g_j \exp\left(\frac{-m\overline{\mathbf{V}}_j^2}{2kT}\right) \qquad (12.54)$$

The degeneracy g_j can be estimated from the uncertainty principle. For a particle with energy in the range $\varepsilon_j \pm \Delta\varepsilon_j/2$, the mean square speed must be in a range given by

$$\frac{m\overline{\mathbf{V}}_j^2}{2} \pm \frac{m}{2}\Delta\overline{\mathbf{V}}_j^2 = \varepsilon_j \pm \frac{\Delta\varepsilon_j}{2} \qquad (12.55)$$

If we can determine the number of indistinguishable particles that have speeds in the range $\Delta\mathbf{V}_j = (\Delta\overline{\mathbf{V}}_j^2)^{1/2}$, then that is the number of particles with energies in the corresponding range given by Eq. (12.55). The number of particles within $\Delta\varepsilon_j$ around ε_j is just the degeneracy g_j.

The product of momentum and position in space for all particles with energy $\varepsilon_j \pm \Delta\varepsilon_j$ is $m\,\Delta\mathbf{V}_x \cdot m\,\Delta\mathbf{V}_y \cdot m\,\Delta\mathbf{V}_z \cdot \Delta x\,\Delta y\,\Delta z$. This six-dimensional quantity is a part of what is called *momentum space*. The momentum space occupied by a single particle is given by Eq.(12.33) as $m\,\Delta\mathbf{V}_x \cdot m\,\Delta\mathbf{V}_y \cdot m\,\Delta\mathbf{V}_z \cdot \Delta x\,\Delta y\,\Delta z \approx h^3$. The degeneracy is then the momentum space occupied by all particles with energies within $\Delta\varepsilon_j$ divided by the momentum space per particle, or

$$g_j = \frac{m\,\Delta\mathbf{V}_x \cdot m\,\Delta\mathbf{V}_y \cdot m\,\Delta\mathbf{V}_z \cdot \Delta x\,\Delta y\,\Delta z}{h^3}$$

$$\approx \left(\frac{m}{h}\right)^3 d\mathbf{V}_x\,d\mathbf{V}_y\,d\mathbf{V}_z\,dx\,dy\,dz \qquad (12.56)$$

The partition function now becomes

$$Z = \left(\frac{m}{h}\right)^3 \sum_{j=0}^{\infty} \left\{ dx\,dy\,dz \exp\left[\frac{-m(\mathbf{V}_x^2 + \mathbf{V}_y^2 + \mathbf{V}_z^2)}{kT}\right] d\mathbf{V}_x\,d\mathbf{V}_y\,d\mathbf{V}_z \right\}_j$$
$$(12.57)$$

As we did for the tennis ball in Sec. 12.3.1, we could show that the difference between energy levels for monatomic gases at most conditions is very small. This allows us to replace the summation in Eq. (12.57) with an integration over speed and direction, so that Z becomes

$$Z = \left(\frac{m}{h}\right)^3 \int_{z=0}^{\infty} \int_{y=0}^{\infty} \int_{x=0}^{\infty} dx\,dy\,dz$$
$$\times \left[\int_{\mathbf{V}_x=-\infty}^{\infty} \exp\left(\frac{-m\mathbf{V}_x^2}{2kT}\right) d\mathbf{V}_x \int_{\mathbf{V}_y=-\infty}^{\infty} \exp\left(\frac{-m\mathbf{V}_y^2}{2kT}\right) d\mathbf{V}_y \right.$$
$$\left. \times \int_{\mathbf{V}_z=-\infty}^{\infty} \exp\left(\frac{-m\mathbf{V}_z^2}{2kT}\right) d\mathbf{V}_z \right] \qquad (12.58)$$

The triple integral over $dx\,dy\,dz$ is just the system volume V. The remaining three integrals can be evaluated with the aid of a table of integrals, and each gives $(2\pi kT/m)^{1/2}$. The partition function for a monatomic gas with no electronic energy states finally becomes

$$Z = \left(\frac{V}{h^3}\right)(2\pi m kT)^{3/2} \qquad (12.59)$$

Example Problem 12.1

Derive the expression for the internal energy u, specific heat at constant volume c_v, and absolute entropy of argon. Compare the prediction from the derived expressions with the results of Table D.3 at 300°C and 1 atm.

Solution
Starting with Eq. (12.52), we have

$$U = N_p kT^2 \frac{d(\ln Z)}{dT}$$

$$= N_p kT^2 \frac{d}{dT}\left(\ln\frac{V}{h^3} + \frac{3}{2}\ln 2\pi mk + \frac{3}{2}\ln T\right)$$

$$= \frac{3N_p kT}{2}$$

Letting $N_p = N_a$ for 1 mol of gas, dividing through by the molecular weight, and noting that $N_a k/M = R$, we get

$$u = \frac{U}{M} = \frac{3RT}{2}$$

Then

$$c_v = \left(\frac{\partial u}{\partial T}\right)_v = \frac{3R}{2}$$

These are the same results that we obtained for an ideal monatomic gas using the classical approach.

Substituting the result for U into Eq. (12.49) gives

$$S = kN_p\left(\ln\frac{Z}{N_p} + \frac{5}{2}\right)$$

Substituting Eq. (12.59) for Z gives

$$S = kN_p\left\{\ln\left[\left(\frac{V}{N_p h^3}\right)(2\pi mkT)^{3/2}\right] + \frac{5}{2}\right\}$$

If we set $N_p = N_a$, then S will be the entropy per mole. Dividing by the molecular weight of the gas M results in the specific entropy, kJ/(kg · K), as

$$s = R\left\{\ln\left[\left(\frac{V}{N_a h^3}\right)(2\pi mkT)^{3/2}\right] + \frac{5}{2}\right\}$$

Note that the Helmholtz function [Eq. (12.53)] can be used to find the equation of state for this gas. The total differential of the Helmholtz function in Eq. (8.5c) gives P as

$$P = -\left(\frac{\partial A}{\partial V}\right)_T$$

$$= -N_p kT\frac{\partial}{\partial V}\left\{\ln V + \ln\left[\frac{(2\pi mkT)^{3/2}}{N_p h^3}\right] + 1\right\}_T$$

$$= \frac{N_p kT}{V}$$

or, with $N_p = N_a$,

$$PV = N_a kT = \bar{R}T$$

which should not be surprising by now. Substituting into the last relation for s to eliminate V results in

$$s = R\left(\tfrac{3}{2}\ln T + \ln\frac{V}{N_a} + \tfrac{3}{2}\ln 2\pi mk - 3\ln h + \tfrac{5}{2}\right)$$

$$= R(\tfrac{5}{2}\ln T - \ln P + \ln N_a k + \tfrac{3}{2}\ln 2\pi mk - 3\ln h + \tfrac{5}{2})$$

$$= c_P\ln T - R\ln P + \text{const}$$

where we have used $c_P = c_v + R = 5R/2$, and

$$\text{Const} = R\left(\tfrac{3}{2}\ln 2\pi m - 3\ln h + \tfrac{5}{2}\ln k + \tfrac{5}{2}\right)$$

$$= R\left[\ln \frac{(2\pi m)^{3/2}k^{5/2}}{h^3} + \frac{5}{2}\right]$$

The final form is called the *Sackur-Tetrode equation,* and it predicts the absolute entropy of an ideal gas. The dependence on T and P is seen to be the same as indicated in Table 5.4 for an ideal gas; now, however, no arbitrary constant has been introduced into the relation for entropy. All constants in the Sackur-Tetrode equation are known for a given monatomic gas.

Now we can compare the predictions with the table values for argon. Using $k = 1.3806 \times 10^{-26}$ kJ/K, $h = 6.6262 \times 10^{-37}$ kJ $\cdot$ s (Table B.1), $R = 0.20815$ kJ/(kg $\cdot$ K), $m = M/N_a = 39.944/(6.023 \times 10^{26}) = 6.6319 \times 10^{-26}$ kg, we find

$$u = \frac{3RT}{2} = 3(0.20815)\,\frac{273.15 + 300}{2} = 178.95 \text{ kJ/kg}$$

which compares with 179.05 from Table D.3. For c_v,

$$c_v = \frac{3R}{2} = \frac{3(0.20815)}{2} = 0.3122 \text{ kJ/(kg} \cdot \text{K)}$$

which compares with 0.3124 kJ/(kg $\cdot$ K) from Table D.3.

Using the Sackur-Tetrode equation, after combining the log terms, we find

$$s = R\left[\ln \frac{(2\pi m)^{3/2}(kT)^{5/2}}{Ph^3} + \frac{5}{2}\right]$$

$$= 0.20813$$

$$\cdot \left\{\ln \frac{[2\pi(6.6319 \times 10^{-26})/1000]^{3/2}[(1.3806 \times 10^{-26})(573.15)]^{5/2}}{(101.32)(6.6262 \times 10^{-37})^3} + \frac{5}{2}\right\}$$

$$= 0.20813[\ln(5.0823 \times 10^7) + \tfrac{5}{2}]$$

$$= 4.2134 \text{ kJ/(kg} \cdot \text{K)}$$

which compares with $s = s_0 = 4.2163$ kJ/(kg $\cdot$ K) from Table D.3.

Comments

The differences between the calculated values and the table values arise for two reasons. Small differences are introduced by the use of slightly different values for the fundamental constants. More importantly, the tables are developed by including the effect of electronic energy states, which increases the values of u and s above those calculated by ignoring the electronic contributions as we did here. At low temperatures, the effect of the electronic states is small for most gases. The tables include the effect of the first 285 electronic states for argon.

In this section, we use the quantum statistics for general classes of particles that obey various types of restrictions (bosons, fermions) and show that the statistics reduce to the Maxwell-Boltzmann statistics for a dilute gas composed of indistinguishable particles. In describing the behavior of matter that is not in the form of a dilute gas, the original Bose-Einstein or Fermi-Dirac statistics must be used, because the assumption that $g_j \gg N_j$ is not valid. In such cases the Maxwell-Boltzmann statistics and distribution give erroneous predictions.

Given the case of a dilute gas, we predicted the values of the thermodynamic properties for the case of a monatomic gas. Note that we made no assumptions about the structure of the gas—only that it obeyed the applicable statistics for the number of particles that exist in a given energy level. This procedure can be continued for a diatomic or polyatomic gas, with allowance for the greater number of energy states that can exist because of the presence of quantized rotational, vibrational, and electronic energies. The gas tables in App. D were constructed from the resulting relations.

12.5 First Law

Equation (12.6) gives a general relation for the internal energy of a system of particles in terms of the energy levels of each individual particle. The internal energy can evidently be changed in two ways: either the energy level distribution ε_j can be altered, or the distribution of particles among the energy levels N_j can be changed. Thus

$$dU = \sum_{j=0}^{\infty} \varepsilon_j \, dN_j + \sum_{j=0}^{\infty} N_j \, d\varepsilon_j \tag{12.60}$$

Suppose we consider a control mass; that is, the total number of particles N_p in the system is fixed, so the total mass remains constant. First, consider a change in U that occurs without a change in the distribution of N_j, that is, $dN_j = 0$. For such a case, the thermodynamic probability w of the system is unchanged by the process [Eq. (12.34), (12.36), or (12.40a), depending on the applicable statistics]. Since $S = k \ln w$, such a process is *isentropic*. We noted in Chap. 5 that a process that changes the internal energy isentropically is classified as work, so we can identify

$$\delta W = \sum_{j=0}^{\infty} N_j \, d\varepsilon_j \tag{12.61}$$

Consider now a case in which U is changed by changing the distribution of particles among the energy states while holding the energy state distribution constant, that is, $d\varepsilon_j = 0$. In that case, w *is* changed, so there is a change in the value of S. If the process is reversible, then this can occur only by the mechanism of heat transfer, so we can identify

$$\delta Q = \sum_{j=0}^{\infty} \varepsilon_j \, dN_j \qquad\qquad (12.62)$$

and we find that Eq. (12.60) is indeed a statement of the first law for a closed system: $\delta Q + \delta W = dU$.

12.6 Concluding Remarks

After careful reading of this chapter, some new insights should be found. First, entropy is directly related to our uncertainty about the microscopic state of a system of particles, since it is directly proportional to ln w, the natural log of the possible number of ways of arranging the particles in the system among the available energy levels and the available geometric positions. Thus, we can see why entropy is called a measure of uncertainty, chaos, etc. These terms simply reflect the fact that the more ways there are to arrange the particles in the system, the less we know about the particular state of the particles and the greater the entropy of the system. We can see why the entropy of a system of N_p particles confined to volume V increases when the particles are allowed to expand to a volume $2V$; there are now more positions for the particles to occupy, and we become less certain about the state of the set of particles.

The relation $S = k \ln w$ is fundamental to much of science and describes not only the behavior of the quantity we know as entropy but also information about many other types of systems, including biological systems and information transfer in communication and computer systems. Ludwig Boltzmann, who first derived the relation, saw it as important enough that he requested to have it engraved on his headstone, and it was. We have observed that different formulations for w may result, depending on the quantum mechanical rules obeyed by the particles under study. We studied briefly only the particular case of a dilute monatomic gas; however, the methods apply to many other types of particles, including electrons, photons, atoms, and molecules constrained in the form of solids and liquids, superfluids, etc. We need only use the correct formulation for w and the partition function for these particles to predict their thermodynamic behavior, although the correct formulation may become quite complex.

We leave this introductory study of thermodynamics at this point, having examined many of the practical applications of the subject to engineering problems as based on the fundamental principles. The broad and powerful applications of the subject to engineering problems have been stressed. The application to diverse fields of science, which, without exception, make use of some of the tools outlined here, also extends to broad fields of the liberal arts. It is particularly interesting to see the impact of such concepts as entropy, the wave equation, and the uncertainty principle on fields such as theology and philosophy, which allow and apply these ideas much as we do in science and engineering.

Problems

12.1 Derive the result $\varepsilon_{mp} = kT/2$ for the most probable energy in the Maxwell-Boltzmann distribution, Eq. (12.9).

12.2 Compare the values of $\bar{c}_v$ for the monatomic and diatomic gases in Table C.2 to the predictions of Eqs. (12.22) and (12.24), respectively.

12.3 Find the uncertainty in position $\Delta x(m)$ relative to the accuracy of the speed measurement of the tennis ball of Sec. 12.3.1 when it moves with a speed of 10 m/s.

12.4 A laboratory experiment is able to measure the speed of an electron within 0.5 percent. The mass of an electron is given in Table B.1. What is the minimum uncertainty in the position of the electron that can be determined in the experiment if the measured electron speed is 3×10^5 m/s?

12.5 If the position of a 1-g pellet can be measured within 10^{-3} cm as it moves through the air, what is the minimum uncertainty possible in a measurement of the pellet speed?

12.6 Using Eq. (12.59) for the partition function and the complete form of the Maxwell-Boltzmann distribution for boltzons, Eq. (12.41), see whether the classical Maxwell-Boltzmann distribution of Eq. (12.9) is obtained for a monatomic gas.

12.7 Develop a relation for the enthalpy of an ideal gas in terms of the partition function Z, T, and V.

12.8 A gas has three particles. There are four energy states, each with degeneracy of 2, available to each particle. The populations of the states are $N_0 = 2$, $N_1 = 1$, $N_2 = N_3 = 0$. Find the thermodynamic probability and the entropy of the gas if the gas particles obey (a) Bose-Einstein statistics, (b) Fermi-Dirac statistics, and (c) Maxwell-Boltzmann statistics.

12.9 (a) What is the value of the partition function for a gas with four energy states, each with degeneracy of 10, available to each particle? The energy states have values of $\varepsilon_0 = 0$, $\varepsilon_1 = kT$, $\varepsilon_2 = 2kT$, $\varepsilon_3 = 3kT$. (b) What are the relative populations of the four energy states.

12.10 An ideal monatomic gas has an atomic weight of 40 and is in thermodynamic equilibrium at $T = 300$ K. For this gas what is the value of (a) the mean speed of the atoms, (b) c_P, (c) u, (d) h, and (e) s?

12.11 Plot the absolute entropy of krypton at $P = 2$ atm in the range $0 < T \leq 2000$ K. Ignore electronic energy states.

12.12 Starting with $S = k \ln w$, derive a relation for the change in entropy of an ideal monatomic gas during a constant-temperature expansion from V_1 to V_2. Compare the result with the classical result for the same process.

APPENDIX A

A Short History of the Development of Thermodynamics

Confound these ancients. They've stolen all our best ideas.
Ben Jonson

The development of classical thermodynamics provides a look into the thought processes of scientists and engineers, and it allows us a glimpse of human failings as well as the triumphs of intellect that were produced.

The history of this subject is usually presented as an orderly progression of ideas, each building on the foundation laid by earlier investigators. This view has some validity. But, as with present-day research, there were often long periods when invalid ideas were tenaciously held in the face of decisive evidence of their falsity. In other cases, a number of investigators almost simultaneously adopted a whole new block of theory and built upon it. It is simply not accurate to describe this history in a linear fashion. In 1889, Samuel P. Langley, no mean researcher in his own right, examined another branch of research in his address as the retiring president of the American Assocation for the Advancement of Science:

> We often hear (the progress of science) likened to the march of an army towards some definite end; but this, it has seemed to me, is not the way science usually does move, but only the way it seems to move in the retrospective view of the compiler, who probably knows almost nothing of the real confusion, diversity and retrograde motion of the individuals comprising the body, and only shows us such parts of it as he, looking backward from his present standpoint, now sees to have been the right direction.

> I believe this comparison of the progress of science to that of an army, which obeys an impulse from one head, has more error than truth in it; and, though all similes are more or less misleading, I would prefer to ask

you to think rather of a moving crowd, where the direction of the whole comes somehow from the independent impulses of the individual members; not wholly unlike a pack of hounds, which, in the long run perhaps catches its game, but where, nevertheless, when at fault, each individual goes his own way, by scent not by sight, some running back and some forward; where the louder-voiced bring many to follow them nearly as often in a wrong path as in a right one; where the entire pack even has been known to move off bodily on a false scent; for this, if a less dignified illustration, would be one which had the merit of having a truth in it, left out of sight by the writers of texbooks.

With this warning in mind, we can embark on a brief historical voyage through the development of classical thermodynamics.

The concept of energy was introduced in the field of mechanics by Galileo Galelei (1564–1642). As early as the middle of the seventeenth century, Sir Isaac Newton and Christian Huygens had used it as a convenient method of solving problems involving the calculation of the height reached by a swinging pendulum. In the more general sense in which energy is used in thermodynamics, however, the early attempts to quantify energy were hampered by the lack of understanding that heat transfer and work were simply different forms of energy transfer. Even more basically, researchers were plagued by confusion between the concepts of heat transfer and temperature. Temperature measurement was necessary before ideas of heat transfer could be developed. The early development of the air thermometer by Galileo (1592), the sealed alcohol thermometer by Ferdinand II, Grand Duke of Tuscany in 1641, the suggestion of an oil-filled thermometer by Newton (1701) (who proposed a temperature scale that began at 0° at the melting point of ice and had a second fixed point of 12° as "the maximum heat that the thermometer can attain by contact with the human body"), and finally the development of the mercury-in-glass thermometer by Gabriel D. Fahrenheit in 1715 — all laid the foundation for later investigations.

Heat Engines and the Theory of Thermodynamics

Engineers were designing various types of heat engines even before the development of careful temperature measurement and thermodynamic theory. Thomas Savery (1650?–1715) patented a steam-operated pump in England in 1698 based on a principle suggested by Edward Somerset, the Marquis of Worcester, around 1663. Savery's pump used a valve system that was controlled by hand, making it cumbersome, slow, and inefficient. Thomas Newcomen (1663–1729), an ironmonger from Dartmouth, put his first coal-burning engine into service in England in 1712. The engine used atmospheric pressure on one side of a piston to drive the piston against the lower pressure of condensing steam on the other side, and in later versions the valves were operated by the pump linkage so that the engine could operate with scant attention (Fig. A.1).

Not until 1760 did Joseph Black (1728–1799), a professor of medicine and chemistry at the University of Glasgow, finally lay the foundation of the quantitative science of heat transfer. In that year, he measured the heat capacity (specific heat) of various materials and noted the difference between temperature and heat transfer. From 1761 to 1764, he demonstrated the concepts of latent heat of fusion and evaporation. These concepts became the basis for what was to be called the *caloric theory,* which envisioned heat transfer as a colorless fluid that migrated from a body at higher temperature to one at lower temperature. This caloric fluid was envisioned at the time to be indestructible.

James Watt (1736–1819), a Scottish instrument maker who worked for some time in Black's laboratory, had in the meantime realized some of the limitations of the Newcomen engine, and he was building engines that used steam pressure on one side of the piston to drive against the lower pressure of condensing steam on the other. He used some of Black's results to design an external condenser (1765), rather than relying on condensation in the cylinder as Newcomen had done. This allowed the cylinder to remain at higher temperatures during the entire cycle and greatly improved the engine efficiency. By 1778, Watt and his partner Matthew Boulton (1728–1809) were in competition with other manufacturers. To compare his engines with others, Watt introduced the concept of *duty,* which is a form of engine efficiency, defined as "the number of pounds raised through a height of 1 foot per bushel of coal used." By 1781–1782, Watt had introduced the double-acting engine that turned a flywheel, allowing continuous rotary motion. This had not been possible with earlier engines, because power was produced only on the downstroke. He defined the horsepower as the rate of work from a mill horse, calculating that value as 33,000 pounds of force per minute exerted through a distance of 1 foot (1782–1783). This value was conservative by a factor of 2, to avoid dissatisfied customers.

Benjamin Thompson (1753–1814) was an American who made the unfortunate choice of backing the British and spying for them at the time of the revolution. So he found it discreet to emigrate to England, where he became a lieutenant colonel in the British army. He invented, among other things, the drip coffee pot. He took a commission to aid in providing armaments for a Bavarian prince, who gave him the title of Count Rumford, and while so engaged (1798), he observed that there was a continuous heat release during the boring of a cannon. If this were so, how could the "caloric fluid" be conserved when it was apparently being continuously produced? Rumford deduced that heat transfer was "a kind of motion." This observation was, of course, available to anyone who observed the frictional production of thermal energy, but Rumford is generally given credit for bringing it to the attention of the scientific community. The caloric theory continued to be widely accepted as correct for

Figure A.1 *Newcomen's engine.*

more than 50 years, and indeed much of the mathematical interpretation of the caloric theory was taken completely into the modern view of thermodynamics. Rumford had earlier (1787–1799) carried out careful experiments which showed that the weight of a given amount of water was unchanged during the freezing process, to an accuracy of 1 part in 1 million. Thus, if there was a caloric fluid, it was essentially weightless. Rumford ended his days in Paris after marrying (1805) the widow of Lavoisier, who had been one of the strongest advocates of the caloric theory.

In the early years of the nineteenth century, Richard Trevithick (1771–1819) in England and Oliver Evans (1755–1819) in the United States were producing steam engines that worked at boiler pressures above atmospheric pressure (Figs. A.2 and A.3), thus considerably increasing cycle efficiency. Trevithick used pressures of about 15 psig.

The Rev. Robert Stirling (1790–1878), an English parish minister, patented a practical heat engine in 1816 that used air as the working fluid. In theory, the cycle used in the Stirling engine approaches the ideal cycle soon to be proposed by Carnot. The engine had practical importance because it operated at low pressure, eliminating the safety problems arising from the increasing use of higher pressures in steam boilers.

In 1824, the brilliant French military engineer Nicolas Leonard Sadi Carnot (1796–1832), son of Napoleon's minister of war, presented his only published paper, "Reflexions on the Motive Power of Fire, and on Machines Fitted to Develop that Power," which presented one form of what we now call the second law of thermodynamics as applied to the amount of work that can be obtained from an engine using a heat transfer as its driving energy. He also stated a reasoned form of the first law. Carnot's work was presented in terms of the caloric theory, and this fact led many later researchers to cling to that theory, since Carnot's predictions were evidently valid and were confirmed by experiment. However, Carnot himself was apparently beginning to question the caloric theory basis of his own work, as evidenced by changes he made in proofs of his manuscript and by some of his unpublished notes. He died at age 36 of cholera while recovering from scarlet fever, having provided probably the single most important individual contribution to classical thermodynamics.

In the early 1840s, James Prescott Joule (1818–1889) and Julius Robert Mayer (1814–1878) almost simultaneously set forth the idea that heat transfer and mechanical work were simply different forms of the same quantity, which we now recognize as energy transfer. Neither of these representations of the "mechanical theory of heat" was accepted at the time.

Joule's ideas were based on a remarkable series of experiments. He immersed the armature of a dynamo in a rotating vessel of water and measured the heating of the water when a current flowed through the armature with (1) the armature stationary, (2) the armature rotating in the

1. Lancashire Boiler.

S. Boiler Shell.
E. Boiler End.
G. Gusset Stays.
I. Internal flues.
X. External flues.
T. Galloway tubes.
B. Bowling rings.
F. Furnace.
R. Bridge.
f. Feed valve.
C. Scum cock.

2. Locomotive Boiler.

S. Boiler Shell.
F. Fire box.
D. Steam dome.
P. Steam pipe.
B. Smoke box.
G. Girders.
Y. Stays.
T. Tubes.
L. Longl stay.

3. Marine Boiler.

S. Boiler Shell.
F. Fire box.
G. Girders.
L. Longl stays.
f. Furnace.
B. Bridge.
T. Tubes.
R. Stays.
H. Mudhole door.

Figure A.2 Early steam boiler designs.

WILLIAMS' PERRAN FOUNDRY CO.

IMPROVED HORIZONTAL HIGH-PRESSURE ENGINES.

These Engines are all made of the very best materials, and constructed after the most approved models, with a view to obtain the greatest results at the least expenditure. They are made sufficiently strong to yield more than double the nominal power if worked at a boiler pressure of, say 60 lbs. to 65 lbs.

It is needless to enter into a long descriptive specification of these Engines, as the high position they have held for so many years has established their reputation for great durability and economy in the consumption of fuel.

SPECIFICATION OF EACH SIZE GIVEN UPON APPLICATION.

Below are a few particulars of the smaller sizes, but **Williams' Perran Foundry Co.** construct Engines on this system up to the largest size made.

PARTICULARS AND PRICE OF ENGINES ONLY.

Nominal Horse Power.	Diameter of Cylinder.	Length of Stroke.	Price at Works.			If fitted with expansion valves extra			If with Feed Water Heater extra.			Cost of Packing.			Approximate weight with Fly Wheel.
	inches.	inches.	£	s.	d.	£	s.	d.	£	s.	d	£	s.	d.	
4	6	10				9	0	0	9	0	0	2	0	0	
6	8	12				9	0	0	9	0	0	2	10	0	
8	9	14				9	0	0	9	0	0	3	0	0	
10	10	18				9	0	0	9	0	0	3	10	0	
12	12	20				14	0	0	14	0	0	4	0	0	
14	14	24							14	0	0	4	10	0	
16	15	24							18	0	0	5	0	0	
18	16	30							18	0	0	5	10	0	
20	18	30				Included in price of Engine.			18	0	0	6	0	0	
25	20	36							25	0	0	6	10	0	
30	22	42							25	0	0	7	10	0	

(For Boilers see next page.)

PERRANARWORTHAL, CORNWALL,
AND
1 & 2 GREAT WINCHESTER STREET BUILDINGS, LONDON, E.C.

Figure A.3 Catalog page for steam engines. Note proposed boiler pressure of 60 to 65 psig.

forward direction, and (3) the armature rotating in the reverse direction. He found that the heating could be increased or decreased depending on the direction of rotation. He measured a work input of 4.60 N · m/°C per kilogram of water. He then designed a series of experiments, including forcing water through holes in a piston, 4.25 N · m/(kg · °C); friction between two surfaces submerged in water or mercury, 4.25 N · m/(kg · °C); air pumped into a closed cylinder, 4.60 N · m/(kg · °C); air in the cylinder allowed to escape slowly, 4.38 N · m/(kg · °C); and air from one cylinder allowed to escape slowly into a second cylinder, with both cylinders submerged in the same tank (no net work done, and no change in temperature, since no energy crosses the system boundary). Joule published these results, and the mechanical theory of heat began to attract more attention. Joule later made even more careful measurements of the temperature change of an insulated canister of water when stirred by the measured work input from a paddle wheel stirrer. While on his honeymoon in Switzerland, he hoped to find a high enough waterfall to provide sufficient energy in the form of work that he could make still more careful measurements.

In the meantime, Mayer, who had conceived the mechanical theory of heat while a ship's physician in the East Indies, had tried to publish similar results based on data obtained by Black. Mayer had great difficulty getting his work into print in the scientific journals; his first submitted manuscript was not even acknowledged. Some of his later work (1842) was ridiculed. His despair was so great that he attempted suicide by jumping from a window, but he only broke both legs. Because attempted suicide was considered proof of insanity, he was placed in an asylum for some time. He was finally given some measure of recognition in later years as having been equal with Joule in establishing the mechanical theory of heat.

Another physician who had difficulty in publishing his work was Herman Ludwig von Helmholtz (1821 – 1894), self-taught in mathematics and physics. He gave the first clear analytical exposition of energy on a generalized basis in 1847 in a privately published paper that he also could not get accepted by professional journals.

In the meantime, Carnot's work had been expanded and clarified by Emile Clapeyron (1799 – 1864). During the later 1840s, many thermodynamicists, including the physicists William Thomson (1824 – 1907) (later Lord Kelvin), Rudolf Julius Emanuel Clausius (1822 – 1888), and the Scottish engineer William John Macquorn Rankine (1820 – 1872), struggled with the problem of how to reconcile Carnot's work, which was based on the caloric theory, with the experimental confirmation of the mechanical theory of heat being provided by Joule and Mayer. Because Joule's results depended on temperature difference measurements of the order of 0.01 °F, there was room for considerable skepticism over his results.

In 1848, Kelvin, a professor of natural philosophy at the University of Glasgow and then 24 years of age, had suggested an absolute temperature scale, based in part on Carnot's results. Kelvin was engaged in work on telegraphy in support of the laying of the Atlantic cable, but he found time to publish work in 1849 in which appeared the first use of the term *thermodynamic*, and the term *mechanical energy* was also introduced. In 1850, he finally abandoned the caloric theory entirely, and from 1852 to 1862 he worked with Joule in a series of experiments to measure the temperature change of a gas during a controlled expansion. This was done to attempt to disprove a supposed assumption of Mayer's that such a temperature change was always zero, which Joule and Thomson (Kelvin) believed had introduced an error into Mayer's value for the mechanical equivalent of heat. (Mayer had actually only assumed the zero value was valid for air in the range of conditions where air acts as an ideal gas, which it is.) The experimental work led to very important results, including determination of the so-called Joule-Thomson coefficient for real gases.

Clausius in the meantime realized that there were two distinct laws at work, the second law, as expounded by Carnot, and the first law, which Clausius first constructed with a logical theory in 1850. In this work, he defined the internal energy U and showed clearly the difference between the specific heats measured at constant volume and constant pressure. He also showed that this formulation agreed with experiment. Although both Clausius and Kelvin had been using the function Q_{rev}/T for some years, Clausius recognized the value of the function as a property and coined the word *entropy* to describe it. He also assigned it the symbol S. His statement of the first law as "*Die Energie der Welt ist konstant*" remains a concise and generally valid one.

Rankine, in applying the theory of thermodynamics to heat engines, in 1853 defined the thermodynamic efficiency of a heat engine and in 1854 showed the usefulness of *P-v* diagrams as related to work. He wrote the first thermodynamics textbook in 1859.

In 1862, the cycle used in modern gasoline-powered internal combustion engines was proposed in a French patent issued to Alphonse Beau de Rochas (1815 – 1893), although no engine is known to have been built as a result. The cycle was introduced in a practical engine by a young merchant in Cologne, Nikolaus August Otto (1832 – 1891), and his partner Eugen Langen (1833 – 1895) in 1876 and was demonstrated at the Paris Exposition of 1878. Otto's work was first applied to engines using illuminating gas. The successful engine followed a series of attempts which were ridiculed at the time, but which led to continual improvement and final success. Otto was unaware of Beau de Rochas' work and fought many legal battles to maintain rights of production of his engines, finally losing.

Captain John Ericsson (1803 – 1889) was a Swedish engineer who spent a long and productive career in the United States. He perfected the screw propeller for ships, replacing the paddle wheel common at that time, and he built the ship *Monitor* for the Union forces in the Civil War, which set the standard for iron-clad ships for the next 50 years. He had earlier invented the shell-and-tube heat exchanger for use as a condenser for marine engines. He, with others, became interested in air engines (Fig. A.4), again because of the number of catastrophic boiler explosions, particularly on ships where they resulted from a combination of higher-pressure boilers and careless operation. In 1850, he built and demonstrated a large hot-air engine for the 2200-ton ship *Ericsson.* The engine was a technical success, but took so much space from the cargo hold with its four 14-ft-diameter cylinders with 6-ft strokes that the ship probably could not have competed economically with steam-powered machinery. The *Ericsson* sank in a storm before conclusive tests could be performed. Ericsson later marketed smaller solar-powered and coal-fired hot-air engines, the latter with some success.

In 1873, George Bailey Brayton (1830–1892), born in East Greenwich, Vermont, introduced and in 1876 marketed an internal combustion engine that operated by injecting compressed air through a hot grating into the combustion chamber. Heavy oil or other liquid fuel was injected directly into the air before it entered the combustion chamber. Because there was no sudden explosion, the combustion occurred at nearly constant pressure. An earlier model had operated with gas as the fuel, but the flame had passed back through the grating into the fuel supply, making the engine dangerous to operate. The heavy-oil engine did not enjoy continued commercial success, but the thermodynamics of the engine have become the basis for modern gas-turbine engines.

Josiah Willard Gibbs (1839–1903) is often noted as the most brilliant but least recognized thermodynamicist in the United States. He received the first doctoral degree in engineering awarded in the United States (usually assumed to be in mechanical engineering since the subject was gearing). He developed the *T-s* diagram as a tool for analyzing heat transfer in thermodynamic systems and provided methods for analyzing thermodynamic equilibrium in the general sense. In 1878 he published work defining the phase rule, which made thermodynamics the basis for the field of physical chemistry. His later contributions in establishing the fundamentals of statistical thermodynamics fall outside the realm of classical thermodynamics, but are equally important.

Gottlieb Daimler (1834–1900) was superintendent of Otto's gas engine works in Deutz, Germany, and realized that the Otto engine must be made to operate on volatile liquid fuel if it was to become practical for use in transportation. In 1879, he obtained a patent for a multicylinder engine operating on a common crank, and in 1883–1884 he, along with his brilliant design engineer Wilhelm Maybach (1846–1929), produced the first commercially successful automotive engine.

Dr. Rudolph Christian Karl Diesel (1858–1913) was born in France but was sent to Germany by his parents as a child, where he later attended the *Technische Hochschule* in Munich as a student in machine design. He designed large steam engines and boilers, but he kept studying to find a replacement for the steam engine that would be more efficient than the 6

THE DIFFERENT USES FOR COMPRESSED AIR.—Drawn by W. Louis Sonntag, Jun.—[See Page 1202.]

1. Car using the Hardie Compressed-Air Motor on 125th Street, New York City.   2. The Mekarski Compressed-Air System used in Paris—recharging a Car *en route*.   3. Motor and Train in Paris (Mekarski System).   4. Compressed-Air Locomotive to be operated on the Elevated Railroad in New York City. 5. Switches moved by Compressed Air and controlled by Electricity in the Pennsylvania Railroad Yards.   6. The Use of Compressed Air for drilling, excavating, etc., on the Chicago Drainage Canal.

Figure A.4   Some proposed and actual uses of compressed air for transportation systems.

to 10 percent then available from the steam cycles. He developed his cycle of operation based on the use of the compression stroke to reach high temperatures, with the combustion process then occurring at constant temperature by metering of the rate of fuel injection, believing this to be the nearest practical approach to the Carnot cycle for internal combustion engines. He demonstrated a modification of his cycle in 1893, but his first attempt resulted in an explosion of the engine which came close to ending his life as well as his experiments. He continued to develop the concept, and finally he demonstrated a practical operating engine in 1897. Failing health, continuing criticism, and serious financial setbacks beset Diesel, and in 1913 he disappeared from a boat crossing the English Channel on a calm moonlit night.

All the engines developed to this point were reciprocating, using a piston in a cylinder to drive a flywheel to provide circular motion. As early as 1791, the Englishman John Barber had patented an engine with all the elements of a contemporary gas turbine, and further paper designs were made by others until practical turbines finally were developed for steam cycles almost simultaneously by Sir Charles A. Parsons (1854–1931) in England and Carl G. P. DeLaval (1845–1913) in Sweden during the 1880s.

Refrigeration and Thermodynamics

To follow the development of refrigeration, we must back up in time to the middle of the seventeenth century, when the Englishman Robert Boyle (1627–1691) observed the reduction of the boiling temperature of water when the pressure was reduced. Dr. William Cullen, a professor of medicine at the University of Glasgow, observed in 1755 that an isolated container of water underwent a drop in temperature during evaporation. In 1844, Thomas Masters of London patented an ice cream maker that used a mixture of ice and salt to lower the brine temperature. Charles E. Monroe of Cambridge, Massachusetts, in 1871 patented a food cooler based on the idea of evaporation of water from the porous lining of a refrigerator. However, the most notable of the entrepreneurs who used the methods of "natural" refrigeration was Frederic Tudor (1783–1864), who cut and stored natural ice (Fig. A.5). By 1804, he was making regular shipments of ice to the South from the North, and by 1834 he had extended his trade to the West Indies, South Africa, and finally to Europe.

Sir John Leslie (1766–1832), professor of mathematics at Edinburgh University, building on the observations by Boyle and Cullen, used sulfuric acid to absorb water vapor from a vessel containing water, and thus produce a vacuum in the closed container. The vacuum in turn caused the saturation temperature of the water producing the vapor to be low enough that ice could be formed. In 1810, 1-lb blocks were made by

ICE HOUSES ON THE HUDSON.

DOWN THE RIVER TO NEW YORK.

DISTRIBUTING TO THE WAGONS.

SUPPLYING AN OCEAN STEAMER.

WEIGHING A LOAD.

SMALL CUSTOMERS.

DELIVERING UP TOWN.

FILLING A CELLAR.

THE ICE INDUSTRY OF NEW YORK.—Drawn by F. Ray.—[See Page 565.]

Figure A.5 Use of natural ice for refrigeration.

this process. By 1881, Franz Windhausen was producing six 672-lb blocks per cycle in a commercial-scale machine in Germany. The sulfuric acid was regenerated for further use after each cycle by heating it with steam to drive off the absorbed water.

In 1858, a system based on the observation that ammonia could reach lower temperatures than water when boiled at the same pressure was marketed by Ferdinand P. E. Carre (1824–?). In the early 1930s, the Crosley "Icyball" system based on exactly Carre's cycle was sold widely in the rural United States. The gas refrigerator operates on the same cycle in a continuous rather than batch fashion.

In 1755, M. Hoell observed that compressed air leaving a pressurized air line cooled when it escaped from the line. In 1828, Trevithick proposed a refrigeration machine based on Hoell's observation, and in 1851 Dr. John Gorrie (1803–1855) was issued a U. S. patent for the first machine to operate successfully on the air compression–expansion cycle.

Jacob Perkins (1766–1849), an American living in London, observed that working fluids other than air could operate more efficiently, especially if they could be condensed easily after compression. He built the first practical vapor-compression machine, which was patented in 1834. The use of ammonia in such a machine was demonstrated successfully by David Boyle of Chicago, who developed the machine between 1869 and 1873 and operated a 1-ton/day ice-making plant in Jefferson, Texas, in 1873. Carl P. G. Linde (1842–1934) of Munich used an advanced cycle with much better mechanical detail which was developed to an experimental stage in 1873 and to commercial use in 1875 (Fig. A.6).

Summary

The work of this diverse group of scientists and engineers drawn from many nations follows Langley's picture referred to at the beginning of this section — a pack of hunting hounds questing after an elusive quarry. Classical thermodynamics presents a beautifully coherent theory that finally joins the many threads that were followed by the early investigators. However, the history of that theory is not a continuous line of advancement, as perhaps best indicated by the fact that formulation of what we now call the second law preceded the formulation of the first law by more than 25 years. The history is a tangled string filled with loops and knots.

For further information on the early history of thermodynamics, see Refs. 1 to 11.

EXAMINING SLAB OF ICE IN PROCESS OF FREEZING.

COMPLETELY FROZEN—WEIGHT, FOUR AND A HALF TONS.

READY FOR THE CIRCULAR SAWS.

SAWING INTO SMALL PIECES.

ARTIFICIAL ICE MANUFACTURE.—From Sketches by Horace Bradley.—[See Page 67.]

Figure A.6 *Early systems for large-scale ice production.*

References

1. R. Bruce Lindsay, *Julius Robert Mayer—Prophet of Energy*, Pergamon, Elmsford, N.Y., 1973.
2. R. Bruce Lindsay, *Energy, Historical Development of the Concept*, vol. 1, *Benchmark Papers on Energy,* Hutchinson & Ross, Stroudsburg, Pa., 1975.
3. R. Bruce Lindsay, *Applications of Energy: Nineteenth Century*, vol. 2, *Benchmark Papers on Energy,* Dowden, Hutchinson and Ross, Stroudsburg, Pa., 1976.
4. William Edgar Knowles Middleton, *A History of the Thermometer and Its Use in Meteorology*, Johns Hopkins, Baltimore, Md., 1966.
5. E. Mendoza, "A Sketch for a History of Early Thermodynamics," *Physics Today,* vol. 14, no. 2, February 1961, pp. 32–42.
6. Charles Singer, *A Short History of Science to the Nineteenth Century*, Oxford at the Clarendon Press, London, 1941.
7. Abbot Payson Usher, *A History of Mechanical Inventions*, rev. ed., Harvard University Press, Cambridge, Mass., 1954.
8. Carl J. Echkardt, *Men and Power*, privately published.
9. C. Lyle Cummins, Jr., *Internal Fire*, Carnot Press, Lake Osweago, Ore., 1976.
10. Arthur W. J. G. Orde-Hume, *Perpetual Motion, The History of an Obsession*, St. Martin's Press, New York, 1977.
11. P. W. Atkins, *The Second Law*, Scientific American Library, New York, 1984.

APPENDIX B

Conversion Factors

TABLE B.1 Fundamental Numerical Values

First Bohr electron radius	$a_0 = 0.5292 \times 10^{-10}$ m
Speed of light in a vacuum	$c_0 = 2.9979 \times 10^8$ m/s
Electronic charge	$e = 1.6022 \times 10^{-19}$ C
Gravitational acceleration	$g = 9.81$ m/s^2
Planck's constant	$h = 6.6262 \times 10^{-34}$ J $\cdot$ s
Boltzmann constant	$k = 1.3806 \times 10^{-23}$ J/K
Electron rest mass	$m_e = 9.1096 \times 10^{-31}$ kg
Classical electron radius	$r_0 = 2.8179 \times 10^{-15}$ m
Electron volt	1 eV $= 1.6022 \times 10^{-19}$ J
Temperature associated with 1 eV	1 eV$/k = 11{,}605$ K
Ionization potential of hydrogen atom	13.606 eV
Avogadro's number	$N_a = 6.023 \times 10^{26}$ particles/kmol

TABLE B.2 Values for the Universal Gas Constant

$\overline{R} = 8.31441$ kPa $\cdot$ m^3/(kmol $\cdot$ K)

$= 8.31441$ kJ/(kmol $\cdot$ K)

$= 0.0820568$ liter $\cdot$ atm/(gmol $\cdot$ K)

$= 0.0820568$ m^3 $\cdot$ atm/(kmol $\cdot$ K)

$= 1.98586$ Btu/(lbmol $\cdot$ °R)

$= 0.730235$ ft^3 $\cdot$ atm/(lbmol $\cdot$ °R)

$= 1545.3$ ft $\cdot$ lbf/(lbmol $\cdot$ °R)

$= 10.73$ psia $\cdot$ ft^3/(lbmol $\cdot$ °R)

TABLE B.3 Conversion Factors for Length

	Mile, mi	Kilometer, km	Meter, m	Foot, ft	Inch, in
1 mile =	1	1.609	1609	5280	6.336×10^4
1 kilometer =	0.6214	1	10^3	3.281×10^3	3.937×10^4
1 meter =	6.214×10^{-4}	10^{-3}	1	3.281	39.37
1 foot =	1.894×10^{-4}	3.048×10^{-4}	0.3048	1	12
1 inch =	1.578×10^{-5}	2.540×10^{-5}	2.540×10^{-2}	8.333×10^{-2}	1
1 centimeter =	6.214×10^{-6}	10^{-5}	10^{-2}	3.281×10^{-2}	0.3937
1 millimeter =	6.214×10^{-7}	10^{-6}	10^{-3}	3.281×10^{-3}	0.03937
1 micrometer =	6.214×10^{-10}	10^{-9}	10^{-6}	3.281×10^{-6}	3.937×10^{-5}
1 nanometer =	6.214×10^{-13}	10^{-12}	10^{-9}	3.281×10^{-9}	3.937×10^{-8}
1 angstrom =	6.214×10^{-14}	10^{-13}	10^{-10}	3.281×10^{-10}	3.937×10^{-9}

	Centimeter, cm	Millimeter, mm	Micrometer, μm	Nanometer, nm	Angstrom, Å
1 mile =	1.609×10^5	1.609×10^6	1.609×10^9	1.609×10^{12}	1.609×10^{13}
1 kilometer =	10^5	10^6	10^9	10^{12}	10^{13}
1 meter =	10^2	10^3	10^6	10^9	10^{10}
1 foot =	30.48	3.048×10^2	3.048×10^5	3.048×10^8	3.048×10^9
1 inch =	2.540	25.40	2.540×10^4	2.540×10^7	2.540×10^8
1 centimeter =	1	10	10^4	10^7	10^8
1 millimeter =	10^{-1}	1	10^3	10^6	10^7
1 micrometer =	10^{-4}	10^{-3}	1	10^3	10^4
1 nanometer =	10^{-7}	10^{-6}	10^{-3}	1	10
1 angstrom =	10^{-8}	10^{-7}	10^{-4}	10^{-1}	1

TABLE B.4 Useful Conversion Factors*

Area
  $1 \text{ m}^2 = 10.764 \text{ ft}^2$
  $1 \text{ ft}^2 = 0.092903 \text{ m}^2$
  $1 \text{ in}^2 = 6.4516 \times 10^{-4} \text{ m}^2$

Mass
  $1 \text{ kg} = 2.2046 \text{ lbm}$
  $1 \text{ lbm} = 0.45359 \text{ kg}$

Volume
  $1 \text{ m}^3 = 35.315 \text{ ft}^3$
  $1 \text{ ft}^3 = 0.028317 \text{ m}^3$
  $1 \text{ gal} = 0.13368 \text{ ft}^3$
  $1 \text{ liter} = 1000.0 \text{ cm}^3$

Density
  $1 \text{ lbm/ft}^3 = 16.019 \text{ kg/m}^3$
  $1 \text{ kg/m}^3 = 0.062428 \text{ lbm/ft}^3$

Force
  $1 \text{ N} = 1 \text{ kg} \cdot \text{m/s}^2 = 0.22481 \text{ lbf}$
  $1 \text{ lbf} = 4.4482 \text{ N}$

TABLE B.4 *Useful Conversion Factors (Continued)*

Pressure

$1\ Pa = 1\ N/m^2 = 1 \times 10^{-5}\ bar = 1.4504 \times 10^{-4}\ psia = 9.8692 \times 10^{-6}\ atm$
$= 0.020886\ lbf/ft^2$
$1\ bar = 10^5\ Pa = 0.98692\ atm = 14.504\ psia = 2088.6\ lbf/ft^2$
$1\ lbf/in^2\ (psia) = 144\ lbf/ft^2 = 6894.8\ Pa = 6.8948 \times 10^{-2}\ bar = 0.068046\ atm$
$1\ atm = 101.325\ kPa = 14.696\ psia = 1.0133\ bar = 2116.2\ lbf/ft^2$

Energy

$1\ J = 1\ N \cdot m = 1\ kg \cdot m^2/s^2$
$1\ kJ = 1\ kW \cdot s = 0.94783\ Btu = 0.23885\ kcal = 737.56\ ft \cdot lbf$
$1\ Btu = 1.0550\ kJ = 0.25200\ kcal = 778.16\ ft \cdot lbf$
$1\ kcal = 4.1868\ kJ = 3.9684\ Btu = 3088.0\ ft \cdot lbf$
$1\ kWh = 3.60 \times 10^3\ kJ = 2655.2 \times 10^3\ ft \cdot lbf = 3412.2\ Btu = 859.86\ kcal/h$
$1\ ft \cdot lbf = 1.2851 \times 10^{-3}\ Btu = 1.3558 \times 10^{-3}\ kJ$

Energy Rate (Power)

$1\ W = 1\ J/s$
$1\ W = 3.4122\ Btu/h = 0.85987\ kcal/h = 1.34102 \times 10^{-3}\ hp = 0.73756\ ft \cdot lbf/s$
$1\ Btu/h = 0.29307\ W = 0.25200\ kcal/h = 3.9300 \times 10^{-4}\ hp = 0.21616\ ft \cdot lbf/s$
$1\ kcal/h = 1.1630\ W = 3.9683\ Btu/h = 1.5595 \times 10^{-3}\ hp = 0.85778\ ft \cdot lbf/s$
$1\ horsepower\ (hp) = 550\ ft \cdot lbf/s = 2544.5\ Btu/h = 745.70\ W$
$1\ ft \cdot lbf/s = 4.6262\ Btu/h = 1.3558\ W = 1.8182 \times 10^{-3}\ hp$
$1\ ton\ (cooling\ capacity) = 12,000\ Btu/h = 3.5168\ kW$

Specific Energy, Specific Enthalpy

$1\ kJ/kg = 0.42992\ Btu/lbm = 0.23885\ kcal/kg = 334.55\ ft \cdot lbf/lbm$
$1\ Btu/lbm = 2.3260\ kJ/kg = 0.55556\ kcal/kg = 778.16\ ft \cdot lbf/lbm$
$1\ kcal/kg = 4.1868\ kJ/kg = 1.8000\ Btu/lbm = 1400.7\ ft \cdot lbf/lbm$
$1\ ft \cdot lbf/lbm = 2.9891 \times 10^{-3}\ kJ/kg = 1.2851 \times 10^{-3}\ Btu/lbm$
$= 7.1394 \times 10^{-4}\ kcal/kg$

Energy Rate per Unit Area

$1\ W/m^2 = 0.31700\ Btu/(h \cdot ft^2) = 0.85986\ kcal/(h \cdot m^2)$
$1\ Btu/(h \cdot ft^2) = 3.1546\ W/m^2 = 2.7125\ kcal/(h \cdot m^2)$
$1\ kcal/(h \cdot m^2) = 1.1630\ W/m^2 = 0.36867\ Btu/(h \cdot ft^2)$

Entropy

$1\ kJ/K = 0.52657\ Btu/°R = 0.23885\ kcal/K$
$1\ Btu/°R = 1.8991\ kJ/K = 0.45359\ kcal/K$
$1\ kcal/K = 4.1868\ kJ/K = 2.2047\ Btu/°R$

Specific Entropy, Specific Heat, Gas Constant

$1\ kJ/(kg \cdot K) = 0.23885\ Btu/(lbm \cdot °R) = 0.23885\ kcal/(kg \cdot K)$
$1\ Btu/(lbm \cdot °R) = 4.1868\ kJ/(kg \cdot K) = 1.0000\ kcal/(kg \cdot K)$
$1\ kcal/(kg \cdot K) = 4.1868\ kJ/(kg \cdot K) = 1.0000\ Btu/(lbm \cdot °R)$

Temperature

$T,\ K = \frac{5}{9}T,°R = \frac{5}{9}(T,°F + 459.67) = T,°C + 273.15$
$T,°R = \frac{9}{5}T,\ K = \frac{9}{5}(T,°C + 273.15) = T,°F + 459.67$
$T,°F = \frac{9}{5}T,°C + 32$
$T,°C = \frac{5}{9}(T,°F - 32)$

* All energy conversions based on International Steam Table values.

APPENDIX C

Thermodynamic Properties in Dimensionless Form or for Both SI and USCS Units

TABLE C.1 Triple-Point Properties for Common Substances

Substance	T, K	T, °R	P, kPa	P, psia
Ammonia (NH_3)	195.4	351.7	6.18	0.896
Carbon dioxide (CO_2)	216.6	389.9	516.6	74.93
Helium 4 (λ point)	2.17	3.91	5.07	0.735
Hydrogen (H_2)	13.84	24.91	7.09	1.03
Nitrogen (N_2)	63.18	113.7	12.56	1.822
Oxygen (O_2)	54.36	97.85	0.152	0.0220
Water (H_2O)	273.16	491.69	0.6113	0.08866

SI values from William Z. Black and James G. Hartley, *Thermodynamics,* Harper & Row, New York, 1985 (used with permission).

TABLE C.2 Gas Constants and Zero-Pressure Specific Heats for Various Ideal Gases at 300 K (540°R)

Gas	Molecular Weight	Gas Constant R kJ/(kg · K)	ft · lbf/(lbm · °R)	c_P^o kJ/(kg · K)	Btu/(lbm · °R)	c_v^o kJ/(kg · K)	Btu/(lbm · °R)	$k = \dfrac{c_P^o}{c_v^o}$
Air	28.97	0.28700	53.34	1.0052	0.2401	0.7180	0.1716	1.400
Argon (Ar)	39.944	0.20813	38.68	0.5207	0.1244	0.3124	0.0746	1.667
Butane (C_4H_{10})	58.120	0.14304	26.58	1.7164	0.415	1.5734	0.381	1.09
Carbon dioxide (CO_2)	44.01	0.18892	35.10	0.8464	0.2021	0.6573	0.1569	1.288
Carbon monoxide (CO)	28.01	0.29683	55.16	1.0411	0.2487	0.7441	0.1777	1.399
Ethane (C_2H_6)	30.07	0.27650	51.38	1.7662	0.427	1.4897	0.361	1.183
Ethylene (C_2H_4)	28.052	0.29637	55.07	1.5482	0.411	1.2518	0.340	1.208
Helium (He)	4.003	2.07703	386.0	5.1926	1.25	3.1156	0.753	1.667
Hydrogen (H_2)	2.016	4.12418	766.4	14.3193	3.4198	10.1919	2.4340	1.405
Methane (CH_4)	16.04	0.51835	96.35	2.2537	0.532	1.7354	0.403	1.32
Neon (Ne)	20.183	0.41195	76.55	1.0299	0.246	0.6179	0.1477	1.667
Nitrogen (N_2)	28.016	0.29680	55.15	1.0404	0.2485	0.7434	0.1776	1.400
Octane (C_8H_{18})	114.14	0.07279	13.53	1.7113	0.409	1.6385	0.392	1.044
Oxygen (O_2)	32.000	0.25983	48.28	0.9190	0.2195	0.6590	0.1574	1.395
Propane (C_3H_8)	44.094	0.18855	35.04	1.6794	0.407	1.4909	0.362	1.124
Water (H_2O)	18.016	0.46152	85.76	1.8649	0.4454	1.4031	0.3351	1.329

From Gordon Van Wylen and Richard Sonntag, *Fundamentals of Classical Thermodynamics,* 2d ed., New York, 1976 (with permission) and *GASPROPS,* Wiley Professional Software, Wiley, New York, 1984.

TABLE C.3   Specific Heats and Specific Volumes of Various Liquids and Solids

Substance	T (K)	T (°R)	c_P kJ/(kg · K)	c_P Btu/(lbm · °R)	v m³/kg	v ft³/lbm
Liquid						
Ammonia						
(sat)	253	455	4.52	1.08	15.04×10^{-4}	24.06×10^{-3}
(sat)	323	581	5.10	1.22	17.76×10^{-4}	28.41×10^{-3}
Benzene						
(1 atm)	288	518	1.80	0.430	11.4×10^{-4}	18.3×10^{-3}
(1 atm)	338	608	1.92	0.459	—	—
Bismuth						
(1 atm)	698	1256	0.144	0.0344	1.02×10^{-4}	1.63×10^{-3}
(1 atm)	1033	1859	0.164	0.0392	—	—
Ethyl alcohol						
(1 atm)	298	536	2.43	0.580	12.7×10^{-4}	20.3×10^{-3}
Glycerin						
(1 atm)	283	509	2.32	0.554	7.94×10^{-4}	12.7×10^{-3}
(1 atm)	323	581	2.58	0.616	—	—
Mercury						
(1 atm)	283	509	0.138	0.0330	0.738×10^{-4}	1.18×10^{-3}
(1 atm)	588	1058	0.134	0.0320	—	—
Refrigerant 12						
(sat)	233	419	0.883	0.211	6.59×10^{-4}	10.66×10^{-3}
(sat)	253	455	0.908	0.217	6.85×10^{-4}	10.97×10^{-3}
(sat)	323	581	1.02	0.244	8.26×10^{-4}	13.20×10^{-3}
Sodium						
(1 atm)	368	662	1.38	0.330	—	—
(1 atm)	813	1463	1.26	0.301	—	—
Water						
(1 atm)	273	492	4.213	1.006	10.00×10^{-4}	16.02×10^{-3}
(1 atm)	298	536	4.177	0.998	10.03×10^{-4}	16.06×10^{-3}
(1 atm)	323	581.	4.178	0.998	10.12×10^{-4}	16.21×10^{-3}
(1 atm)	373	671	4.213	1.006	10.43×10^{-4}	16.70×10^{-3}
Solid						
Aluminum	23	41	0.0163	0.00389	—	—
	73	131	0.318	0.0760	—	—
	173	311	0.699	0.167	—	—
	273	492	0.870	0.208	3.7×10^{-4}	5.9×10^{-3}
	373	671	0.941	0.225	—	—
	573	1031	1.04	0.248	—	—
Carbon						
(diamond)	298	536	0.519	0.124	2.86×10^{-4}	4.58×10^{-3}
(graphite)	298	536	0.711	0.170	5.12×10^{-4}	8.20×10^{-3}
Chromium	298	536	0.448	0.107	1.40×10^{-4}	2.24×10^{-3}
Copper	50	90	0.0967	0.0231	—	—
	100	180	0.252	0.0602	—	—
	173	311	0.328	0.0783	—	—
	223	401	0.361	0.0862	—	—
	273	492	0.381	0.0910	—	—
	300	540	0.385	0.0920	1.12×10^{-4}	1.79×10^{-3}
	373	671	0.393	0.0939	—	—
	473	851	0.403	0.0963	—	—
Gold	298	536	0.129	0.0308	0.518×10^{-4}	0.830×10^{-3}

TABLE C.3 Continued

Substance	T K	T °R	c_P kJ/(kg · K)	c_P Btu/(lbm · °R)	v m³/kg	v ft³/lbm
Solid						
Ice	73	131	0.678	0.162	—	—
	133	239	1.096	0.2618	—	—
	213	383	1.640	0.3917	—	—
	262	472	2.033	0.4856	—	—
	270.8	487.4	1.682	0.4017	11.1×10^{-4}	17.8×10^{-3}
Iron	293	527	0.448	0.107	0.127×10^{-4}	0.203×10^{-3}
Lead	3	5	0.0033	0.00079	—	—
	14	25	0.0305	0.00728	—	—
	173	311	0.118	0.0282	—	—
	273	492	0.124	0.0296	0.882×10^{-4}	1.41×10^{-3}
	373	671	0.134	0.0320	—	—
	573	1031	0.149	0.0356	—	—
Nickel	298	536	0.444	0.1060	1.12×10^{-4}	1.79×10^{-3}
Silver	293	527	0.233	0.0557	0.952×10^{-4}	1.52×10^{-3}
	773	1391	0.243	0.0580	—	—
Sodium	298	536	1.226	0.2928	10.3×10^{-4}	16.5×10^{-3}
Tungsten	298	536	0.134	0.0320	0.518×10^{-4}	0.830×10^{-3}
Zinc	298	536	0.385	0.0920	1.40×10^{-4}	2.24×10^{-3}

Values of c_P from Kenneth G. Wark, *Thermodynamics,* 4th ed., McGraw-Hill, New York, 1983 (used with permission). Values of specific volume from various sources.

TABLE C.4 Critical Constants

Gas	Formula	Molecular Weight	Critical Temperature T_{cr} K	Critical Temperature T_{cr} °R	Critical Pressure P_{cr} atm	Critical Pressure P_{cr} MPa	Critical Volume v_{cr} m³/kmol	Critical Volume v_{cr} ft³/lbmol	Acentric Factor ω	Critical Compressibility Factor z_{cr}
Acetic acid	$C_2H_4O_2$	60.05	594.8	1070.6	57.2	5.79	0.1711	2.740	0.454	0.200
Acetone	C_3H_6O	58.08	508.7	915.6	46.6	4.72	0.213	3.41	—	0.238
Acetonitrile	CH_3CN	41.05	547.9	986.2	47.7	4.83	0.173	2.77	—	0.184
Acetylene	C_2H_2	26.02	309	557	61.6	6.28	0.1130	1.810	0.184	0.274
Air	—	28.97	132.5	238.5	37.2	3.77	—	—	—	—
Ammonia	NH_3	17.03	405.5	729.8	111.3	11.27	0.0724	1.16	0.250	0.242
Argon	Ar	39.944	151	272	48.0	4.86	0.0752	1.20	−0.004	0.291
Benzene	C_6H_6	78.11	562	1012	48.6	4.92	0.2604	4.171	0.212	0.274
Bromine	Br_2	159.832	584	1052	102	10.33	0.1355	2.171	0.132	0.288
n-Butane	C_4H_{10}	58.120	425.2	765.2	37.5	3.80	0.2547	4.080	0.193	0.274
Carbon dioxide	CO_2	44.01	304.2	547.5	72.9	7.38	0.0943	1.51	0.225	0.274
Carbon disulfide	CS_2	76.13	552	994	78	7.90	0.170	2.8	—	0.293

TABLE C.4 **Continued**

Gas	Formula	Molecular Weight	Critical Temperature T_{cr}		Critical Pressure P_{cr}		Critical Volume v_{cr}		Acentric Factor ω	Critical Compressibility Factor z_{cr}
			K	°R	atm	MPa	m³/kmol	ft³/lbmol		
Carbon monoxide	CO	28.01	133	240	34.5	3.49	0.0930	1.49	0.049	0.294
Carbon tetrachloride	CCl₄	153.84	556.4	1001.5	45.0	4.56	0.2760	4.421	0.194	0.272
Chlorine	Cl₂	70.914	417	751	76.1	7.71	0.1243	1.991	0.073	0.276
Chloroform	CHCl₃	119.39	536.6	965.8	54.0	5.47	0.2404	3.851	0.216	0.294
Cyclohexane	C₆H₁₂	84.16	553	996	40	4.05	0.308	4.93	—	0.272
Decane	C₁₀H₂₂	142.17	619.4	1115	21.24	2.152	0.6113	9.792	0.490	0.255
Deuterium (normal)	D₂	4.00	38.4	69.1	16.4	1.66	—	—	−0.130	—
Dichlorodifluoromethane (refrigerant 12)	CCl₂F₂	120.92	384.7	692.4	39.6	4.01	0.2179	3.490	0.176	0.273
Dichlorofluoromethane (refrigerant 21)	CHCl₂F	102.93	451.7	813.0	51.0	5.17	0.197	3.16	—	0.272
Diethyl ether	C₄H₁₀O	74.08	466.0	838.8	35.5	3.60	0.2822	4.520	0.281	0.262
Dioxane	C₄H₈O₂	88.10	585	1054	50.7	5.14	0.240	3.9	—	0.253
Ethane	C₂H₆	30.068	305.5	549.8	48.2	4.88	0.1480	2.371	0.098	0.284
Ethyl acetate	C₄H₈O₂	88.10	523.3	941.9	37.8	3.83	0.286	4.58	—	0.252
Ethyl alcohol	C₂H₅OH	46.07	516.0	929.0	63.0	6.38	0.1673	2.680	0.635	0.248
Ethyl chloride	C₂H₅Cl	64.50	460.4	828.7	52	5.27	0.1961	3.141	0.190	0.270
Ethylene	C₂H₄	28.052	282.4	508.3	50.5	5.12	0.1243	1.991	0.085	0.268
Ethylene oxide	C₂H₄O	44.05	468	843	71.0	7.19	0.138	2.2	—	0.255
Ethyl methyl ketone	C₄H₈O	72.10	533	960	46.6	4.72	0.213	3.41	—	0.227
Helium	He⁴	4.003	5.3	9.5	2.26	0.229	0.0578	0.926	−0.387	0.300
Helium 3	He³	3.00	3.34	6.01	1.15	0.116	—	—	—	—
Heptane	C₇H₁₆	100.12	540.17	972.31	27.00	2.735	0.4108	6.580	0.351	0.250
n-Hexane	C₆H₁₄	86.172	507.9	914.2	29.9	3.03	0.3678	5.892	0.296	0.264
Hydrazine	N₂H₄	32.05	653	1176	145	14.7	—	—	—	—
Hydrogen (normal)	H₂	2.016	33.3	59.9	12.8	1.30	0.0650	1.04	−0.220	0.304
Hydrogen chloride	HCl	36.47	324.6	584.3	81.5	8.26	0.048	0.76	—	0.147
Hydrogen cyanide	HCN	27.03	456.7	822.0	53.2	5.07	0.1349	2.161	0.407	0.197
Hydrogen sulfide	H₂S	34.08	373.6	672.5	88.9	9.01	0.977	1.57	—	0.283
Isobutane	C₄H₁₀	58.12	408.1	734.6	36.0	3.65	0.263	4.21	—	0.283
Isopropyl alcohol	C₃H₇OH	60.09	508.8	915.8	53	5.4	0.219	3.51	—	0.278
Krypton	Kr	83.7	209.4	376.9	54.3	5.50	0.0922	1.48	−0.002	0.291
Methane	CH₄	16.042	191.1	343.9	45.8	4.64	0.0990	1.59	0.008	0.289
Methyl alcohol	CH₃OH	32.04	513.2	923.7	78.5	7.95	0.1180	1.890	0.559	0.220
Methyl chloride	CH₃Cl	50.49	416.3	749.3	65.9	6.68	0.1430	2.291	0.156	0.276
Neon	Ne	20.183	44.5	80.1	26.9	2.72	0.0417	0.668	0.0	0.307
Nitric oxide	NO	30.01	179	323	65	6.58	0.0578	0.926	0.607	0.251
Nitrogen	N₂	28.016	126.2	227.1	33.5	3.39	0.0901	1.44	0.040	0.291
Nitrogen peroxide	NO₂	46.01	431	776	100	10.1	0.082	1.3	—	0.232
Nitrous oxide	N₂O	44.02	309.7	557.4	71.7	7.26	0.0962	1.54	0.160	0.272
Nonane	C₉H₂₀	128.16	596	1072	22.86	2.316	0.5532	8.861	0.444	0.258
Octane	C₈H₁₈	114.14	569.4	1024.9	24.66	2.498	0.4901	7.851	0.394	0.259
Oxygen	O₂	32.00	154.8	278.6	50.1	5.08	0.0780	1.25	0.021	0.308
Pentane	C₅H₁₂	72.09	470.3	846.6	33.04	3.347	0.3103	4.971	0.251	0.269
Propane	C₃H₈	44.094	370.0	665.9	42.0	4.25	0.1998	3.200	0.152	0.277
Propene	C₃H₆	42.078	365.0	656.9	45.6	4.62	0.1811	2.901	0.148	0.276
n-Propyl alcohol	C₃H₇OH	60.09	537	967	50.2	5.09	0.220	3.52	—	0.251
Propyne	C₃H₄	40.062	401	722	52.8	5.35	—	—	0.218	—

TABLE C.4 Continued

Gas	Formula	Molecular Weight	Critical Temperature T_{cr}		Critical Pressure P_{cr}		Critical Volume v_{cr}		Acentric Factor ω	Critical Compressibility Factor z_{cr}
			K	°R	atm	MPa	m³/kmol	ft³/lbmol		
Sulfur dioxide	SO_2	64.06	430.7	775.2	77.8	7.88	0.1218	1.951	0.251	0.268
Sulfur trioxide	SO_3	80.06	491.5	884.7	83.6	8.47	0.1268	2.031	0.410	0.262
Toluene	C_7H_8	92.06	593.8	1068.8	41.6	4.21	0.3153	5.051	0.257	0.276
Trichlorofluoromethane (refrigerant 11)	CCl_3F	137.38	471.2	848.1	43.2	4.38	0.2479	3.971	0.188	0.277
Water	H_2O	18.016	647.29	1165.1	218.0	22.09	0.0568	0.900	0.344	0.230
Xenon	Xe	131.3	289.75	521.55	58.0	5.88	0.1188	1.902	0.002	0.289

Data adapted with permission from K. A. Kobe and R. E. Lynn, Jr., "The Critical Properties of Elements and Compounds," *Chemical Reviews,* vol. 52, 1953, pp. 117–236 (Table 22), copyright 1953 American Chemical Society, and from Ernest G. Cravalho and Joseph L. Smith, *Engineering Thermodynamics,* Pitman, Marshfield, Mass., 1981 (used with permission).

TABLE C.5 Empirical Constants for the Benedict-Webb-Rubin Equation

Gas	Formula	A_0		B_0		C_0	
		N · m⁴/kg²	lbf · ft⁴/lbm²	m³/kg	ft³/lbm	N · m⁴ · K²/kg²	lbf · ft⁴ · °R²/lbm²
Methane	CH_4	731.195	3918.49	2.65735×10^{-3}	4.25667×10^{-2}	0.889635×10^7	1.54469×10^8
Ethylene	C_2H_4	430.550	2307.33	1.98649×10^{-3}	3.18205×10^{-2}	1.69071×10^7	2.93562×10^8
Ethane	C_2H_6	466.269	2498.75	2.08914×10^{-3}	3.34648×10^{-2}	2.01509×10^7	3.49885×10^8
Propylene	C_3H_6	350.217	1876.82	2.02308×10^{-3}	3.24066×10^{-2}	2.51642×10^7	4.36932×10^8
Propane	C_3H_8	358.575	1921.61	2.20855×10^{-3}	3.53776×10^{-2}	2.65194×10^7	4.60462×10^8
i-Butane	C_4H_{10}	307.308	1646.87	2.36826×10^{-3}	3.79359×10^{-2}	2.55256×10^7	4.43207×10^8
i-Butylene	C_4H_8	288.571	1546.46	2.06958×10^{-3}	3.31515×10^{-2}	2.98871×10^7	5.18937×10^8
n-Butane	C_4H_{10}	302.865	1623.06	2.14127×10^{-3}	3.43000×10^{-2}	2.98168×10^7	5.17716×10^8
i-Pentane	C_5H_{12}	249.391	1336.49	2.22006×10^{-3}	3.55620×10^{-2}	3.40357×10^7	5.90970×10^8
n-Pentane	C_5H_{12}	237.376	1272.10	2.17426×10^{-3}	3.48283×10^{-2}	4.13424×10^7	7.17838×10^8
n-Hexane	C_6H_{14}	197.242	1057.02	2.06498×10^{-3}	3.30778×10^{-2}	4.53487×10^7	7.87400×10^8
n-Heptane	C_7H_{16}	177.041	948.77	1.98756×10^{-3}	3.18377×10^{-2}	4.79543×10^7	8.32642×10^8

Gas	Formula	a		b		c	
		N · m⁷/kg³	lbf · ft⁷/lbm³	m⁶/kg²	ft⁶/lbm²	N · m⁷ · K²/kg³	lbf · ft⁷ · °R²/lbm³
Methane	CH_4	1.21466	104.270	1.31523×10^{-5}	3.37476×10^{-3}	0.62577×10^5	1.74047×10^7
Ethylene	C_2H_4	1.19119	102.256	1.09451×10^{-5}	2.80842×10^{-3}	0.97139×10^5	2.70175×10^7
Ethane	C_2H_6	1.28892	110.645	1.23191×10^{-5}	3.16097×10^{-3}	1.22361×10^5	3.40325×10^7
Propylene	C_3H_6	1.05482	90.5491	1.05806×10^{-5}	2.71489×10^{-3}	1.39829×10^5	3.88910×10^7
Propane	C_3H_8	1.12224	96.3367	1.15892×10^{-5}	2.97369×10^{-3}	1.52759×10^5	4.24872×10^7
i-Butane	C_4H_{10}	1.00195	86.0106	1.25806×10^{-5}	3.22807×10^{-3}	1.47891×10^5	4.11333×10^7
i-Butylene	C_4H_8	0.97316	83.539	1.10774×10^{-5}	2.84236×10^{-3}	1.58056×10^5	4.39605×10^7
n-Butane	C_4H_{10}	0.97334	83.555	1.18582×10^{-5}	3.04271×10^{-3}	1.63610×10^5	4.55052×10^7
i-Pentane	C_5H_{12}	1.01546	87.1704	1.28545×10^{-5}	3.29835×10^{-3}	1.87887×10^5	5.22574×10^7
n-Pentane	C_5H_{12}	1.10159	94.5640	1.28545×10^{-5}	3.29835×10^{-3}	2.22807×10^5	6.19698×10^7
n-Hexane	C_6H_{14}	1.12913	96.9282	1.47181×10^{-5}	3.77654×10^{-3}	2.40013×10^5	6.67554×10^7
n-Heptane	C_7H_{16}	1.04602	89.7937	1.51575×10^{-5}	3.88928×10^{-3}	2.49275×10^5	6.93314×10^7

TABLE C.5 Continued

Gas	Formula	α		γ	
		m^9/kg^3	ft^9/lbm^3	m^6/kg^2	ft^6/lbm^2
Methane	CH_4	30.1853×10^{-9}	12.4068×10^{-4}	23.3469×10^{-6}	5.99061×10^{-3}
Ethylene	C_2H_4	8.08173×10^{-9}	3.32175×10^{-4}	11.7469×10^{-6}	3.01415×10^{-3}
Ethane	C_2H_6	8.97220×10^{-9}	3.68775×10^{-4}	13.0701×10^{-6}	3.35367×10^{-3}
Propylene	C_3H_6	6.13014×10^{-9}	2.51961×10^{-4}	10.3453×10^{-6}	2.65451×10^{-3}
Propane	C_3H_8	7.09776×10^{-9}	2.91732×10^{-4}	11.3317×10^{-6}	2.90761×10^{-3}
i-Butane	C_4H_{10}	5.48279×10^{-9}	2.25353×10^{-4}	10.0799×10^{-6}	2.58641×10^{-3}
i-Butylene	C_4H_8	5.16963×10^{-9}	2.12482×10^{-4}	9.41616×10^{-6}	2.41610×10^{-3}
n-Butane	C_4H_{10}	5.62184×10^{-9}	2.31069×10^{-4}	10.0799×10^{-6}	2.58641×10^{-3}
i-Pentane	C_5H_{12}	4.53682×10^{-9}	1.86472×10^{-4}	8.90805×10^{-6}	2.28573×10^{-3}
n-Pentane	C_5H_{12}	4.83038×10^{-9}	1.98538×10^{-4}	9.13893×10^{-6}	2.34497×10^{-3}
n-Hexane	C_6H_{14}	4.40244×10^{-9}	1.80949×10^{-4}	8.99353×10^{-6}	2.30766×10^{-3}
n-Heptane	C_7H_{16}	4.33982×10^{-9}	1.78375×10^{-4}	8.97754×10^{-6}	2.30356×10^{-3}

For SI units, R is Pa $\cdot$ m³/(kg $\cdot$ K) or J/(kg $\cdot$ K), P in Pa, v in m³/kg, and T in K.

For USCS units, R is in ft $\cdot$ lbf/(lbm $\cdot$ °R), P in lbf/ft², v in ft³/lbm, and T in °R.

SI data, as corrected, from Ernest Cravalho and Joseph L. Smith, *Engineering Thermodynamics,* Pitman, Marshfield, Mass., 1981 (used with permission).

TABLE C.6 Lee-Kesler Values for the Compressibility Factor

(a) Values for $z^{(0)}$

T_r	P_r						
	0.010	0.050	0.100	0.200	0.400	0.600	0.800
0.30	0.0029	0.0145	0.0290	0.0579	0.1158	0.1737	0.2315
0.35	0.0026	0.0130	0.0261	0.0522	0.1043	0.1564	0.2084
0.40	0.0024	0.0119	0.0239	0.0477	0.0953	0.1429	0.1904
0.45	0.0022	0.0110	0.0221	0.0442	0.0882	0.1322	0.1762
0.50	0.0021	0.0103	0.0207	0.0413	0.0825	0.1236	0.1647
0.55	0.9804	0.0098	0.0195	0.0390	0.0778	0.1166	0.1553
0.60	0.9849	0.0093	0.0186	0.0371	0.0741	0.1109	0.1476
0.65	0.9881	0.9377	0.0178	0.0356	0.0710	0.1063	0.1415
0.70	0.9904	0.9504	0.8958	0.0344	0.0687	0.1027	0.1366
0.75	0.9922	0.9598	0.9165	0.0336	0.0670	0.1001	0.1330
0.80	0.9935	0.9669	0.9319	0.8539	0.0661	0.0985	0.1307
0.85	0.9946	0.9725	0.9436	0.8810	0.0661	0.0983	0.1301
0.90	0.9954	0.9768	0.9528	0.9015	0.7800	0.1006	0.1321
0.93	0.9959	0.9790	0.9573	0.9115	0.8059	0.6635	0.1359
0.95	0.9961	0.9803	0.9600	0.9174	0.8206	0.6967	0.1410
0.97	0.9963	0.9815	0.9625	0.9227	0.8338	0.7240	0.5580
0.98	0.9965	0.9821	0.9637	0.9253	0.8398	0.7360	0.5887
0.99	0.9966	0.9826	0.9648	0.9277	0.8455	0.7471	0.6138
1.00	0.9967	0.9832	0.9659	0.9300	0.8509	0.7574	0.6353
1.01	0.9968	0.9837	0.9669	0.9322	0.8561	0.7671	0.6542
1.02	0.9969	0.9842	0.9679	0.9343	0.8610	0.7761	0.6710
1.05	0.9971	0.9855	0.9707	0.9401	0.8743	0.8002	0.7130
1.10	0.9975	0.9874	0.9747	0.9485	0.8930	0.8323	0.7649
1.15	0.9978	0.9891	0.9780	0.9554	0.9081	0.8576	0.8032
1.20	0.9981	0.9904	0.9808	0.9611	0.9205	0.8779	0.8330
1.30	0.9985	0.9926	0.9852	0.9702	0.9396	0.9083	0.8764
1.40	0.9988	0.9942	0.9884	0.9768	0.9534	0.9298	0.9062
1.50	0.9991	0.9954	0.9909	0.9818	0.9636	0.9456	0.9278
1.60	0.9993	0.9964	0.9928	0.9856	0.9714	0.9575	0.9439
1.70	0.9994	0.9971	0.9943	0.9886	0.9775	0.9667	0.9563
1.80	0.9995	0.9977	0.9955	0.9910	0.9823	0.9739	0.9659
1.90	0.9996	0.9982	0.9964	0.9929	0.9861	0.9796	0.9735
2.00	0.9997	0.9986	0.9972	0.9944	0.9892	0.9842	0.9796
2.20	0.9998	0.9992	0.9983	0.9967	0.9937	0.9910	0.9886
2.40	0.9999	0.9996	0.9991	0.9983	0.9969	0.9957	0.9948
2.60	1.0000	0.9998	0.9997	0.9994	0.9991	0.9990	0.9990
2.80	1.0000	1.0000	1.0001	1.0002	1.0007	1.0013	1.0021
3.00	1.0000	1.0002	1.0004	1.0008	1.0018	1.0030	1.0043
3.50	1.0001	1.0004	1.0008	1.0017	1.0035	1.0055	1.0075
4.00	1.0001	1.0005	1.0010	1.0021	1.0043	1.0066	1.0090

TABLE C.6 Continued

(a) Values for $z^{(0)}$							
P_r							
1.000	1.200	1.500	2.000	3.000	5.000	7.000	10.000
0.2892	0.3470	0.4335	0.5775	0.8648	1.4366	2.0048	2.8507
0.2604	0.3123	0.3901	0.5195	0.7775	1.2902	1.7987	2.5539
0.2379	0.2853	0.3563	0.4744	0.7095	1.1758	1.6373	2.3211
0.2200	0.2638	0.3294	0.4384	0.6551	1.0841	1.5077	2.1338
0.2056	0.2465	0.3077	0.4092	0.6110	1.0094	1.4017	1.9801
0.1939	0.2323	0.2899	0.3853	0.5747	0.9475	1.3137	1.8520
0.1842	0.2207	0.2753	0.3657	0.5446	0.8959	1.2398	1.7440
0.1765	0.2113	0.2634	0.3495	0.5197	0.8526	1.1773	1.6519
0.1703	0.2038	0.2538	0.3364	0.4991	0.8161	1.1241	1.5729
0.1656	0.1981	0.2464	0.3260	0.4823	0.7854	1.0787	1.5047
0.1626	0.1942	0.2411	0.3182	0.4690	0.7598	1.0400	1.4456
0.1614	0.1924	0.2382	0.3132	0.4591	0.7388	1.0071	1.3943
0.1630	0.1935	0.2383	0.3114	0.4527	0.7220	0.9793	1.3496
0.1664	0.1963	0.2405	0.3122	0.4507	0.7138	0.9648	1.3257
0.1705	0.1998	0.2432	0.3138	0.4501	0.7092	0.9561	1.3108
0.1779	0.2055	0.2474	0.3164	0.4504	0.7052	0.9480	1.2968
0.1844	0.2097	0.2503	0.3182	0.4508	0.7035	0.9442	1.2901
0.1959	0.2154	0.2538	0.3204	0.4514	0.7018	0.9406	1.2835
0.2901	0.2237	0.2583	0.3229	0.4522	0.7004	0.9372	1.2772
0.4648	0.2370	0.2640	0.3260	0.4533	0.6991	0.9339	1.2710
0.5146	0.2629	0.2715	0.3297	0.4547	0.6980	0.9307	1.2650
0.6026	0.4437	0.3131	0.3452	0.4604	0.6956	0.9222	1.2481
0.6880	0.5984	0.4580	0.3953	0.4770	0.6950	0.9110	1.2232
0.7443	0.6803	0.5798	0.4760	0.5042	0.6987	0.9033	1.2021
0.7858	0.7363	0.6605	0.5605	0.5425	0.7069	0.8990	1.1844
0.8438	0.8111	0.7624	0.6908	0.6344	0.7358	0.8998	1.1580
0.8827	0.8595	0.8256	0.7753	0.7202	0.7761	0.9112	1.1419
0.9103	0.8933	0.8689	0.8328	0.7887	0.8200	0.9297	1.1339
0.9308	0.9180	0.9000	0.8738	0.8410	0.8617	0.9518	1.1320
0.9463	0.9367	0.9234	0.9043	0.8809	0.8984	0.9745	1.1343
0.9583	0.9511	0.9413	0.9275	0.9118	0.9297	0.9961	1.1391
0.9678	0.9624	0.9552	0.9456	0.9359	0.9557	1.0157	1.1452
0.9754	0.9715	0.9664	0.9599	0.9550	0.9772	1.0328	1.1516
0.9865	0.9847	0.9826	0.9806	0.9827	1.0094	1.0600	1.1635
0.9941	0.9936	0.9935	0.9945	1.0011	1.0313	1.0793	1.1728
0.9993	0.9998	1.0010	1.0040	1.0137	1.0463	1.0926	1.1792
1.0031	1.0042	1.0063	1.0106	1.0223	1.0565	1.1016	1.1830
1.0057	1.0074	1.0101	1.0153	1.0284	1.0635	1.1075	1.1848
1.0097	1.0120	1.0156	1.0221	1.0368	1.0723	1.1138	1.1834
1.0115	1.0140	1.0179	1.0249	1.0401	1.0747	1.1136	1.1773

TABLE C.6 Continued

				P_r			

(b) Values for $z^{(1)}$

T_r	0.010	0.050	0.100	0.200	0.400	0.600	0.800
0.30	−0.0008	−0.0040	−0.0081	−0.0161	−0.0323	−0.0484	−0.0645
0.35	−0.0009	−0.0046	−0.0093	−0.0185	−0.0370	−0.0554	−0.0738
0.40	−0.0010	−0.0048	−0.0095	−0.0190	−0.0380	−0.0570	−0.0758
0.45	−0.0009	−0.0047	−0.0094	−0.0187	−0.0374	−0.0560	−0.0745
0.50	−0.0009	−0.0045	−0.0090	−0.0181	−0.0360	−0.0539	−0.0716
0.55	−0.0314	−0.0043	−0.0086	−0.0172	−0.0343	−0.0513	−0.0682
0.60	−0.0205	−0.0041	−0.0082	−0.0164	−0.0326	−0.0487	−0.0646
0.65	−0.0137	−0.0772	−0.0078	−0.0156	−0.0309	−0.0461	−0.0611
0.70	−0.0093	−0.0507	−0.1161	−0.0148	−0.0294	−0.0438	−0.0579
0.75	−0.0064	−0.0339	−0.0744	−0.0143	−0.0282	−0.0417	−0.0550
0.80	−0.0044	−0.0228	−0.0487	−0.1160	−0.0272	−0.0401	−0.0526
0.85	−0.0029	−0.0152	−0.0319	−0.0715	−0.0268	−0.0391	−0.0509
0.90	−0.0019	−0.0099	−0.0205	−0.0442	−0.1118	−0.0396	−0.0503
0.93	−0.0015	−0.0075	−0.0154	−0.0326	−0.0763	−0.1662	−0.0514
0.95	−0.0012	−0.0062	−0.0126	−0.0262	−0.0589	−0.1110	−0.0540
0.97	−0.0010	−0.0050	−0.0101	−0.0208	−0.0450	−0.0770	−0.1647
0.98	−0.0009	−0.0044	−0.0090	−0.0184	−0.0390	−0.0641	−0.1100
0.99	−0.0008	−0.0039	−0.0079	−0.0161	−0.0335	−0.0531	−0.0796
1.00	−0.0007	−0.0034	−0.0069	−0.0140	−0.0285	−0.0435	−0.0588
1.01	−0.0006	−0.0030	−0.0060	−0.0120	−0.0240	−0.0351	−0.0429
1.02	−0.0005	−0.0026	−0.0051	−0.0102	−0.0198	−0.0277	−0.0303
1.05	−0.0003	−0.0015	−0.0029	−0.0054	−0.0092	−0.0097	−0.0032
1.10	−0.0000	0.0000	0.0001	0.0007	0.0038	0.0106	0.0236
1.15	0.0002	0.0011	0.0023	0.0052	0.0127	0.0237	0.0396
1.20	0.0004	0.0019	0.0039	0.0084	0.0190	0.0326	0.0499
1.30	0.0006	0.0030	0.0061	0.0125	0.0267	0.0429	0.0612
1.40	0.0007	0.0036	0.0072	0.0147	0.0306	0.0477	0.0661
1.50	0.0008	0.0039	0.0078	0.0158	0.0323	0.0497	0.0677
1.60	0.0008	0.0040	0.0080	0.0162	0.0330	0.0501	0.0677
1.70	0.0008	0.0040	0.0081	0.0163	0.0329	0.0497	0.0667
1.80	0.0008	0.0040	0.0081	0.0162	0.0325	0.0488	0.0652
1.90	0.0008	0.0040	0.0079	0.0159	0.0318	0.0477	0.0635
2.00	0.0008	0.0039	0.0078	0.0155	0.0310	0.0464	0.0617
2.20	0.0007	0.0037	0.0074	0.0147	0.0293	0.0437	0.0579
2.40	0.0007	0.0035	0.0070	0.0139	0.0276	0.0411	0.0544
2.60	0.0007	0.0033	0.0066	0.0131	0.0260	0.0387	0.0512
2.80	0.0006	0.0031	0.0062	0.0124	0.0245	0.0365	0.0483
3.00	0.0006	0.0029	0.0059	0.0117	0.0232	0.0345	0.0456
3.50	0.0005	0.0026	0.0052	0.0103	0.0204	0.0303	0.0401
4.00	0.0005	0.0023	0.0046	0.0091	0.0182	0.0270	0.0357

TABLE C.6 Continued

(b) Values for $z^{(1)}$

			P_r				
1.000	1.200	1.500	2.000	3.000	5.000	7.000	10.000
−0.0806	−0.0966	−0.1207	−0.1608	−0.2407	−0.3996	−0.5572	−0.7915
−0.0921	−0.1105	−0.1379	−0.1834	−0.2738	−0.4523	−0.6279	−0.8863
−0.0946	−0.1134	−0.1414	−0.1879	−0.2799	−0.4603	−0.6365	−0.8936
−0.0929	−0.1113	−0.1387	−0.1840	−0.2734	−0.4475	−0.6162	−0.8606
−0.0893	−0.1069	−0.1330	−0.1762	−0.2611	−0.4253	−0.5831	−0.8099
−0.0849	−0.1015	−0.1263	−0.1669	−0.2465	−0.3991	−0.5446	−0.7521
−0.0803	−0.0960	−0.1192	−0.1572	−0.2312	−0.3718	−0.5047	−0.6928
−0.0759	−0.0906	−0.1122	−0.1476	−0.2160	−0.3447	−0.4653	−0.6346
−0.0718	−0.0855	−0.1057	−0.1385	−0.2013	−0.3184	−0.4270	−0.5785
−0.0681	−0.0808	−0.0996	−0.1298	−0.1872	−0.2929	−0.3901	−0.5250
−0.0648	−0.0767	−0.0940	−0.1217	−0.1736	−0.2682	−0.3545	−0.4740
−0.0622	−0.0731	−0.0888	−0.1138	−0.1602	−0.2439	−0.3201	−0.4254
−0.0604	−0.0701	−0.0840	−0.1059	−0.1463	−0.2195	−0.2862	−0.3788
−0.0602	−0.0687	−0.0810	−0.1007	−0.1374	−0.2045	−0.2661	−0.3516
−0.0607	−0.0678	−0.0788	−0.0967	−0.1310	−0.1943	−0.2526	−0.3339
−0.0623	−0.0669	−0.0759	−0.0921	−0.1240	−0.1837	−0.2391	−0.3163
−0.0641	−0.0661	−0.0740	−0.0893	−0.1202	−0.1783	−0.2322	−0.3075
−0.0680	−0.0646	−0.0715	−0.0861	−0.1162	−0.1728	−0.2254	−0.2989
−0.0879	−0.0609	−0.0678	−0.0824	−0.1118	−0.1672	−0.2185	−0.2902
−0.0223	−0.0473	−0.0621	0.0778	−0.1072	−0.1615	−0.2116	−0.2816
−0.0062	0.0227	−0.0524	−0.0722	−0.1021	−0.1556	−0.2047	−0.2731
0.0220	0.1059	0.0451	−0.0432	−0.0838	−0.1370	−0.1835	−0.2476
0.0476	0.0897	0.1630	0.0698	−0.0373	−0.1021	−0.1469	−0.2056
0.0625	0.0943	0.1548	0.1667	0.0332	−0.0611	−0.1084	−0.1642
0.0719	0.0991	0.1477	0.1990	0.1095	−0.0141	−0.0678	−0.1231
0.0819	0.1048	0.1420	0.1991	0.2079	0.0875	0.0176	−0.0423
0.0857	0.1063	0.1383	0.1894	0.2397	0.1737	0.1008	0.0350
0.0864	0.1055	0.1345	0.1806	0.2433	0.2309	0.1717	0.1058
0.0855	0.1035	0.1303	0.1729	0.2381	0.2631	0.2255	0.1673
0.0838	0.1008	0.1259	0.1658	0.2305	0.2788	0.2628	0.2179
0.0816	0.0978	0.1216	0.1593	0.2224	0.2846	0.2871	0.2576
0.0792	0.0947	0.1173	0.1532	0.2144	0.2848	0.3017	0.2876
0.0767	0.0916	0.1133	0.1476	0.2069	0.2819	0.3097	0.3096
0.0719	0.0857	0.1057	0.1374	0.1932	0.2720	0.3135	0.3355
0.0675	0.0803	0.0989	0.1285	0.1812	0.2602	0.3089	0.3459
0.0634	0.0754	0.0929	0.1207	0.1706	0.2484	0.3009	0.3475
0.0598	0.0711	0.0876	0.1138	0.1613	0.2372	0.2915	0.3443
0.0565	0.0672	0.0828	0.1076	0.1529	0.2268	0.2817	0.3385
0.0497	0.0591	0.0728	0.0949	0.1356	0.2042	0.2584	0.3194
0.0443	0.0527	0.0651	0.0849	0.1219	0.1857	0.2378	0.2994

From Ernest G. Cravalho and Joseph L. Smith, Jr., *Engineering Thermodynamics,* Pitman, Marshfield, Mass., 1981 (used with permission), as calculated from data in Byung Ik Lee and Michael G. Kesler, "A Generalized Thermodynamic Correlation Based on Three-Parameter Corresponding States," *American Institute of Chemical Engineering Journal,* vol. 21, no. 3, May 1975, pp. 510–527.

TABLE C.7 **Lee-Kesler Values for Residual Enthalpy**

(a) Values of $[(h_{ideal} - h_{actual})/(RT_{cr})]^{(0)}$

T_r	P_r						
	0.010	0.050	0.100	0.200	0.400	0.600	0.800
0.30	6.045	6.043	6.040	6.034	6.022	6.011	5.999
0.35	5.906	5.904	5.901	5.895	5.882	5.870	5.858
0.40	5.763	5.761	5.757	5.751	5.738	5.726	5.713
0.45	5.615	5.612	5.609	5.603	5.590	5.577	5.564
0.50	5.465	5.463	4.459	5.453	5.440	5.427	5.414
0.55	0.032	5.312	5.309	5.303	5.290	5.278	5.265
0.60	0.027	5.162	5.159	5.153	5.141	5.129	5.116
0.65	0.023	0.118	5.008	5.002	4.991	4.980	4.968
0.70	0.020	0.101	0.213	4.848	4.838	4.828	4.818
0.75	0.017	0.088	0.183	4.687	4.679	4.672	4.664
0.80	0.015	0.078	0.160	0.345	4.507	4.504	4.499
0.85	0.014	0.069	0.141	0.300	4.309	4.313	4.316
0.90	0.012	0.062	0.126	0.264	0.596	4.074	4.094
0.93	0.011	0.058	0.118	0.246	0.545	0.960	3.920
0.95	0.011	0.056	0.113	0.235	0.516	0.885	3.763
0.97	0.011	0.054	0.109	0.225	0.490	0.824	1.356
0.98	0.010	0.053	0.107	0.221	0.478	0.797	1.273
0.99	0.010	0.052	0.105	0.216	0.466	0.773	1.206
1.00	0.010	0.051	0.103	0.212	0.455	0.750	1.151
1.01	0.010	0.050	0.101	0.208	0.445	0.728	1.102
1.02	0.010	0.049	0.099	0.203	0.434	0.708	1.060
1.05	0.009	0.046	0.094	0.192	0.407	0.654	0.955
1.10	0.008	0.042	0.086	0.175	0.367	0.581	0.827
1.15	0.008	0.039	0.079	0.160	0.334	0.523	0.732
1.20	0.007	0.036	0.073	0.148	0.305	0.474	0.657
1.30	0.006	0.031	0.063	0.127	0.259	0.399	0.545
1.40	0.005	0.027	0.055	0.110	0.224	0.341	0.463
1.50	0.005	0.024	0.048	0.097	0.196	0.297	0.400
1.60	0.004	0.021	0.043	0.086	0.173	0.261	0.350
1.70	0.004	0.019	0.038	0.076	0.153	0.231	0.309
1.80	0.003	0.017	0.034	0.068	0.137	0.206	0.275
1.90	0.003	0.015	0.031	0.062	0.123	0.185	0.246
2.00	0.003	0.014	0.028	0.056	0.111	0.167	0.222
2.20	0.002	0.012	0.023	0.046	0.092	0.137	0.182
2.40	0.002	0.010	0.019	0.038	0.076	0.114	0.150
2.60	0.002	0.008	0.016	0.032	0.064	0.095	0.125
2.80	0.001	0.007	0.014	0.027	0.054	0.080	0.105
3.00	0.001	0.006	0.011	0.023	0.045	0.067	0.088
3.50	0.001	0.004	0.007	0.015	0.029	0.043	0.056
4.00	0.000	0.002	0.005	0.009	0.017	0.026	0.033

TABLE C.7   Continued

(a) Values of $[(h_{ideal} - h_{actual})/(RT_{cr})]^{(0)}$							
			P_r				
1.000	**1.200**	**1.500**	**2.000**	**3.000**	**5.000**	**7.000**	**10.000**
5.987	5.975	5.957	5.927	5.868	5.748	5.628	5.446
5.845	5.833	5.814	5.783	5.721	5.595	5.469	5.278
5.700	5.687	5.668	5.636	5.572	5.442	5.311	5.113
5.551	5.538	5.519	5.486	5.421	5.288	5.154	4.950
5.401	5.388	5.369	5.336	5.270	5.135	4.999	4.791
5.252	5.239	5.220	5.187	5.121	4.986	4.849	4.638
5.104	5.091	5.073	5.041	4.976	4.842	4.704	4.492
4.956	4.945	4.927	4.896	4.833	4.702	4.565	4.353
4.808	4.797	4.781	4.752	4.693	4.566	4.432	4.221
4.655	4.646	4.632	4.607	4.554	4.434	4.303	4.095
4.494	4.488	4.478	4.459	4.413	4.303	4.178	3.974
4.316	4.316	4.312	4.302	4.269	4.173	4.056	3.857
4.108	4.118	4.127	4.132	4.119	4.043	3.935	3.744
3.953	3.976	4.000	4.020	4.024	3.963	3.863	3.678
3.825	3.865	3.904	3.940	3.958	3.910	3.815	3.634
3.658	3.732	3.796	3.853	3.890	3.856	3.767	3.591
3.544	3.652	3.736	3.806	3.854	3.829	3.743	3.569
3.376	3.558	3.670	3.758	3.818	3.801	3.719	3.548
2.584	3.441	3.598	3.706	3.782	3.774	3.695	3.526
1.796	3.283	3.516	3.652	3.744	3.746	3.671	3.505
1.627	3.039	3.442	3.595	3.705	3.718	3.647	3.484
1.359	2.034	3.030	3.398	3.583	3.632	3.575	3.420
1.120	1.487	2.203	2.965	3.353	3.484	3.453	3.315
0.968	1.239	1.719	2.479	3.091	3.329	3.329	3.211
0.857	1.076	1.443	2.079	2.807	3.166	3.202	3.107
0.698	0.860	1.116	1.560	2.274	2.825	2.942	2.899
0.588	0.716	0.915	1.253	1.857	2.486	2.679	2.692
0.505	0.611	0.774	1.046	1.549	2.175	2.421	2.486
0.440	0.531	0.667	0.894	1.318	1.904	2.177	2.285
0.387	0.466	0.583	0.777	1.139	1.672	1.953	2.091
0.344	0.413	0.515	0.683	0.996	1.476	1.751	1.908
0.307	0.368	0.458	0.606	0.880	1.309	1.571	1.736
0.276	0.330	0.411	0.541	0.782	1.167	1.411	1.577
0.226	0.269	0.334	0.437	0.629	0.937	1.143	1.295
0.187	0.222	0.275	0.359	0.513	0.761	0.929	1.058
0.155	0.185	0.228	0.297	0.422	0.621	0.756	0.858
0.130	0.154	0.190	0.246	0.348	0.508	0.614	0.689
0.109	0.129	0.159	0.205	0.288	0.415	0.495	0.545
0.069	0.081	0.099	0.127	0.174	0.239	0.270	0.264
0.041	0.048	0.058	0.072	0.095	0.116	0.110	0.061

TABLE C.7 **Continued**

				(b) Values of $[(h_{ideal} - h_{actual})/(RT_{cr})]^{(1)}$			
				P_r			
T_r	0.010	0.050	0.100	0.200	0.400	0.600	0.800
0.30	11.098	11.096	11.095	11.091	11.083	11.076	11.069
0.35	10.656	10.655	10.654	10.653	10.650	10.646	10.643
0.40	10.121	10.121	10.121	10.120	10.121	10.121	10.121
0.45	9.515	9.515	9.515	9.517	9.519	9.521	9.523
0.50	8.868	8.869	8.870	8.872	8.876	8.880	8.884
0.55	0.080	8.211	8.212	8.215	8.221	8.226	8.232
0.60	0.059	7.568	7.570	7.573	7.579	7.585	7.591
0.65	0.045	0.247	6.949	6.952	6.959	6.966	6.973
0.70	0.034	0.185	0.415	6.360	6.367	6.373	6.381
0.75	0.027	0.142	0.306	5.796	5.802	5.809	5.816
0.80	0.021	0.110	0.234	0.542	5.266	5.271	5.278
0.85	0.017	0.087	0.182	0.401	4.753	4.754	4.758
0.90	0.014	0.070	0.144	0.308	0.751	4.254	4.248
0.93	0.012	0.061	0.126	0.265	0.612	1.236	3.942
0.95	0.011	0.056	0.115	0.241	0.542	0.994	3.737
0.97	0.010	0.052	0.105	0.219	0.483	0.837	1.616
0.98	0.010	0.050	0.101	0.209	0.457	0.776	1.324
0.99	0.009	0.048	0.097	0.200	0.433	0.722	1.154
1.00	0.009	0.046	0.093	0.191	0.410	0.675	1.034
1.01	0.009	0.044	0.089	0.183	0.389	0.632	0.940
1.02	0.008	0.042	0.085	0.175	0.370	0.594	0.863
1.05	0.007	0.037	0.075	0.153	0.318	0.498	0.691
1.10	0.006	0.030	0.061	0.123	0.251	0.381	0.507
1.15	0.005	0.025	0.050	0.099	0.199	0.296	0.385
1.20	0.004	0.020	0.040	0.080	0.158	0.232	0.297
1.30	0.003	0.013	0.026	0.052	0.100	0.142	0.177
1.40	0.002	0.008	0.016	0.032	0.060	0.083	0.100
1.50	0.001	0.005	0.009	0.018	0.032	0.042	0.048
1.60	0.000	0.002	0.004	0.007	0.012	0.013	0.011
1.70	0.000	0.000	0.000	-0.000	-0.003	-0.009	-0.017
1.80	-0.000	-0.001	-0.003	-0.006	-0.015	-0.025	-0.037
1.90	-0.001	-0.003	-0.005	-0.011	-0.023	-0.037	-0.053
2.00	-0.001	-0.003	-0.007	-0.015	-0.030	-0.047	-0.065
2.20	-0.001	-0.005	-0.010	-0.020	-0.040	-0.062	-0.083
2.40	-0.001	-0.006	-0.012	-0.023	-0.047	-0.071	-0.095
2.60	-0.001	-0.006	-0.013	-0.026	-0.052	-0.078	-0.104
2.80	-0.001	-0.007	-0.014	-0.028	-0.055	-0.082	-0.110
3.00	-0.001	-0.007	-0.014	-0.029	-0.058	-0.086	-0.114
3.50	-0.002	-0.008	-0.016	-0.031	-0.062	-0.092	-0.122
4.00	-0.002	-0.008	-0.016	-0.032	-0.064	-0.096	-0.127

TABLE C.7 Continued

From Ernest G. Cravalho and Joseph L. Smith, Jr., *Engineering Thermodynamics,* Pitman, Marshfield, Mass., 1981 (used with permission), as calculated from data in Byung Ik Lee and Michael G. Kesler, "A Generalized Thermodynamic Correlation Based on Three-Parameter Corresponding States," *American Institute of Chemical Engineering Journal,* vol. 21, no. 3, May 1975, pp. 510–527.

(b) Values of $[(h_{\text{ideal}} - h_{\text{actual}})/(RT_{\text{cr}})]^{(1)}$

P_r							
1.000	1.200	1.500	2.000	3.000	5.000	7.000	10.000
11.062	11.055	11.044	11.027	10.992	10.935	10.872	10.781
10.640	10.637	10.632	10.624	10.609	10.581	10.554	10.529
10.121	10.121	10.121	10.122	10.123	10.128	10.135	10.150
9.525	9.527	9.531	9.537	9.549	9.576	9.611	9.663
8.888	8.892	8.899	8.909	8.932	8.978	9.030	9.111
8.238	8.243	8.252	8.267	8.298	8.360	8.425	8.531
7.596	7.603	7.614	7.632	7.669	7.745	7.824	7.950
6.980	6.987	6.997	7.017	7.059	7.147	7.239	7.381
6.388	6.395	6.407	6.429	6.475	6.574	6.677	6.837
5.824	5.832	5.845	5.868	5.918	6.027	6.142	6.318
5.285	5.293	5.306	5.330	5.385	5.506	5.632	5.824
4.763	4.771	4.784	4.810	4.872	5.008	5.149	5.358
4.249	4.255	4.268	4.298	4.371	4.530	4.688	4.916
3.934	3.937	3.951	3.987	4.073	4.251	4.422	4.662
3.712	3.713	3.730	3.773	3.873	4.068	4.248	4.497
3.470	3.467	3.492	3.551	3.670	3.885	4.077	4.336
3.332	3.327	3.363	3.434	3.568	3.795	3.992	4.257
3.164	3.164	3.223	3.313	3.464	3.705	3.909	4.178
2.471	2.952	3.065	3.186	3.358	3.615	3.825	4.100
1.375	2.595	2.880	3.051	3.251	3.525	3.742	4.023
1.180	1.723	2.650	2.906	3.142	3.435	3.661	3.947
0.877	0.878	1.496	2.381	2.800	3.167	3.418	3.722
0.617	0.673	0.617	1.261	2.167	2.720	3.023	3.362
0.459	0.503	0.487	0.604	1.497	2.275	2.641	3.019
0.349	0.381	0.381	0.361	0.934	1.840	2.273	2.692
0.203	0.218	0.218	0.178	0.300	1.066	1.592	2.086
0.111	0.115	0.108	0.070	0.044	0.504	1.012	1.547
0.049	0.046	0.032	−0.008	−0.078	0.142	0.556	1.080
0.005	−0.004	−0.023	−0.065	−0.151	−0.082	0.217	0.689
−0.027	−0.040	−0.063	−0.109	−0.202	−0.223	−0.028	0.369
−0.051	−0.067	−0.094	−0.143	−0.241	−0.317	−0.203	0.112
−0.070	−0.088	−0.117	−0.169	−0.271	−0.381	−0.330	−0.092
−0.085	−0.105	−0.136	−0.190	−0.295	−0.428	−0.424	−0.255
−0.106	−0.128	−0.163	−0.221	−0.331	−0.493	−0.551	−0.489
−0.120	−0.144	−0.181	−0.242	−0.356	−0.535	−0.631	−0.645
−0.130	−0.156	−0.194	−0.257	−0.376	−0.567	−0.687	−0.754
−0.137	−0.164	−0.204	−0.269	−0.391	−0.591	−0.729	−0.836
−0.142	−0.170	−0.211	−0.278	−0.403	−0.611	−0.763	−0.899
−0.152	−0.181	−0.224	−0.294	−0.425	−0.650	−0.827	−1.015
−0.158	−0.188	−0.233	−0.306	−0.442	−0.680	−0.874	−1.097

TABLE C.8   Lee-Kesler Values for Residual Entropy

(a) Values of $[(s_{ideal} - s_{actual})/R]^{(0)}$

T_r	P_r						
	0.010	0.050	0.100	0.200	0.400	0.600	0.800
0.30	11.614	10.008	9.319	8.635	7.961	7.574	7.304
0.35	11.185	9.579	8.890	8.205	7.529	7.140	6.869
0.40	10.802	9.196	8.506	7.821	7.144	6.755	6.483
0.45	10.453	8.847	8.157	7.472	6.794	6.404	6.132
0.50	10.137	8.531	7.841	7.156	6.479	6.089	5.816
0.55	0.038	8.245	7.555	6.870	6.193	5.803	5.531
0.60	0.029	7.983	7.294	6.610	5.933	5.544	5.273
0.65	0.023	0.122	7.052	6.368	5.694	5.306	5.036
0.70	0.018	0.096	0.206	6.140	5.467	5.082	4.814
0.75	0.015	0.078	0.164	5.917	5.248	4.866	4.600
0.80	0.013	0.064	0.134	0.294	5.026	4.649	4.388
0.85	0.011	0.054	0.111	0.239	4.785	4.418	4.166
0.90	0.009	0.046	0.094	0.199	0.463	4.145	3.912
0.93	0.008	0.042	0.085	0.179	0.408	0.750	3.723
0.95	0.008	0.039	0.080	0.168	0.377	0.671	3.556
0.97	0.007	0.037	0.075	0.157	0.350	0.607	1.056
0.98	0.007	0.036	0.073	0.153	0.337	0.580	0.971
0.99	0.007	0.035	0.071	0.148	0.326	0.555	0.903
1.00	0.007	0.034	0.069	0.144	0.315	0.532	0.847
1.01	0.007	0.033	0.067	0.139	0.304	0.510	0.799
1.02	0.006	0.032	0.065	0.135	0.294	0.491	0.757
1.05	0.006	0.030	0.060	0.124	0.267	0.439	0.656
1.10	0.005	0.026	0.053	0.108	0.230	0.371	0.537
1.15	0.005	0.023	0.047	0.096	0.201	0.319	0.452
1.20	0.004	0.021	0.042	0.085	0.177	0.277	0.389
1.30	0.003	0.017	0.033	0.068	0.140	0.217	0.298
1.40	0.003	0.014	0.027	0.056	0.114	0.174	0.237
1.50	0.002	0.011	0.023	0.046	0.094	0.143	0.194
1.60	0.002	0.010	0.019	0.039	0.079	0.120	0.162
1.70	0.002	0.008	0.017	0.033	0.067	0.102	0.137
1.80	0.001	0.007	0.014	0.029	0.058	0.088	0.117
1.90	0.001	0.006	0.013	0.025	0.051	0.076	0.102
2.00	0.001	0.006	0.011	0.022	0.044	0.067	0.089
2.20	0.001	0.004	0.009	0.018	0.035	0.053	0.070
2.40	0.001	0.004	0.007	0.014	0.028	0.042	0.056
2.60	0.001	0.003	0.006	0.012	0.023	0.035	0.046
2.80	0.000	0.002	0.005	0.010	0.020	0.029	0.039
3.00	0.000	0.002	0.004	0.008	0.017	0.025	0.033
3.50	0.000	0.001	0.003	0.006	0.012	0.017	0.023
4.00	0.000	0.001	0.002	0.004	0.009	0.013	0.017

TABLE C.8 Continued

(a) Values of $[(s_{ideal} - s_{actual})/R]^{(0)}$

			P_r				
1.000	**1.200**	**1.500**	**2.000**	**3.000**	**5.000**	**7.000**	**10.000**
7.099	6.935	6.740	6.497	6.182	5.847	5.683	5.578
6.663	6.497	6.299	6.052	5.728	5.376	5.194	5.060
6.275	6.109	5.909	5.660	5.330	4.967	4.772	4.619
5.924	5.757	5.557	5.306	4.974	4.603	4.401	4.234
5.608	5.441	5.240	4.989	4.656	4.282	4.074	3.899
5.324	5.157	4.956	4.706	4.373	3.998	3.788	3.607
5.066	4.900	4.700	4.451	4.120	3.747	3.537	3.353
4.830	4.665	4.467	4.220	3.892	3.523	3.315	3.131
4.610	4.446	4.250	4.007	3.684	3.322	3.117	2.935
4.399	4.238	4.045	3.807	3.491	3.138	2.939	2.761
4.191	4.034	3.846	3.615	3.310	2.970	2.777	2.605
3.976	3.825	3.646	3.425	3.135	2.812	2.629	2.463
3.738	3.599	3.434	3.231	2.964	2.663	2.491	2.334
3.569	3.444	3.295	3.108	2.860	2.577	2.412	2.262
3.433	3.326	3.193	3.023	2.790	2.520	2.362	2.215
3.259	3.188	3.081	2.932	2.719	2.463	2.312	2.170
3.142	3.106	3.019	2.884	2.682	2.436	2.287	2.148
2.972	3.010	2.953	2.835	2.646	2.408	2.263	2.126
2.178	2.893	2.879	2.784	2.609	2.380	2.239	2.105
1.391	2.736	2.798	2.730	2.571	2.352	2.215	2.083
1.225	2.495	2.706	2.673	2.533	2.325	2.191	2.062
0.965	1.523	2.328	2.483	2.415	2.242	2.121	2.001
0.742	1.012	1.557	2.081	2.202	2.104	2.007	1.903
0.607	0.790	1.126	1.649	1.968	1.966	1.897	1.810
0.512	0.651	0.890	1.308	1.727	1.827	1.789	1.722
0.385	0.478	0.628	0.891	1.299	1.554	1.581	1.556
0.303	0.372	0.478	0.663	0.990	1.303	1.386	1.402
0.246	0.299	0.381	0.520	0.777	1.088	1.208	1.260
0.204	0.247	0.312	0.421	0.628	0.913	1.050	1.130
0.172	0.208	0.261	0.350	0.519	0.773	0.915	1.013
0.147	0.177	0.222	0.296	0.438	0.661	0.799	0.908
0.127	0.153	0.191	0.255	0.375	0.570	0.702	0.815
0.111	0.134	0.167	0.221	0.325	0.497	0.620	0.733
0.087	0.105	0.130	0.172	0.251	0.388	0.492	0.599
0.070	0.084	0.104	0.138	0.201	0.311	0.399	0.496
0.058	0.069	0.086	0.113	0.164	0.255	0.329	0.416
0.048	0.058	0.072	0.094	0.137	0.213	0.277	0.353
0.041	0.049	0.061	0.080	0.116	0.181	0.236	0.303
0.029	0.034	0.042	0.056	0.081	0.126	0.166	0.216
0.021	0.025	0.031	0.041	0.059	0.093	0.123	0.162

TABLE C.8   Continued

(b) Values of $[(s_{ideal} - s_{actual})/R]^{(1)}$

T_r	P_r						
	0.010	0.050	0.100	0.200	0.400	0.600	0.800
0.30	16.782	16.774	16.764	16.744	16.705	16.665	16.626
0.35	15.413	15.408	15.401	15.387	15.359	15.333	15.305
0.40	13.990	13.986	13.981	13.972	13.953	13.934	13.915
0.45	12.564	12.561	12.558	12.551	12.537	12.523	12.509
0.50	11.202	11.200	11.197	11.192	11.182	11.172	11.162
0.55	0.115	9.948	9.946	9.942	9.935	9.928	9.921
0.60	0.078	8.828	8.826	8.823	8.817	8.811	8.806
0.65	0.055	0.309	7.832	7.829	7.824	7.819	7.815
0.70	0.040	0.216	0.491	6.951	6.945	6.941	6.937
0.75	0.029	0.156	0.340	6.173	6.167	6.162	6.158
0.80	0.022	0.116	0.246	0.578	5.475	5.468	5.462
0.85	0.017	0.088	0.183	0.408	4.853	4.841	4.832
0.90	0.013	0.068	0.140	0.301	0.744	4.269	4.249
0.93	0.011	0.058	0.120	0.254	0.593	1.219	3.914
0.95	0.010	0.053	0.109	0.228	0.517	0.961	3.697
0.97	0.010	0.048	0.099	0.206	0.456	0.797	1.570
0.98	0.009	0.046	0.094	0.196	0.429	0.734	1.270
0.99	0.009	0.044	0.090	0.186	0.405	0.680	1.098
1.00	0.008	0.042	0.086	0.177	0.382	0.632	0.977
1.01	0.008	0.040	0.082	0.169	0.361	0.590	0.883
1.02	0.008	0.039	0.078	0.161	0.342	0.552	0.807
1.05	0.007	0.034	0.069	0.140	0.292	0.460	0.642
1.10	0.005	0.028	0.055	0.112	0.229	0.350	0.470
1.15	0.005	0.023	0.045	0.091	0.183	0.275	0.361
1.20	0.004	0.019	0.037	0.075	0.149	0.220	0.286
1.30	0.003	0.013	0.026	0.052	0.102	0.148	0.190
1.40	0.002	0.010	0.019	0.037	0.072	0.104	0.133
1.50	0.001	0.007	0.014	0.027	0.053	0.076	0.097
1.60	0.001	0.005	0.011	0.021	0.040	0.057	0.073
1.70	0.001	0.004	0.008	0.016	0.031	0.044	0.056
1.80	0.001	0.003	0.006	0.013	0.024	0.035	0.044
1.90	0.001	0.003	0.005	0.010	0.019	0.028	0.036
2.00	0.000	0.002	0.004	0.008	0.016	0.023	0.029
2.20	0.000	0.001	0.003	0.006	0.011	0.016	0.021
2.40	0.000	0.001	0.002	0.004	0.008	0.012	0.015
2.60	0.000	0.001	0.002	0.003	0.006	0.009	0.012
2.80	0.000	0.001	0.001	0.003	0.005	0.008	0.010
3.00	0.000	0.001	0.001	0.002	0.004	0.006	0.008
3.50	0.000	0.000	0.001	0.001	0.003	0.004	0.006
4.00	0.000	0.000	0.001	0.001	0.002	0.003	0.005

TABLE C.8 Continued

(b) Values of $[(s_{ideal} - s_{actual})/R]^{(1)}$

P_r

1.000	1.200	1.500	2.000	3.000	5.000	7.000	10.000
16.586	16.547	16.488	16.390	16.195	15.837	15.468	14.925
15.278	15.251	15.211	15.144	15.011	14.751	14.496	14.153
13.896	13.877	13.849	13.803	13.714	13.541	13.376	13.144
12.496	12.482	12.462	12.430	12.367	12.248	12.145	11.999
11.153	11.143	11.129	11.107	11.063	10.985	10.920	10.836
9.914	9.907	9.897	9.882	9.853	9.806	9.769	9.732
8.799	8.794	8.787	8.777	8.760	8.736	8.723	8.720
7.810	7.807	7.801	7.794	7.784	7.779	7.785	7.811
6.933	6.930	6.926	6.922	6.919	6.929	6.952	7.002
6.155	6.152	6.149	6.147	6.149	6.174	6.213	6.285
5.458	5.455	5.453	5.452	5.461	5.501	5.555	5.648
4.826	4.822	4.820	4.822	4.839	4.898	4.969	5.082
4.238	4.232	4.230	4.236	4.267	4.351	4.442	4.578
3.894	3.885	3.884	3.896	3.941	4.046	4.151	4.300
3.658	3.647	3.648	3.669	3.728	3.851	3.966	4.125
3.406	3.391	3.401	3.437	3.517	3.661	3.788	3.957
3.264	3.247	3.268	3.318	3.412	3.569	3.701	3.875
3.093	3.082	3.126	3.195	3.306	3.477	3.616	3.796
2.399	2.868	2.967	3.067	3.200	3.387	3.532	3.717
1.306	2.513	2.784	2.933	3.094	3.297	3.450	3.640
1.113	1.655	2.557	2.790	2.986	3.209	3.369	3.565
0.820	0.831	1.443	2.283	2.655	2.949	3.134	3.348
0.577	0.640	0.618	1.241	2.067	2.534	2.767	3.013
0.437	0.489	0.502	0.654	1.471	2.138	2.428	2.708
0.343	0.385	0.412	0.447	0.991	1.767	2.115	2.430
0.226	0.254	0.282	0.300	0.481	1.147	1.569	1.944
0.158	0.178	0.200	0.220	0.290	0.730	1.138	1.544
0.115	0.130	0.147	0.166	0.206	0.479	0.823	1.222
0.086	0.098	0.112	0.129	0.159	0.334	0.604	0.969
0.067	0.076	0.087	0.102	0.127	0.248	0.456	0.775
0.053	0.060	0.070	0.083	0.105	0.195	0.355	0.628
0.043	0.049	0.057	0.069	0.089	0.160	0.286	0.518
0.035	0.040	0.048	0.058	0.077	0.136	0.238	0.434
0.025	0.029	0.035	0.043	0.060	0.105	0.178	0.322
0.019	0.022	0.027	0.034	0.048	0.086	0.143	0.254
0.015	0.018	0.021	0.028	0.041	0.074	0.120	0.210
0.012	0.014	0.018	0.023	0.035	0.065	0.104	0.188
0.010	0.012	0.015	0.020	0.031	0.058	0.093	0.158
0.007	0.009	0.011	0.015	0.024	0.046	0.073	0.122
0.006	0.007	0.009	0.012	0.020	0.038	0.060	0.100

From Ernest G. Cravalho and Joseph L. Smith, Jr., *Engineering Thermodynamics,* Pitman, Marshfield, Mass., 1981 (used with permission), as calculated from data in Byung Ik Lee and Michael G. Kesler, "A Generalized Thermodynamic Correlation Based on Three-Parameter Corresponding States," *American Institute of Chemical Engineering Journal,* vol. 21, no. 3, May 1975, pp. 510–527.

TABLE C.9   Enthalpies and Gibbs Function of Formation of Selected Organic Compounds at 298.15 K (77°F)

Substance	Formula	Molecular Weight	State	$\Delta \bar{h}_f^o$ MJ/kmol	$\Delta \bar{h}_f^o$ Btu/lbmol	$\Delta \bar{g}_f^o$ MJ/kmol	$\Delta \bar{g}_f^o$ Btu/lbmol
Methane	CH_4	16.04	g	−74.90	−32,200	−50.83	−21,850
Ethane	C_2H_6	30.07	g	−84.72	−36,420	−32.9	−14.100
Propane	C_3H_8	44.09	g	−103.9	−44,670	−23.5	−10,100
n-Butane	C_4H_{10}	58.12	g	−126.2	−54,260	−17.2	−7,390
i-Butane	C_4H_{10}	58.12	g	−134.6	−57,870	−20.9	−8,990
n-Pentane	C_5H_{12}	72.15	g	−146.5	−62,980	−8.37	−3,600
n-Hexane	C_6H_{14}	86.17	g	−167.3	−71,930	−0.29	−120
n-Heptane	C_7H_{16}	100.20	g	−187.9	−80,780	8.12	3,490
n-Octane	C_8H_{18}	114.22	g	−208.6	−89,680	16.5	7,090
Ethylene	C_2H_4	28.05	g	52.32	22,490	68.17	29,310
Propylene	C_3H_6	42.08	g	20.4	8,770	62.76	26,980
Acetylene	C_2H_2	26.04	g	226.9	97,550	209.3	89,980
Benzene	C_6H_6	78.11	g	82.98	35,670	129.7	55,760
			l	49.07	21,100	124.4	53,480
Methanol	CH_4O	32.04	g	−201.3	−86,540	−162.6	−69,910
			l	−238.7	−102,600	−166.3	−71,500
Ethanol	C_2H_6O	46.07	g	−235.0	−101,000	−168.4	−72,400
			l	−277.2	−119,200	−174.3	−74,940
Ammonia	NH_3	17.03	g	−46.22	−19,870	−16.6	−7,140
			aq	−80.89	−34,780	−26.7	−11,500
Carbon monoxide	CO	28.01	g	−110.6	−47,550	−137.4	−59,070
Carbon dioxide	CO_2	44.01	g	−393.8	−169,300	−394.6	−169,600
Hydrogen chloride	HCl	36.47	g	−92.37	−39,710	−95.33	−40,980
			aq	−167.6	−72,060	−131.3	−56,450
Nitric oxide	NO	30.01	g	90.44	38,880	86.75	37,300
Nitrogen dioxide	NO_2	46.01	g	33.9	14,600	51.87	22,300
Nitrous oxide	N_2O	44.02	g	81.60	35,100	103.7	44,580
Sulfur dioxide	SO_2	64.07	g	−297.1	−127,700	−300.6	−129,200
Sulfur trioxide	SO_3	80.07	g	−395.4	−170,000	−370.6	−159,300
Sulfuric acid	H_2SO_4	98.08	aq	−908.12	−390,420	−742.49	−319,210
			l	−814.12	−350,010	−690.11	−296,690
Water	H_2O	18.02	g	−242.0	−104,000	−228.7	−90,320
			l	−286.0	−123,000	−237.4	−102,100

SI values from B. G. Kyle, *Chemical and Process Thermodynamics,* Prentice-Hall, Englewood Cliffs, N.J., 1984 (with permission) from data in:

Selected Values of Chemical Thermodynamic Properties, National Bureau of Standards Circular 500, 1952.

Selected Values of Properties of Hydrocarbons, National Bureau of Standards Circular 461, 1947.

D. R. Stull, E. F. Westrum, Jr., and G. C. Sinke, *The Chemical Thermodynamics of Organic Compounds,* Wiley, New York, 1969.

TABLE C.10   Values of Absolute Entropy at the Standard Reference State of 25°C (77°F) and 1 atm

Substance	Formula	State	Molecular Weight	$\bar{s}^o$ kJ/(kmol · K)	$\bar{s}^o$ Btu/(lbmol · °R)
Acetylene	C_2H_2	g	26.02	200.8	48.00
Benzene	C_6H_6	l	78.11	173.3	41.41
		g		269.2	64.34
n-Butane	C_4H_{10}	l	58.120	231.	55.2
		g		310.1	74.12
Carbon (graphite)		s	14.01	5.694	1.361
Carbon dioxide	CO_2	g	44.01	213.7	51.07
Carbon monoxide	CO	g	28.01	197.9	47.30
Ethane	C_2H_6	g	30.068	229.5	54.85
Ethylene	C_2H_4	g	28.052	219.5	52.45
Hydrogen	H_2	g	2.016	130.58	31.211
Hydrogen chloride	HCl	g	36.465	186.8	44.64
Methane	CH_4	g	16.042	186.3	44.52
Nitrogen	N_2	g	28.016	191.5	45.767
Oxygen	O_2	g	32.00	205.03	49.003
Propane	C_3H_8	g	44.094	269.9	64.51
Sulfur (rhombic crystals)	S_2		32.006	31.93	7.631
Water	H_2O	l	18.016	69.96	16.72
		v		188.7	45.11

Data from Daniel R. Stull, Edgar F. Westrum, Jr., and Gerard C. Sinke, *The Chemical Thermodynamics of Organic Compounds*, Wiley, New York, 1969 (as gathered from various sources).

TABLE C.11   Enthalpy of Combustion and Adiabatic Flame Temperatures (*Stoichiometric combustion with dry air, fuel and air at* 298 K, *lower values*)

Substance	Formula	Enthalpy of Combustion kJ/kmol	Enthalpy of Combustion Btu/lbmol	Adiabatic Flame Temperature K	Adiabatic Flame Temperature °R
Acetylene	C_2H_2	-12.5×10^5	-5.37×10^5	2859	5146
Benzene	C_6H_6	-31.7×10^5	-13.6×10^5	2484	4471
n-Butane	C_4H_{10}	-26.5×10^5	-11.4×10^5	2357	4243
Carbon monoxide	CO	-2.83×10^5	-1.22×10^5	2615	4707
Ethane	C_2H_6	-14.3×10^5	-6.15×10^5	2338	4208
Ethylene	C_2H_4	-13.2×10^5	-5.67×10^5	2523	4541
Hydrogen	H_2	-2.42×10^5	-1.04×10^5	2490	4482
Methane	CH_4	-8.03×10^5	-3.45×10^5	2285	4113
Propane	C_3H_8	-20.5×10^5	-8.81×10^5	2629	4732

Selected data from A. G. Gaydon and H. G. Wolfhard, *Flames, Their Structure, Radiation and Temperature*, 2d ed., Macmillan, New York, 1960, and from Henry C. Barnett and Robert R. Hibbard (eds.), "Basic Considerations in the Combustion of Hydrocarbon Fuels," NACA Report 1300, 1957.

TABLE C.12 Logarithms to Base 10 of the Equilibrium Constant K

Temperature, K	Temperature, °R	$\frac{1}{2}O_2 \leftrightarrow O$	$\frac{1}{2}H_2 \leftrightarrow H$	$\frac{1}{2}N_2 \leftrightarrow N$	$H_2 + \frac{1}{2}O_2 \leftrightarrow H_2O$	$C + O_2 \leftrightarrow CO_2$	$C + \frac{1}{2}O_2 \leftrightarrow CO$
600	1080	−18.574	−16.336	−38.081	18.633	34.405	14.318
700	1260	−15.449	−13.599	−32.177	15.583	29.506	12.946
800	1440	−13.101	−11.539	−27.744	13.289	25.830	11.914
900	1620	−11.272	−9.934	−24.292	11.498	22.970	11.108
1000	1800	−9.807	−8.646	−21.528	10.062	20.680	10.459
1100	1980	−8.606	−7.589	−19.265	8.883	18.806	9.926
1200	2160	−7.604	−6.707	−17.377	7.899	17.243	9.479
1300	2340	−6.755	−5.958	−15.778	7.064	15.920	9.099
1400	2520	−6.027	−5.315	−14.406	6.347	14.785	8.771
1500	2700	−5.395	−4.756	−13.217	5.725	13.801	8.485
1600	2880	−4.842	−4.266	−12.175	5.180	12.940	8.234
1700	3060	−4.353	−3.833	−11.256	4.699	12.180	8.011
1800	3240	−3.918	−3.448	−10.437	4.270	11.504	7.811
1900	3420	−3.529	−3.102	−9.705	3.886	10.898	7.631
2000	3600	−3.178	−2.790	−9.046	3.540	10.353	7.469
2100	3780	−2.860	−2.508	−8.449	3.227	9.860	7.321
2200	3960	−2.571	−2.251	−7.905	2.942	9.411	7.185
2300	4140	−2.307	−2.016	−7.409	2.682	9.001	7.061
2400	4320	−2.065	−1.800	−6.954	2.443	8.625	6.946
2500	4500	−1.842	−1.601	−6.535	2.224	8.280	6.840
2600	4680	−1.636	−1.417	−6.149	2.021	7.960	6.741
2700	4860	−1.446	−1.247	−5.790	1.833	7.664	6.649
2800	5040	−1.268	−1.089	−5.457	1.658	7.388	6.563
2900	5220	−1.103	−0.941	−5.147	1.495	7.132	6.483
3000	5400	−0.949	−0.803	−4.858	1.343	6.892	6.407
3100	5580	−0.805	−0.674	−4.587	1.201	6.668	6.336
3200	5760	−0.670	−0.553	−4.332	1.067	6.458	6.269
3300	5940	−0.543	−0.439	−4.093	0.942	6.260	6.206
3400	6120	−0.423	−0.332	−3.868	0.824	6.074	6.145
3500	6300	−0.310	−0.231	−3.656	0.712	5.898	6.088
3600	6480	−0.204	−0.135	−3.455	0.607	5.732	6.034
3700	6660	−0.103	−0.044	−3.265	0.507	5.574	5.982
3800	6840	−0.007	0.042	−3.086	0.413	5.425	5.933
3900	7020	0.084	0.123	−2.915	0.323	5.283	5.886
4000	7200	0.170	0.201	−2.752	0.238	5.149	5.841

Temperature, K	Temperature, °R	$\frac{1}{2}O_2 + \frac{1}{2}N_2 \leftrightarrow NO$	$O_2 + \frac{1}{2}N_2 \leftrightarrow NO_2$	$\frac{1}{2}O_2 + \frac{1}{2}H_2 \leftrightarrow OH$	$\frac{1}{2}O_2 + N_2 \leftrightarrow N_2O$	$C + 2H_2 \leftrightarrow CH_4$	$C + \frac{3}{2}H_2 \rightarrow CH_3$
600	1080	−7.210	−6.111	−2.568	−11.040	2.001	−13.212
700	1260	−6.086	−5.714	−2.085	−10.021	0.951	−11.458
800	1440	−5.243	−5.417	−1.724	−9.253	0.146	−10.152
900	1620	−4.587	−5.185	−1.444	−8.654	−0.493	−9.145
1000	1800	−4.062	−5.000	−1.222	−8.171	−1.011	−8.344
1100	1980	−3.633	−4.848	−1.041	−7.774	−1.440	−7.693
1200	2160	−3.275	−4.721	−0.890	−7.442	−1.801	−7.153
1300	2340	−2.972	−4.612	−0.764	−7.158	−2.107	−6.698
1400	2520	−2.712	−4.519	−0.656	−6.914	−2.372	−6.309
1500	2700	−2.487	−4.438	−0.563	−6.701	−2.602	−5.974
1600	2880	−2.290	−4.367	−0.482	−6.514	−2.803	−5.681
1700	3060	−2.116	−4.304	−0.410	−6.347	−2.981	−5.423
1800	3240	−1.962	−4.248	−0.347	−6.198	−3.139	−5.195
1900	3420	−1.823	−4.198	−0.291	−6.065	−3.281	−4.991
2000	3600	−1.699	−4.152	−0.240	−5.943	−3.408	−4.808
2100	3780	−1.586	−4.111	−0.195	−5.833	−3.523	−4.642
2200	3960	−1.484	−4.074	−0.153	−5.732	−3.627	−4.492
2300	4140	−1.391	−4.040	−0.116	−5.639	−3.722	−4.355
2400	4320	−1.305	−4.008	−0.082	−5.554	−3.809	−4.230
2500	4500	−1.227	−3.979	−0.050	−5.475	−3.889	−4.115
2600	4680	−1.154	−3.953	−0.021	−5.401	−3.962	−4.009
2700	4860	−1.087	−3.928	0.005	−5.333	−4.030	−3.911
2800	5040	−1.025	−3.905	0.030	−5.270	−4.093	−3.820
2900	5220	−0.967	−3.884	0.053	−5.210	−4.152	−3.736
3000	5400	−0.913	−3.864	0.074	−5.154	−4.206	−3.659
3100	5580	−0.863	−3.846	0.094	−5.102	−4.257	−3.584
3200	5760	−0.815	−3.828	0.112	−5.052	−4.304	−3.515
3300	5940	−0.771	−3.812	0.129	−5.006	−4.349	−3.451
3400	6120	−0.729	−3.797	0.145	−4.962	−4.391	−3.391
3500	6300	−0.690	−3.783	0.160	−4.920	−4.430	−3.334
3600	6480	−0.653	−3.770	0.174	−4.881	−4.467	−3.280
3700	6660	−0.618	−3.757	0.188	−4.843	−4.503	−3.230
3800	6840	−0.585	−3.746	0.200	−4.807	−4.536	−3.182
3900	7020	−0.554	−3.734	0.212	−4.773	−4.568	−3.137
4000	7200	−0.524	−3.724	0.223	−4.741	−4.598	−3.095

TABLE C.12 Continued

Temperature, K	Temperature, °R	$C + H_2 \leftrightarrow CH_2$	$C + \frac{1}{2}H_2 \leftrightarrow CH$	$2C + H_2 \leftrightarrow C_2H_2$	$C_2 + 2H_2 \leftrightarrow C_2H_4$	$\frac{1}{2}N_2 + \frac{3}{2}H_2 \leftrightarrow NH_3$	$\frac{1}{2}N_2 + H_2 \leftrightarrow NH_2$	$\frac{1}{2}N_2 + \frac{1}{2}H_2 \leftrightarrow NH$
600	1080	−30.678	−45.842	−16.687	−7.652	−1.377	−18.326	−31.732
700	1260	−25.898	−38.448	−13.882	−7.114	−2.023	−15.996	−27.049
800	1440	−22.319	−32.905	−11.784	−6.728	−2.518	−14.255	−23.537
900	1620	−19.540	−28.597	−10.155	−6.438	−2.910	−12.905	−20.806
1000	1800	−17.321	−25.152	−8.856	−6.213	−3.228	−11.827	−18.621
1100	1980	−15.508	−22.336	−7.795	−6.034	−3.490	−10.948	−16.834
1200	2160	−14.000	−19.991	−6.913	−5.889	−3.710	−10.216	−15.345
1300	2340	−12.726	−18.008	−6.168	−5.766	−3.897	−9.598	−14.084
1400	2520	−11.635	−16.310	−5.531	−5.664	−4.058	−9.069	−13.004
1500	2700	−10.691	−14.838	−4.979	−5.575	−4.197	−8.610	−12.068
1600	2880	−9.866	−13.551	−4.497	−5.497	−4.319	−8.210	−11.249
1700	3060	−9.139	−12.417	−4.072	−5.430	−4.426	−7.856	−10.526
1800	3240	−8.493	−11.409	−3.695	−5.369	−4.521	−7.542	−9.883
1900	3420	−7.916	−10.507	−3.358	−5.316	−4.605	−7.261	−9.308
2000	3600	−7.397	−9.696	−3.055	−5.267	−4.681	−7.009	−8.790
2100	3780	−6.929	−8.963	−2.782	−5.223	−4.749	−6.780	−8.322
2200	3960	−6.503	−8.296	−2.532	−5.183	−4.810	−6.572	−7.896
2300	4140	−6.115	−7.687	−2.306	−5.146	−4.866	−6.382	−7.507
2400	4320	−5.760	−7.130	−2.098	−5.113	−4.916	−6.208	−7.151
2500	4500	−5.433	−6.617	−1.906	−5.081	−4.963	−6.048	−6.823
2600	4680	−5.133	−6.144	−1.730	−5.052	−5.005	−5.899	−6.520
2700	4860	−4.854	−5.706	−1.566	−5.025	−5.044	−5.762	−6.240
2800	5040	−4.596	−5.300	−1.415	−5.000	−5.079	−5.635	−5.979
2900	5220	−4.356	−4.922	−1.274	−4.977	−5.112	−5.516	−5.737
3000	5400	−4.132	−4.569	−1.142	−4.955	−5.143	−5.405	−5.511
3100	5580	−3.923	−4.239	−1.019	−4.934	−5.171	−5.300	−5.299
3200	5760	−3.728	−3.930	−0.903	−4.915	−5.197	−5.203	−5.100
3300	5940	−3.544	−3.639	−0.795	−4.897	−5.221	−5.111	−4.914
3400	6120	−3.372	−3.366	−0.693	−4.880	−5.244	−5.024	−4.738
3500	6300	−3.210	−3.108	−0.597	−4.864	−5.265	−4.942	−4.572
3600	6480	−3.056	−2.865	−0.506	−4.848	−5.285	−4.865	−4.416
3700	6660	−2.912	−2.636	−0.420	−4.834	−5.304	−4.791	−4.267
3800	6840	−2.775	−2.418	−0.339	−4.821	−5.321	−4.721	−4.127
3900	7020	−2.646	−2.212	−0.262	−4.808	−5.338	−4.655	−3.994
4000	7200	−2.523	−2.016	−0.189	−4.796	−5.353	−4.592	−3.867

Values taken from Roger A. Strehlow, *Combustion Fundamentals*, McGraw-Hill, New York, 1984 (with permission). The original source is the JANAF thermodynamic tables including supplements.

554

Figure C.1 *Generalized compressibility charts. (Used with permission of Dr. Edward E. Obert, University of Wisconsin.)*

(b) $0 < P_r < 7$

Figure C.1 (Continued)

Figure C.1 (Continued)

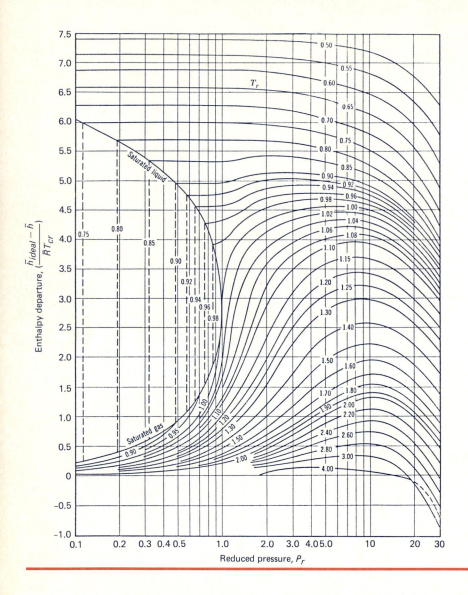

Figure C.2 Generalized enthalpy departure chart. [*Redrawn from Gordon van Wylen and Richard Sontag, Fundamentals of Classical Thermodynamics, (SI version), 2d ed., Wiley, New York, 1976. Reprinted by permission of John Wiley & Sons, Inc.*]

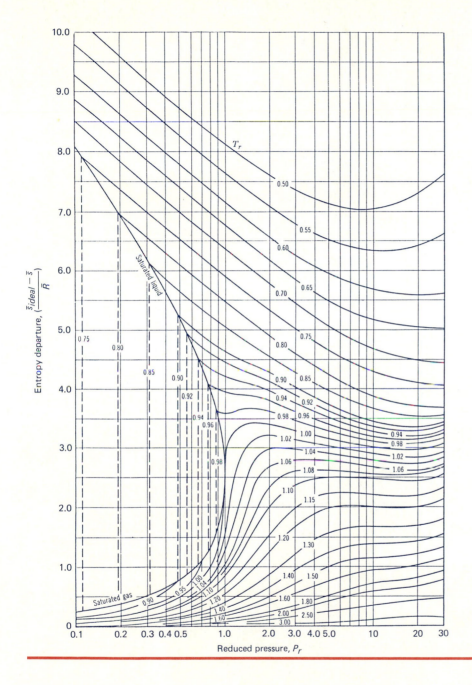

Entropy departure, $\left(\dfrac{\bar{s}_{ideal} - \bar{s}}{\bar{R}}\right)$

T_r

0.50
0.55
0.60
0.65
0.70
0.75
0.80
0.85
0.90
0.94
0.98
1.00
1.02
1.04
1.06
1.08
1.10
1.15
1.20
1.30
1.40
1.50
1.60
1.80
2.00
2.50
3.00

0.94
0.98
1.02
1.06

Saturated liquid

Saturated gas

0.75
0.80
0.85
0.90
0.92
0.94
0.96
0.98

0.90
0.95
1.00
1.04
1.10
1.20
1.40
1.60

Reduced pressure, P_r

Figure C.3 Generalized entropy departure chart. [*Redrawn from Gordon van Wylen and Richard Sontag, Fundamentals of Classical Thermodynamics, (SI version), 2d ed., Wiley, New York, 1976. Reprinted by permission of John Wiley & Sons, Inc.*]

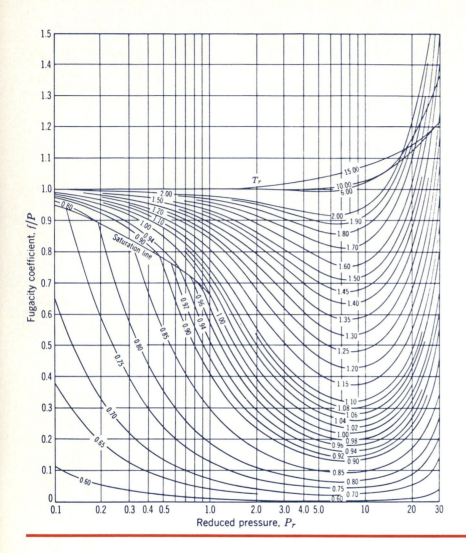

Figure C.4 Generalized fugacity departure chart. [*Redrawn from Gordon van Wylen and Richard Sontag, Fundamentals of Classical Thermodynamics, (SI version), 2d ed., Wiley, New York, 1976. Reprinted by permission of John Wiley & Sons, Inc.*]

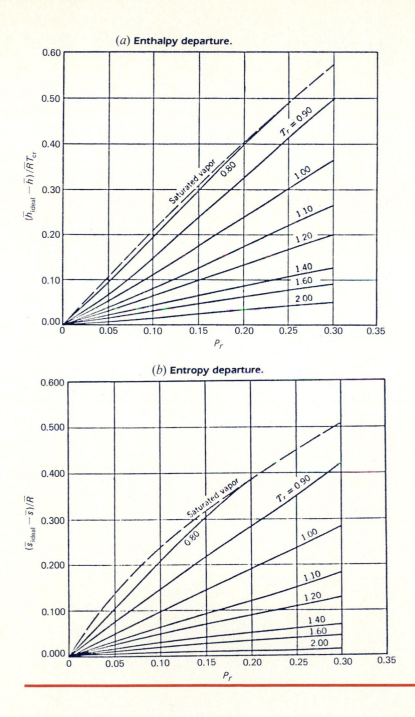

(a) Enthalpy departure.

(b) Entropy departure.

Figure C.5 Generalized departure charts at low pressures. [*Redrawn from Gordon van Wylen and Richard Sontag, Fundamentals of Classical Thermodynamics, [SI version], 2d ed., Wiley, New York, 1976. Reprinted by permission of John Wiley & Sons, Inc.*]

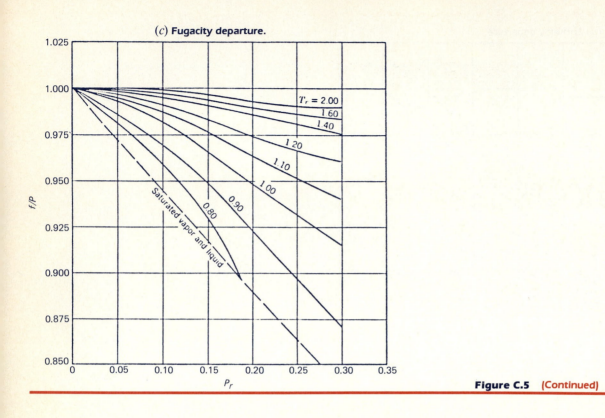

(*c*) **Fugacity departure.**

Figure C.5 (Continued)

APPENDIX D

Thermodynamic Data for Various Substances—SI Units

TABLE D.1 Temperature-Dependent Molar Heat Capacities of Gases at Zero Pressure (SI Units)

$$\bar{c}_p^o = a + bT + cT^2 + dT^3$$
$$[T \text{ in K}, \bar{c}_p^o \text{ in kJ/(kmol} \cdot \text{K)}]$$

Substance	Formula	a	b	c	d	Temperature Range, K	Error, % Max.	Error, % Avg.
Nitrogen	N_2	28.90	-0.1571×10^{-2}	0.8081×10^{-5}	-2.873×10^{-9}	273–1800	0.59	0.34
Oxygen	O_2	25.48	1.520×10^{-2}	-0.7155×10^{-5}	1.312×10^{-9}	273–1800	1.19	0.28
Air		28.11	0.1967×10^{-2}	0.4802×10^{-5}	-1.966×10^{-9}	273–1800	0.72	0.33
Hydrogen	H_2	29.11	-0.1916×10^{-2}	0.4003×10^{-5}	-0.8704×10^{-9}	273–1800	1.01	0.26
Carbon monoxide	CO	28.16	0.1675×10^{-2}	0.5372×10^{-5}	-2.222×10^{-9}	273–1800	0.89	0.37
Carbon dioxide	CO_2	22.26	5.981×10^{-2}	-3.501×10^{-5}	7.469×10^{-9}	273–1800	0.67	0.22
Water vapor	H_2O	32.24	0.1923×10^{-2}	1.055×10^{-5}	-3.595×10^{-9}	273–1800	0.53	0.24
Nitric oxide	NO	29.34	-0.09395×10^{-2}	0.9747×10^{-5}	-4.187×10^{-9}	273–1500	0.97	0.36
Nitrous oxide	N_2O	24.11	5.8632×10^{-2}	-3.562×10^{-5}	10.58×10^{-9}	273–1500	0.59	0.26
Nitrogen dioxide	NO_2	22.9	5.715×10^{-2}	-3.52×10^{-5}	7.87×10^{-9}	273–1500	0.46	0.18
Ammonia	NH_3	27.568	2.5630×10^{-2}	0.99072×10^{-5}	-6.6909×10^{-9}	273–1500	0.91	0.36
Sulfur	S_2	27.21	2.218×10^{-2}	-1.628×10^{-5}	3.986×10^{-9}	273–1800	0.99	0.38
Sulfur dioxide	SO_2	25.78	5.795×10^{-2}	-3.812×10^{-5}	8.612×10^{-9}	273–1800	0.45	0.24
Sulfur trioxide	SO_3	16.40	14.58×10^{-2}	-11.20×10^{-5}	32.42×10^{-9}	273–1300	0.29	0.13
Acetylene	C_2H_2	21.8	9.2143×10^{-2}	-6.527×10^{-5}	18.21×10^{-9}	273–1500	1.46	0.59
Benzene	C_6H_6	-36.22	48.475×10^{-2}	-31.57×10^{-5}	77.62×10^{-9}	273–1500	0.34	0.20
Methanol	CH_4O	19.0	9.152×10^{-2}	-1.22×10^{-5}	-8.039×10^{-9}	273–1000	0.18	0.08
Ethanol	C_2H_6O	19.9	20.96×10^{-2}	-10.38×10^{-5}	20.05×10^{-9}	273–1500	0.40	0.22
Hydrogen chloride	HCl	30.33	-0.7620×10^{-2}	1.327×10^{-5}	-4.338×10^{-9}	273–1500	0.22	0.08
Methane	CH_4	19.89	5.024×10^{-2}	1.269×10^{-5}	-11.01×10^{-9}	273–1500	1.33	0.57
Ethane	C_2H_6	6.900	17.27×10^{-2}	-6.406×10^{-5}	7.285×10^{-9}	273–1500	0.83	0.28
Propane	C_3H_8	-4.04	30.48×10^{-2}	-15.72×10^{-5}	31.74×10^{-9}	273–1500	0.40	0.12
n-Butane	C_4H_{10}	3.96	37.15×10^{-2}	-18.34×10^{-5}	35.00×10^{-9}	273–1500	0.54	0.24
i-Butane	C_4H_{10}	-7.913	41.60×10^{-2}	-23.01×10^{-5}	49.91×10^{-9}	273–1500	0.25	0.13
n-Pentane	C_5H_{12}	6.774	45.43×10^{-2}	-22.46×10^{-5}	42.29×10^{-9}	273–1500	0.56	0.21
n-Hexane	C_6H_{14}	6.938	55.22×10^{-2}	-28.65×10^{-5}	57.69×10^{-9}	273–1500	0.72	0.20
Ethylene	C_2H_4	3.95	15.64×10^{-2}	-8.344×10^{-5}	17.67×10^{-9}	273–1500	0.54	0.13
Propylene	C_3H_6	3.15	23.83×10^{-2}	-12.18×10^{-5}	24.62×10^{-9}	273–1500	0.73	0.17

Data from B. G. Kyle, *Chemical and Process Thermodynamics,* Prentice-Hall, Englewood Cliffs, N.J., 1984 (used with permission).

TABLE D.2 Ideal Gas Properties of Air at Low Pressure — SI Units

T, °C	u, kJ/kg	h, kJ/kg	s_0, kJ/(kg · K)	P_0	v_0	c_P, kJ/(kg · K)	c_v, kJ/(kg · K)	k
−200	52.32	73.33	5.2906	0.0101	20,488.	1.0025	0.7152	1.4016
−150	88.09	123.46	5.8128	0.0624	5,592.1	1.0025	0.7153	1.4015
−100	123.85	173.58	6.1544	0.2051	2,391.5	1.0026	0.7153	1.4015
−50	159.65	223.74	6.4090	0.4979	1,269.6	1.0030	0.7158	1.4013
0	195.46	273.92	6.6119	1.0095	766.52	1.0041	0.7169	1.4007
25	213.40	299.03	6.6999	1.3715	615.80	1.0051	0.7179	1.4001
50	231.36	324.18	6.7808	1.8186	503.36	1.0065	0.7193	1.3993
100	267.42	374.60	6.9259	3.0147	350.63	1.0107	0.7235	1.3970
150	303.74	425.28	7.0533	4.6998	255.05	1.0169	0.7296	1.3937
200	340.42	476.32	7.1673	6.9913	191.71	1.0249	0.7376	1.3894
250	377.54	527.80	7.2707	10.024	147.85	1.0345	0.7472	1.3844
300	415.17	579.79	7.3656	13.952	116.37	1.0452	0.7580	1.3789
350	453.36	632.34	7.4535	18.950	93.152	1.0568	0.7695	1.3732
400	492.13	685.48	7.5356	25.218	75.617	1.0687	0.7815	1.3675
450	531.51	739.22	7.6126	32.977	62.120	1.0808	0.7935	1.3620
500	571.49	793.56	7.6852	42.475	51.563	1.0927	0.8054	1.3566
550	612.06	848.49	7.7540	53.987	43.192	1.1042	0.8170	1.3516
600	653.19	903.98	7.8195	67.812	36.475	1.1154	0.8281	1.3468
650	694.87	960.02	7.8819	84.280	31.028	1.1260	0.8388	1.3424
700	737.07	1,016.6	7.9415	103.75	26.571	1.1360	0.8488	1.3384
750	779.75	1,073.6	7.9987	126.61	22.892	1.1455	0.8583	1.3346
800	822.89	1,131.1	8.0536	153.28	19.883	1.1544	0.8672	1.3312
850	866.47	1,189.1	8.1063	184.21	17.271	1.1628	0.8756	1.3280
900	910.45	1,247.4	8.1571	219.89	15.113	1.1706	0.8834	1.3251
950	954.81	1,306.1	8.2062	260.84	13.284	1.1779	0.8907	1.3225
1,000	999.52	1,365.2	8.2535	307.60	11.725	1.1848	0.8976	1.3200
1,100	1,088.6	1,483.0	8.3436	420.98	9.2400	1.1973	0.9101	1.3156
1,200	1,181.5	1,604.6	8.4281	565.19	7.3836	1.2083	0.9210	1.3118
1,300	1,274.1	1,726.0	8.5078	746.01	5.9736	1.2180	0.9308	1.3086
1,400	1,367.7	1,848.2	8.5831	969.91	4.8867	1.2267	0.9395	1.3057
1,500	1,462.0	1,971.3	8.6546	1,244.0	4.0377	1.2345	0.9473	1.3032
1,600	1,557.1	2,095.1	8.7225	1,576.1	3.3666	1.2416	0.9544	1.3010
1,700	1,652.9	2,219.6	8.7872	1,974.9	2.8302	1.2480	0.9608	1.2989
1,800	1,749.3	2,344.7	8.8491	2,449.8	2.3972	1.2539	0.9667	1.2971
1,900	1,846.2	2,470.4	8.9083	3,011.0	2.0446	1.2593	0.9721	1.2955
2,000	1,943.7	2,596.6	8.9650	3,669.4	1.7549	1.2644	0.9771	1.2939
2,100	2,041.7	2,723.3	9.0196	4,437.2	1.5150	1.2690	0.9818	1.2925
2,200	2,140.1	2,850.5	9.0720	5,327.2	1.3151	1.2734	0.9862	1.2913
2,300	2,239.0	2,978.0	9.1226	6,353.2	1.1473	1.2775	0.9902	1.2901
2,400	2,338.2	3,106.0	9.1714	7,530.1	1.0056	1.2813	0.9940	1.2889
2,500	2,437.8	3,234.3	9.2185	8,873.5	0.8853	1.2848	0.9976	1.2879
2,600	2,537.7	3,363.0	9.2641	10,401.	0.7826	1.2882	1.0010	1.2869
2,700	2,638.0	3,492.0	9.3082	12,129.	0.6944	1.2914	1.0042	1.2860
2,800	2,738.6	3,621.3	9.3510	14,078.	0.6184	1.2944	1.0071	1.2852
2,900	2,839.5	3,750.9	9.3925	16,267.	0.5526	1.2972	1.0100	1.2844
3,000	2,940.6	3,880.8	9.4328	18,719.	0.4953	1.2999	1.0126	1.2836

Properties generated from the program *GASPROPS*, Wiley Professional Software, Wiley, New York, 1984.

TABLE D.3 Ideal Gas Properties of Argon at Low Pressure — SI Units

T, °C	u, kJ/kg	h, kJ/kg	s_0, kJ/(kg · K)	P_0	v_0	c_P, kJ/(kg · K)	c_v, kJ/(kg · K)	k
−200	22.85	38.09	3.1444	0.3641	412.66	0.5207	0.3124	1.667
−150	38.47	64.12	3.4156	1.3402	188.75	0.5207	0.3124	1.667
−100	54.09	90.15	3.5930	3.1433	113.15	0.5207	0.3124	1.667
−50	69.71	116.19	3.7251	5.9294	77.305	0.5207	0.3124	1.667
0	85.33	142.22	3.8304	9.8325	57.063	0.5207	0.3124	1.667
25	93.14	155.24	3.8760	12.241	50.031	0.5207	0.3124	1.667
50	100.95	168.26	3.9179	14.973	44.333	0.5207	0.3124	1.667
100	116.57	194.29	3.9928	21.458	35.719	0.5207	0.3124	1.667
150	132.19	220.32	4.0583	29.391	29.573	0.5207	0.3124	1.667
200	147.81	246.36	4.1165	38.865	25.007	0.5207	0.3124	1.667
250	163.43	272.39	4.1688	49.969	21.505	0.5207	0.3124	1.667
300	179.05	298.42	4.2163	62.787	18.751	0.5207	0.3124	1.667
350	194.68	324.46	4.2598	77.400	16.538	0.5207	0.3124	1.667
400	210.30	350.49	4.3000	93.885	14.728	0.5207	0.3124	1.667
450	225.92	376.53	4.3373	112.32	13.225	0.5207	0.3124	1.667
500	241.54	402.56	4.3721	132.76	11.962	0.5207	0.3124	1.667
550	257.16	428.59	4.4048	155.20	10.888	0.5207	0.3124	1.667
600	272.78	454.63	4.4355	179.98	9.9651	0.5207	0.3124	1.667
650	288.40	480.66	4.4645	206.88	9.1657	0.5207	0.3124	1.667
700	304.02	506.69	4.4919	236.07	8.4677	0.5207	0.3124	1.667
750	319.64	532.73	4.5180	267.59	7.8540	0.5207	0.3124	1.667
800	335.26	558.76	4.5429	301.51	7.3109	0.5207	0.3124	1.667
850	350.88	584.80	4.5666	337.90	6.8277	0.5207	0.3124	1.667
900	366.50	610.83	4.5892	376.79	6.3954	0.5207	0.3124	1.667
950	382.12	636.86	4.6110	418.26	6.0069	0.5207	0.3124	1.667
1,000	397.74	662.90	4.6318	462.36	5.6562	0.5207	0.3124	1.667
1,100	428.98	714.96	4.6712	558.64	5.0490	0.5207	0.3124	1.667
1,200	460.22	767.03	4.7078	666.04	4.5432	0.5207	0.3124	1.667
1,300	491.46	819.10	4.7420	784.98	4.1165	0.5207	0.3124	1.667
1,400	522.70	871.17	4.7741	915.83	3.7526	0.5207	0.3124	1.667
1,500	553.94	923.23	4.8043	1,059.0	3.4394	0.5207	0.3124	1.667
1,600	585.18	975.30	4.8329	1,214.8	3.1674	0.5207	0.3124	1.667
1,700	616.42	1,027.4	4.8600	1,383.6	2.9294	0.5207	0.3124	1.667
1,800	647.66	1,079.4	4.8857	1,565.7	2.7198	0.5207	0.3124	1.667
1,900	678.90	1,131.5	4.9102	1,761.5	2.5341	0.5207	0.3124	1.667
2,000	710.14	1,183.6	4.9337	1,971.4	2.3685	0.5207	0.3124	1.667
2,100	741.38	1,235.6	4.9561	2,195.6	2.2202	0.5207	0.3124	1.667
2,200	772.62	1,287.7	4.9776	2,434.4	2.0868	0.5207	0.3124	1.667
2,300	803.86	1,339.8	4.9982	2,688.1	1.9662	0.5207	0.3124	1.667
2,400	835.11	1,391.8	5.0181	2,957.2	1.8568	0.5207	0.3124	1.667
2,500	866.35	1,443.9	5.0372	3,241.7	1.7572	0.5207	0.3124	1.667
2,600	897.59	1,496.0	5.0556	3,542.2	1.6661	0.5207	0.3124	1.667
2,700	928.83	1,548.0	5.0734	3,858.7	1.5827	0.5207	0.3124	1.667
2,800	960.07	1,600.1	5.0907	4,191.6	1.5060	0.5207	0.3124	1.667
2,900	991.31	1,652.2	5.1073	4,541.2	1.4353	0.5207	0.3124	1.667
3,000	1,022.6	1,704.3	5.1235	4,907.7	1.3700	0.5207	0.3124	1.667

Properties generated from the program *GASPROPS,* Wiley Professional Software, Wiley, New York, 1984.

TABLE D.4 Ideal Gas Properties of Nitrogen at Low Pressure — SI Units

T, °C	u, kJ/kg	h, kJ/kg	s_0, kJ/(kg · K)	P_0	v_0	c_P, kJ/(kg · K)	c_v, kJ/(kg · K)	k
−200	54.31	76.04	5.3795	0.0074	28,801.	1.0395	0.7425	1.4000
−150	91.44	128.01	5.9210	0.0461	7,822.2	1.0395	0.7425	1.4000
−100	128.56	179.99	6.2752	0.1521	3,334.4	1.0395	0.7425	1.4000
−50	165.72	231.99	6.5392	0.3702	1,765.7	1.0398	0.7428	1.3998
0	202.86	283.99	6.7494	0.7517	1,064.4	1.0400	0.7430	1.3997
25	221.44	309.99	6.8405	1.0218	854.74	1.0403	0.7433	1.3996
50	240.03	336.01	6.9243	1.3550	698.56	1.0409	0.7439	1.3993
100	277.27	388.10	7.0742	2.2452	486.83	1.0430	0.7460	1.3981
150	314.66	440.34	7.2055	3.4953	354.62	1.0468	0.7498	1.3961
200	352.29	492.82	7.3227	5.1880	267.15	1.0527	0.7557	1.3930
250	390.26	545.64	7.4289	7.4179	206.58	1.0605	0.7635	1.3890
300	428.67	598.89	7.5261	10.293	163.11	1.0699	0.7729	1.3842
350	467.58	652.65	7.6160	13.935	130.99	1.0806	0.7836	1.3790
400	507.05	706.97	7.6998	18.484	106.68	1.0922	0.7952	1.3735
450	547.10	761.88	7.7785	24.094	87.916	1.1042	0.8072	1.3679
500	587.77	817.39	7.8527	30.940	73.197	1.1163	0.8193	1.3625
550	629.03	873.51	7.9231	39.213	61.489	1.1284	0.8314	1.3572
600	670.90	930.22	7.9900	49.124	52.064	1.1401	0.8431	1.3523
650	713.34	987.52	8.0538	60.906	44.398	1.1515	0.8545	1.3476
700	756.34	1,045.4	8.1148	74.808	38.105	1.1623	0.8653	1.3432
750	799.86	1,103.7	8.1733	91.104	32.897	1.1726	0.8756	1.3392
800	843.89	1,162.6	8.2294	110.09	28.554	1.1824	0.8854	1.3355
850	888.39	1,222.0	8.2835	132.08	24.909	1.1915	0.8945	1.3320
900	933.34	1,281.8	8.3356	157.42	21.830	1.2002	0.9032	1.3288
950	978.70	1,342.0	8.3859	186.47	19.215	1.2082	0.9112	1.3259
1,000	1,024.5	1,402.6	8.4344	219.61	16.982	1.2158	0.9188	1.3232
1,100	1,117.0	1,524.9	8.5269	299.87	13.413	1.2295	0.9325	1.3185
1,200	1,210.9	1,648.4	8.6137	401.81	10.739	1.2415	0.9445	1.3145
1,300	1,305.9	1,773.1	8.6956	529.47	8.7032	1.2520	0.9550	1.3110
1,400	1,401.8	1,898.8	8.7730	687.32	7.1306	1.2612	0.9642	1.3080
1,500	1,498.7	2,025.3	8.8465	880.31	5.9001	1.2693	0.9723	1.3055
1,600	1,596.3	2,152.6	8.9163	1,113.9	4.9260	1.2765	0.9795	1.3032
1,700	1,694.6	2,280.6	8.9827	1,393.9	4.1466	1.2829	0.9859	1.3013
1,800	1,793.4	2,409.2	9.0465	1,726.8	3.5168	1.2886	0.9916	1.2995
1,900	1,892.9	2,538.3	9.1073	2,119.5	3.0033	1.2937	0.9967	1.2980
2,000	1,992.8	2,667.9	9.1656	2,579.6	2.5812	1.2983	1.0013	1.2966
2,100	2,093.1	2,797.9	9.2216	3,115.0	2.2315	1.3025	1.0055	1.2954
2,200	2,193.8	2,928.4	9.2754	3,734.6	1.9398	1.3062	1.0092	1.2943
2,300	2,294.9	3,059.2	9.3273	4,447.4	1.6947	1.3097	1.0127	1.2933
2,400	2,396.4	3,190.3	9.3773	5,263.4	1.4877	1.3129	1.0159	1.2924
2,500	2,498.1	3,321.7	9.4255	6,193.0	1.3117	1.3158	1.0188	1.2915
2,600	2,600.1	3,453.4	9.4722	7,247.3	1.1613	1.3184	1.0214	1.2908
2,700	2,702.4	3,585.4	9.5173	8,438.0	1.0321	1.3209	1.0239	1.2901
2,800	2,804.9	3,717.6	9.5611	9,777.7	0.9207	1.3232	1.0262	1.2894
2,900	2,907.6	3,850.0	9.6035	11,279.	0.8241	1.3254	1.0284	1.2888
3,000	3,010.6	3,982.7	9.6446	12,957.	0.7400	1.3274	1.0304	1.2882

Properties generated from the program *GASPROPS*, Wiley Professional Software, Wiley, New York, 1984.

TABLE D.5   Ideal Gas Properties of Oxygen at Low Pressure — SI Units

T, °C	u, kJ/kg	h, kJ/kg	s_0, kJ/(kg · K)	P_0	v_0	c_P, kJ/(kg · K)	c_v, kJ/(kg · K)	k
−200	47.55	66.57	5.1302	0.0376	4,993.6	0.9100	0.6500	1.4000
−150	80.06	112.08	5.6045	0.2330	1,355.1	0.9102	0.6502	1.3999
−100	112.58	157.60	5.9147	0.7689	577.44	0.9104	0.6504	1.3998
−50	145.12	203.14	6.1457	1.8711	305.83	0.9116	0.6516	1.3991
0	177.78	248.80	6.3303	3.8077	183.96	0.9154	0.6554	1.3967
25	194.20	271.72	6.4107	5.1865	147.41	0.9187	0.6587	1.3947
50	210.72	294.74	6.4848	6.8990	120.12	0.9230	0.6630	1.3922
100	244.14	341.16	6.6183	11.534	82.963	0.9342	0.6742	1.3856
150	278.19	388.21	6.7366	18.186	59.668	0.9482	0.6882	1.3778
200	312.98	436.01	6.8434	27.425	44.241	0.9637	0.7037	1.3695
250	348.57	484.59	6.9410	39.928	33.599	0.9797	0.7197	1.3613
300	384.95	533.97	7.0311	56.485	26.021	0.9954	0.7354	1.3536
350	422.09	584.12	7.1150	78.005	20.486	1.0103	0.7503	1.3465
400	459.96	634.99	7.1935	105.53	16.358	1.0243	0.7643	1.3402
450	498.50	686.53	7.2673	140.22	13.226	1.0372	0.7772	1.3345
500	537.66	738.69	7.3371	183.39	10.811	1.0490	0.7890	1.3295
550	577.39	791.41	7.4032	236.48	8.9260	1.0597	0.7997	1.3251
600	617.62	844.65	7.4659	301.12	7.4359	1.0695	0.8095	1.3212
650	658.32	898.34	7.5257	379.04	6.2455	1.0783	0.8183	1.3177
700	699.44	952.47	7.5828	472.19	5.2850	1.0864	0.8264	1.3146
750	740.94	1,007.0	7.6374	582.64	4.5031	1.0937	0.8337	1.3119
800	782.80	1,061.8	7.6898	712.68	3.8614	1.1005	0.8404	1.3094
850	824.98	1,117.0	7.7400	864.76	3.3306	1.1066	0.8466	1.3071
900	867.46	1,172.5	7.7884	1,041.5	2.8885	1.1124	0.8524	1.3050
950	910.21	1,228.2	7.8349	1,245.8	2.5177	1.1177	0.8577	1.3032
1,000	953.22	1,284.3	7.8798	1,480.7	2.2049	1.1227	0.8627	1.3014
1,100	1,040.0	1,397.0	7.9650	2,055.6	1.7130	1.1319	0.8719	1.2982
1,200	1,127.6	1,510.6	8.0449	2,795.2	1.3515	1.1403	0.8803	1.2954
1,300	1,216.0	1,625.0	8.1200	3,732.7	1.0808	1.1481	0.8881	1.2928
1,400	1,305.2	1,740.2	8.1910	4,905.3	0.8747	1.1556	0.8956	1.2903
1,500	1,395.1	1,856.1	8.2583	6,355.3	0.7155	1.1629	0.9028	1.2880
1,600	1,485.7	1,972.8	8.3223	8,130.1	0.5908	1.1699	0.9099	1.2857
1,700	1,577.1	2,090.1	8.3833	10,283.	0.4921	1.1769	0.9169	1.2836
1,800	1,669.1	2,208.2	8.4417	12,872.	0.4130	1.1838	0.9238	1.2815
1,900	1,761.8	2,326.9	8.4976	15,963.	0.3491	1.1906	0.9306	1.2794
2,000	1,855.2	2,446.3	8.5513	19,629.	0.2970	1.1973	0.9373	1.2774
2,100	1,949.3	2,566.3	8.6030	23,949.	0.2541	1.2039	0.9438	1.2755
2,200	2,044.0	2,687.0	8.6528	29,010.	0.2186	1.2103	0.9503	1.2736
2,300	2,139.4	2,808.4	8.7009	34,910.	0.1890	1.2166	0.9566	1.2718
2,400	2,235.3	2,930.4	8.7474	41,752.	0.1642	1.2228	0.9628	1.2701
2,500	2,331.9	3,052.9	8.7925	49,650.	0.1432	1.2288	0.9688	1.2684
2,600	2,429.1	3,176.1	8.8361	58,728.	0.1255	1.2346	0.9746	1.2668
2,700	2,526.8	3,299.9	8.8784	69,121.	0.1103	1.2402	0.9802	1.2653
2,800	2,625.1	3,424.1	8.9195	80,971.	0.0973	1.2456	0.9856	1.2638
2,900	2,723.9	3,549.0	8.9595	94,435.	0.0862	1.2508	0.9908	1.2624
3,000	2,823.2	3,674.3	8.9984	109,680.	0.0765	1.2557	0.9957	1.2611

Properties generated from the program *GASPROPS,* Wiley Professional Software, Wiley, New York, 1984.

TABLE D.6 Ideal Gas Properties of Water Vapor at Low Pressure — SI Units

T, °C	u, kJ/kg	h, kJ/kg	s_0, kJ/(kg · K)	P_0	v_0	c_P, kJ/(kg · K)	c_v, kJ/(kg · K)	k
0	377.07	503.22	10.310	0.0503	24,725.	1.8590	1.3972	1.3305
25	412.05	549.75	10.423	0.0643	21,130.	1.8644	1.4026	1.3292
50	447.22	596.46	10.470	0.0712	20,660.	1.8713	1.4095	1.3277
100	518.11	690.45	10.751	0.1309	12,988.	1.8897	1.4279	1.3234
150	590.08	785.51	11.135	0.3009	6,405.3	1.9132	1.4514	1.3182
200	663.32	881.84	11.350	0.4795	4,494.2	1.9401	1.4782	1.3124
250	737.94	979.55	11.547	0.7338	3,247.2	1.9691	1.5073	1.3064
300	814.07	1,078.8	11.728	1.0864	2,403.1	1.9995	1.5377	1.3003
350	891.69	1,179.5	11.896	1.5650	1,813.6	2.0311	1.5693	1.2943
400	971.00	1,281.9	12.054	2.2036	1,391.4	2.0636	1.6018	1.2883
450	1,051.9	1,385.9	12.203	3.0441	1,082.0	2.0969	1.6351	1.2825
500	1,134.5	1,491.6	12.345	4.1338	851.91	2.1308	1.6689	1.2767
550	1,218.8	1,599.0	12.479	5.5338	677.53	2.1654	1.7035	1.2711
600	1,304.9	1,708.1	12.608	7.3140	543.76	2.2006	1.7388	1.2656
650	1,392.7	1,819.0	12.731	9.5579	439.93	2.2361	1.7743	1.2603
700	1,482.3	1,931.7	12.850	12.367	358.43	2.2717	1.8099	1.2552
750	1,573.7	2,046.2	12.965	15.860	293.84	2.3072	1.8454	1.2503
800	1,666.8	2,162.5	13.076	20.161	242.46	2.3424	1.8806	1.2456
850	1,761.7	2,280.4	13.183	25.446	201.04	2.3772	1.9154	1.2411
900	1,858.3	2,400.1	13.288	31.904	167.49	2.4114	1.9496	1.2369
950	1,956.7	2,521.6	13.389	39.729	140.23	2.4448	1.9830	1.2329
1,000	2,056.7	2,644.6	13.488	49.192	117.89	2.4777	2.0159	1.2291
1,100	2,261.4	2,895.6	13.677	74.225	84.264	2.5402	2.0784	1.2222
1,200	2,472.2	3,152.5	13.858	109.78	61.122	2.5988	2.1370	1.2161
1,300	2,688.6	3,415.2	14.030	159.48	44.931	2.6533	2.1915	1.2107
1,400	2,910.3	3,683.1	14.195	228.04	33.419	2.7034	2.2416	1.2060
1,500	3,136.8	3,955.7	14.354	321.26	25.140	2.7498	2.2880	1.2018
1,600	3,367.8	4,232.9	14.503	444.36	19.201	2.7923	2.3304	1.1982
1,700	3,602.8	4,514.0	14.619	570.41	15.756	2.8310	2.3692	1.1949
1,800	3,841.5	4,799.0	14.793	832.17	11.347	2.8667	2.4048	1.1920
1,900	4,083.7	5,087.3	14.929	1,116.9	8.8623	2.8996	2.4378	1.1894
2,000	4,329.0	5,378.8	15.060	1,483.8	6.9780	2.9299	2.4681	1.1871
2,100	4,577.2	5,673.2	15.187	1,952.8	5.5353	2.9577	2.4959	1.1850
2,200	4,828.2	5,970.3	15.309	2,547.2	4.4225	2.9837	2.5218	1.1831
2,300	5,081.6	6,269.9	15.428	3,294.7	3.5574	3.0080	2.5461	1.1814
2,400	5,337.3	6,571.9	15.543	4,227.8	2.8799	3.0306	2.5687	1.1798
2,500	5,595.3	6,876.0	15.655	5,384.9	2.3457	3.0519	2.5901	1.1783
2,600	5,855.3	7,182.2	15.763	6,814.4	1.9205	3.0721	2.6103	1.1769
2,700	6,117.3	7,490.4	15.869	8,560.8	1.5819	3.0911	2.6293	1.1756
2,800	6,383.9	7,803.1	15.971	10,674.	1.3114	3.1095	2.6477	1.1744
2,900	6,651.2	8,116.7	16.069	13,213.	1.0939	3.1271	2.6653	1.1733
3,000	6,918.7	8,430.2	16.164	16,247.	0.9176	—	—	—

Properties generated from the program *GASPROPS,* Wiley Professional Software, Wiley, New York, 1984.

TABLE D.7 Ideal Gas Properties of Carbon Dioxide at Low Pressure — SI Units

T, °C	u, kJ/kg	h, kJ/kg	s_0, kJ/(kg · K)	P_0	v_0	c_P, kJ/(kg · K)	c_v, kJ/(kg · K)	k
−200	34.56	48.39	3.8607	0.0075	18,191.	0.6618	0.4728	1.3999
−150	58.63	81.91	4.2065	0.0468	4,911.	0.6712	0.4821	1.3921
−100	73.96	106.69	4.4403	0.1612	2,002.4	0.7078	0.5188	1.3644
−50	110.29	152.47	4.6262	0.4313	964.6	0.7616	0.5725	1.3302
0	140.51	192.15	4.7857	1.0033	507.6	0.8176	0.6286	1.3008
25	156.57	212.93	4.8585	1.4747	376.97	0.8443	0.6553	1.2884
50	173.27	234.36	4.9275	2.1249	283.55	0.8697	0.6807	1.2777
100	208.50	279.04	5.0560	4.1943	165.8	0.9166	0.7275	1.2599
150	245.94	325.94	5.1738	7.8275	100.79	0.9584	0.7694	1.2457
200	285.37	374.82	5.2830	13.948	63.25	0.9960	0.8070	1.2343
250	326.58	425.48	5.3847	23.903	40.80	1.0299	0.8409	1.2248
300	369.40	477.75	5.4801	39.606	26.98	1.0606	0.8716	1.2169
350	413.69	531.50	5.5700	63.737	18.23	1.0886	0.8996	1.2102
400	459.32	586.57	5.6550	99.962	12.55	1.1141	0.9251	1.2044
450	506.16	642.87	5.7357	153.20	8.8011	1.1375	0.9484	1.1993
500	554.12	700.29	5.8125	230.00	6.2676	1.1588	0.9697	1.1949
550	603.10	758.72	5.8857	338.87	4.5292	1.1782	0.9892	1.1911
600	653.02	818.08	5.9557	490.89	3.3164	1.1961	1.0071	1.1877
650	703.78	878.30	6.0228	700.12	2.4585	1.2124	1.0234	1.1847
700	755.33	939.30	6.0871	984.37	1.8433	1.2273	1.0383	1.1821
750	807.58	1,001.0	6.1472	1,352.6	1.4103	1.2410	1.0520	1.1797
800	860.48	1,063.4	6.2153	1,939.5	1.0317	1.2535	1.0645	1.1776
850	913.99	1,126.3	6.2345	2,147.4	0.9752	1.2650	1.0760	1.1757
900	968.07	1,189.9	6.2842	2,794.0	0.7829	1.2756	1.0866	1.1740
950	1,022.7	1,253.9	6.3821	4,689.7	0.4863	1.2855	1.0964	1.1724
1,000	1,077.7	1,318.4	6.4243	5,865.1	0.4047	1.2946	1.1055	1.1710
1,100	1,189.1	1,448.6	6.5248	9,983.1	0.2565	1.3105	1.1214	1.1686
1,200	1,301.9	1,580.4	6.6174	16,299.	0.1685	1.3240	1.1349	1.1666
1,300	1,416.0	1,713.4	6.7047	25,872.	0.1134	1.3359	1.1469	1.1648
1,400	1,531.3	1,847.6	6.7873	40,053.	0.0779	1.3462	1.1571	1.1634
1,500	1,647.4	1,982.6	6.8657	60,656.	0.0545	1.3551	1.1661	1.1621
1,600	1,764.8	2,118.9	6.9403	90,050.	0.0388	1.3633	1.1742	1.1610
1,700	1,882.5	2,255.5	7.0114	131,170.	0.0280	1.3704	1.1813	1.1600
1,800	1,999.7	2,391.6	7.0793	187,920.	0.0206	1.3767	1.1877	1.1592
1,900	2,122.2	2,533.0	7.1443	265,110.	0.0153	1.3824	1.1934	1.1584
2,000	2,223.2	2,653.0	7.2067	368,730.	0.0115	1.3876	1.1986	1.1577
2,100	2,354.0	2,802.6	7.2665	506,000.	0.0087	1.3923	1.2033	1.1571
2,200	2,481.3	2,948.9	7.3241	686,580.	0.0067	1.3965	1.2075	1.1566
2,300	2,600.5	3,087.0	7.3795	920,640.	0.0052	1.4006	1.2115	1.1560
2,400	2,722.4	3,227.7	7.4329	1,221,400.	0.0041	1.4043	1.2152	1.1556
2,500	2,843.9	3,368.2	7.4846	1,605.700.	0.0032	1.4077	1.2187	1.1551
2,600	2,966.0	3,509.1	7.5346	2,092,100.	0.0026	1.4110	1.2219	1.1547
2,700	3,088.3	3,650.4	7.5829	2,701,300.	0.0021	1.4140	1.2250	1.1543
2,800	3,211.0	3,792.0	7.6296	3,458,500.	0.0017	1.4169	1.2279	1.1540
2,900	3,333.9	3,933.8	7.6751	4,399,900.	0.0013	1.4196	1.2306	1.1536
3,000	3,457.1	4,075.9	7.7193	5,561,300.	0.0011	1.4222	1.2331	1.1533

Properties generated from the program *GASPROPS*, Wiley Professional Software, Wiley, New, York, 1984.

TABLE D.8   Properties of Saturated Water—Temperature Table, SI Units

T_{sat}, °C	P_{sat}, kPa	Specific Volume, m³/kg			Internal Energy, kJ/kg			Enthalpy, kJ/kg			Entropy, kJ/(kg · K)		
		v_l	v_{lg}	v_g	u_l	u_{lg}	u_g	h_l	h_{lg}	h_g	s_l	s_{lg}	s_g
0	0.61	0.001000	206.13	206.13	0.00	2,373.9	2,373.9	0.0	2,500.0	2,500.0	−0.0012	9.1590	9.1578
5	0.87	0.001000	147.20	147.20	21.04	2,361.1	2,382.1	21.0	2,489.6	2,510.6	0.0757	8.9510	9.0267
10	1.23	0.001000	106.36	106.36	42.02	2,347.8	2,389.8	42.0	2,478.4	2,520.4	0.1509	8.7511	8.9020
15	1.71	0.001001	78.036	78.037	62.95	2,333.7	2,396.7	63.0	2,466.8	2,529.7	0.2244	8.5582	8.7827
20	2.34	0.001002	57.801	57.802	83.86	2,319.9	2,403.7	83.9	2,455.0	2,538.9	0.2965	8.3718	8.6684
25	3.17	0.001003	43.446	43.447	104.75	2,305.5	2,410.3	104.8	2,443.1	2,547.9	0.3672	8.1919	8.5591
30	4.24	0.001004	32.907	32.908	125.63	2,291.6	2,417.2	125.6	2,431.2	2,556.8	0.4367	8.0180	8.4546
35	5.62	0.001006	25.250	25.251	146.50	2,277.3	2,423.8	146.5	2,419.2	2,565.7	0.5049	7.8496	8.3545
40	7.37	0.001008	19.536	19.537	167.37	2,263.2	2,430.6	167.4	2,407.3	2,574.6	0.5720	7.6864	8.2584
45	9.58	0.001010	15.262	15.263	188.24	2,249.1	2,437.3	188.3	2,395.3	2,583.5	0.6381	7.5281	8.1662
50	12.33	0.001012	12.046	12.047	209.12	2,234.7	2,443.8	209.1	2,383.2	2,592.3	0.7031	7.3745	8.0776
55	15.74	0.001014	9.5771	9.5781	230.01	2,220.4	2,450.4	230.0	2,371.1	2,601.1	0.7672	7.2253	7.9925
60	19.92	0.001017	7.6776	7.6786	250.91	2,206.0	2,456.9	250.9	2,358.9	2,609.8	0.8303	7.0804	7.9107
65	25.00	0.001020	6.1996	6.2006	271.83	2,191.6	2,463.4	271.9	2,346.6	2,618.4	0.8926	6.9394	7.8320
70	31.15	0.001023	5.0452	5.0462	292.76	2,177.0	2,469.7	292.8	2,334.2	2,626.9	0.9540	6.8023	7.7563
75	38.54	0.001026	4.1328	4.1338	313.70	2,162.3	2,476.0	313.7	2,321.6	2,635.4	1.0146	6.6687	7.6834
80	47.35	0.001029	3.4074	3.4085	334.67	2,147.6	2,482.3	334.7	2,309.0	2,643.7	1.0744	6.5387	7.6131
85	57.80	0.001032	2.8276	2.8286	355.65	2,132.8	2,488.4	355.7	2,296.2	2,651.9	1.1335	6.4118	7.5453
90	70.10	0.001036	2.3604	2.3614	376.66	2,117.8	2,494.5	376.7	2,283.3	2,660.0	1.1917	6.2881	7.4798
95	84.52	0.001039	1.9806	1.9817	397.69	2,102.8	2,500.5	397.8	2,270.2	2,668.0	1.2493	6.1673	7.4166
100	101.32	0.001043	1.6689	1.6699	418.75	2,087.9	2,506.6	418.9	2,257.0	2,675.8	1.3062	6.0492	7.3554
105	120.80	0.001047	1.4142	1.4152	439.83	2,072.8	2,512.6	440.0	2,243.6	2,683.6	1.3624	5.9338	7.2962
110	143.27	0.001051	1.2063	1.2074	460.95	2,057.2	2,518.2	461.1	2,230.0	2,691.1	1.4179	5.8209	7.2388
115	169.07	0.001056	1.0350	1.0361	482.10	2,041.3	2,523.4	482.3	2,216.3	2,698.6	1.4728	5.7105	7.1833
120	198.55	0.001060	0.89100	0.8921	503.28	2,025.4	2,528.7	503.5	2,202.3	2,705.8	1.5271	5.6023	7.1293
125	232.11	0.001065	0.76938	0.7704	524.51	2,009.6	2,534.1	524.8	2,188.2	2,712.9	1.5807	5.4962	7.0770
130	270.15	0.001070	0.66702	0.6681	545.78	1,993.6	2,539.4	546.1	2,173.8	2,719.9	1.6338	5.3922	7.0261
135	313.09	0.001075	0.58074	0.5818	567.09	1,977.3	2,544.4	567.4	2,159.2	2,726.6	1.6864	5.2902	6.9766
140	361.39	0.001080	0.50739	0.5085	588.46	1,960.9	2,549.3	588.8	2,144.3	2,733.1	1.7384	5.1900	6.9284
145	415.53	0.001085	0.44462	0.4457	609.88	1,944.3	2,554.2	610.3	2,129.1	2,739.4	1.7899	5.0916	6.8815
150	475.99	0.001091	0.39100	0.3921	631.35	1,927.5	2,558.8	631.9	2,113.6	2,745.5	1.8409	4.9948	6.8358
155	543.30	0.001096	0.34514	0.3462	652.89	1,910.3	2,563.2	653.5	2,097.8	2,751.3	1.8915	4.8996	6.7911
160	618.00	0.001102	0.30566	0.3068	674.50	1,892.8	2,567.3	675.2	2,081.7	2,756.9	1.9416	4.8059	6.7475
165	700.66	0.001108	0.27131	0.2724	696.18	1,875.1	2,571.3	697.0	2,065.2	2,762.2	1.9912	4.7135	6.7048
170	791.86	0.001114	0.24141	0.2425	717.93	1,857.2	2,575.2	718.8	2,048.4	2,767.2	2.0405	4.6224	6.6630
175	892.20	0.001121	0.21538	0.2165	739.77	1,839.0	2,578.8	740.8	2,031.2	2,772.0	2.0894	4.5325	6.6220
180	1,002.3	0.001127	0.19266	0.1938	761.69	1,820.5	2,582.1	762.8	2,013.6	2,776.4	2.1380	4.4437	6.5817
185	1,122.9	0.001134	0.17272	0.1739	783.70	1,801.6	2,585.3	785.0	1,995.5	2,780.5	2.1862	4.3559	6.5421
190	1,254.5	0.001141	0.15513	0.1563	805.80	1,782.4	2,588.3	807.2	1,977.1	2,784.3	2.2341	4.2691	6.5032
195	1,398.0	0.001148	0.13964	0.1408	828.01	1,762.9	2,590.9	829.6	1,958.1	2,787.8	2.2817	4.1834	6.4651
200	1,553.9	0.001156	0.12597	0.1271	850.32	1,743.0	2,593.3	852.1	1,938.8	2,790.9	2.3290	4.0986	6.4276
205	1,723.1	0.001164	0.11386	0.1150	872.74	1,722.7	2,595.4	874.7	1,918.9	2,793.6	2.3761	4.0147	6.3908
210	1,906.3	0.001172	0.10307	0.1042	895.28	1,702.0	2,597.3	897.5	1,898.5	2,796.0	2.4230	3.9314	6.3544
215	2,104.3	0.001180	0.09345	0.0946	917.94	1,681.0	2,598.9	920.4	1,877.6	2,798.0	2.4696	3.8485	6.3181
220	2,317.8	0.001189	0.08486	0.0860	940.73	1,659.5	2,600.2	943.5	1,856.2	2,799.7	2.5161	3.7661	6.2821
225	2,547.8	0.001198	0.07716	0.0784	963.66	1,637.6	2,601.3	966.7	1,834.2	2,800.9	2.5623	3.6841	6.2464
230	2,795.0	0.001208	0.07022	0.0714	986.73	1,615.4	2,602.2	990.1	1,811.7	2,801.8	2.6084	3.6025	6.2109
235	3,060.3	0.001218	0.06400	0.0652	1,010.0	1,592.7	2,602.7	1,013.7	1,788.6	2,802.3	2.6544	3.5213	6.1757
240	3,344.7	0.001228	0.05851	0.0597	1,033.6	1,569.1	2,602.5	1,037.5	1,764.8	2,802.3	2.7002	3.4404	6.1406
245	3,649.0	0.001239	0.05353	0.0548	1,056.9	1,545.2	2,602.1	1,061.4	1,740.5	2,801.9	2.7460	3.3597	6.1057

T_{sat}, °C	P_{sat}, kPa	Specific Volume, m³/kg			Internal Energy, kJ/kg			Enthalpy, kJ/kg			Entropy, kJ/(kg · K)		
		v_l	v_{lg}	v_g	u_l	u_{lg}	u_g	h_l	h_{lg}	h_g	s_l	s_{lg}	s_g
250	3,974.2	0.001250	0.04893	0.0502	1,080.7	1,521.0	2,601.7	1,085.6	1,715.5	2,801.2	2.7917	3.2792	6.0708
255	4,321.3	0.001262	0.04471	0.0460	1,104.6	1,496.7	2,601.3	1,110.1	1,689.9	2,800.0	2.8373	3.1986	6.0359
260	4,691.2	0.001275	0.04086	0.0421	1,128.8	1,471.9	2,600.7	1,134.8	1,663.5	2,798.3	2.8829	3.1180	6.0009
265	5,085.0	0.001288	0.03738	0.0387	1,153.2	1,446.4	2,599.6	1,159.8	1,636.5	2,796.3	2.9286	3.0372	5.9657
270	5,503.8	0.001302	0.03424	0.0355	1,177.9	1,420.3	2,598.1	1,185.1	1,608.7	2,793.7	2.9743	2.9560	5.9303
275	5,948.6	0.001317	0.03139	0.0327	1,202.8	1,393.4	2,596.3	1,210.7	1,580.1	2,790.8	3.0200	2.8745	5.8945
280	6,420.5	0.001333	0.02878	0.0301	1,228.1	1,366.0	2,594.0	1,236.6	1,550.8	2,787.4	3.0660	2.7924	5.8584
285	6,920.8	0.001349	0.02639	0.0277	1,253.7	1,337.9	2,591.6	1,263.0	1,520.6	2,783.6	3.1121	2.7097	5.8218
290	7,450.6	0.001366	0.02418	0.0255	1,279.6	1,297.7	2,577.3	1,289.8	1,477.9	2,767.7	3.1585	2.6262	5.7847
295	8,011.1	0.001385	0.02214	0.0235	1,306.0	1,265.5	2,571.5	1,317.1	1,442.8	2,759.9	3.2052	2.5417	5.7469
300	8,603.7	0.001404	0.02025	0.0217	1,332.8	1,232.0	2,564.8	1,344.9	1,406.2	2,751.1	3.2523	2.4560	5.7083
305	9,214.4	0.001425	0.01850	0.0199	1,360.2	1,197.5	2,557.6	1,373.3	1,367.9	2,741.2	3.3000	2.3688	5.6687
310	9,869.4	0.001447	0.01688	0.0183	1,388.0	1,161.1	2,549.2	1,402.3	1,327.8	2,730.1	3.3483	2.2797	5.6279
315	10,561.	0.001470	0.01538	0.0169	1,416.5	1,123.0	2,539.6	1,432.1	1,285.5	2,717.6	3.3973	2.1884	5.5858
320	11,289.	0.001499	0.01398	0.0155	1,445.7	1,083.0	2,528.7	1,462.6	1,240.9	2,703.5	3.4473	2.0947	5.5420
325	12,056.	0.001528	0.01267	0.0142	1,475.5	1,040.9	2,516.4	1,494.0	1,193.6	2,687.5	3.4984	1.9979	5.4962
330	12,862.	0.001561	0.01143	0.0130	1,506.2	996.3	2,502.5	1,526.3	1,143.3	2,669.6	3.5507	1.8973	5.4480
335	13,712.	0.001598	0.01026	0.0119	1,537.8	949.0	2,486.8	1,559.7	1,089.6	2,649.3	3.6045	1.7922	5.3967
340	14,605.	0.001639	0.00914	0.0108	1,570.4	898.7	2,469.0	1,594.3	1,032.2	2,626.5	3.6601	1.6820	5.3420
345	15,545.	0.001686	0.00808	0.0098	1,606.3	842.1	2,448.4	1,632.5	967.7	2,600.2	3.7176	1.5658	5.2834
350	16,535.	0.001741	0.00706	0.0088	1,643.0	780.5	2,423.5	1,671.8	897.2	2,569.0	3.7775	1.4416	5.2191
355	17,577.	0.001808	0.00605	0.0079	1,682.1	710.9	2,393.0	1,713.9	817.3	2,531.2	3.8400	1.3054	5.1454
360	18,675.	0.001896	0.00504	0.0069	1,726.2	629.5	2,355.7	1,761.6	723.7	2,485.3	3.9056	1.1531	5.0587
365	19,833.	0.002016	0.00400	0.0060	1,777.9	531.0	2,308.9	1,817.8	610.3	2,428.1	3.9746	0.9822	4.9569
370	21,054.	0.002225	0.00274	0.0050	1,843.3	394.1	2,237.3	1,890.1	451.9	2,342.0	4.0476	0.7555	4.8030
374.4	22,090.	0.00315	0.00000	0.00315	2,029.6	0.0	2,029.6	2,099.3	0.0	2,099.3	4.4298	0.0	4.4298

Properties generated from the program *STEAMCALC*, Wiley Professional Software, Wiley, New York, 1984.

TABLE D.9 Properties of Saturated Water — Pressure Table, SI Units

P_{sat}, kPa	T_{sat}, °C	Specific Volume, m³/kg			Internal Energy, kJ/kg			Enthalpy, kJ/kg			Entropy, kJ/(kg · K)		
		v_l	v_{lg}	v_g	u_l	u_{lg}	u_g	h_l	h_{lg}	h_g	s_l	s_{lg}	s_g
1.00	7.0	0.001000	129.08	129.08	29.40	2,356.1	2,385.5	29.4	2,485.2	2,514.6	0.1058	8.8704	8.9763
1.50	13.0	0.001001	88.067	88.068	54.68	2,339.3	2,394.0	54.7	2,471.4	2,526.1	0.1956	8.6337	8.8292
2.00	17.5	0.001001	67.073	67.074	73.41	2,326.8	2,400.2	73.4	2,460.9	2,534.3	0.2607	8.4642	8.7249
2.50	21.1	0.001002	54.290	54.291	88.41	2,316.7	2,405.1	88.4	2,452.4	2,540.8	0.3120	8.3322	8.6442
3.00	24.1	0.001003	45.751	45.752	100.96	2,308.0	2,409.0	101.0	2,445.3	2,546.2	0.3545	8.2240	8.5786
3.50	26.7	0.001003	39.483	39.484	111.81	2,300.9	2,412.7	111.8	2,439.1	2,550.9	0.3908	8.1324	8.5233
4.00	29.0	0.001004	34.779	34.780	121.37	2,294.5	2,415.9	121.4	2,433.6	2,555.0	0.4226	8.0529	8.4756
4.50	31.0	0.001005	31.128	31.129	129.95	2,288.6	2,418.6	130.0	2,428.7	2,558.7	0.4509	7.9827	8.4336
5.00	32.9	0.001005	28.194	28.195	137.73	2,283.3	2,421.0	137.7	2,424.3	2,562.0	0.4764	7.9197	8.3961
5.50	34.6	0.001006	25.773	25.774	144.86	2,278.4	2,423.3	144.9	2,420.2	2,565.0	0.4996	7.8626	8.3622
6.00	36.2	0.001006	23.742	23.743	151.45	2,274.0	2,425.4	151.5	2,416.4	2,567.9	0.5209	7.8104	8.3313
6.50	37.7	0.001007	22.013	22.014	157.58	2,269.8	2,427.4	157.6	2,412.9	2,570.5	0.5407	7.7623	8.3030
7.00	39.0	0.001007	20.522	20.523	163.31	2,266.0	2,429.3	163.3	2,409.6	2,572.9	0.5590	7.7177	8.2768
7.50	40.3	0.001008	19.225	19.226	168.70	2,262.3	2,431.0	168.7	2,406.5	2,575.2	0.5763	7.6762	8.2524
8.00	41.5	0.001008	18.086	18.087	173.79	2,258.9	2,432.7	173.8	2,403.6	2,577.4	0.5924	7.6372	8.2296
8.50	42.7	0.001009	17.080	17.081	178.61	2,255.6	2,434.2	178.6	2,400.8	2,579.4	0.6077	7.6006	8.2083
9.00	43.8	0.001009	16.185	16.186	183.19	2,252.5	2,435.7	183.2	2,398.2	2,581.4	0.6222	7.5660	8.1881
9.50	44.8	0.001010	15.383	15.384	187.56	2,249.5	2,437.1	187.6	2,395.7	2,583.2	0.6359	7.5332	8.1691
10.00	45.8	0.001010	14.660	14.661	191.74	2,246.7	2,438.4	191.7	2,393.3	2,585.0	0.6490	7.5021	8.1511
15.00	54.0	0.001014	10.020	10.021	225.83	2,223.2	2,449.0	225.8	2,373.5	2,599.4	0.7544	7.2548	8.0092
20.00	60.1	0.001017	7.6483	7.6493	251.28	2,205.7	2,457.0	251.3	2,358.7	2,610.0	0.8314	7.0779	7.9093
25.00	65.0	0.001020	6.2015	6.2025	271.80	2,191.6	2,463.3	271.8	2,346.6	2,618.4	0.8925	6.9396	7.8321
30.00	69.1	0.001022	5.2277	5.2287	289.09	2,179.5	2,468.6	289.1	2,336.3	2,625.5	0.9433	6.8260	7.7693
35.00	72.7	0.001024	4.5249	4.5259	304.11	2,169.0	2,473.1	304.1	2,327.4	2,631.5	0.9869	6.7295	7.7164
40.00	75.9	0.001026	3.9918	3.9929	317.42	2,159.7	2,477.1	317.5	2,319.4	2,636.8	1.0253	6.6455	7.6707
45.00	78.7	0.001028	3.5744	3.5755	329.40	2,151.3	2,480.7	329.4	2,312.2	2,641.6	1.0595	6.5710	7.6305
50.00	81.3	0.001030	3.2389	3.2398	340.31	2,143.6	2,483.9	340.4	2,305.5	2,645.9	1.0904	6.5042	7.5946
60.00	86.0	0.001033	2.7305	2.7316	359.66	2,129.9	2,489.6	359.7	2,293.7	2,653.5	1.1446	6.3880	7.5326
70.00	90.0	0.001036	2.3638	2.3648	376.49	2,117.9	2,494.4	376.6	2,283.4	2,659.9	1.1913	6.2891	7.4804
80.00	93.5	0.001038	2.0859	2.0869	391.43	2,107.2	2,498.7	391.5	2,274.1	2,665.6	1.2323	6.2029	7.4352
90.00	96.7	0.001041	1.8667	1.8678	404.90	2,097.7	2,502.6	405.0	2,265.7	2,670.7	1.2689	6.1265	7.3954
100.00	99.6	0.001043	1.6898	1.6908	417.20	2,089.0	2,506.2	417.3	2,258.0	2,675.3	1.3020	6.0578	7.3598
101.32	100.0	0.001043	1.66895	1.6700	418.74	2,087.9	2,506.6	418.8	2,257.0	2,675.8	1.3062	6.0493	7.3554
125.00	106.0	0.001048	1.36965	1.3707	444.01	2,069.7	2,513.7	444.1	2,240.9	2,685.1	1.3734	5.9113	7.2847
150.00	111.4	0.001053	1.15612	1.1572	466.74	2,052.9	2,519.6	466.9	2,226.3	2,693.2	1.4330	5.7904	7.2234
175.00	116.1	0.001057	1.00248	1.0035	486.58	2,037.9	2,524.5	486.8	2,213.4	2,700.1	1.4844	5.6873	7.1717
200.00	120.2	0.001060	0.88498	0.8860	504.25	2,024.7	2,529.0	504.5	2,201.7	2,706.2	1.5295	5.5974	7.1269
225.00	124.0	0.001064	0.79229	0.7934	520.22	2,012.8	2,533.0	520.5	2,191.1	2,711.5	1.5700	5.5175	7.0874
250.00	127.4	0.001067	0.71751	0.7186	534.82	2,001.9	2,536.7	535.1	2,181.2	2,716.3	1.6066	5.4455	7.0521
275.00	130.6	0.001070	0.65602	0.6571	548.30	1,991.7	2,540.0	548.6	2,172.1	2,720.7	1.6401	5.3800	7.0201
300.00	133.5	0.001073	0.60457	0.6056	560.83	1,982.1	2,542.9	561.2	2,163.5	2,724.6	1.6710	5.3199	6.9910
325.00	136.3	0.001076	0.56082	0.5619	572.57	1,973.1	2,545.7	572.9	2,155.4	2,728.3	1.6998	5.2643	6.9641
350.00	138.9	0.001079	0.52305	0.5241	583.60	1,964.6	2,548.2	584.0	2,147.7	2,731.6	1.7266	5.2126	6.9392
375.00	141.3	0.001081	0.49007	0.4911	594.03	1,956.6	2,550.6	594.4	2,140.3	2,734.8	1.7519	5.1642	6.9161
400.00	143.6	0.001084	0.46105	0.4621	603.93	1,948.9	2,552.8	604.4	2,133.3	2,737.7	1.7757	5.1187	6.8944
425.00	145.8	0.001086	0.43534	0.4364	613.36	1,941.6	2,554.9	613.8	2,126.6	2,740.4	1.7982	5.0758	6.8740
450.00	147.9	0.001088	0.41242	0.4135	622.35	1,934.5	2,556.9	622.8	2,120.1	2,743.0	1.8196	5.0352	6.8548
475.00	149.9	0.001091	0.39188	0.3930	630.97	1,927.7	2,558.7	631.5	2,113.9	2,745.4	1.8400	4.9966	6.8366
500.00	151.8	0.001093	0.37336	0.3745	639.24	1,921.2	2,560.4	639.8	2,107.9	2,747.6	1.8595	4.9598	6.8193
550.00	155.5	0.001097	0.34129	0.3424	654.86	1,908.7	2,563.5	655.5	2,096.4	2,751.8	1.8960	4.8910	6.7871

TABLE D.9   Continued

P_{sat}, kPa	T_{sat}, °C	Specific Volume, m³/kg			Internal Energy, kJ/kg			Enthalpy, kJ/kg			Entropy, kJ/(kg · K)		
		v_l	v_{lg}	v_g	u_l	u_{lg}	u_g	h_l	h_{lg}	h_g	s_l	s_{lg}	s_g
600.00	158.8	0.001101	0.31443	0.3155	669.42	1,896.9	2,566.3	670.1	2,085.5	2,755.6	1.9298	4.8278	6.7576
650.00	162.0	0.001104	0.29151	0.2926	683.06	1,885.8	2,568.8	683.8	2,075.2	2,759.0	1.9613	4.7692	6.7305
700.00	164.9	0.001108	0.27168	0.2728	695.93	1,875.3	2,571.2	696.7	2,065.4	2,762.1	1.9907	4.7146	6.7053
750.00	167.7	0.001111	0.25439	0.2555	708.11	1,865.3	2,573.4	708.9	2,056.0	2,765.0	2.0183	4.6634	6.6817
800.00	170.4	0.001115	0.23919	0.2403	719.68	1,855.7	2,575.4	720.6	2,047.0	2,767.6	2.0445	4.6152	6.6597
850.00	172.9	0.001118	0.22572	0.2268	730.71	1,846.5	2,577.2	731.7	2,038.4	2,770.0	2.0692	4.5696	6.6388
900.00	175.3	0.001121	0.21372	0.2148	741.27	1,837.7	2,578.9	742.3	2,030.0	2,772.3	2.0928	4.5264	6.6192
950.00	177.7	0.001124	0.20295	0.2041	751.39	1,829.1	2,580.5	752.5	2,021.9	2,774.3	2.1152	4.4853	6.6005
1,000	179.9	0.001127	0.19322	0.1943	761.11	1,820.8	2,581.9	762.2	2,014.0	2,776.3	2.1367	4.4460	6.5827
1,100	184.1	0.001133	0.17631	0.1774	779.52	1,805.1	2,584.6	780.8	1,999.0	2,779.8	2.1771	4.3725	6.5495
1,200	187.9	0.001138	0.16209	0.1632	796.71	1,790.2	2,586.9	798.1	1,984.7	2,782.8	2.2145	4.3047	6.5191
1,300	191.6	0.001143	0.14998	0.1511	812.87	1,776.1	2,589.0	814.4	1,971.1	2,785.4	2.2493	4.2417	6.4910
1,400	195.0	0.001148	0.13956	0.1407	828.13	1,762.7	2,590.8	829.7	1,958.0	2,787.8	2.2820	4.1829	6.4649
1,500	198.3	0.001153	0.13050	0.1317	842.61	1,749.8	2,592.4	844.3	1,945.5	2,789.8	2.3127	4.1277	6.4405
1,600	201.4	0.001158	0.12254	0.1237	856.39	1,737.3	2,593.7	858.2	1,933.4	2,791.7	2.3418	4.0757	6.4176
1,700	204.3	0.001163	0.11549	0.1167	869.56	1,725.4	2,595.0	871.5	1,921.7	2,793.3	2.3695	4.0265	6.3960
1,800	207.1	0.001167	0.10918	0.1104	882.18	1,713.9	2,596.1	884.3	1,910.4	2,794.7	2.3958	3.9797	6.3755
1,900	209.8	0.001172	0.10351	0.1047	894.30	1,702.7	2,597.0	896.5	1,899.4	2,795.9	2.4210	3.9350	6.3559
2,000	212.4	0.001176	0.09839	0.0996	905.97	1,691.9	2,597.9	908.3	1,888.7	2,797.0	2.4450	3.8922	6.3372
2,250	218.4	0.001186	0.08751	0.0887	933.41	1,666.2	2,599.6	936.1	1,863.1	2,799.2	2.5012	3.7925	6.2936
2,500	223.9	0.001196	0.07873	0.0799	958.75	1,642.1	2,600.9	961.7	1,839.0	2,800.7	2.5525	3.7015	6.2540
2,750	229.1	0.001206	0.07147	0.0727	982.37	1,619.4	2,601.8	985.7	1,816.0	2,801.7	2.5997	3.6178	6.2176
3,000	233.8	0.001215	0.06538	0.0666	1,004.5	1,597.9	2,602.4	1,008.2	1,794.0	2,802.2	2.6437	3.5402	6.1839
3,250	238.3	0.001225	0.06028	0.0615	1,025.4	1,577.0	2,602.5	1,029.4	1,772.9	2,802.3	2.6848	3.4676	6.1524
3,500	242.5	0.001234	0.05594	0.0572	1,045.3	1,556.8	2,602.1	1,049.6	1,752.6	2,802.2	2.7234	3.3995	6.1229
3,750	246.5	0.001242	0.05208	0.0533	1,064.2	1,537.6	2,601.8	1,068.8	1,732.9	2,801.7	2.7600	3.3350	6.0950
4,000	250.3	0.001251	0.04864	0.0499	1,082.2	1,519.3	2,601.5	1,087.2	1,713.9	2,801.1	2.7947	3.2739	6.0686
5,000	263.9	0.001285	0.03811	0.0394	1,147.9	1,451.9	2,599.8	1,154.3	1,642.4	2,796.7	2.9186	3.0548	5.9734
6,000	275.5	0.001319	0.03109	0.0324	1,205.6	1,390.4	2,596.0	1,213.5	1,576.9	2,790.5	3.0251	2.8655	5.8906
7,000	285.8	0.001352	0.02603	0.0274	1,257.8	1,333.5	2,591.3	1,267.2	1,515.7	2,782.9	3.1194	2.6965	5.8159
8,000	295.0	0.001385	0.02214	0.0235	1,305.9	1,265.8	2,571.7	1,317.0	1,442.9	2,759.9	3.2050	2.5421	5.7471
9,000	303.3	0.001418	0.01907	0.0205	1,351.0	1,209.3	2,560.3	1,363.7	1,380.9	2,744.7	3.2840	2.3981	5.6821
10,000	311.0	0.001452	0.01658	0.0180	1,393.7	1,153.8	2,547.5	1,408.2	1,319.5	2,727.7	3.3580	2.2616	5.6196
11,000	318.1	0.001489	0.01450	0.0160	1,434.6	1,098.6	2,533.1	1,451.0	1,258.0	2,709.0	3.4283	2.1305	5.5588
12,000	324.7	0.001527	0.01273	0.0143	1,474.0	1,043.3	2,517.3	1,492.4	1,196.0	2,688.4	3.4958	2.0028	5.4986
13,000	331.0	0.001568	0.01119	0.0128	1,512.4	987.6	2,499.9	1,532.8	1,133.1	2,665.8	3.5611	1.8771	5.4382
14,000	336.9	0.001612	0.00984	0.0114	1,549.8	931.1	2,480.9	1,572.4	1,068.8	2,641.2	3.6249	1.7520	5.3769
15,000	342.4	0.001661	0.00862	0.0103	1,586.6	873.4	2,460.0	1,611.5	1,002.8	2,614.3	3.6877	1.6265	5.3142
16,000	347.7	0.001715	0.00752	0.0092	1,626.1	810.1	2,436.2	1,653.6	930.4	2,584.0	3.7498	1.4996	5.2494
17,000	352.3	0.001769	0.00660	0.0084	1,660.2	750.3	2,410.5	1,690.3	862.5	2,552.8	3.8054	1.3819	5.1872
18,000	357.0	0.001839	0.00566	0.0075	1,698.6	680.8	2,379.4	1,731.7	782.6	2,514.3	3.8652	1.2481	5.1134
19,000	361.4	0.001926	0.00475	0.0067	1,740.1	603.6	2,343.6	1,776.7	693.9	2,470.5	3.9249	1.1063	5.0312
20,000	365.7	0.002037	0.00384	0.0059	1,785.8	515.1	2,300.9	1,826.6	591.9	2,418.5	3.9846	0.9568	4.9414
21,000	369.8	0.002208	0.00281	0.0050	1,839.7	401.7	2,241.4	1,886.0	460.8	2,346.8	4.0443	0.7681	4.8124
22,000	373.7	0.002623	0.00114	0.0038	1,944.6	174.0	2,118.6	2,002.3	199.0	2,201.3	4.1042	0.4563	4.5605
22,090	374.4	0.00315	0.00000	0.00315	2,029.6	0.0	2,029.6	2,099.3	0.0	2,099.3	4.4298	0.0	4.4298

Properties generated from the program *STEAMCALC*, Wiley Professional Software, Wiley, New York, 1984.

TABLE D.10 Properties of Superheated Steam — SI Units

P, kPa	T, °C	v, m³/kg	u, kJ/kg	h, kJ/kg	s, kJ/(kg · K)	P, kPa	T, °C	v, m³/kg	u, kJ/kg	h, kJ/kg	s, kJ/(kg · K)
10 (T_{sat} = 45.8°C)						**101.33** (T_{sat} = 100.0°C)					
	100	17.196	2516.2	2688.1	8.4498		150	1.9109	2583.1	2776.7	7.6084
	150	19.513	2588.2	2783.3	8.6893		200	2.1439	2658.0	2875.3	7.8286
	200	21.826	2661.2	2879.5	8.9040		250	2.3747	2733.3	2973.9	8.0268
	250	24.136	2735.5	2976.8	9.0996		300	2.6043	2809.6	3073.4	8.2085
	300	26.446	2811.2	3075.6	9.2799		350	2.8334	2887.1	3174.2	8.3770
	350	28.754	2888.3	3175.9	9.4476		400	3.0621	2966.1	3276.3	8.5347
	400	31.063	2967.1	3277.7	9.6048		450	3.2905	3046.6	3380.0	8.6832
	450	33.371	3047.4	3381.1	9.7530		500	3.5189	3128.7	3485.3	8.8240
	500	35.679	3129.4	3486.2	9.8935		550	3.7471	3212.5	3592.2	8.9580
	550	37.987	3213.1	3593.0	10.027		600	3.9752	3298.0	3700.8	9.0860
	600	40.295	3298.5	3701.5	10.155		650	4.2033	3385.2	3811.1	9.2089
	650	42.603	3385.7	3811.7	10.278		700	4.4313	3474.1	3923.1	9.3271
	700	44.911	3474.5	3923.7	10.396		750	4.6593	3564.8	4036.9	9.4411
	750	47.219	3565.2	4037.4	10.510		800	4.8872	3657.3	4152.4	9.5513
	800	49.526	3657.6	4152.9	10.620		850	5.1152	3751.5	4269.8	9.6582
	850	51.834	3751.8	4270.2	10.727						
50 (T_{sat} = 81.3°C)						**200** (T_{sat} = 120.2°C)					
	100	3.4182	2512.0	2682.9	7.6959		150	0.9596	2577.2	2769.1	7.2804
	150	3.8894	2586.0	2780.5	7.9413		200	1.0804	2654.5	2870.6	7.5072
	200	4.3561	2659.8	2877.6	8.1583		250	1.1989	2730.9	2970.7	7.7084
	250	4.8206	2734.5	2975.6	8.3551		300	1.3162	2807.8	3071.1	7.8916
	300	5.2840	2810.5	3074.7	8.5360		350	1.4329	2885.8	3172.4	8.0610
	350	5.7468	2887.8	3175.1	8.7040		400	1.5492	2965.0	3274.9	8.2192
	400	6.2092	2966.6	3277.1	8.8614		450	1.6653	3045.7	3378.8	8.3682
	450	6.6715	3047.1	3380.6	9.0098		500	1.7813	3128.0	3484.3	8.5092
	500	7.1336	3129.1	3485.8	9.1504		550	1.8971	3211.9	3591.3	8.6434
	550	7.5956	3212.9	3592.6	9.2843		600	2.0129	3297.4	3700.0	8.7716
	600	8.0575	3298.3	3701.2	9.4123		650	2.1286	3384.7	3810.4	8.8945
	650	8.5193	3385.5	3811.4	9.5351		700	2.2442	3473.7	3922.5	9.0128
	700	8.9811	3474.4	3923.4	9.6532		750	2.3598	3564.4	4036.4	9.1268
	750	9.4428	3565.0	4037.2	9.7672		800	2.4754	3656.9	4152.0	9.2372
	800	9.9045	3657.5	4152.7	9.8775		850	2.5909	3751.1	4269.3	9.3441
	850	10.366	3751.7	4270.0	9.9843						
100 (T_{sat} = 99.6°C)						**300** (T_{sat} = 133.5°C)					
	100	1.6956	2506.4	2676.0	7.3610		150	0.6338	2570.8	2760.9	7.0779
	150	1.9363	2583.1	2776.8	7.6146		200	0.7164	2650.8	2865.7	7.3122
	200	2.1724	2658.1	2875.3	7.8347		250	0.7965	2728.5	2967.4	7.5165
	250	2.4062	2733.3	2974.0	8.0329		300	0.8753	2806.1	3068.7	7.7014
	300	2.6388	2809.6	3073.5	8.2146		350	0.9535	2884.4	3170.5	7.8717
	350	2.8708	2887.1	3174.2	8.3831		400	1.0314	2963.9	3273.4	8.0305
	400	3.1025	2966.1	3276.4	8.5407		450	1.1091	3044.8	3377.6	8.1798
	450	3.3340	3046.6	3380.0	8.6893		500	1.1866	3127.2	3483.2	8.3211
	500	3.5654	3128.8	3485.3	8.8300		550	1.2639	3211.2	3590.4	8.4554
	550	3.7966	3212.5	3592.2	8.9640		600	1.3413	3296.9	3699.2	8.5838
	600	4.0277	3298.0	3700.8	9.0921		650	1.4185	3384.2	3809.7	8.7068
	650	4.2588	3385.2	3811.1	9.2149		700	1.4957	3473.2	3921.9	8.8252
	700	4.4898	3474.1	3923.1	9.3331		750	1.5728	3564.0	4035.8	8.9393
	750	4.7208	3564.8	4036.9	9.4471		800	1.6499	3656.5	4151.5	9.0497
	800	4.9518	3657.3	4152.4	9.5574		850	1.7270	3750.8	4268.9	9.1566
	850	5.1827	3751.5	4269.8	9.6643						

$P,$ kPa	$T,$ °C	$v,$ m³/kg	$u,$ kJ/kg	$h,$ kJ/kg	$s,$ kJ/(kg · K)	$P,$ kPa	$T,$ °C	$v,$ m³/kg	$u,$ kJ/kg	$h,$ kJ/kg	$s,$ kJ/(kg · K)
400 (T_{sat} = 143.6°C)						1000 (T_{sat} = 179.9°C)					
	150	0.4707	2563.9	2752.2	6.9287		200	0.2059	2621.4	2827.3	6.6930
	200	0.5343	2647.0	2860.7	7.1712		250	0.2328	2710.0	2942.8	6.9251
	250	0.5952	2726.0	2964.1	7.3789		300	0.2580	2793.1	3051.1	7.1229
	300	0.6549	2804.3	3066.2	7.5655		350	0.2824	2874.7	3157.1	7.3003
	350	0.7139	2883.1	3168.6	7.7367		400	0.3065	2956.3	3262.7	7.4633
	400	0.7725	2962.9	3271.9	7.8961		450	0.3303	3038.5	3368.8	7.6154
	450	0.8309	3044.0	3376.3	8.0458		500	0.3540	3121.9	3475.9	7.7585
	500	0.8892	3126.5	3482.2	8.1873		550	0.3775	3206.6	3584.2	7.8942
	550	0.9474	3210.6	3589.5	8.3219		600	0.4010	3292.8	3693.8	8.0235
	600	1.0054	3296.3	3698.5	8.4504		650	0.4244	3380.6	3805.0	8.1473
	650	1.0635	3383.7	3809.0	8.5735		700	0.4477	3470.0	3917.7	8.2662
	700	1.1214	3472.7	3921.3	8.6919		750	0.4710	3561.0	4032.0	8.3808
	750	1.1793	3563.5	4035.3	8.8062		800	0.4943	3653.8	4148.1	8.4916
	800	1.2372	3656.1	4151.0	8.9166		850	0.5175	3748.3	4265.8	8.5988
	850	1.2951	3750.4	4268.4	9.0236						
600 (T_{sat} = 158.8°C)						1500 (T_{sat} = 198.3°C)					
	200	0.3521	2639.0	2850.2	6.9669		250	0.15199	2695.4	2923.4	6.7093
	250	0.3939	2720.8	2957.2	7.1819		300	0.16971	2783.3	3037.8	6.9183
	300	0.4344	2800.6	3061.3	7.3719		350	0.18654	2867.4	3147.2	7.1014
	350	0.4742	2880.3	3164.8	7.5451		400	0.20292	2950.6	3255.0	7.2677
	400	0.5136	2960.7	3268.9	7.7057		450	0.21906	3034.0	3362.5	7.4219
	450	0.5528	3042.2	3373.8	7.8562		500	0.23503	3118.1	3470.6	7.5664
	500	0.5919	3125.0	3480.1	7.9982		550	0.25089	3203.3	3579.7	7.7030
	550	0.6308	3209.3	3587.7	8.1332		600	0.26666	3289.9	3689.9	7.8331
	600	0.6696	3295.1	3696.9	8.2619		650	0.28237	3378.0	3801.5	7.9574
	650	0.7084	3382.6	3807.7	8.3853		700	0.29803	3467.6	3914.7	8.0767
	700	0.7471	3471.8	3920.1	8.5039		750	0.31364	3558.9	4029.4	8.1917
	750	0.7858	3562.7	4034.2	8.6182		800	0.32921	3651.8	4145.7	8.3027
	800	0.8245	3655.3	4150.0	8.7287		850	0.34475	3746.5	4263.6	8.4101
	850	0.8631	3749.7	4267.6	8.8358						
800 (T_{sat} = 170.4°C)						2000 (T_{sat} = 212.4°C)					
	200	0.2608	2630.4	2839.1	6.8156		250	0.11145	2679.5	2902.4	6.5451
	250	0.2932	2715.5	2950.1	7.0388		300	0.12550	2772.9	3023.9	6.7671
	300	0.3241	2796.9	3056.2	7.2326		350	0.13856	2860.0	3137.1	6.9565
	350	0.3544	2877.5	3161.0	7.4079		400	0.15113	2944.8	3247.1	7.1263
	400	0.3842	2958.5	3265.8	7.5697		450	0.16343	3029.3	3356.1	7.2826
	450	0.4137	3040.4	3371.4	7.7209		500	0.17556	3114.2	3465.3	7.4286
	500	0.4432	3123.5	3478.0	7.8635		550	0.18757	3200.0	3575.1	7.5662
	550	0.4725	3208.0	3586.0	7.9988		600	0.19950	3287.0	3686.0	7.6970
	600	0.5017	3294.0	3695.4	8.1278		650	0.21137	3375.4	3798.1	7.8218
	650	0.5309	3381.6	3806.3	8.2514		700	0.22318	3465.3	3911.6	7.9416
	700	0.5600	3470.9	3918.9	8.3702		750	0.23494	3556.8	4026.7	8.0569
	750	0.5891	3561.9	4033.1	8.4846		800	0.24667	3649.9	4143.2	8.1681
	800	0.6181	3654.6	4149.1	8.5953		850	0.25836	3744.7	4261.5	8.2758
	850	0.6471	3749.0	4266.7	8.7024						

P, kPa	T, °C	v, m³/kg	u, kJ/kg	h, kJ/kg	s, kJ/(kg · K)	P, kPa	T, °C	v, m³/kg	u, kJ/kg	h, kJ/kg	s, kJ/(kg · K)
2500						5000					
(T_{sat} = 223.9°C)						(T_{sat} = 263.9°C)					
	250	0.08699	2662.2	2879.7	6.4076		300	0.045302	2698.2	2924.7	6.2085
	300	0.09893	2762.0	3009.3	6.6446		350	0.051943	2809.9	3069.6	6.4512
	350	0.10975	2852.2	3126.6	6.8409		400	0.057792	2907.7	3196.6	6.6474
	400	0.12004	2938.9	3239.1	7.0145		450	0.063252	3000.0	3316.2	6.8188
	450	0.13005	3024.6	3349.7	7.1730		500	0.068495	3090.0	3432.5	6.9743
	500	0.13987	3110.3	3459.9	7.3205		550	0.073603	3179.5	3547.5	7.1184
	550	0.14958	3196.6	3570.6	7.4592		600	0.078617	3269.2	3662.3	7.2538
	600	0.15921	3284.1	3682.1	7.5906		650	0.083560	3359.6	3777.4	7.3820
	650	0.16876	3372.8	3794.7	7.7161		700	0.088447	3451.1	3893.4	7.5044
	700	0.17827	3462.9	3908.6	7.8362		750	0.093289	3543.9	4010.4	7.6216
	750	0.18772	3554.6	4024.0	7.9518		800	0.098094	3638.2	4128.7	7.7345
	800	0.19714	3648.0	4140.8	8.0633		850	0.102867	3734.0	4248.3	7.8435
	850	0.20653	3742.9	4259.3	8.1712						
3000						6000					
(T_{sat} = 233.8°C)						(T_{sat} = 275.6°C)					
	250	0.07055	2643.2	2854.9	6.2855		300	0.036146	2667.1	2884.0	6.0669
	300	0.08116	2750.6	2994.1	6.5399		350	0.042223	2790.9	3044.2	6.3354
	350	0.09053	2844.3	3115.9	6.7437		400	0.047380	2894.3	3178.6	6.5429
	400	0.09931	2932.9	3230.9	6.9213		450	0.052104	2989.7	3302.3	6.7202
	450	0.10779	3019.8	3343.1	7.0822		500	0.056592	3081.7	3421.3	6.8793
	500	0.11608	3106.3	3454.5	7.2312		550	0.060937	3172.5	3538.1	7.0258
	550	0.12426	3193.2	3566.0	7.3709		600	0.065185	3263.1	3654.2	7.1627
	600	0.13234	3281.1	3678.1	7.5031		650	0.069360	3354.3	3770.4	7.2922
	650	0.14036	3370.2	3791.3	7.6291		700	0.073479	3446.4	3887.2	7.4154
	700	0.14832	3460.6	3905.6	7.7497		750	0.077552	3539.6	4004.9	7.5333
	750	0.15624	3552.5	4021.2	7.8656		800	0.081588	3634.3	4123.8	7.6467
	800	0.16412	3646.0	4138.4	7.9774		850	0.085592	3730.4	4243.9	7.7562
	850	0.17197	3741.1	4257.1	8.0855						
4000						7000					
(T_{sat} = 250.3°C)						(T_{sat} = 285.8°C)					
	300	0.058835	2725.8	2961.2	6.3622		300	0.029459	2631.9	2838.1	5.9299
	350	0.066448	2827.6	3093.4	6.5835		350	0.035234	2770.6	3017.2	6.2301
	400	0.073377	2920.6	3214.1	6.7699		400	0.039922	2880.4	3159.8	6.4504
	450	0.079959	3010.0	3329.8	6.9358		450	0.044132	2979.1	3288.1	6.6343
	500	0.086343	3098.2	3443.6	7.0879		500	0.048087	3073.2	3409.8	6.7971
	550	0.092599	3186.4	3556.8	7.2298		550	0.051890	3165.4	3528.6	6.9460
	600	0.098764	3275.2	3670.2	7.3636		600	0.055591	3257.0	3646.1	7.0846
	650	0.10486	3364.9	3784.3	7.4907		650	0.059218	3348.9	3763.4	7.2153
	700	0.11090	3455.9	3899.5	7.6121		700	0.062788	3441.6	3881.1	7.3394
	750	0.11690	3548.2	4015.8	7.7287		750	0.066312	3535.3	3999.5	7.4580
	800	0.12285	3642.1	4133.5	7.8411		800	0.069798	3630.3	4118.9	7.5720
	850	0.12878	3737.6	4252.7	7.9496		850	0.073253	3726.7	4239.5	7.6818

TABLE D.10 Continued

P, kPa	T, °C	v, m³/kg	u, kJ/kg	h, kJ/kg	s, kJ/(kg · K)	P, kPa	T, °C	v, m³/kg	u, kJ/kg	h, kJ/kg	s, kJ/(kg · K)
8000						**25,000**					
(T_{sat} = 295.0°C)											
	300	0.024265	2592.0	2786.2	5.7926		500	0.011128	2894.0	3172.2	5.9731
	350	0.029949	2748.7	2988.3	6.1316		550	0.012721	3024.0	3342.1	6.1861
	400	0.034311	2865.8	3140.3	6.3665		600	0.014126	3139.0	3492.2	6.3632
	450	0.038145	2968.3	3273.5	6.5574		650	0.015416	3247.0	3632.4	6.5194
	500	0.041705	3064.6	3398.2	6.7243		700	0.016631	3351.6	3767.4	6.6618
	550	0.045103	3158.2	3519.0	6.8757		750	0.017790	3454.6	3899.4	6.7941
	600	0.048395	3250.8	3638.0	7.0160		800	0.018907	3557.2	4029.8	6.9186
	650	0.051612	3343.5	3756.4	7.1479		850	0.019990	3659.9	4159.6	7.0368
	700	0.054770	3436.8	3874.9	7.2729						
	750	0.057883	3531.0	3994.0	7.3922						
	800	0.060957	3626.4	4114.0	7.5068						
	850	0.064000	3723.1	4235.1	7.6170						
10,000						**30,000**					
(T_{sat} = 311.0°C)											
	350	0.022422	2699.5	2923.7	5.9448		500	0.008681	2833.1	3093.6	5.8080
	400	0.026409	2834.8	3098.9	6.2156		550	0.010166	2980.2	3285.2	6.0484
	450	0.029743	2945.8	3243.2	6.4226		600	0.011437	3103.9	3447.0	6.2393
	500	0.032760	3046.9	3374.5	6.5982		650	0.012582	3217.3	3594.7	6.4039
	550	0.035598	3143.6	3499.6	6.7549		700	0.013647	3325.6	3735.0	6.5519
	600	0.038320	3238.4	3621.6	6.8988		750	0.014655	3431.4	3871.1	6.6882
	650	0.040964	3332.6	3742.2	7.0332		800	0.015619	3536.2	4004.8	6.8158
	700	0.043547	3427.0	3862.5	7.1601		850	0.016549	3640.7	4137.2	6.9364
	750	0.046083	3522.2	3983.0	7.2809						
	800	0.048581	3618.4	4104.2	7.3965						
	850	0.051047	3715.8	4226.3	7.5077						
15,000						**35,000**					
(T_{sat} = 342.4°C)											
	350	0.011460	2536.0	2707.9	5.4667		600	0.009520	3068.0	3401.2	6.1278
	400	0.015662	2742.8	2977.8	5.8845		650	0.010562	3187.1	3556.7	6.3011
	450	0.018452	2884.1	3160.8	6.1472		700	0.011521	3299.2	3702.5	6.4548
	500	0.020796	3000.2	3312.1	6.3496		750	0.012420	3408.0	3842.6	6.5953
	550	0.022909	3105.7	3449.3	6.5216		800	0.013275	3515.0	3979.6	6.7260
	600	0.024884	3206.3	3579.6	6.6753		850	0.014095	3621.4	4114.7	6.8491
	650	0.026768	3304.8	3706.3	6.8164						
	700	0.028587	3402.4	3831.2	6.9482						
	750	0.030356	3500.1	3955.4	7.0727						
	800	0.032086	3598.3	4079.6	7.1912						
	850	0.033783	3697.4	4204.2	7.3046						
20,000						**40,000**					
(T_{sat} = 365.7°C)											
	400	0.009948	2622.8	2821.8	5.5595		650	0.009053	3156.5	3518.7	6.2077
	450	0.012707	2812.4	3066.5	5.9110		700	0.009930	3272.6	3669.9	6.3672
	500	0.014772	2949.4	3244.8	6.1497		750	0.010748	3384.2	3814.2	6.5118
	550	0.016548	3065.8	3396.8	6.3402		800	0.011521	3493.6	3954.4	6.6456
	600	0.018162	3173.2	3536.4	6.5049		850	0.012259	3601.8	4092.2	6.7712
	650	0.019672	3276.2	3669.7	6.6533						
	700	0.021112	3377.2	3799.5	6.7093						
	750	0.022499	3477.5	3927.5	6.9186						
	800	0.023845	3577.9	4054.8	7.0400						
	850	0.025159	3678.7	4181.9	7.1558						

Properties generated from the program *STEAMCALC*, Wiley Professional Software, Wiley, New York, 1984.

Figure D.1 Temperature-entropy diagram for steam. (From Lester Haar, John S. Gallagher, and George S. Kell, NBS/NRC Steam Tables, 1984. With permission from Hemisphere Publishing Corporation, New York.)

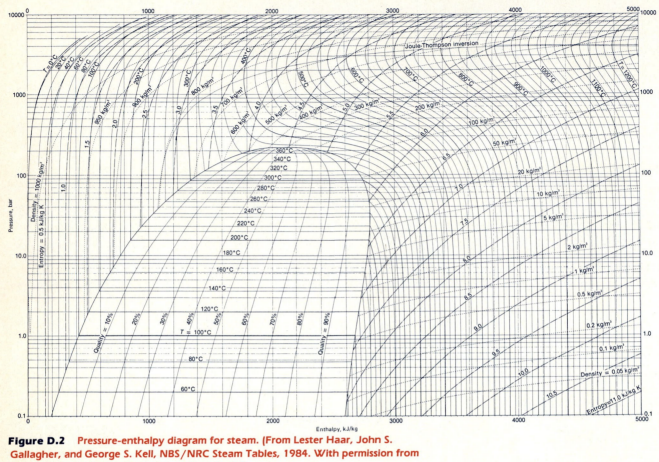

Figure D.2 *Pressure-enthalpy diagram for steam. (From Lester Haar, John S. Gallagher, and George S. Kell, NBS/NRC Steam Tables, 1984. With permission from Hemisphere Publishing Corporation, New York.)*

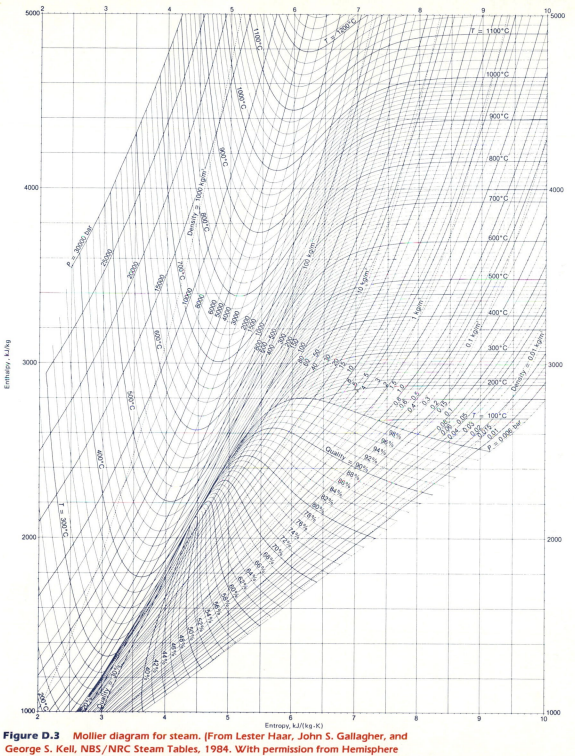

Figure D.3   Mollier diagram for steam. (From Lester Haar, John S. Gallagher, and George S. Kell, NBS/NRC Steam Tables, 1984. With permission from Hemisphere Publishing Corporation, New York.)

TABLE D.11   Properties of Saturated Refrigerant 12 — Temperature Table, SI Units

Reference state: $h_l = 200$ kJ/kg and $s_l = 1$ kJ/(kg · K) at 0°C

T_{sat}, °C	P_{sat}, kPa	Specific Volume, m³/kg			Internal Energy, kJ/kg			Enthalpy, kJ/kg			Entropy, kJ/(kg · K)		
		v_l	v_{lg}	v_g	u_l	u_{lg}	u_g	h_l	h_{lg}	h_g	s_l	s_{lg}	s_g
−100	1.2	0.0005991	10.114	10.115	112.09	181.96	294.05	112.09	193.85	305.94	0.60086	1.1195	1.7204
−90	2.8	0.0006078	4.4206	4.4211	120.64	177.21	297.85	120.64	189.76	310.40	0.64889	1.0360	1.6849
−80	6.2	0.0006169	2.1402	2.1408	129.20	172.56	301.76	129.20	185.75	314.96	0.69439	0.96171	1.6561
−70	12.3	0.0006266	1.1279	1.1285	137.78	167.97	305.75	137.79	181.79	319.58	0.73770	0.89485	1.6326
−60	22.6	0.0006369	0.63783	0.63847	146.43	163.37	309.80	146.45	177.78	324.23	0.77930	0.83407	1.6134
−50	39.1	0.0006478	0.38275	0.38340	155.13	158.77	313.89	155.15	173.74	328.89	0.81920	0.77857	1.5978
−40	64.1	0.0006594	0.24142	0.24208	163.89	154.12	318.01	163.93	169.60	333.53	0.85765	0.72744	1.5851
−30	100.3	0.0006720	0.15881	0.15948	172.73	149.41	322.14	172.80	165.34	338.14	0.89483	0.68001	1.5748
−20	150.8	0.0006854	0.10823	0.10891	181.65	144.61	326.25	181.75	160.93	342.68	0.93085	0.63571	1.5666
−10	219.0	0.0007000	0.075992	0.076692	190.66	139.69	330.35	190.81	156.33	347.14	0.96586	0.59408	1.5599
0	308.4	0.0007159	0.054710	0.055426	199.77	134.64	334.41	200.00	151.51	351.50	1.0000	0.55469	1.5547
10	423.1	0.0007332	0.040200	0.040934	209.05	129.37	338.41	209.36	146.37	355.73	1.0334	0.51695	1.5504
20	567.0	0.0007524	0.030042	0.030794	218.46	123.89	342.34	218.89	140.92	359.81	1.0663	0.48071	1.5470
30	744.6	0.0007738	0.022745	0.023519	228.07	118.11	346.18	228.65	135.05	363.70	1.0988	0.44548	1.5443
40	960.3	0.0007980	0.017383	0.018181	237.94	111.95	349.90	238.71	128.65	367.35	1.1311	0.41082	1.5419
50	1218.9	0.0008257	0.013350	0.014176	248.18	105.26	353.43	249.18	121.53	370.71	1.1636	0.37608	1.5397
60	1525.4	0.0008581	0.010259	0.011117	258.83	97.90	356.74	260.14	113.55	373.69	1.1965	0.34084	1.5374
70	1885.2	0.0008971	0.007832	0.008729	270.11	89.58	359.70	271.81	104.35	376.15	1.2304	0.30409	1.5344
80	2303.9	0.0009460	0.005879	0.006825	282.24	79.92	362.16	284.42	93.46	377.88	1.2658	0.26466	1.5304
90	2787.7	0.0010117	0.004249	0.005261	295.63	68.16	363.79	298.45	80.01	378.46	1.3040	0.22031	1.5243
100	3343.2	0.0011129	0.002793	0.003906	311.26	52.54	363.80	314.98	61.87	376.86	1.3476	0.16582	1.5134
111.6	4010	0.001792	0	0.001792	340.0	0	340.0	347.37	0	347.37	1.4267	0	1.4267

Properties generated from the program *REFRIG*, Wiley Professional Software, Wiley, New York, 1984.

TABLE D.12 Properties of Saturated Refrigerant 12 — Pressure Table, SI Units

Reference state: $h_l = 200$ kJ/kg and $s_l = 1$ kJ/(kg · K) at 0°C

P_{sat}, kPa	T_{sat}, °C	Specific Volume, m³/kg			Internal Energy, kJ/kg			Enthalpy, kJ/kg			Entropy, kJ/(kg · K)		
		v_l	v_{lg}	v_g	u_l	u_{lg}	u_g	h_l	h_{lg}	h_g	s_l	s_{lg}	s_g
2	−94.1	0.0006041	6.1375	6.1381	117.11	179.16	296.27	117.11	191.43	308.55	0.62939	1.0693	1.6987
4	−85.7	0.0006116	3.2067	3.2073	124.29	175.21	299.51	124.30	188.04	312.34	0.66860	1.0033	1.6719
6	−80.4	0.0006166	2.1954	2.1960	128.88	172.73	301.61	128.88	185.90	314.79	0.69273	0.96434	1.6571
8	−76.3	0.0006204	1.6785	1.6791	132.33	170.88	303.20	132.33	184.30	316.64	0.71044	0.93649	1.6469
10	−73.1	0.0006235	1.3632	1.3638	135.13	169.38	304.51	135.13	183.01	318.15	0.72455	0.91476	1.6393
20	−62.1	0.0006347	0.71472	0.71535	144.62	164.32	308.95	144.64	178.62	323.25	0.77078	0.84628	1.6171
30	−55.0	0.0006423	0.49002	0.49067	150.80	161.06	311.85	150.82	175.76	326.57	0.79956	0.80555	1.6051
40	−49.6	0.0006483	0.37487	0.37552	155.50	158.57	314.07	155.53	173.56	329.09	0.82089	0.77627	1.5972
50	−45.2	0.0006533	0.30449	0.30514	159.36	156.52	315.88	159.39	171.75	331.14	0.83798	0.75332	1.5913
60	−41.4	0.0006578	0.25688	0.25754	162.66	154.78	317.43	162.69	170.19	332.88	0.85233	0.73438	1.5867
70	−38.1	0.0006617	0.22245	0.22311	165.55	153.24	318.79	165.60	168.81	334.41	0.86475	0.71824	1.5830
80	−35.2	0.0006654	0.19634	0.19701	168.14	151.86	320.00	168.19	167.57	335.76	0.87573	0.70415	1.5799
90	−32.5	0.0006687	0.17585	0.17652	170.50	150.60	321.10	170.56	166.43	336.99	0.88558	0.69164	1.5772
100	−30.1	0.0006719	0.15932	0.15999	172.66	149.45	322.10	172.73	165.38	338.10	0.89454	0.68037	1.5749
200	−12.5	0.0006962	0.082849	0.083545	188.38	140.94	329.32	188.52	157.51	346.03	0.95715	0.60432	1.5615
300	−0.8	0.0007145	0.056197	0.056912	199.00	135.07	334.07	199.21	151.93	351.14	0.99712	0.55793	1.5551
400	8.2	0.0007299	0.042478	0.043208	207.34	130.34	337.68	207.63	147.34	354.97	1.0274	0.52373	1.5511
500	15.6	0.0007437	0.034075	0.034819	214.31	126.32	340.63	214.68	143.36	358.04	1.0520	0.49647	1.5484
600	22.0	0.0007565	0.028376	0.029133	220.38	122.75	343.13	220.84	139.77	360.61	1.0729	0.47354	1.5465
700	27.7	0.0007686	0.024244	0.025013	225.81	119.49	345.30	226.35	136.46	362.81	1.0913	0.45362	1.5449
800	32.8	0.0007802	0.021105	0.021886	230.76	116.46	347.22	231.39	133.34	364.73	1.1077	0.43589	1.5436
900	37.4	0.0007913	0.018635	0.019426	235.33	113.60	348.94	236.05	130.38	366.42	1.1227	0.41984	1.5425
1000	41.7	0.0008023	0.016637	0.017439	239.60	110.89	350.49	240.40	127.53	367.93	1.1365	0.40510	1.5416
1500	59.2	0.0008554	0.010470	0.011326	258.00	98.50	356.49	259.28	114.20	373.48	1.1940	0.34359	1.5375
2000	72.9	0.0009100	0.007226	0.008136	273.52	86.96	360.47	275.34	101.41	376.75	1.2404	0.29305	1.5335
2500	84.2	0.0009711	0.005159	0.006130	287.70	75.28	362.98	290.13	88.18	378.31	1.2815	0.24674	1.5282
3000	94.0	0.0010459	0.003659	0.004704	301.50	62.57	364.07	304.64	73.55	378.19	1.3205	0.20032	1.5208
3500	102.6	0.0011515	0.002416	0.003568	315.96	47.32	363.29	319.99	55.78	375.77	1.3606	0.14844	1.5090
4000	110.3	0.0013844	0.001023	0.002408	334.75	22.56	357.31	340.29	26.65	366.94	1.4126	0.06949	1.4821
4010	111.6	0.001792	0	0.001792	340.0	0	340.0	347.37	0	347.37	1.4267	0	1.4267

Properties generated from the program *REFRIG,* Wiley Professional Software, Wiley, New York, 1984.

TABLE D.13 Properties of Superheated Refrigerant 12 — SI Units

Reference state: $h_l = 200$ kJ/kg and $s_l = 1$ kJ/(kg · K) at 0°C

P, kPa	T, °C	v, m³/kg	u, kJ/kg	h, kJ/kg	s, kJ/(kg · K)	P, kPa	T, °C	v, m³/kg	u, kJ/kg	h, kJ/kg	s, kJ/(kg · K)
20 ($T_{sat} = -62.1$°C)						100 ($T_{sat} = -30.1$°C)					
	−50.0	0.7585	314.36	329.53	1.6460		−25.0	0.1639	324.65	341.04	1.5868
	−25.0	0.8466	326.09	343.02	1.7032		0.0	0.1826	337.45	355.71	1.6432
	0.0	0.9339	338.52	357.20	1.7576		25.0	0.2009	350.78	370.87	1.6962
	25.0	1.0209	351.60	372.02	1.8096		50.0	0.2188	364.64	386.53	1.7466
	50.0	1.1075	365.29	387.44	1.8592		75.0	0.2366	379.01	402.67	1.7947
	75.0	1.1940	379.53	403.41	1.9068		100.0	0.2542	393.84	419.27	1.8408
	100.0	1.2804	394.28	419.89	1.9525		125.0	0.2717	409.12	436.29	1.8849
	125.0	1.3666	409.49	436.83	1.9964		150.0	0.2892	424.80	453.72	1.9273
	150.0	1.4528	425.13	454.18	2.0387		175.0	0.3066	440.85	471.51	1.9682
	175.0	1.5390	441.14	471.92	2.0794		200.0	0.3240	457.25	489.64	2.0075
	200.0	1.6251	457.51	490.01	2.1187						
40 ($T_{sat} = -49.6$°C)						200 ($T_{sat} = -12.5$°C)					
	−25.0	0.4200	325.74	342.53	1.6542		0.0	0.08861	336.04	353.76	1.5904
	0.0	0.4645	338.26	356.83	1.7091		25.0	0.09833	349.72	369.38	1.6451
	25.0	0.5084	351.40	371.74	1.7612		50.0	0.10771	363.81	385.35	1.6965
	50.0	0.5521	365.13	387.22	1.8111		75.0	0.11687	378.34	401.71	1.7452
	75.0	0.5956	379.40	403.23	1.8588		100.0	0.12590	393.29	418.47	1.7917
	100.0	0.6390	394.17	419.73	1.9046		125.0	0.13483	408.65	435.62	1.8362
	125.0	0.6823	409.40	436.69	1.9485		150.0	0.14370	424.39	453.13	1.8788
	150.0	0.7255	425.05	454.07	1.9909		175.0	0.15254	440.49	471.00	1.9198
	175.0	0.7687	441.07	471.82	2.0316		200.0	0.16131	456.92	489.18	1.9593
	200.0	0.8119	457.44	489.92	2.0709						
60 ($T_{sat} = -41.4$°C)						400 ($T_{sat} = 8.2$°C)					
	−25.0	0.2777	325.38	342.04	1.6249		25.0	0.04692	347.46	366.22	1.5900
	0.0	0.3079	337.99	356.46	1.6802		50.0	0.05207	362.08	382.91	1.6436
	25.0	0.3376	351.20	371.45	1.7327		75.0	0.05698	376.96	399.75	1.6938
	50.0	0.3670	364.97	386.99	1.7827		100.0	0.06173	392.15	416.85	1.7412
	75.0	0.3962	379.27	403.04	1.8306		125.0	0.06638	407.69	434.24	1.7863
	100.0	0.4252	394.06	419.58	1.8764		150.0	0.07095	423.56	451.94	1.8294
	125.0	0.4542	409.31	436.56	1.9205		175.0	0.07548	439.76	469.95	1.8707
	150.0	0.4831	424.96	453.95	1.9628		200.0	0.07997	456.26	488.25	1.9104
	175.0	0.5120	441.00	471.72	2.0036						
	200.0	0.5408	457.38	489.83	2.0429						
80 ($T_{sat} = -35.2$°C)						600 ($T_{sat} = 22.0$°C)					
	−25.0	0.2066	325.01	341.54	1.6037		25.0	0.02962	344.97	362.74	1.5536
	0.0	0.2296	337.72	356.09	1.6595		50.0	0.03345	360.23	380.30	1.6101
	25.0	0.2522	350.99	371.16	1.7122		75.0	0.03698	375.52	397.70	1.6619
	50.0	0.2744	364.81	386.76	1.7625		100.0	0.04032	390.98	415.17	1.7103
	75.0	0.2965	379.14	402.86	1.8104		125.0	0.04355	406.71	432.84	1.7561
	100.0	0.3183	393.95	419.42	1.8564		150.0	0.04670	422.72	450.74	1.7997
	125.0	0.3402	409.21	436.43	1.9005		175.0	0.04980	439.01	468.89	1.8413
	150.0	0.3619	424.88	453.84	1.9429		200.0	0.05286	455.59	487.31	1.8813
	175.0	0.3836	440.93	471.62	1.9837						
	200.0	0.4053	457.31	489.74	2.0230						

						Reference state: $h_l = 200$ kJ/kg and $s_l = 1$ kJ/(kg · K) at 0°C					
P, kPa	T, °C	v, m³/kg	u, kJ/kg	h, kJ/kg	s, kJ/(kg · K)	P, kPa	T, °C	v, m³/kg	u, kJ/kg	h, kJ/kg	s, kJ/(kg · K)
800 ($T_{sat} = 32.8$°C)						2000 ($T_{sat} = 72.9$°C)					
	50.0	0.02407	358.25	377.50	1.5842		75.0	0.00831	362.23	378.86	1.5395
	75.0	0.02693	374.00	395.55	1.6379		100.0	0.01003	381.30	401.37	1.6017
	100.0	0.02959	389.77	413.45	1.6874		125.0	0.01142	399.02	421.86	1.6547
	125.0	0.03212	405.70	431.40	1.7339		150.0	0.01265	416.31	441.61	1.7027
	150.0	0.03456	421.86	449.51	1.7780		175.0	0.01379	433.50	461.08	1.7472
	175.0	0.03695	438.26	467.82	1.8200		200.0	0.01488	450.73	480.48	1.7893
	200.0	0.03930	454.92	486.36	1.8602						
1000 ($T_{sat} = 41.7$°C)						4000 ($T_{sat} = 110.3$°C)					
	50.0	0.01837	356.09	374.45	1.5620		125.0	0.004056	382.49	398.71	1.5633
	75.0	0.02087	372.41	393.28	1.6180		150.0	0.005172	404.82	425.51	1.6281
	100.0	0.02313	388.52	411.65	1.6688		175.0	0.005999	424.38	448.38	1.6803
	125.0	0.02525	404.67	429.92	1.7162		200.0	0.006708	443.05	469.89	1.7267
	150.0	0.02728	420.98	448.26	1.7607						
	175.0	0.02924	437.50	466.74	1.8031						
	200.0	0.03116	454.24	485.40	1.8436						

Properties generated from the program *REFRIG*, Wiley Professional Software, Wiley, New York, 1984.

Figure D.4   Pressure-enthalpy diagram for refrigerant 12. (Freon® 12 is the DuPont trademark for refrigerant 12.)

TABLE D.14 Properties of Saturated Ammonia—Temperature Table, SI Units

Reference state: $h_l = 200$ kJ/kg and $s_l = 1$ kJ/(kg · K) at 0°C and 1 atm

T_{sat}, °C	P_{sat}, kPa	Specific Volume, m³/kg			Internal Energy, kJ/kg			Enthalpy, kJ/kg			Entropy, kJ/(kg · K)		
		v_l	v_{lg}	v_g	u_l	u_{lg}	u_g	h_l	h_{lg}	h_g	s_l	s_{lg}	s_g
−50	40.7	0.001424	2.6357	2.6371	−33.62	1318.5	1284.9	−33.6	1425.8	1392.3	0.0566	6.3895	6.4461
−48	45.8	0.001429	2.3606	2.3621	−23.45	1310.9	1287.4	−23.4	1419.0	1395.6	0.1020	6.3023	6.4043
−46	51.4	0.001434	2.1192	2.1206	−13.47	1303.3	1289.9	−13.4	1412.3	1398.9	0.1461	6.2173	6.3635
−44	57.5	0.001439	1.9067	1.9081	−3.65	1296.0	1292.3	−3.6	1405.7	1402.1	0.1892	6.1344	6.3236
−42	64.3	0.001444	1.7192	1.7207	6.03	1288.7	1294.7	6.1	1399.2	1405.3	0.2312	6.0533	6.2845
−40	71.6	0.001449	1.5530	1.5545	15.86	1281.3	1297.1	16.0	1392.5	1408.5	0.2736	5.9727	6.2463
−38	79.7	0.001454	1.4060	1.4075	25.33	1274.1	1299.5	25.4	1386.2	1411.6	0.3140	5.8948	6.2089
−36	88.4	0.001460	1.2753	1.2768	34.71	1267.1	1301.8	34.8	1379.9	1414.7	0.3538	5.8185	6.1723
−34	97.9	0.001465	1.1590	1.1604	44.01	1260.1	1304.1	44.2	1373.6	1417.7	0.3928	5.7436	6.1365
−32	108.3	0.001470	1.0551	1.0566	53.25	1253.1	1306.3	53.4	1367.3	1420.7	0.4313	5.6701	6.1014
−30	119.5	0.001476	0.9623	0.9637	62.44	1246.1	1308.6	62.6	1361.1	1423.7	0.4692	5.5977	6.0670
−28	131.6	0.001481	0.8790	0.8805	71.58	1239.2	1310.8	71.8	1354.8	1426.6	0.5067	5.5265	6.0332
−26	144.6	0.001487	0.8043	0.8058	80.70	1232.2	1312.9	80.9	1348.5	1429.5	0.5437	5.4564	6.0001
−24	158.7	0.001492	0.7371	0.7386	89.79	1225.3	1315.0	90.0	1342.2	1432.3	0.5804	5.3873	5.9677
−22	173.9	0.001498	0.6765	0.6780	98.87	1218.3	1317.1	99.1	1335.9	1435.0	0.6167	5.3192	5.9358
−20	190.2	0.001504	0.6219	0.6234	107.94	1211.3	1319.2	108.2	1329.5	1437.7	0.6527	5.2519	5.9046
−18	207.7	0.001509	0.5724	0.5739	117.02	1204.2	1321.2	117.3	1323.1	1440.4	0.6884	5.1855	5.8739
−16	226.4	0.001515	0.5277	0.5292	126.09	1197.1	1323.2	126.4	1316.6	1443.0	0.7238	5.1199	5.8437
−14	246.5	0.001521	0.4870	0.4885	135.18	1190.0	1325.1	135.6	1310.0	1445.6	0.7590	5.0551	5.8141
−12	268.0	0.001527	0.4501	0.4516	144.28	1182.8	1327.1	144.7	1303.4	1448.1	0.7940	4.9910	5.7850
−10	290.9	0.001534	0.4165	0.4180	153.39	1175.5	1328.9	153.8	1296.7	1450.5	0.8288	4.9276	5.7564
−8	315.3	0.001540	0.3859	0.3874	162.53	1168.2	1330.8	163.0	1289.9	1452.9	0.8634	4.8649	5.7282
−6	341.3	0.001546	0.3579	0.3595	171.68	1160.9	1332.6	172.2	1283.1	1455.3	0.8978	4.8028	5.7005
−4	369.0	0.001553	0.3324	0.3340	180.86	1153.5	1334.3	181.4	1276.1	1457.6	0.9320	4.7413	5.6733
−2	398.4	0.001559	0.3090	0.3106	190.06	1146.0	1336.0	190.7	1269.1	1459.8	0.9661	4.6804	5.6465
0	429.6	0.001566	0.2876	0.2892	199.29	1138.4	1337.7	200.0	1262.0	1461.9	1.0000	4.6201	5.6201
2	462.7	0.001573	0.2679	0.2695	208.54	1130.8	1339.3	209.3	1254.8	1464.0	1.0337	4.5603	5.5941
4	497.7	0.001579	0.2498	0.2514	217.81	1123.1	1340.9	218.6	1247.5	1466.1	1.0673	4.5011	5.5684
6	534.8	0.001586	0.2332	0.2348	227.11	1115.4	1342.5	228.0	1240.1	1468.1	1.1008	4.4424	5.5432
8	573.9	0.001594	0.2179	0.2195	236.44	1107.5	1344.0	237.4	1232.6	1470.0	1.1341	4.3842	5.5183
10	615.3	0.001601	0.2038	0.2054	245.79	1099.6	1345.4	246.8	1225.0	1471.8	1.1672	4.3264	5.4937
12	658.9	0.001608	0.1907	0.1923	255.16	1091.7	1346.8	256.2	1217.4	1473.6	1.2002	4.2692	5.4694
14	704.9	0.001615	0.1787	0.1803	264.55	1083.7	1348.2	265.7	1209.6	1475.3	1.2331	4.2124	5.4455
16	753.2	0.001623	0.1675	0.1691	273.97	1075.5	1349.5	275.2	1201.7	1476.9	1.2658	4.1560	5.4218
18	804.1	0.001631	0.1572	0.1588	283.41	1067.4	1350.8	284.7	1193.7	1478.5	1.2984	4.1001	5.3984
20	857.6	0.001638	0.1476	0.1492	292.87	1059.1	1352.0	294.3	1185.7	1479.9	1.3308	4.0446	5.3753
22	913.8	0.001646	0.1387	0.1403	302.35	1050.8	1353.1	303.9	1177.5	1481.3	1.3630	3.9894	5.3524
24	972.7	0.001655	0.1304	0.1320	311.86	1042.4	1354.2	313.5	1169.2	1482.7	1.3951	3.9347	5.3298
26	1034.5	0.001663	0.1227	0.1243	321.38	1033.9	1355.3	323.1	1160.8	1483.9	1.4271	3.8803	5.3074
28	1099.3	0.001671	0.1155	0.1172	330.92	1025.3	1356.2	332.8	1152.3	1485.0	1.4589	3.8263	5.2852
30	1167.1	0.001680	0.1088	0.1105	340.48	1016.7	1357.2	342.4	1143.7	1486.1	1.4906	3.7726	5.2632
32	1238.0	0.001689	0.1026	0.1043	350.07	1007.9	1358.0	352.2	1134.9	1487.1	1.5222	3.7193	5.2414
34	1312.2	0.001698	0.0967	0.0984	359.67	999.1	1358.8	361.9	1126.1	1488.0	1.5536	3.6662	5.2198
36	1389.6	0.001707	0.0913	0.0930	369.30	990.2	1359.5	371.7	1117.1	1488.8	1.5849	3.6134	5.1983
38	1470.5	0.001716	0.0862	0.0879	378.96	981.2	1360.2	381.5	1108.0	1489.4	1.6160	3.5609	5.1769
40	1554.9	0.001725	0.0814	0.0831	388.64	972.1	1360.8	391.3	1098.7	1490.0	1.6471	3.5086	5.1557
42	1642.9	0.001735	0.0769	0.0787	398.34	962.9	1361.3	401.2	1089.3	1490.5	1.6780	3.4565	5.1346
44	1734.7	0.001745	0.0727	0.0745	408.08	953.6	1361.7	411.1	1079.8	1490.9	1.7089	3.4047	5.1135
46	1830.2	0.001755	0.0688	0.0706	417.84	944.2	1362.0	421.1	1070.1	1491.2	1.7396	3.3530	5.0926
48	1929.6	0.001766	0.0651	0.0669	427.64	934.7	1362.3	431.0	1060.3	1491.3	1.7703	3.3015	5.0717
50	2033.1	0.001776	0.0616	0.0634	437.47	925.0	1362.5	441.1	1050.3	1491.3	1.8009	3.2501	5.0509

Properties generated from the program *REFRIG*, Wiley Professional Software, Wiley, New York, 1984.

TABLE D.15 Properties of Saturated Ammonia—Pressure Table, SI Units

Reference state: $h_l = 200$ kJ/kg and $s_l = 1$ kJ/(kg · K) at 0°C

P_{sat}, kPa	T_{sat}, °C	Specific Volume, m³/kg			Internal Energy, kJ/kg			Enthalpy, kJ/kg			Entropy, kJ/(kg · K)		
		v_l	v_{lg}	v_g	u_l	u_{lg}	u_g	h_l	h_{lg}	h_g	s_l	s_{lg}	s_g
40.0	−50.3	0.001424	2.6792	2.6806	−35.13	1319.7	1284.5	−35.1	1426.8	1391.8	0.0498	6.4025	6.4523
45.0	−48.3	0.001429	2.3997	2.4012	−24.97	1312.0	1287.0	−24.9	1420.0	1395.1	0.0952	6.3153	6.4105
50.0	−46.5	0.001433	2.1745	2.1759	−15.85	1305.1	1289.3	−15.8	1413.9	1398.1	0.1356	6.2376	6.3732
55.0	−44.8	0.001437	1.9892	1.9906	−7.59	1298.9	1291.3	−7.5	1408.3	1400.8	0.1720	6.1676	6.3396
60.0	−43.2	0.001441	1.8336	1.8351	−0.00	1293.2	1293.2	0.1	1403.3	1403.3	0.2051	6.1037	6.3088
65.0	−41.8	0.001445	1.7014	1.7028	7.01	1288.0	1295.0	7.1	1398.6	1405.7	0.2355	6.0451	6.2806
70.0	−40.4	0.001448	1.5871	1.5886	13.80	1282.8	1296.6	13.9	1393.9	1407.8	0.2648	5.9897	6.2544
75.0	−39.1	0.001451	1.4879	1.4893	19.93	1278.2	1298.1	20.0	1389.8	1409.8	0.2910	5.9391	6.2301
80.0	−37.9	0.001455	1.4007	1.4021	25.69	1273.9	1299.6	25.8	1385.9	1411.7	0.3156	5.8919	6.2075
85.0	−36.8	0.001458	1.3235	1.3249	31.14	1269.8	1300.9	31.3	1382.3	1413.5	0.3387	5.8475	6.1862
90.0	−35.7	0.001460	1.2545	1.2560	36.30	1265.9	1302.2	36.4	1378.8	1415.2	0.3605	5.8057	6.1661
95.0	−34.6	0.001463	1.1925	1.1940	41.22	1262.2	1303.4	41.4	1375.5	1416.8	0.3812	5.7660	6.1471
100.0	−33.6	0.001466	1.1367	1.1381	45.92	1258.6	1304.6	46.1	1372.3	1418.4	0.4008	5.7284	6.1292
120.0	−29.9	0.001476	0.9582	0.9596	62.87	1245.8	1308.7	63.0	1360.8	1423.8	0.4710	5.5943	6.0654
140.0	−26.7	0.001485	0.8292	0.8307	77.55	1234.6	1312.2	77.8	1350.7	1428.5	0.5310	5.4805	6.0115
160.0	−23.8	0.001493	0.7315	0.7330	90.59	1224.6	1315.2	90.8	1341.7	1432.5	0.5836	5.3813	5.9649
180.0	−21.2	0.001500	0.6549	0.6564	102.36	1215.6	1317.9	102.6	1333.5	1436.1	0.6306	5.2932	5.9237
200.0	−18.9	0.001507	0.5931	0.5946	113.12	1207.2	1320.4	113.4	1325.9	1439.3	0.6731	5.2139	5.8870
220.0	−16.7	0.001513	0.5422	0.5437	123.05	1199.5	1322.5	123.4	1318.8	1442.2	0.7120	5.1418	5.8538
240.0	−14.6	0.001519	0.4995	0.5010	132.30	1192.2	1324.5	132.7	1312.1	1444.8	0.7479	5.0755	5.8234
260.0	−12.7	0.001525	0.4631	0.4646	140.97	1185.4	1326.4	141.4	1305.8	1447.2	0.7813	5.0142	5.7955
280.0	−10.9	0.001531	0.4318	0.4333	149.14	1178.9	1328.1	149.6	1299.8	1449.4	0.8126	4.9571	5.7697
300.0	−9.2	0.001536	0.4045	0.4061	156.87	1172.8	1329.6	157.3	1294.1	1451.5	0.8420	4.9036	5.7456
320.0	−7.6	0.001541	0.3805	0.3821	164.21	1166.9	1331.1	164.7	1288.7	1453.4	0.8697	4.8534	5.7231
340.0	−6.1	0.001546	0.3593	0.3608	171.23	1161.2	1332.5	171.8	1283.4	1455.2	0.8961	4.8058	5.7019
360.0	−4.6	0.001551	0.3403	0.3419	177.93	1155.8	1333.8	178.5	1278.3	1456.8	0.9211	4.7608	5.6819
380.0	−3.2	0.001555	0.3233	0.3248	184.36	1150.6	1335.0	184.9	1273.5	1458.4	0.9450	4.7181	5.6631
400.0	−1.9	0.001560	0.3078	0.3094	190.55	1145.6	1336.1	191.2	1268.7	1459.9	0.9679	4.6772	5.6451
450.0	1.2	0.001570	0.2751	0.2767	205.05	1133.7	1338.7	205.8	1257.5	1463.3	1.0210	4.5828	5.6038
500.0	4.1	0.001580	0.2488	0.2503	218.40	1122.6	1341.0	219.2	1247.0	1466.2	1.0694	4.4974	5.5668
550.0	6.8	0.001589	0.2270	0.2286	230.80	1112.3	1343.1	231.7	1237.1	1468.8	1.1140	4.4193	5.5333
600.0	9.3	0.001598	0.2088	0.2104	242.39	1102.5	1344.9	243.3	1227.8	1471.1	1.1552	4.3474	5.5026
650.0	11.6	0.001607	0.1933	0.1949	253.28	1093.3	1346.6	254.3	1218.9	1473.2	1.1937	4.2806	5.4742
700.0	13.8	0.001615	0.1799	0.1815	263.59	1084.5	1348.1	264.7	1210.4	1475.1	1.2297	4.2182	5.4479
750.0	15.9	0.001622	0.1682	0.1698	273.36	1076.1	1349.4	274.6	1202.2	1476.8	1.2637	4.1597	5.4233
800.0	17.8	0.001630	0.1579	0.1596	282.67	1068.0	1350.7	284.0	1194.4	1478.3	1.2958	4.1045	5.4003
850.0	19.7	0.001637	0.1489	0.1505	291.55	1060.3	1351.8	292.9	1186.8	1479.7	1.3263	4.0523	5.3785
900.0	21.5	0.001645	0.1407	0.1424	300.07	1052.9	1352.9	301.5	1179.5	1481.0	1.3553	4.0027	5.3579
950.0	23.2	0.001651	0.1335	0.1351	308.24	1045.6	1353.8	309.8	1172.4	1482.2	1.3829	3.9555	5.3384
1000.0	24.9	0.001658	0.1269	0.1285	316.10	1038.6	1354.7	317.8	1165.5	1483.2	1.4094	3.9104	5.3198
1050.0	26.5	0.001665	0.1209	0.1225	323.69	1031.8	1355.5	325.4	1158.7	1484.2	1.4348	3.8672	5.3020
1100.0	28.0	0.001671	0.1154	0.1171	331.02	1025.2	1356.3	332.9	1152.2	1485.1	1.4592	3.8257	5.2850
1150.0	29.5	0.001678	0.1104	0.1121	338.12	1018.8	1356.9	340.0	1145.8	1485.9	1.4828	3.7858	5.2686
1200.0	30.9	0.001684	0.1058	0.1075	344.98	1012.6	1357.6	347.0	1139.6	1486.6	1.5054	3.7475	5.2530
1250.0	32.3	0.001690	0.1016	0.1033	351.65	1006.5	1358.2	353.8	1133.5	1487.2	1.5273	3.7105	5.2378
1300.0	33.7	0.001696	0.0977	0.0993	358.13	1000.6	1358.7	360.3	1127.5	1487.8	1.5485	3.6747	5.2232
1350.0	35.0	0.001702	0.0940	0.0957	364.42	994.7	1359.2	366.7	1121.7	1488.4	1.5690	3.6401	5.2091
1400.0	36.3	0.001708	0.0906	0.0923	370.57	989.0	1359.6	373.0	1115.9	1488.9	1.5889	3.6065	5.1955
1450.0	37.5	0.001714	0.0874	0.0891	376.55	983.5	1360.0	379.0	1110.2	1489.3	1.6083	3.5739	5.1822
1500.0	38.7	0.001719	0.0845	0.0862	382.39	978.0	1360.4	385.0	1104.7	1489.7	1.6271	3.5423	5.1694

TABLE D.15 Continued

Reference state: $h_l = 200$ kJ/kg and $s_l = 1$ kJ/(kg $\cdot$ K) at 0°C

P_{sat}, kPa	T_{sat}, °C	Specific Volume, m³/kg			Internal Energy, kJ/kg			Enthalpy, kJ/kg			Entropy, kJ/(kg $\cdot$ K)		
		v_l	v_{lg}	v_g	u_l	u_{lg}	u_g	h_l	h_{lg}	h_g	s_l	s_{lg}	s_g
1600.0	41.0	0.001730	0.0791	0.0808	393.65	967.4	1361.0	396.4	1093.9	1490.3	1.6631	3.4817	5.1447
1700.0	43.3	0.001741	0.0743	0.0760	404.44	957.1	1361.5	407.4	1083.4	1490.8	1.6974	3.4240	5.1214
1800.0	45.4	0.001752	0.0700	0.0718	414.79	947.2	1361.9	417.9	1073.1	1491.1	1.7300	3.3691	5.0991
1900.0	47.4	0.001763	0.0662	0.0679	424.75	937.5	1362.2	428.1	1063.2	1491.3	1.7613	3.3166	5.0779
2000.0	49.4	0.001773	0.0627	0.0645	434.37	928.1	1362.4	437.9	1053.4	1491.3	1.7913	3.2662	5.0575

Properties generated from the program *REFRIG*, Wiley Professional Software, Wiley, New York, 1984.

TABLE D.16 Properties of Superheated Ammonia—SI Units

Reference state: $h_l = 200$ kJ/kg and $s_l = 1$ kJ/(kg $\cdot$ K) at 0°C

P, kPa	T, °C	v, m³/kg	u, kJ/kg	h, kJ/kg	s, kJ/(kg $\cdot$ K)	P, kPa	T, °C	v, m³/kg	u, kJ/kg	h, kJ/kg	s, kJ/(kg $\cdot$ K)
50 ($T_{sat} = -46.5$°C)						100 ($T_{sat} = -33.6$°C)					
	−20	2.4464	1332.0	1454.3	6.6077		−20	1.2102	1327.5	1448.6	6.2518
	−10	2.5471	1348.1	1475.5	6.6898		−10	1.2622	1344.3	1470.5	6.3369
	0	2.6474	1364.3	1496.6	6.7687		0	1.3137	1361.0	1492.4	6.4184
	10	2.7472	1380.5	1517.8	6.8449		10	1.3647	1377.7	1514.1	6.4966
	20	2.8466	1396.7	1539.1	6.9187		20	1.4153	1394.3	1535.8	6.5719
	30	2.9458	1413.1	1560.4	6.9902		30	1.4657	1411.0	1557.5	6.6447
	40	3.0447	1429.6	1581.8	7.0596		40	1.5158	1427.7	1579.3	6.7152
	50	3.1435	1446.1	1603.3	7.1273		50	1.5658	1444.5	1601.0	6.7837
	60	3.2417	1462.8	1624.9	7.1931		60	1.6153	1461.3	1622.9	6.8502
	70	3.3406	1479.7	1646.7	7.2576		70	1.6652	1478.3	1644.8	6.9152
	80	3.4389	1496.7	1668.6	7.3205		80	1.7148	1495.4	1666.9	6.9786
	90	3.5373	1513.8	1690.7	7.3821		90	1.7643	1512.7	1689.1	7.0406
	100	3.6355	1531.1	1712.9	7.4425		100	1.8137	1530.1	1711.5	7.1013
75 ($T_{sat} = -39.1$°C)						125 ($T_{sat} = -29.1$°C)					
	−20	1.6223	1329.8	1451.4	6.4011		−20	0.96270	1325.3	1445.6	6.1337
	−10	1.6906	1346.2	1473.0	6.4846		−10	1.0052	1342.4	1468.0	6.2206
	0	1.7583	1362.6	1494.5	6.5648		0	1.0469	1359.4	1490.2	6.3034
	10	1.8255	1379.1	1516.0	6.6420		10	1.0881	1376.2	1512.3	6.3826
	20	1.8924	1395.5	1537.5	6.7166		20	1.1290	1393.1	1534.2	6.4588
	30	1.9591	1412.0	1559.0	6.7887		30	1.1696	1409.9	1556.1	6.5322
	40	2.0255	1428.6	1580.5	6.8587		40	1.2100	1426.7	1578.0	6.6032
	50	2.0917	1445.3	1602.2	6.9267		50	1.2502	1443.6	1599.9	6.6721
	60	2.1574	1462.1	1623.9	6.9929		60	1.2903	1460.6	1621.9	6.7390
	70	2.2237	1479.0	1645.8	7.0576		70	1.3302	1477.6	1643.9	6.8042
	80	2.2895	1496.0	1667.8	7.1208		80	1.3700	1494.8	1666.0	6.8679
	90	2.3553	1513.3	1689.9	7.1826		90	1.4097	1512.1	1688.3	6.9300
	100	2.4210	1530.6	1712.2	7.2431		100	1.4494	1529.6	1710.7	6.9909

Reference state: $h_l = 200$ **kJ/kg and** $s_l = 1$ **kJ/(kg · K) at 0°C**

$P,$ kPa	$T,$ °C	$v,$ m³/kg	$u,$ kJ/kg	$h,$ kJ/kg	$s,$ kJ/(kg · K)	$P,$ kPa	$T,$ °C	$v,$ m³/kg	$u,$ kJ/kg	$h,$ kJ/kg	$s,$ kJ/(kg · K)
150						300					
($T_{sat} = -25.2$°F)						($T_{sat} = -9.2$°C)					
	−20	0.79774	1323.0	1442.6	6.0355		0	0.4238	1347.3	1474.4	5.8312
	−10	0.83380	1340.4	1465.5	6.1241		10	0.4425	1365.9	1498.7	5.9183
	0	0.86901	1357.7	1488.0	6.2082		20	0.4608	1384.2	1522.4	6.0008
	10	0.90377	1374.8	1510.4	6.2885		30	0.4787	1402.1	1545.7	6.0790
	20	0.93817	1391.8	1532.6	6.3655		40	0.4964	1419.9	1568.8	6.1539
	30	0.97227	1408.8	1554.6	6.4396		50	0.5138	1437.6	1591.7	6.2259
	40	1.0062	1425.8	1576.7	6.5111		60	0.5311	1455.2	1614.5	6.2953
	50	1.0398	1442.8	1598.7	6.5804		70	0.5483	1472.8	1637.2	6.3626
	60	1.0734	1459.8	1620.8	6.6477		80	0.5653	1490.4	1660.0	6.4279
	70	1.1068	1476.9	1643.0	6.7132		90	0.5823	1508.0	1682.7	6.4914
	80	1.1401	1494.2	1665.2	6.7771		100	0.5992	1525.8	1705.6	6.5534
	90	1.1733	1511.5	1687.5	6.8394						
	100	1.2065	1529.0	1710.0	6.9005						
200						350					
($T_{sat} = -18.9$°C)						($T_{sat} = -5.4$°C)					
	−10	0.6192	1336.4	1460.3	5.9683		0	0.3601	1343.7	1469.7	5.7424
	0	0.6466	1354.3	1483.6	6.0553		10	0.3765	1362.8	1494.6	5.8320
	10	0.6733	1371.9	1506.5	6.1377		20	0.3926	1381.5	1518.9	5.9164
	20	0.6995	1389.3	1529.2	6.2164		30	0.4082	1399.9	1542.7	5.9962
	30	0.7255	1406.6	1551.7	6.2919		40	0.4235	1417.9	1566.1	6.0722
	40	0.7513	1423.8	1574.1	6.3645		50	0.4386	1435.8	1589.3	6.1451
	50	0.7768	1441.0	1596.4	6.4347		60	0.4536	1453.6	1612.4	6.2153
	60	0.8023	1458.3	1618.7	6.5027		70	0.4685	1471.3	1635.3	6.2832
	70	0.8275	1475.6	1641.1	6.5687		80	0.4832	1489.1	1658.2	6.3490
	80	0.8527	1492.9	1663.5	6.6330		90	0.4978	1506.9	1681.1	6.4129
	90	0.8778	1510.4	1685.9	6.6958		100	0.5124	1524.7	1704.1	6.4753
	100	0.9028	1528.0	1708.5	6.7572						
250						400					
($T_{sat} = -13.7$°C)						($T_{sat} = -1.9$°C)					
	−10	0.4905	1332.4	1455.0	5.8436		0	0.3123	1340.0	1464.9	5.6634
	0	0.5129	1350.8	1479.1	5.9334		10	0.3270	1359.7	1490.5	5.7556
	10	0.5349	1368.9	1502.7	6.0182		20	0.3413	1378.8	1515.4	5.8418
	20	0.5563	1386.8	1525.8	6.0987		30	0.3552	1397.5	1539.6	5.9233
	30	0.5775	1404.4	1548.8	6.1755		40	0.3688	1415.9	1563.4	6.0005
	40	0.5983	1421.9	1571.5	6.2493		50	0.3823	1434.0	1586.9	6.0744
	50	0.6190	1439.3	1594.1	6.3203		60	0.3954	1452.0	1610.2	6.1452
	60	0.6396	1456.7	1616.6	6.3891		70	0.4086	1469.9	1633.4	6.2138
	70	0.6600	1474.2	1639.2	6.4557		80	0.4216	1487.8	1656.4	6.2801
	80	0.6803	1491.6	1661.7	6.5205		90	0.4345	1505.7	1679.5	6.3445
	90	0.7005	1509.2	1684.3	6.5837		100	0.4473	1523.6	1702.6	6.4072
	100	0.7206	1526.9	1707.0	6.6453						

TABLE D.16 Continued

Reference state: $h_l = 200$ kJ/kg and $s_l = 1$ kJ/(kg · K) at 0°C

P, kPa	T, °C	v, m³/kg	u, kJ/kg	h, kJ/kg	s, kJ/(kg · K)	P, kPa	T, °C	v, m³/kg	u, kJ/kg	h, kJ/kg	s, kJ/(kg · K)
450 (T_{sat} = 1.3°C)						800 (T_{sat} = 17.8°C)					
	10	0.2885	1356.5	1486.3	5.6865		20	0.1614	1355.6	1484.7	5.4222
	20	0.3014	1376.1	1511.8	5.7749		40	0.1772	1398.8	1540.6	5.6065
	30	0.3141	1395.2	1536.5	5.8579		60	0.1919	1438.8	1592.3	5.7668
	40	0.3263	1413.9	1560.7	5.9364		80	0.2059	1477.2	1641.9	5.9113
	50	0.3384	1432.2	1584.5	6.0113		100	0.2195	1514.8	1690.6	6.0449
	60	0.3502	1450.4	1608.0	6.0829		120	0.2328	1552.2	1738.5	6.1704
	70	0.3620	1468.5	1631.4	6.1521		140	0.2459	1589.8	1786.5	6.2897
	80	0.3737	1486.5	1654.7	6.2189		160	0.2589	1627.8	1834.9	6.4040
	90	0.3852	1504.5	1677.9	6.2837		180	0.2717	1666.4	1883.8	6.5142
	100	0.3967	1522.5	1701.1	6.3467						
500 (T_{sat} = 4.1°C)						1000 (T_{sat} = 24.9°C)					
	20	0.2695	1373.3	1508.1	5.7138		40	0.1387	1389.6	1528.3	5.4672
	40	0.2923	1411.8	1557.9	5.8783		60	0.1511	1431.8	1582.9	5.6365
	60	0.3141	1448.8	1605.9	6.0267		80	0.1627	1471.6	1634.3	5.7864
	80	0.3354	1485.2	1652.9	6.1637		100	0.1739	1510.2	1684.1	5.9236
	100	0.3562	1521.5	1699.6	6.2923		120	0.1848	1548.3	1733.1	6.0515
	120	0.3768	1557.9	1746.3	6.4144		140	0.1955	1586.4	1781.9	6.1725
	140	0.3973	1594.8	1793.4	6.5313		160	0.2060	1624.8	1830.8	6.2881
	160	0.4176	1632.2	1841.0	6.6437		180	0.2164	1663.7	1880.1	6.3994
	180	0.4377	1670.3	1889.2	6.7524						
600 (T_{sat} = 9.3°C)						1200 (T_{sat} = 30.9°C)					
	20	0.2215	1367.7	1500.6	5.6049		40	0.1129	1379.8	1515.2	5.3458
	40	0.2412	1407.6	1552.3	5.7756		60	0.1238	1424.6	1573.1	5.5251
	60	0.2598	1445.5	1601.4	5.9277		80	0.1339	1465.9	1626.6	5.6810
	80	0.2778	1482.6	1649.3	6.0672		100	0.1435	1505.5	1677.7	5.8219
	100	0.2955	1519.2	1696.5	6.1974		120	0.1528	1544.4	1727.7	5.9523
	120	0.3129	1556.0	1743.7	6.3206		140	0.1618	1583.0	1777.2	6.0751
	140	0.3300	1593.1	1791.1	6.4382		160	0.1707	1621.8	1826.7	6.1921
	160	0.3470	1630.8	1839.0	6.5513		180	0.1795	1661.0	1876.5	6.3044
	180	0.3639	1669.0	1887.4	6.6605						
700 (T_{sat} = 13.8°C)						1400 (T_{sat} = 36.3°C)					
	20	0.1872	1361.8	1492.8	5.5090		40	0.0943	1369.3	1501.4	5.2357
	40	0.2047	1403.3	1546.5	5.6863		60	0.1042	1417.0	1563.0	5.4265
	60	0.2210	1442.2	1596.9	5.8423		80	0.1132	1460.1	1618.6	5.5888
	80	0.2367	1479.9	1645.6	5.9842		100	0.1217	1500.8	1671.2	5.7337
	100	0.2521	1517.0	1693.5	6.1161		120	0.1299	1540.4	1722.2	5.8667
	120	0.2671	1554.1	1741.1	6.2405		140	0.1378	1579.5	1772.4	5.9914
	140	0.2820	1591.5	1788.9	6.3589		160	0.1455	1618.8	1822.5	6.1098
	160	0.2967	1629.3	1837.0	6.4726		180	0.1532	1658.3	1872.8	6.2232
	180	0.3113	1667.7	1885.6	6.5824						

Reference state: $h_l = 200$ kJ/kg and $s_l = 1$ kJ/(kg · K) at 0°C

P, kPa	T, °C	v, m³/kg	u, kJ/kg	h, kJ/kg	s, kJ/(kg · K)	P, kPa	T, °C	v, m³/kg	u, kJ/kg	h, kJ/kg	s, kJ/(kg · K)
1600 (T_{sat} = 41.0°C)						2000 (T_{sat} = 49.4°C)					
	60	0.0895	1409.1	1552.3	5.3366		60	0.0687	1392.1	1529.6	5.1742
	80	0.0977	1454.0	1610.4	5.5060		80	0.0760	1441.4	1593.3	5.3600
	100	0.1054	1495.9	1664.6	5.6552		100	0.0825	1485.9	1650.9	5.5187
	120	0.1127	1536.3	1716.6	5.7911		120	0.0886	1528.0	1705.2	5.6606
	140	0.1198	1576.0	1767.7	5.9177		140	0.0945	1568.9	1757.9	5.7913
	160	0.1266	1615.7	1818.3	6.0375		160	0.1002	1609.5	1809.9	5.9141
	180	0.1334	1655.6	1869.1	6.1520		180	0.1057	1650.1	1861.6	6.0308
1800 (T_{sat} = 45.4°C)											
	60	0.0780	1400.8	1541.2	5.2531						
	80	0.0857	1447.8	1602.0	5.4303						
	100	0.0927	1491.0	1657.8	5.5841						
	120	0.0993	1532.2	1711.0	5.7229						
	140	0.1057	1572.5	1762.8	5.8515						
	160	0.1119	1612.6	1814.1	5.9728						
	180	0.1180	1652.9	1865.3	6.0884						

Properties generated from the program *REFRIG*, Wiley Professional Software, Wiley, New York, 1984.

Figure D.5 Psychrometric chart. *(From American Society of Heating, Refrigerating and Air-Conditioning Engineers, used with permission).*

APPENDIX E

Thermodynamic Data
for Various
Substances — USCS
Units

TABLE E.1 **Temperature-Dependent Molar Heat Capacities of Gases at Zero Pressure — USCS Units**

$$\bar{c}_P^o = a + bT + cT^2 + dT^3$$
$$[T \text{ in } °R, \bar{c}_P^o \text{ in Btu/(lbmol} \cdot °R)]$$

Substance	Formula	a	b	c	d	Temperature Range, °R	Error, % Max.	Error, % Avg.
Nitrogen	N_2	6.903	-0.02085×10^{-2}	0.05957×10^{-5}	-0.1176×10^{-9}	491–3240	0.59	0.34
Oxygen	O_2	6.085	0.2017×10^{-2}	-0.05275×10^{-5}	0.05372×10^{-9}	491–3240	1.19	0.28
Air		6.713	0.02609×10^{-2}	0.03540×10^{-5}	-0.08052×10^{-9}	491–3240	0.72	0.33
Hydrogen	H_2	6.952	-0.02542×10^{-2}	0.02952×10^{-5}	-0.03565×10^{-9}	491–3240	1.01	0.26
Carbon monoxide	CO	6.726	0.02222×10^{-2}	0.03960×10^{-5}	-0.09100×10^{-9}	491–3240	0.89	0.37
Carbon dioxide	CO_2	5.316	0.79361×10^{-2}	-0.2581×10^{-5}	0.3059×10^{-9}	491–3240	0.67	0.22
Water vapor	H_2O	7.700	0.02552×10^{-2}	0.07781×10^{-5}	-0.1472×10^{-9}	491–3240	0.53	0.24
Nitric oxide	NO	7.008	-0.01247×10^{-2}	0.07185×10^{-5}	-0.1715×10^{-9}	491–2700	0.97	0.36
Nitrous oxide	N_2O	5.758	0.7780×10^{-2}	-0.2596×10^{-5}	0.4331×10^{-9}	491–2700	0.59	0.26
Nitrogen dioxide	NO_2	5.48	0.7583×10^{-2}	-0.260×10^{-5}	0.322×10^{-9}	491–2700	0.46	0.18
Ammonia	NH_3	6.5846	0.34028×10^{-2}	0.073034×10^{-5}	-0.27402×10^{-9}	491–2700	0.91	0.36
Sulfur	S_2	6.499	0.2943×10^{-2}	-0.1200×10^{-5}	0.1632×10^{-9}	491–3240	0.99	0.38
Sulfur dioxide	SO_2	6.157	0.7689×10^{-2}	-0.2810×10^{-5}	0.3527×10^{-9}	491–3240	0.45	0.24
Sulfur trioxide	SO_3	3.918	1.935×10^{-2}	-0.8256×10^{-5}	1.328×10^{-9}	491–2340	0.29	0.13
Acetylene	C_2H_2	5.21	1.2227×10^{-2}	-0.4812×10^{-5}	0.7457×10^{-9}	491–2700	1.46	0.59
Benzene	C_6H_6	-8.650	6.4322×10^{-2}	-2.327×10^{-5}	3.179×10^{-9}	491–2700	0.34	0.20
Methanol	CH_4O	4.55	1.214×10^{-2}	-0.0898×10^{-5}	-0.329×10^{-9}	491–1800	0.18	0.08
Ethanol	C_2H_6O	4.75	2.781×10^{-2}	-0.7651×10^{-5}	0.821×10^{-9}	491–2700	0.40	0.22
Hydrogen chloride	HCl	7.244	-0.1011×10^{-2}	0.09783×10^{-5}	-0.1776×10^{-9}	491–2740	0.22	0.08
Methane	CH_4	4.750	0.6666×10^{-2}	0.09352×10^{-5}	-0.4510×10^{-9}	491–2740	1.33	0.57
Ethane	C_2H_6	1.648	2.291×10^{-2}	-0.4722×10^{-5}	0.2984×10^{-9}	491–2740	0.83	0.28
Propane	C_3H_8	-0.966	4.044×10^{-2}	-1.159×10^{-5}	1.300×10^{-9}	491–2740	0.40	0.12
n-Butane	C_4H_{10}	0.945	4.929×10^{-2}	-1.352×10^{-5}	1.433×10^{-9}	491–2740	0.54	0.24
i-Butane	C_4H_{10}	-1.890	5.520×10^{-2}	-1.696×10^{-5}	2.044×10^{-9}	491–2740	0.25	0.13
n-Pentane	C_5H_{12}	1.618	6.028×10^{-2}	-1.656×10^{-5}	1.732×10^{-9}	491–2740	0.56	0.21
n-Hexane	C_6H_{14}	1.657	7.328×10^{-2}	-2.112×10^{-5}	2.363×10^{-9}	491–2740	0.72	0.20
Ethylene	C_2H_4	0.944	2.075×10^{-2}	-0.6151×10^{-5}	0.7326×10^{-9}	491–2740	0.54	0.13
Propylene	C_3H_6	0.753	3.162×10^{-2}	-0.8981×10^{-5}	1.008×10^{-9}	491–2740	0.73	0.17

Data from B. G. Kyle, *Chemical and Process Thermodynamics,* Prentice-Hall, Englewood Cliffs, N.J., 1984 (used with permission).

TABLE E.2 Ideal Gas Properties of Air at Low Pressure—USCS Units

T, °F	u, Btu/lbm	h, Btu/lbm	s_0, Btu/(lbm · °R)	P_0	v_0	c_P, Btu/(lbm · °R)	c_v, Btu/(lbm · °R)	k
−300	27.28	38.231	1.3098	0.1958	17,248.	0.2394	0.1708	1.402
−200	44.36	62.176	1.4262	1.0692	5,137.2	0.2394	0.1708	1.402
−100	61.46	86.130	1.5043	3.3357	2,280.8	0.2395	0.1709	1.401
0	78.56	110.09	1.5631	7.8574	1,237.5	0.2397	0.1711	1.401
77	91.72	128.56	1.6002	13.504	840.65	0.2401	0.1715	1.400
100	95.69	134.08	1.6103	15.641	756.91	0.2402	0.1716	1.400
200	112.89	158.15	1.6499	27.843	501.15	0.2412	0.1726	1.397
300	130.23	182.35	1.6840	45.807	350.80	0.2428	0.1742	1.394
400	147.77	206.74	1.7142	71.093	255.78	0.2450	0.1764	1.389
500	165.53	231.37	1.7413	105.52	192.37	0.2476	0.1790	1.383
600	183.57	256.27	1.7660	151.21	148.23	0.2505	0.1819	1.377
700	201.92	281.47	1.7887	210.60	116.48	0.2536	0.1850	1.371
800	220.58	306.99	1.8098	286.47	93.013	0.2568	0.1882	1.365
900	239.56	332.83	1.8295	381.97	75.295	0.2600	0.1914	1.358
1,000	258.85	358.99	1.8481	500.64	61.673	0.2631	0.1945	1.353
1,100	278.45	385.45	1.8656	646.40	51.038	0.2661	0.1975	1.347
1,200	298.34	412.20	1.8822	823.62	42.624	0.2689	0.2003	1.343
1,300	318.50	439.22	1.8980	1,037.1	35.891	0.2715	0.2029	1.338
1,400	338.93	466.50	1.9131	1,292.0	30.446	0.2740	0.2054	1.334
1,500	359.59	494.03	1.9275	1,594.1	26.003	0.2764	0.2078	1.330
1,600	380.48	521.78	1.9413	1,949.6	22.347	0.2785	0.2099	1.327
1,700	401.57	549.73	1.9546	2,365.0	19.316	0.2805	0.2119	1.324
1,800	422.87	577.88	1.9673	2,847.7	16.785	0.2824	0.2138	1.321
1,900	444.34	606.22	1.9796	3,405.3	14.657	0.2842	0.2156	1.318
2,000	465.98	634.72	1.9914	4,046.1	12.859	0.2858	0.2172	1.316
2,100	487.77	663.37	2.0029	4,778.9	11.330	0.2873	0.2187	1.314
2,200	509.71	692.18	2.0139	5,613.0	10.023	0.2887	0.2201	1.312
2,300	531.79	721.11	2.0246	6,558.6	8.9005	0.2900	0.2214	1.310
2,400	554.00	750.18	2.0349	7,626.1	7.9320	0.2913	0.2227	1.308
2,500	576.33	779.37	2.0449	8,826.8	7.0926	0.2924	0.2238	1.307
2,600	598.77	808.67	2.0547	10,173.	6.3622	0.2935	0.2249	1.305
2,800	642.99	866.61	2.0733	13,350.	5.1648	0.2955	0.2269	1.302
3,000	689.53	926.87	2.0910	17,268.	4.2381	0.2973	0.2287	1.300
3,200	735.45	986.51	2.1077	22,045.	3.5116	0.2989	0.2303	1.298
3,400	781.67	1,046.5	2.1237	27,813.	2.9354	0.3004	0.2318	1.296
3,600	828.18	1,106.7	2.1389	34,718.	2.4735	0.3018	0.2332	1.294
3,800	874.95	1,167.1	2.1534	42,916.	2.0995	0.3030	0.2344	1.293
4,000	921.97	1,227.9	2.1674	52,578.	1.7942	0.3042	0.2356	1.291
4,200	969.20	1,288.9	2.1807	63,891.	1.5427	0.3053	0.2367	1.290
4,400	1,016.6	1,350.0	2.1936	77,052.	1.3341	0.3063	0.2377	1.289
4,600	1,064.3	1,411.4	2.2060	92,278.	1.1598	0.3072	0.2386	1.288
4,800	1,112.1	1,472.9	2.2179	109,800.	1.0133	0.3081	0.2395	1.286
5,000	1,160.1	1,534.6	2.2294	129,860.	0.8893	0.3089	0.2403	1.286

Properties generated from the program *GASPROPS*, Wiley Professional Software, Wiley, New York, 1984.

TABLE E.3 Ideal Gas Properties of Argon at Low Pressure — USCS Units

T, °F	u, Btu/lbm	h, Btu/lbm	s_0, Btu/(lbm · °R)	P_0	v_0	c_P, Btu/(lbm · °R)	c_v, Btu/(lbm · °R)	k
−300	11.914	19.857	0.7750	0.5837	7,979.2	0.1244	0.0746	1.667
−200	19.376	32.293	0.8355	1.9689	3,847.4	0.1244	0.0746	1.667
−100	26.837	44.729	0.8760	4.4455	2,360.1	0.1244	0.0746	1.667
0	34.299	57.165	0.9065	8.2088	1,633.5	0.1244	0.0746	1.667
77	40.046	66.744	0.9258	12.090	1,294.9	0.1244	0.0746	1.667
100	41.761	69.601	0.9310	13.427	1,215.9	0.1244	0.0746	1.667
200	49.222	82.037	0.9514	20.253	950.16	0.1244	0.0746	1.667
300	56.684	94.473	0.9690	28.822	768.88	0.1244	0.0746	1.667
400	64.146	106.91	0.9844	39.264	638.70	0.1244	0.0746	1.667
500	71.607	119.35	0.9980	51.697	541.52	0.1244	0.0746	1.667
600	79.069	131.78	1.0104	66.235	466.70	0.1244	0.0746	1.667
700	86.531	144.22	1.0216	82.985	407.66	0.1244	0.0746	1.667
800	93.992	156.65	1.0319	102.05	360.09	0.1244	0.0746	1.667
900	101.45	169.09	1.0414	123.52	321.10	0.1244	0.0746	1.667
1,000	108.92	181.53	1.0502	147.50	288.68	0.1244	0.0746	1.667
1,100	116.38	193.96	1.0584	174.08	261.36	0.1244	0.0746	1.667
1,200	123.84	206.40	1.0662	203.34	238.10	0.1244	0.0746	1.667
1,300	131.30	218.83	1.0734	235.36	218.10	0.1244	0.0746	1.667
1,400	138.76	231.27	1.0803	270.24	200.74	0.1244	0.0746	1.667
1,500	146.22	243.71	1.0868	308.05	185.58	0.1244	0.0746	1.667
1,600	153.68	256.14	1.0930	348.86	172.23	0.1244	0.0746	1.667
1,700	161.15	268.58	1.0989	392.76	160.40	0.1244	0.0746	1.667
1,800	168.61	281.01	1.1045	439.82	149.87	0.1244	0.0746	1.667
1,900	176.07	293.45	1.1099	490.11	140.45	0.1244	0.0746	1.667
2,000	183.53	305.89	1.1151	543.69	131.97	0.1244	0.0746	1.667
2,100	190.99	318.32	1.1201	600.65	124.31	0.1244	0.0746	1.667
2,200	198.46	330.76	1.1248	661.05	117.37	0.1244	0.0746	1.667
2,300	205.92	343.20	1.1294	724.95	111.05	0.1244	0.0746	1.667
2,400	213.38	355.63	1.1338	792.41	105.27	0.1244	0.0746	1.667
2,500	220.84	368.07	1.1381	863.52	99.984	0.1244	0.0746	1.667
2,600	228.30	380.50	1.1422	938.32	95.123	0.1244	0.0746	1.667
2,800	243.23	405.38	1.1501	1,099.2	86.504	0.1244	0.0746	1.667
3,000	258.15	430.25	1.1575	1,275.7	79.112	0.1244	0.0746	1.667
3,200	273.07	455.12	1.1645	1,468.1	72.716	0.1244	0.0746	1.667
3,400	288.00	479.99	1.1711	1,677.0	67.138	0.1244	0.0746	1.667
3,600	302.92	504.87	1.1774	1,902.8	62.239	0.1244	0.0746	1.667
3,800	317.84	529.74	1.1834	2,145.9	57.907	0.1244	0.0746	1.667
4,000	332.77	554.61	1.1891	2,406.7	54.056	0.1244	0.0746	1.667
4,200	347.69	579.48	1.1946	2,685.7	50.613	0.1244	0.0746	1.667
4,400	362.61	604.35	1.1998	2,983.2	47.521	0.1244	0.0746	1.667
4,600	377.54	629.23	1.2048	3,299.6	44.731	0.1244	0.0746	1.667
4,800	392.46	654.10	1.2096	3,635.5	42.204	0.1244	0.0746	1.667
5,000	407.33	678.97	1.2143	3,991.0	39.907	0.1244	0.0746	1.667

Properties generated from the program *GASPROPS*, Wiley Professional Software, Wiley, New York, 1984.

TABLE E.4 Ideal Gas Properties of Nitrogen at Low Pressure — USCS Units

T, °F	u, Btu/lbm	h, Btu/lbm	s_0, Btu/(lbm · °R)	P_0	v_0	c_P, Btu/(lbm · °R)	c_v, Btu/(lbm · °R)	k
−300	28.32	39.643	1.3327	0.1443	22,629.	0.2483	0.1773	1.400
−200	46.05	64.471	1.4535	0.7917	6,709.1	0.2483	0.1773	1.400
−100	63.80	89.309	1.5344	2.4781	2,969.0	0.2483	0.1774	1.400
0	81.54	114.15	1.5954	5.8497	1,607.5	0.2484	0.1774	1.400
77	95.20	133.27	1.6338	10.062	1,091.1	0.2485	0.1775	1.400
100	99.29	138.99	1.6443	11.655	982.28	0.2485	0.1776	1.399
200	117.07	163.86	1.6851	20.743	650.56	0.2490	0.1781	1.398
300	134.92	188.81	1.7204	34.075	456.06	0.2500	0.1791	1.396
400	152.90	213.89	1.7514	52.754	333.35	0.2516	0.1806	1.393
500	171.07	239.15	1.7792	78.059	251.49	0.2537	0.1828	1.388
600	189.47	264.64	1.8044	111.47	194.47	0.2563	0.1854	1.383
700	208.16	290.42	1.8277	154.68	153.36	0.2593	0.1883	1.377
800	227.14	316.50	1.8492	209.66	122.91	0.2624	0.1914	1.371
900	246.45	342.90	1.8694	278.59	99.838	0.2656	0.1947	1.364
1,000	266.08	369.62	1.8884	363.96	82.041	0.2688	0.1979	1.359
1,100	286.02	396.66	1.9063	468.53	68.096	0.2719	0.2010	1.353
1,200	306.27	424.01	1.9233	595.35	57.026	0.2750	0.2040	1.348
1,300	326.82	451.65	1.9394	747.79	48.137	0.2778	0.2069	1.343
1,400	347.65	479.57	1.9549	929.52	40.926	0.2806	0.2096	1.338
1,500	368.74	507.75	1.9696	1,144.5	35.025	0.2831	0.2122	1.334
1,600	390.08	536.18	1.9838	1,397.2	30.156	0.2855	0.2145	1.331
1,700	411.64	564.84	1.9974	1,692.2	26.108	0.2877	0.2168	1.327
1,800	433.42	593.72	2.0104	2,034.5	22.721	0.2898	0.2188	1.324
1,900	455.40	622.79	2.0230	2,429.6	19.868	0.2917	0.2207	1.321
2,000	477.57	652.05	2.0352	2,883.2	17.452	0.2935	0.2225	1.319
2,100	499.90	681.48	2.0469	3,401.5	15.394	0.2951	0.2242	1.316
2,200	522.40	711.06	2.0582	3,991.1	13.632	0.2966	0.2257	1.314
2,300	545.04	740.80	2.0692	4,658.9	12.117	0.2981	0.2271	1.312
2,400	567.82	770.67	2.0799	5,412.3	10.808	0.2994	0.2285	1.311
2,500	590.73	800.68	2.0902	6,259.2	9.6729	0.3006	0.2297	1.309
2,600	613.75	830.80	2.1002	7,207.6	8.6838	0.3018	0.2308	1.307
2,800	660.13	891.36	2.1194	9,444.5	7.0603	0.3038	0.2329	1.305
3,000	706.90	952.32	2.1375	12,198.	5.8020	0.3057	0.2347	1.302
3,200	754.00	1,013.6	2.1547	15,550.	4.8144	0.3072	0.2363	1.300
3,400	801.41	1,075.2	2.1711	19,591.	4.0302	0.3087	0.2377	1.298
3,600	849.08	1,137.1	2.1867	24,418.	3.4010	0.3099	0.2390	1.297
3,800	896.99	1,199.2	2.2017	30,138.	2.8912	0.3110	0.2401	1.295
4,000	945.11	1,261.5	2.2160	36,866.	2.4746	0.3120	0.2411	1.294
4,200	993.42	1,324.0	2,2297	44,726.	2.1312	0.3129	0.2420	1.293
4,400	1,041.9	1,386.6	2.2428	53,850.	1.8461	0.3138	0.2428	1.292
4,600	1,090.5	1,449.5	2.2555	64,380.	1.6077	0.3145	0.2436	1.291
4,800	1,139.3	1,512.4	2.2677	76,468.	1.4070	0.3152	0.2443	1.290
5,000	1,188.2	1,575.5	2.2795	90,275.	1.2372	0.3158	0.2449	1.290

Properties generated from the program *GASPROPS*, Wiley Professional Software, Wiley, New York, 1984.

TABLE E.5 Ideal Gas Properties of Oxygen at Low Pressure — USCS Units

T, °F	u, Btu/lbm	h, Btu/lbm	s_0, Btu/(lbm · °R)	P_0	v_0	c_P, Btu/(lbm · °R)	c_v, Btu/(lbm · °R)	k
−300	24.79	34.705	1.2672	0.7280	5,124.7	0.2174	0.1553	1.400
−200	40.32	56.449	1.3730	3.9972	1,518.0	0.2174	0.1553	1.400
−100	55.86	78.195	1.4438	12.508	671.94	0.2175	0.1554	1.400
0	71.43	99.976	1.4973	29.568	363.26	0.2182	0.1561	1.398
77	83.49	116.82	1.5312	51.020	245.79	0.2194	0.1573	1.395
100	87.12	121.87	1.5404	59.183	220.97	0.2199	0.1578	1.394
200	103.03	144.00	1.5767	106.29	145.02	0.2227	0.1606	1.387
300	119.27	166.45	1.6084	177.04	100.27	0.2264	0.1643	1.378
400	135.91	189.29	1.6367	278.99	72.001	0.2305	0.1684	1.369
500	152.96	212.56	1.6623	421.29	53.227	0.2347	0.1726	1.360
600	170.43	236.24	1.6857	614.79	40.275	0.2389	0.1768	1.351
700	188.30	260.32	1.7074	872.14	31.070	0.2428	0.1807	1.344
800	206.55	284.78	1.7277	1,207.9	24.368	0.2463	0.1842	1.337
900	225.14	309.58	1.7466	1,638.7	19.387	0.2496	0.1875	1.331
1,000	244.04	334.68	1.7644	2,183.3	15.622	0.2525	0.1904	1.326
1,100	263.21	360.07	1.7813	2,862.4	12.732	0.2551	0.1930	1.322
1,200	282.64	385.70	1.7972	3,699.5	10.483	0.2575	0.1954	1.318
1,300	302.28	411.56	1.8123	4,720.0	8.7114	0.2596	0.1975	1.314
1,400	322.14	437.63	1.8267	5,952.2	7.3006	0.2616	0.1995	1.311
1,500	342.17	463.87	1.8405	7,426.9	6.1656	0.2633	0.2012	1.309
1,600	362.37	490.28	1.8536	9,177.6	5.2440	0.2649	0.2028	1.306
1,700	382.73	516.85	1.8662	11,241.	4.4893	0.2664	0.2043	1.304
1,800	403.22	543.55	1.8783	13,656.	3.8664	0.2677	0.2056	1.302
1,900	423.85	570.39	1.8899	16,467.	3.3484	0.2690	0.2069	1.300
2,000	444.60	597.35	1.9011	19,718.	2.9148	0.2702	0.2081	1.298
2,100	465.47	624.43	1.9119	23,460.	2.5495	0.2713	0.2092	1.297
2,200	486.45	651.62	1.9223	27,745.	2.2399	0.2724	0.2103	1.295
2,300	507.53	678.92	1.9324	32,632.	1.9761	0.2735	0.2114	1.294
2,400	528.72	706.32	1.9421	38,181.	1.7501	0.2745	0.2124	1.292
2,500	550.01	733.82	1.9516	44,458.	1.5556	0.2755	0.2134	1.291
2,600	571.40	761.42	1.9608	51,533.	1.3873	0.2765	0.2144	1.290
2,800	614.47	816.90	1.9783	68,381.	1.1139	0.2784	0.2163	1.287
3,000	657.92	872.77	1.9950	89,382.	0.9044	0.2803	0.2182	1.285
3,200	701.73	929.00	2.0108	115,280.	0.7418	0.2821	0.2200	1.282
3,400	745.91	985.60	2.0258	146,910.	0.6139	0.2839	0.2218	1.280
3,600	790.45	1,042.6	2.0402	185,210.	0.5122	0.2857	0.2236	1.278
3,800	835.34	1,099.9	2.0540	231,230.	0.4305	0.2874	0.2253	1.276
4,000	880.58	1,157.5	2.0672	286,110.	0.3642	0.2891	0.2270	1.274
4,200	926.16	1,215.5	2.0799	351,160.	0.3101	0.2908	0.2287	1.272
4,400	972.06	1,273.9	2.0922	427,770.	0.2655	0.2924	0.2303	1.270
4,600	1,018.3	1,332.5	2.1040	517,510.	0.2285	0.2940	0.2319	1.268
4,800	1,064.8	1,391.5	2.1154	622,060.	0.1976	0.2955	0.2334	1.266
5,000	1,111.7	1,450.7	2.1265	743,290.	0.1716	0.2970	0.2349	1.264

Properties generated from the program *GASPROPS,* Wiley Professional Software, Wiley, New York, 1984.

TABLE E.6 Ideal Gas Properties of Water Vapor at Low Pressure — USCS Units

T, °F	u, Btu/lbm	h, Btu/lbm	s_0, Btu/(lbm · °R)	P_0	v_0	c_P, Btu/(lbm · °R)	c_v, Btu/(lbm · °R)	k
50	168.12	224.34	2.4802	0.0582	115,150.	0.4445	0.3342	1.330
77	177.15	236.35	2.4895	0.0610	115,720.	0.4453	0.3350	1.329
100	184.86	246.60	2.4896	0.0634	116,210.	0.4461	0.3358	1.328
200	218.66	291.43	2.5595	0.1194	72,654.	0.4507	0.3404	1.324
300	253.00	336.79	2.6584	0.2929	34,120.	0.4568	0.3465	1.318
400	288.00	382.83	2.7153	0.4907	23,049.	0.4640	0.3537	1.312
500	323.75	429.61	2.7668	0.7825	16,134.	0.4717	0.3614	1.305
600	360.30	477.19	2.8140	1.1997	11,620.	0.4799	0.3696	1.298
700	397.68	525.60	2.8576	1.7820	8,561.0	0.4884	0.3781	1.292
800	435.92	574.87	2.8984	2.5790	6,425.6	0.4971	0.3868	1.285
900	475.04	625.02	2.9367	3.6491	4,901.7	0.5060	0.3957	1.279
1,000	515.07	676.08	2.9729	5.0681	3,788.9	0.5152	0.4048	1.272
1,100	556.02	728.06	3.0073	6.9246	2,963.1	0.5245	0.4142	1.266
1,200	597.91	780.98	3.0402	9.3293	2,340.3	0.5339	0.4236	1.260
1,300	640.74	834.84	3.0717	12.415	1,864.6	0.5433	0.4330	1.255
1,400	684.51	889.64	3.1020	16.341	1,497.1	0.5528	0.4424	1.249
1,500	729.23	945.40	3.1312	21.284	1,211.3	0.5621	0.4518	1.244
1,600	774.84	1,002.0	3.1594	27.490	985.67	0.5713	0.4610	1.239
1,700	821.39	1,059.6	3.1867	35.027	806.98	0.5802	0.4699	1.235
1,800	868.85	1,118.1	3.2132	44.744	664.38	0.5890	0.4787	1.230
1,900	917.14	1,177.4	3.2389	56.491	549.51	0.5975	0.4872	1.226
2,000	966.27	1,237.6	3.2639	70.853	456.69	0.6057	0.4954	1.223
2,100	1,016.2	1,298.6	3.2882	88.311	381.30	0.6137	0.5034	1.219
2,200	1,066.9	1,360.3	3.3118	109.45	319.69	0.6213	0.5110	1.216
2,300	1,118.4	1,422.8	3.3349	134.89	269.14	0.6286	0.5183	1.213
2,400	1,170.0	1,486.0	3.3574	165.37	227.48	0.6357	0.5254	1.210
2,500	1,223.1	1,549.9	3.3793	201.79	192.95	0.6424	0.5320	1.207
2,600	1,277.0	1,614.5	3.4008	245.17	164.18	0.6488	0.5384	1.205
2,700	1,331.1	1,679.7	3.4217	296.32	140.28	0.6549	0.5446	1.203
2,800	1,385.9	1,745.4	3.4424	357.48	119.96	0.6607	0.5504	1.200
2,900	1,441.2	1,811.8	3.4617	425.76	103.81	0.6663	0.5560	1.198
3,000	1,497.1	1,878.7	3.4842	522.10	87.172	0.6716	0.5612	1.197
3,200	1,610.3	2,014.0	3.5199	722.06	66.676	0.6814	0.5711	1.193
3,400	1,725.5	2,151.2	3.5564	1,005.0	50.522	0.6904	0.5800	1.190
3,600	1,842.3	2,290.1	3.5915	1,381.5	38.658	0.6986	0.5882	1.188
3,800	1,960.7	2,430.6	3.6253	1,876.4	29.865	0.7060	0.5957	1.185
4,000	2,080.6	2,572.5	3.6578	2,520.5	23.277	0.7129	0.6026	1.183
4,200	2,201.7	2,715.7	3.6892	3,350.9	18.294	0.7193	0.6090	1.181
4,400	2,324.1	2,860.2	3.7196	4,413.5	14.485	0.7252	0.6149	1.179
4,600	2,447.7	3,005.8	3.7490	5,757.7	11.560	0.7308	0.6205	1.178
4,800	2,572.3	3,152.5	3.7774	7,452.8	9.2841	0.7360	0.6257	1.176
5,000	2,698.6	3,300.8	3.8049	9,559.1	7.5136	0.7409	0.6306	1.175

Properties generated from the program *GASPROPS*, Wiley Professional Software, Wiley, New York, 1984.

TABLE E.7 Ideal Gas Properties of Carbon Dioxide at Low Pressure — USCS Units

T, °F	u, Btu/lbm	h, Btu/lbm	s_0, Btu/(lbm · °R)	P_0	v_0	c_P, Btu/(lbm · °R)	c_v, Btu/(lbm · °R)	k
−300	18.07	25.278	0.9525	0.0145	353,760.	0.1583	0.1131	1.399
−200	30.10	41.823	1.0303	0.0812	102,820.	0.1632	0.1181	1.382
−100	41.87	58.109	1.0852	0.2741	42,172.	0.1757	0.1305	1.346
0	55.70	76.456	1.1301	0.7400	19,964.	0.1906	0.1454	1.310
77	67.31	91.543	1.1604	1.4496	11,898.	0.2017	0.1565	1.289
100	70.95	96.221	1.1690	1.7510	10,272.	0.2048	0.1596	1.283
200	87.56	117.35	1.2037	3.7764	5,613.9	0.2175	0.1723	1.262
300	105.37	139.67	1.2351	7.5842	3,219.1	0.2287	0.1836	1.246
400	124.23	163.03	1.2640	14.383	1,920.9	0.2386	0.1935	1.233
500	144.03	187.36	1.2908	26.010	1,185.8	0.2475	0.2024	1.223
600	164.67	212.52	1.3157	45.174	753.87	0.2555	0.2103	1.215
700	186.07	238.43	1.3391	75.792	491.72	0.2627	0.2175	1.208
800	208.15	265.03	1.3611	123.36	328.17	0.2691	0.2240	1.202
900	230.85	292.24	1.3819	195.47	223.55	0.2750	0.2299	1.196
1,000	254.10	320.01	1.4016	302.42	155.12	0.2803	0.2352	1.192
1,100	277.86	348.29	1.4203	457.95	109.45	0.2851	0.2400	1.188
1,200	302.08	377.02	1.4382	680.12	78.424	0.2895	0.2443	1.185
1,300	326.72	406.17	1.4552	992.79	56.962	0.2934	0.2483	1.182
1,400	351.72	435.69	1.4711	1,409.0	42.417	0.2970	0.2519	1.179
1,500	377.07	465.56	1.4888	1,752.9	36.992	0.3003	0.2551	1.177
1,600	402.73	495.73	1.4890	2,096.9	31.568	0.3032	0.2581	1.175
1,700	428.69	526.20	1.5166	3,860.8	17.977	0.3060	0.2608	1.173
1,800	454.90	556.93	1.5303	5,235.1	13.872	0.3084	0.2633	1.171
1,900	481.35	587.89	1.5441	7,098.7	10.683	0.3107	0.2656	1.170
2,000	507.99	619.05	1.5569	9,432.3	8.3805	0.3128	0.2676	1.169
2,100	534.85	650.43	1.5694	12,446.	6.6096	0.3146	0.2695	1.168
2,200	561.88	681.98	1.5815	16,264.	5.2556	0.3164	0.2712	1.166
2,300	589.08	713.69	1.5932	21,082.	4.2068	0.3180	0.2728	1.166
2,400	616.45	745.57	1.6045	27,080.	3.3937	0.3195	0.2743	1.165
2,500	643.97	777.60	1.6155	34,544.	2.7535	0.3209	0.2757	1.164
2,600	671.60	809.76	1.6262	43,774.	2.2463	0.3221	0.2770	1.163
2,700	699.33	842.00	1.6366	55,083.	1.8435	0.3233	0.2781	1.162
2,800	727.24	874.43	1.6467	68,876.	1.5210	0.3244	0.2793	1.162
2,900	755.34	907.04	1.6565	85,636.	1.2608	0.3255	0.2803	1.161
3,000	783.48	939.70	1.6661	105,844.	1.0505	0.3265	0.2813	1.161
3,200	839.25	1,004.5	1.6845	159,051.	0.7395	0.3282	0.2831	1.160
3,400	897.53	1,071.8	1.7020	234,410.	0.5292	0.3298	0.2847	1.159
3,600	948.22	1,131.5	1.7187	339,340.	0.3845	0.3312	0.2861	1.158
3,800	1,007.9	1,200.2	1.7346	483,080.	0.2834	0.3325	0.2873	1.157
4,000	1,069.0	1,270.4	1.7499	678,010.	0.2114	0.3336	0.2884	1.157
4,200	1,126.2	1,336.6	1.7646	937,850.	0.1597	0.3347	0.2895	1.156
4,400	1,184.3	1,403.7	1.7787	1,280,900.	0.1219	0.3356	0.2905	1.155
4,600	1,242.5	1,470.9	1.7922	1,730,000.	0.0940	0.3365	0.2914	1.155
4,800	1,300.8	1,538.3	1.8053	2,310,700.	0.0732	0.3374	0.2922	1.155
5,000	1,359.4	1,605.9 ·	1.8179	3,052,700.	0.0575	0.3382	0.2930	1.154

Properties generated from the program *GASPROPS,* Wiley Professional Software, Wiley, New York, 1984.

TABLE E.8 Properties of Saturated Water — Temperature Table, USCS Units

T_{sat}, °F	P_{sat}, psia	Specific Volume, ft³/lbm			Internal Energy, Btu/lbm			Enthalpy, Btu/lbm			Entropy, Btu/(lbm · °R)		
		v_l	v_{lg}	v_g	u_l	u_{lg}	u_g	h_l	h_{lg}	h_g	s_l	s_{lg}	s_g
40	0.12	0.01602	2,446.9	2,446.9	8.04	1,015.7	1,023.7	8.0	1,070.9	1,078.9	0.0161	2.1433	2.1594
50	0.18	0.01602	1,703.7	1,703.7	18.06	1,009.4	1,027.4	18.1	1,065.5	1,083.6	0.0360	2.0902	2.1262
60	0.26	0.01603	1,208.5	1,208.5	28.06	1,002.7	1,030.7	28.1	1,060.0	1,088.0	0.0555	2.0391	2.0946
70	0.36	0.01605	868.38	868.40	38.05	996.0	1,034.0	38.1	1,054.3	1,092.4	0.0746	1.9899	2.0645
80	0.51	0.01607	633.28	633.30	48.03	989.2	1,037.3	48.0	1,048.6	1,096.7	0.0933	1.9426	2.0359
90	0.70	0.01610	468.08	468.10	58.00	982.5	1,040.5	58.0	1,042.9	1,100.9	0.1116	1.8970	2.0086
100	0.95	0.01613	350.38	350.40	67.97	975.7	1,043.7	68.0	1,037.2	1,105.2	0.1295	1.8530	1.9826
110	1.27	0.01616	265.12	265.13	77.94	969.0	1,046.9	77.9	1,031.5	1,109.4	0.1472	1.8105	1.9577
120	1.69	0.01620	203.24	203.26	87.91	962.1	1,050.0	87.9	1,025.7	1,113.7	0.1645	1.7694	1.9339
130	2.22	0.01625	157.31	157.32	97.89	955.3	1,053.2	97.9	1,020.0	1,117.9	0.1815	1.7296	1.9112
140	2.89	0.01629	122.98	123.00	107.87	948.4	1,056.3	107.9	1,014.1	1,122.0	0.1983	1.6911	1.8894
150	3.72	0.01634	97.022	97.038	117.86	941.5	1,059.4	117.9	1,008.3	1,126.1	0.2148	1.6538	1.8686
160	4.74	0.01639	77.274	77.290	127.86	934.5	1,062.4	127.9	1,002.3	1,130.2	0.2311	1.6175	1.8486
170	5.99	0.01645	62.028	62.044	137.87	927.5	1,065.4	137.9	996.3	1,134.2	0.2471	1.5824	1.8295
180	7.51	0.01650	50.204	50.220	147.89	920.5	1,068.4	147.9	990.2	1,138.2	0.2629	1.5482	1.8111
190	9.34	0.01656	40.939	40.956	157.92	913.4	1,071.3	157.9	984.1	1,142.1	0.2785	1.5149	1.7934
200	11.53	0.01663	33.622	33.639	167.96	906.2	1,074.1	168.0	977.9	1,145.9	0.2938	1.4826	1.7764
210	14.12	0.01670	27.758	27.774	178.02	899.1	1,077.1	178.1	971.6	1,149.7	0.3090	1.4510	1,7600
220	17.19	0.01677	23.066	23.083	188.09	891.9	1,079.9	188.1	965.2	1,153.4	0.3239	1.4203	1.7442
230	20.78	0.01684	19.323	19.340	198.17	884.4	1,082.6	198.2	958.7	1,157.0	0.3387	1.3903	1.7290
240	24.97	0.01692	16.304	16.321	208.28	876.8	1,085.1	208.4	952.2	1,160.5	0.3532	1.3610	1.7142
250	29.83	0.01700	13.810	13.827	218.40	869.3	1,087.7	218.5	945.5	1,164.0	0.3676	1.3324	1.7000
260	35.43	0.01708	11.746	11.763	228.54	861.7	1,090.2	228.7	938.7	1,167.4	0.3818	1.3044	1.6862
270	41.86	0.01717	10.042	10.059	238.71	854.0	1,092.7	238.8	931.8	1,170.6	0.3958	1.2770	1.6729
280	49.20	0.01726	8.6273	8.6446	248.91	846.2	1,095.1	249.1	924.7	1,173.8	0.4097	1.2502	1.6599
290	57.55	0.01735	7.4400	7.4573	259.13	838.3	1,097.4	259.3	917.5	1,176.8	0.4234	1.2239	1.6473
300	67.00	0.01745	6.4426	6.4600	269.38	830.3	1,099.7	269.6	910.2	1,179.8	0.4370	1.1981	1.6351
310	77.66	0.01755	5.6047	5.6223	279.66	822.1	1,101.8	279.9	902.7	1,182.6	0.4504	1.1728	1.6232
320	89.63	0.01765	4.8961	4.9138	289.98	813.8	1,103.7	290.3	895.0	1,185.2	0.4637	1.1479	1.6116
330	103.03	0.01776	4.2894	4.3072	300.34	805.3	1,105.7	300.7	887.1	1,187.8	0.4769	1.1234	1.6003
340	117.97	0.01787	3.7692	3.7871	310.74	796.7	1,107.5	311.1	879.0	1,190.2	0.4900	1.0993	1.5892
350	134.56	0.01799	3.3233	3.3412	321.18	788.0	1,109.2	321.6	870.7	1,192.4	0.5029	1.0755	1.5784
360	152.95	0.01811	2.9392	2.9573	331.67	779.1	1,110.7	332.2	862.3	1,194.4	0.5158	1.0520	1.5678
370	173.26	0.01823	2.6058	2.6240	342.20	770.0	1,112.2	342.8	853.5	1,196.3	0.5285	1.0288	1.5574
380	195.62	0.01836	2.3161	2.3345	352.79	760.7	1,113.5	353.5	844.6	1,198.0	0.5412	1.0060	1.5472
390	220.19	0.01849	2.0643	2.0827	363.44	751.3	1,114.7	364.2	835.4	1,199.6	0.5538	0.9834	1.5372
400	247.09	0.01863	1.8444	1.8630	374.14	741.6	1,115.7	375.0	825.9	1,200.9	0.5663	0.9611	1.5274
410	276.48	0.01877	1.6511	1.6698	384.90	731.7	1,116.6	385.9	816.2	1,202.1	0.5787	0.9390	1.5177
420	308.52	0.01892	1.4808	1.4998	395.73	721.7	1,117.4	396.8	806.2	1,203.0	0.5911	0.9170	1.5081
430	343.37	0.01908	1.3307	1.3498	406.63	711.4	1,118.0	407.8	795.9	1,203.8	0.6034	0.8952	1.4986
440	381.19	0.01924	1.1976	1.2169	417.60	700.9	1,118.5	419.0	785.4	1,204.3	0.6157	0.8734	1.4891
450	422.15	0.01942	1.0791	1.0985	428.65	690.2	1,118.8	430.2	774.5	1,204.7	0.6279	0.8518	1.4797
460	466.43	0.01960	0.9746	0.9942	439.78	679.2	1,119.0	441.5	763.3	1,204.8	0.6401	0.8303	1.4704
470	514.20	0.01979	0.8834	0.9032	451.01	667.7	1,118.7	452.9	751.8	1,204.7	0.6522	0.8089	1.4611
480	565.66	0.01999	0.7997	0.8196	462.33	656.2	1,118.6	464.4	739.9	1,204.4	0.6644	0.7875	1.4518
490	620.99	0.02020	0.7234	0.7436	473.76	644.6	1,118.4	476.1	727.8	1,203.8	0.6765	0.7661	1.4426
500	680.40	0.02042	0.6545	0.6749	485.30	632.8	1,118.1	487.9	715.2	1,203.1	0.6886	0.7447	1.4333
510	744.09	0.02066	0.5929	0.6136	496.97	620.6	1,117.6	499.8	702.2	1,202.1	0.7007	0.7233	1.4240
520	812.26	0.02091	0.5379	0.5588	508.77	608.1	1,116.8	511.9	688.9	1,200.8	0.7128	0.7017	1.4145
530	885.14	0.02118	0.4885	0.5096	520.73	595.2	1,115.9	524.2	675.2	1,199.4	0.7250	0.6800	1.4050
540	962.95	0.02146	0.4437	0.4651	532.85	581.9	1,114.8	530.7	661.0	1,197.7	0.7372	0.6582	1.3954

TABLE E.8 (Continued)

T_{sat}, °F	P_{sat}, psia	Specific Volume, ft³/lbm			Internal Energy, Btu/lbm			Enthalpy, Btu/lbm			Entropy, Btu/(lbm · °R)		
		v_l	v_{lg}	v_g	u_l	u_{lg}	u_g	h_l	h_{lg}	h_g	s_l	s_{lg}	s_g
550	1,045.9	0.02176	0.4027	0.4245	545.17	568.4	1,113.6	549.4	646.3	1,195.7	0.7495	0.6361	1.3856
560	1,134.3	0.02208	0.3652	0.3873	557.68	548.7	1,106.4	562.3	625.4	1,187.7	0.7618	0.6138	1.3757
570	1,228.4	0.02243	0.3309	0.3533	570.43	532.9	1,103.3	575.5	608.1	1,183.6	0.7743	0.5912	1.3655
580	1,328.3	0.02279	0.2994	0.3222	583.44	516.4	1,099.8	589.0	590.0	1,179.0	0.7869	0.5681	1.3550
590	1,431.4	0.02318	0.2705	0.2936	596.75	499.2	1,096.0	602.9	570.8	1,173.7	0.7997	0.5445	1.3442
600	1,543.2	0.02359	0.2438	0.2674	610.38	481.0	1,091.3	617.1	550.6	1,167.7	0.8128	0.5202	1.3330
610	1,661.6	0.02411	0.2191	0.2433	624.35	461.7	1,086.0	631.8	529.1	1,160.8	0.8261	0.4952	1.3213
620	1,786.9	0.02465	0.1962	0.2208	638.73	441.2	1,080.0	646.9	506.1	1,153.0	0.8397	0.4693	1.3090
630	1,919.5	0.02526	0.1746	0.1998	653.55	419.4	1,073.0	662.5	481.5	1,144.0	0.8537	0.4421	1.2959
640	2,059.9	0.02595	0.1543	0.1802	668.86	396.1	1,065.0	678.8	454.9	1,133.7	0.8682	0.4136	1.2818
650	2,208.4	0.02674	0.1350	0.1618	684.72	371.0	1,055.7	695.6	426.2	1,121.9	0.8833	0.3834	1.2667
660	2,365.7	0.02768	0.1166	0.1443	702.78	341.7	1,044.5	714.9	392.8	1,107.7	0.8990	0.3511	1.2501
670	2,532.2	0.02883	0.0987	0.1275	721.23	309.2	1,030.4	734.7	355.4	1,090.2	0.9155	0.3156	1.2311
680	2,708.6	0.03037	0.0808	0.1112	742.14	270.6	1,012.8	757.4	311.1	1,068.5	0.9328	0.2754	1.2083
690	2,895.7	0.03256	0.0621	0.0946	767.05	222.9	989.9	784.5	256.1	1,040.6	0.9512	0.2298	1.1810
700	3,094.3	0.03697	0.0382	0.0752	801.23	150.9	952.1	822.4	172.8	995..2	0.9708	0.1651	1.1359
705.6	3,204.1	0.0500	0	0.0500	872.57	0	872.6	902.5	0	902.5	1.0581	0	1.0581

Properties generated from the program *STEAMCALC*, Wiley Professional Software, Wiley, New York, 1984.

TABLE E.9 Properties of Saturated Water — Pressure Table, USCS Units

P_{sat}, psia	T_{sat}, °F	Specific Volume, ft³/lbm			Internal Energy, Btu/lbm			Enthalpy, Btu/lbm			Entropy, Btu/(lbm · °R)		
		v_l	v_{lg}	v_g	u_l	u_{lg}	u_g	h_l	h_{lg}	h_g	s_l	s_{lg}	s_g
1	101.7	0.01613	333.47	333.48	69.71	974.5	1,044.2	69.7	1,036.2	1,105.9	0.1326	1.8455	1.9782
3	141.5	0.01630	118.70	118.72	109.34	947.4	1,056.7	109.4	1,013.3	1,122.6	0.2008	1.6855	1.8863
5	162.2	0.01640	73.516	73.533	130.10	933.0	1,063.1	130.1	1,001.0	1,131.1	0.2347	1.6096	1.8443
7	176.8	0.01648	53.623	53.639	144.73	922.7	1,067.4	144.7	992.2	1,136.9	0.2579	1.5589	1.8168
9	188.3	0.01655	42.380	42.398	156.19	914.6	1,070.8	156.2	985.2	1,141.4	0.2758	1.5206	1.7964
11	197.8	0.01661	35.126	35.143	165.70	907.8	1,073.5	165.7	979.3	1,145.0	0.2904	1.4898	1.7802
13	205.9	0.01667	30.017	30.033	173.88	902.0	1,075.9	173.9	974.2	1,148.1	0.3028	1.4639	1.7667
14.70	212.0	0.01671	26.732	26.748	180.03	897.6	1,077.7	180.1	970.3	1,150.4	0.3120	1.4448	1.7568
15	213.0	0.01672	26.220	26.236	181.07	896.9	1,078.0	181.1	969.7	1,150.8	0.3135	1.4416	1.7552
20	228.0	0.01683	20.020	20.037	196.11	886.0	1,082.1	196.2	960.1	1,156.2	0.3357	1.3964	1.7320
25	240.1	0.01692	16.285	16.302	208.35	876.8	1,085.1	208.4	952.1	1,160.6	0.3533	1.3608	1.7141
30	250.3	0.01700	13.735	13.752	218.74	869.0	1,087.8	218.8	945.3	1,164.1	0.3681	1.3315	1.6995
35	259.3	0.01708	11.882	11.899	227.81	862.2	1,090.0	227.9	939.2	1,167.1	0.3808	1.3064	1.6872
40	267.2	0.01715	10.480	10.497	235.91	856.1	1,092.0	236.0	933.7	1,169.7	0.3920	1.2845	1.6765
45	274.4	0.01721	9.3829	9.4001	243.23	850.5	1,093.8	243.4	928.7	1,172.0	0.4020	1.2651	1.6671
50	281.0	0.01727	8.4986	8.5158	249.93	845.4	1,095.3	250.1	924.0	1,174.1	0.4111	1.2475	1.6586
55	287.1	0.01733	7.7673	7.7847	256.12	840.6	1,096.7	256.3	919.7	1,176.0	0.4194	1.2316	1.6510
60	292.7	0.01738	7.1535	7.1709	261.89	836.1	1,098.0	262.1	915.6	1,177.6	0.4271	1.2169	1.6440
65	298.0	0.01743	6.6317	6.6492	267.29	831.9	1,099.2	267.5	911.7	1,179.2	0.4342	1.2033	1.6376
70	302.9	0.01748	6.1832	6.2007	272.37	827.9	1,100.3	272.6	908.0	1,180.6	0.4409	1.1907	1.6316
75	307.6	0.01753	5.7937	5.8112	277.18	824.1	1,101.3	277.4	904.5	1,181.9	0.4472	1.1788	1.6260
80	312.0	0.01757	5.4520	5.4696	281.75	820.4	1,102.2	282.0	901.1	1,183.1	0.4531	1.1677	1.6208
85	316.2	0.01761	5.1495	5.1671	286.10	816.9	1,103.0	286.4	897.9	1,184.3	0.4587	1.1572	1.6159
90	320.3	0.01766	4.8791	4.8967	290.25	813.5	1,103.8	290.5	894.8	1,185.3	0.4641	1.1472	1.6113
95	324.1	0.01770	4.6355	4.6532	294.24	810.3	1,104.5	294.5	891.8	1,186.3	0.4692	1.1378	1.6069

TABLE E.9 (Continued)

P_{sat}, psia	T_{sat}, °F	Specific Volume, ft³/lbm			Internal Energy, Btu/lbm			Enthalpy, Btu/lbm			Entropy, Btu/(lbm · °R)		
		v_l	v_{lg}	v_g	u_l	u_{lg}	u_g	h_l	h_{lg}	h_g	s_l	s_{lg}	s_g
100	327.8	0.01774	4.4150	4.4327	298.06	807.1	1,105.2	298.4	888.8	1,187.2	0.4740	1.1287	1.6028
110	334.8	0.01781	4.0315	4.0493	305.29	801.2	1,106.5	305.7	883.3	1,188.9	0.4832	1.1118	1.5950
120	341.2	0.01789	3.7100	3.7279	312.03	795.6	1,107.7	312.4	878.0	1,190.4	0.4916	1.0963	1.5879
130	347.3	0.01796	3.4367	3.4547	318.36	790.3	1,108.7	318.8	873.0	1,191.8	0.4994	1.0818	1.5813
140	353.0	0.01802	3.2015	3.2196	324.34	785.3	1,109.6	324.8	868.2	1,193.0	0.5068	1.0684	1.5752
150	358.4	0.01809	2.9967	3.0148	330.00	780.4	1,110.4	330.5	863.6	1,194.1	0.5137	1.0557	1.5695
160	363.5	0.01815	2.8163	2.8344	335.38	775.8	1,111.2	335.9	859.2	1,195.1	0.5203	1.0438	1.5641
170	368.4	0.01821	2.6561	2.6743	340.51	771.4	1,111.9	341.1	854.9	1,196.0	0.5265	1.0325	1.5590
180	373.1	0.01827	2.5130	2.5313	345.43	767.1	1,112.6	346.0	850.8	1,196.9	0.5324	1.0218	1.5542
190	377.5	0.01832	2.3846	2.4029	350.15	763.0	1,113.1	350.8	846.8	1,197.6	0.5380	1.0117	1.5497
200	381.8	0.01838	2.2686	2.2870	354.68	759.0	1,113.7	355.4	843.0	1,198.3	0.5434	1.0019	1.5454
250	400.9	0.01864	1.8250	1.8436	375.16	740.6	1,115.7	376.0	825.0	1,201.0	0.5675	0.9590	1.5265
300	417.3	0.01888	1.5242	1.5431	392.83	724.3	1,117.1	393.9	808.9	1,202.8	0.5878	0.9229	1.5107
350	431.7	0.01911	1.3068	1.3259	408.50	709.5	1,118.0	409.7	794.2	1,203.9	0.6055	0.8914	1.4969
400	444.6	0.01932	1.1415	1.1608	422.66	695.9	1,118.6	424.1	780.4	1,204.5	0.6213	0.8635	1.4848
450	456.3	0.01953	1.0119	1.0314	435.62	683.3	1,118.9	437.5	767.5	1,204.8	0.6355	0.8383	1.4738
500	467.0	0.01973	0.9100	0.9297	447.63	671.1	1,118.7	449.5	755.3	1,204.7	0.6486	0.8153	1.4639
550	476.9	0.01993	0.8245	0.8445	458.85	659.7	1,118.5	460.9	743.6	1,204.5	0.6606	0.7940	1.4547
600	486.2	0.02012	0.7515	0.7716	469.40	649.0	1,118.4	471.6	732.4	1,204.1	0.6719	0.7742	1.4461
650	494.9	0.02031	0.6888	0.7091	479.38	638.8	1,118.2	481.8	721.7	1,203.5	0.6824	0.7557	1.4381
700	503.1	0.02050	0.6348	0.6553	488.88	629.0	1,117.9	491.5	711.2	1,202.8	0.6923	0.7381	1.4304
750	510.8	0.02068	0.5881	0.6088	497.95	619.5	1,117.5	500.8	701.1	1,202.0	0.7017	0.7215	1.4232
800	518.2	0.02087	0.5473	0.5682	506.65	610.3	1,117.0	509.7	691.3	1,201.1	0.7106	0.7056	1.4162
850	525.2	0.02105	0.5114	0.5324	515.02	601.3	1,116.3	518.3	681.8	1,200.1	0.7192	0.6904	1.4096
900	532.0	0.02123	0.4793	0.5006	523.09	592.6	1,115.7	526.6	672.4	1,199.1	0.7274	0.6758	1.4031
950	538.4	0.02141	0.4506	0.4720	530.91	584.1	1,115.0	534.7	663.3	1,198.0	0.7352	0.6617	1.3969
1,000	544.6	0.02160	0.4244	0.4460	538.48	575.8	1,114.3	542.5	654.3	1,196.8	0.7428	0.6481	1.3909
1,100	556.3	0.02196	0.3787	0.4007	553.02	554.5	1,107.5	557.5	631.6	1,189.1	0.7572	0.6221	1.3794
1,200	567.2	0.02233	0.3401	0.3625	566.86	537.5	1,104.3	571.8	613.0	1,184.8	0.7708	0.5975	1.3683
1,300	577.5	0.02270	0.3071	0.3298	580.13	520.8	1,100.9	585.6	594.7	1,180.3	0.7837	0.5740	1.3577
1,400	587.1	0.02307	0.2785	0.3016	592.91	504.3	1,097.2	598.9	576.4	1,175.3	0.7960	0.5513	1.3473
1,500	596.3	0.02344	0.2535	0.2769	605.29	487.9	1,093.2	611.8	558.2	1,170.0	0.8079	0.5293	1.3372
1,600	605.0	0.02387	0.2313	0.2551	617.31	471.5	1,088.8	624.4	540.0	1,164.4	0.8194	0.5079	1.3272
1,700	613.3	0.02428	0.2114	0.2357	629.05	455.2	1,084.2	636.7	521.7	1,158.4	0.8305	0.4868	1.3173
1,800	621.2	0.02472	0.1934	0.2181	640.55	438.7	1,079.3	648.8	503.1	1,151.9	0.8414	0.4660	1.3074
1,900	628.9	0.02519	0.1770	0.2022	651.84	422.1	1,074.0	660.7	484.4	1,145.1	0.8521	0.4453	1.2974
2,000	636.2	0.02567	0.1619	0.1876	662.95	405.4	1,068.4	672.5	465.3	1,137.8	0.8626	0.4247	1.2873
2,100	643.2	0.02619	0.1480	0.1742	673.92	388.5	1,062.4	684.1	446.0	1,130.1	0.8730	0.4041	1.2771
2,200	650.0	0.02674	0.1350	0.1618	685.52	370.5	1,056.0	696.4	425.4	1,121.9	0.8833	0.3834	1.2667
2,300	656.6	0.02734	0.1229	0.1502	696.79	352.1	1,048.9	708.4	404.4	1,112.9	0.8935	0.3625	1.2560
2,400	662.9	0.02799	0.1114	0.1394	708.07	333.0	1,041.0	720.5	382.4	1,102.9	0.9037	0.3411	1.2449
2,500	668.1	0.02859	0.1021	0.1307	717.59	315.7	1,033.3	730.8	363.0	1,093.8	0.9123	0.3226	1.2349
2,600	673.9	0.02938	0.0917	0.1211	729.03	295.0	1,024.0	743.2	339.1	1,082.3	0.9222	0.3005	1.2227
2,700	679.5	0.03029	0.0817	0.1119	741.07	272.6	1,013.7	756.2	313.4	1,069.6	0.9320	0.2775	1.2094
2,800	685.0	0.03134	0.0717	0.1030	753.91	248.5	1,002.4	770.2	285.7	1,055.8	0.9418	0.2535	1.1953
2,900	690.2	0.03262	0.0616	0.0942	767.67	221.6	989.3	785.2	254.7	1,039.9	0.9516	0.2287	1.1803
3,000	695.3	0.03426	0.0508	0.0850	782.90	190.0	972.9	801.9	218.1	1,020.1	0.9615	0.2002	1.1616
3,204.1	705.6	0.0500	0	0.0500	872.57	0	872.6	902.5	0	902.5	1.0581	0	1.0581

Properties generated from the program *STEAMCALC*, Wiley Professional Software, Wiley, New York, 1984.

TABLE E.10 Properties of Superheated Steam — USCS Units

P, psia	T, °F	v, ft³/lbm	u, Btu/lbm	h, Btu/lbm	s, Btu/(lbm · °R)	P, psia	T, °F	v, ft³/lbm	u, Btu/lbm	h, Btu/lbm	s, Btu/(lbm · °R)
1						14.70					
(T_{sat} = 101.7°F)						(T_{sat} = 212.0°F)					
	200	392.52	1,077.8	1,150.4	2.0512		300	30.524	1,109.8	1,192.8	1.8160
	300	452.26	1,112.1	1,195.8	2.1153		400	34.670	1,145.6	1,239.9	1.8742
	400	511.92	1,147.0	1,241.7	2.1720		500	38.774	1,181.6	1,287.1	1.9261
	500	571.54	1,182.5	1,288.3	2.2233		600	42.857	1,218.2	1,334.8	1.9734
	600	631.14	1,218.8	1,335.6	2.2702		700	46.931	1,255.5	1,383.1	2.0170
	700	690.72	1,256.0	1,383.8	2.3137		800	50.999	1,293.6	1,432.3	2.0576
	800	750.31	1,294.0	1,432.8	2.3542		900	55.064	1,332.5	1,482.2	2.0958
	900	809.89	1,332.8	1,482.7	2.3923		1,000	59.126	1,372.3	1,533.1	2.1319
	1,000	869.46	1,372.5	1,533.4	2.4283		1,100	63.187	1,412.9	1,584.8	2.1662
	1,100	929.04	1,413.2	1,585.1	2.4625		1,200	67.246	1,454.5	1,637.4	2.1989
	1,200	988.61	1,454.7	1,637.1	2.4952		1,300	71.304	1,497.0	1,691.0	2.2302
	1,300	1,048.2	1,497.2	1,691.2	2.5265		1,400	75.361	1,540.5	1,745.4	2.2603
	1,400	1,107.8	1,540.6	1,745.6	2.5566		1,500	79.418	1,584.9	1,800.8	2.2893
	1,500	1,167.3	1,585.0	1,801.0	2.5856		1,600	83.474	1,630.2	1,857.2	2.3174
	1,600	1,226.9	1,630.3	1,857.4	2.6137						
5						25					
(T_{sat} = 162.2°F)						(T_{sat} = 240.1°F)					
	200	78.146	1,076.4	1,148.8	1.8718		300	17.829	1,108.0	1,190.4	1.7550
	300	90.241	1,111.4	1,194.9	1.9370		400	20.307	1,144.6	1,238.5	1.8144
	400	102.24	1,146.6	1,241.2	1.9942		500	22.741	1,180.9	1,286.2	1.8668
	500	114.21	1,182.3	1,287.9	2.0456		600	25.154	1,217.7	1,334.1	1.9144
	600	126.15	1,218.7	1,335.4	2.0927		700	27.557	1,255.1	1,382.6	1.9581
	700	138.09	1,255.8	1,383.6	2.1361		800	29.955	1,293.3	1,431.8	1.9988
	800	150.01	1,293.8	1,432.6	2.1767		900	32.349	1,332.2	1,481.9	2.0371
	900	161.94	1,332.7	1,482.5	2.2148		1,000	34.740	1,372.1	1,532.8	2.0732
	1,000	173.86	1,372.5	1,533.3	2.2509		1,100	37.130	1,412.8	1,584.5	2.1075
	1,100	185.78	1,413.1	1,585.0	2.2851		1,200	39.519	1,454.4	1,637.2	2.1402
	1,200	197.70	1,454.7	1,637.6	2.3178		1,300	41.906	1,496.9	1,690.8	2.1715
	1,300	209.62	1,497.2	1,691.1	2.3491		1,400	44.293	1,540.4	1,745.3	2.2017
	1,400	221.54	1,540.6	1,745.6	2.3792		1,500	46.679	1,584.7	1,800.7	2.2307
	1,500	233.45	1,585.0	1,801.0	2.4082		1,600	49.064	1,630.1	1,857.1	2.2588
	1,600	245.37	1,630.3	1,857.3	2.4363						
10						50					
(T_{sat} = 193.2°F)						(T_{sat} = 281.0°F)					
	200	38.844	1,074.7	1,146.6	1.7927						
	300	44.986	1,110.6	1,193.9	1.8595		300	8.7695	1,103.1	1,184.3	1.6721
	400	51.034	1,146.1	1,240.5	1.9172		400	10.062	1,141.9	1,235.0	1.7349
	500	57.041	1,181.9	1,287.5	1.9689		500	11.307	1,179.2	1,283.9	1.7887
	600	63.028	1,218.4	1,335.1	2.0160		600	12.529	1,216.5	1,332.4	1.8368
	700	69.005	1,255.7	1,383.4	2.0596		700	13.741	1,254.2	1,381.4	1.8809
	800	74.977	1,293.7	1,432.4	2.1002		800	14.947	1,292.6	1,430.9	1.9219
	900	80.945	1,332.6	1,482.4	2.1383		900	16.150	1,331.6	1,481.1	1.9602
	1,000	86.910	1,372.4	1,533.2	2.1744		1,000	17.350	1,371.6	1,532.1	1.9964
	1,100	92.874	1,413.0	1,584.9	2.2086		1,100	18.549	1,412.3	1,584.0	2.0308
	1,200	98.837	1,454.6	1,637.5	2.2413		1,200	19.746	1,454.0	1,636.7	2.0636
	1,300	104.80	1,497.1	1,691.0	2.2727		1,300	20.942	1,496.6	1,690.3	2.0949
	1,400	110.76	1,540.5	1,745.5	2.3028		1,400	22.137	1,540.0	1,744.9	2.1251
	1,500	116.72	1,584.9	1,800.9	2.3318		1,500	23.332	1,584.5	1,800.3	2.1542
	1,600	122.68	1,630.2	1,857.2	2.3598		1,600	24.526	1,629.8	1,856.8	2.1822

TABLE E.10 (Continued)

P, psia	T, °F	v, ft³/lbm	u, Btu/lbm	h, Btu/lbm	s, Btu/(lbm · °R)	P, psia	T, °F	v, ft³/lbm	u, Btu/lbm	h, Btu/lbm	s, Btu/(lbm · °R)
75 (T_{sat} = 307.6°F)						150 (T_{sat} = 358.4°F)					
	400	6.6452	1,139.2	1,231.4	1.6869		400	3.2209	1,129.9	1,219.3	1.5995
	500	7.4944	1,177.5	1,281.5	1.7421		500	3.6800	1,172.0	1,274.1	1.6599
	600	8.3205	1,215.3	1,330.8	1.7910		600	4.1112	1,211.5	1,325.6	1.7109
	700	9.1356	1,253.3	1,380.1	1.8354		700	4.5299	1,250.5	1,376.2	1.7566
	800	9.9447	1,291.8	1,429.9	1.8766		800	4.9422	1,289.6	1,426.8	1.7984
	900	10.750	1,331.1	1,480.3	1.9151		900	5.3508	1,329.3	1,477.8	1.8374
	1,000	11.554	1,371.1	1,531.4	1.9514		1,000	5.7569	1,369.5	1,529.3	1.8739
	1,100	12.355	1,411.9	1,583.4	1.9858		1,100	6.1612	1,410.6	1,581.6	1.9086
	1,200	13.155	1,453.6	1,636.2	2.0186		1,200	6.5642	1,452.5	1,634.7	1.9416
	1,300	13.954	1,496.2	1,689.9	2.0501		1,300	6.9661	1,495.2	1,688.6	1.9731
	1,400	14.752	1,539.7	1,744.5	2.0802		1,400	7.3672	1,538.8	1,743.3	2.0033
	1,500	15.550	1,584.2	1,800.0	2.1093		1,500	7.7675	1,583.3	1,799.0	2.0325
	1,600	16.347	1,629.6	1,856.4	2.1374		1,600	8.1673	1,628.8	1,855.5	2.0606
100 (T_{sat} = 327.8°F)						175 (T_{sat} = 370.8°F)					
	400	4.9349	1,136.2	1,227.5	1.6518		400	2.7295	1,126.5	1,214.9	1.5784
	500	5.5878	1,175.7	1,279.1	1.7085		500	3.1343	1,170.0	1,271.5	1.6408
	600	6.2161	1,214.0	1,329.1	1.7581		600	3.5097	1,210.2	1,323.9	1.6927
	700	6.8328	1,252.4	1,378.8	1.8029		700	3.8719	1,249.5	1,374.9	1.7388
	800	7.4435	1,291.1	1,428.8	1.8443		800	4.2275	1,288.9	1,425.8	1.7808
	900	8.0506	1,330.5	1,479.4	1.8829		900	4.5794	1,328.7	1,477.0	1.8199
	1,000	8.6552	1,370.6	1,530.7	1.9193		1,000	4.9288	1,369.0	1,528.7	1.8566
	1,100	9.2581	1,411.5	1,582.8	1.9538		1,100	5.2764	1,410.2	1,581.0	1.8913
	1,200	9.8597	1,453.2	1,635.7	1.9867		1,200	5.6227	1,452.1	1,634.2	1.9243
	1,300	10.460	1,495.9	1,689.4	2.0182		1,300	5.9679	1,494.9	1,688.1	1.9559
	1,400	11.060	1,539.4	1,744.1	2.0484		1,400	6.3122	1,538.5	1,742.9	1.9862
	1,500	11.659	1,583.9	1,799.7	2.0775		1,500	6.6558	1,583.1	1,798.6	2.0154
	1,600	12.257	1,629.3	1,856.1	2.1056		1,600	6.9988	1,628.6	1,855.2	2.0435
125 (T_{sat} = 344.3°F)						200 (T_{sat} = 381.8°F)					
	400	3.9073	1,133.1	1,223.5	1.6235		400	2.3599	1,123.0	1,210.3	1.5594
	500	4.4434	1,173.9	1,276.6	1.6820		500	2.7247	1,168.0	1,268.9	1.6240
	600	4.9533	1,212.8	1,327.4	1.7323		600	3.0584	1,208.9	1,322.1	1.6767
	700	5.4511	1,251.4	1,377.5	1.7775		700	3.3783	1,248.6	1,373.6	1.7232
	800	5.9427	1,290.4	1,427.8	1.8191		800	3.6915	1,288.1	1,424.8	1.7655
	900	6.4307	1,329.9	1,478.6	1.8579		900	4.0009	1,328.0	1,476.1	1.8047
	1,000	6.9162	1,370.1	1,530.0	1.8944		1,000	4.3077	1,368.5	1,528.0	1.8415
	1,100	7.4000	1,411.0	1,582.2	1.9290		1,100	4.6128	1,409.7	1,580.4	1.8763
	1,200	7.8824	1,452.8	1,635.2	1.9619		1,200	4.9165	1,451.7	1,633.7	1.9094
	1,300	8.3637	1,495.5	1,689.0	1.9934		1,300	5.2191	1,494.5	1,687.7	1.9410
	1,400	8.8442	1,539.1	1,743.7	2.0236		1,400	5.5209	1,538.2	1,742.5	1.9713
	1,500	9.3240	1,583.6	1,799.3	2.0527		1,500	5.8220	1,582.8	1,798.3	2.0005
	1,600	9.8031	1,629.1	1,855.8	2.0809		1,600	6.1224	1,628.3	1,854.9	2.0287

TABLE E.10 (Continued)

P, psia	T, °F	v, ft³/lbm	u, Btu/lbm	h, Btu/lbm	s, Btu/(lbm · °R)	P, psia	T, °F	v, ft³/lbm	u, Btu/lbm	h, Btu/lbm	s, Btu/(lbm · °R)
225 (T_{sat} = 391.8°F)						300 (T_{sat} = 417.3°F)					
	400	2.0715	1,119.2	1,205.5	1.5419		500	1.7665	1,159.5	1,257.6	1.5701
	500	2.4058	1,166.0	1,266.1	1.6088		600	2.0045	1,203.4	1,314.7	1.6268
	600	2.7072	1,207.5	1,320.3	1.6625		700	2.2264	1,244.6	1,368.2	1.6751
	700	2.9944	1,247.6	1,372.3	1.7094		800	2.4407	1,285.1	1,420.6	1.7184
	800	3.2746	1,287.4	1,423.7	1.7519		900	2.6509	1,325.6	1,472.8	1.7582
	900	3.5509	1,327.4	1,475.3	1.7913		1,000	2.8586	1,366.5	1,525.2	1.7954
	1,000	3.8247	1,368.0	1,527.3	1.8282		1,100	3.0644	1,408.0	1,578.1	1.8305
	1,100	4.0967	1,409.3	1,579.8	1.8630		1,200	3.2688	1,450.2	1,631.6	1.8638
	1,200	4.3673	1,451.3	1,633.1	1.8962		1,300	3.4722	1,493.1	1,685.9	1.8955
	1,300	4.6368	1,494.2	1,687.2	1.9278		1,400	3.6747	1,537.0	1,741.0	1.9260
	1,400	4.9055	1,537.9	1,742.1	1.9582		1,500	3.8764	1,581.7	1,796.9	1.9552
	1,500	5.1735	1,582.5	1,797.9	1.9874		1,600	4.0776	1,627.3	1,853.6	1.9835
	1,600	5.4408	1,628.0	1,854.6	2.0156						
250 (T_{sat} = 400.9°F)						350 (T_{sat} = 431.7°F)					
	500	2.1504	1,163.9	1,263.4	1.5949		500	1.4914	1,154.9	1,251.5	1.5481
	600	2.4262	1,206.2	1,318.4	1.6495		600	1.7028	1,200.5	1,310.8	1.6070
	700	2.6872	1,246.6	1,370.9	1.6969		700	1.8971	1,242.6	1,365.5	1.6563
	800	2.9411	1,286.6	1,422.7	1.7397		800	2.0833	1,283.6	1,418.5	1.7002
	900	3.1909	1,326.8	1,474.4	1.7793		900	2.2652	1,324.4	1,471.1	1.7403
	1,000	3.4382	1,367.5	1,526.6	1.8162		1,000	2.4445	1,365.4	1,523.8	1.7777
	1,100	3.6837	1,408.8	1,579.3	1.8512		1,100	2.6220	1,407.1	1,576.9	1.8129
	1,200	3.9279	1,450.9	1,632.6	1.8843		1,200	2.7981	1,449.4	1,630.6	1.8463
	1,300	4.1710	1,493.8	1,686.8	1.9160		1,300	2.9730	1,492.4	1,685.0	1.8781
	1,400	4.4132	1,537.6	1,741.7	1.9464		1,400	3.1472	1,536.3	1,740.2	1.9086
	1,500	4.6546	1,582.2	1,797.6	1.9756		1,500	3.3206	1,581.1	1,796.2	1.9380
	1,600	4.8956	1,627.8	1,854.3	2.0038		1,600	3.4934	1,626.8	1,853.0	1.9663
275 (T_{sat} = 409.4°F)						400 (T_{sat} = 444.6°F)					
	500	1.9411	1,161.7	1,260.5	1.5821		500	1.2842	1,150.0	1,245.1	1.5281
	600	2.1962	1,204.8	1,316.6	1.6377		600	1.4763	1,197.6	1,306.8	1.5894
	700	2.4359	1,245.6	1,369.6	1.6855		700	1.6500	1,240.5	1,362.7	1.6398
	800	2.6681	1,285.9	1,421.7	1.7286		800	1.8152	1,282.0	1,416.4	1.6842
	900	2.8964	1,326.2	1,473.6	1.7683		900	1.9759	1,323.1	1,469.4	1.7247
	1,000	3.1220	1,367.0	1,525.9	1.8054		1,000	2.1340	1,364.4	1,522.4	1.7623
	1,100	3.3459	1,408.4	1,578.7	1.8404		1,100	2.2902	1,406.2	1,575.7	1.7976
	1,200	3.5684	1,450.5	1,632.1	1.8736		1,200	2.4450	1,448.6	1,629.6	1.8311
	1,300	3.7898	1,493.5	1,686.3	1.9053		1,300	2.5987	1,491.8	1,684.1	1.8630
	1,400	4.0103	1,537.3	1,741.4	1.9357		1,400	2.7515	1,535.7	1,739.4	1.8936
	1,500	4.2302	1,582.0	1,797.2	1.9650		1,500	2.9037	1,580.5	1,795.5	1.9230
	1,600	4.4494	1,627.5	1,854.0	1.9932		1,600	3.0552	1,626.2	1,852.4	1.9513

TABLE E.10 (Continued)

P, psia	T, °F	v, ft³/lbm	u, Btu/lbm	h, Btu/lbm	s, Btu/(lbm · °R)	P, psia	T, °F	v, ft³/lbm	u, Btu/lbm	h, Btu/lbm	s, Btu/(lbm · °R)
450 (T_{sat} = 456.3°F)						550 (T_{sat} = 476.9°F)					
	500	1.1222	1,144.9	1,238.3	1.5095		500	0.8846	1,133.6	1,223.7	1.4750
	600	1.2999	1,194.5	1,302.8	1.5735		600	1.0425	1,188.2	1,294.3	1.5451
	700	1.4577	1,238.4	1,359.8	1.6250		700	1.1777	1,234.1	1,354.0	1.5990
	800	1.6066	1,280.4	1,414.2	1.6699		800	1.3031	1,277.2	1,409.8	1.6452
	900	1.7509	1,321.9	1,467.7	1.7108		900	1.4236	1,319.3	1,464.2	1.6868
	1,000	1.8924	1,363.4	1,521.0	1.7486		1,000	1.5411	1,361.3	1,518.1	1.7250
	1,100	2.0321	1,405.3	1,574.5	1.7841		1,100	1.6567	1,403.5	1,572.1	1.7608
	1,200	2.1704	1,447.8	1,628.6	1.8177		1,200	1.7709	1,446.3	1,626.5	1.7946
	1,300	2.3075	1,491.1	1,683.2	1.8497		1,300	1.8841	1,489.7	1,681.4	1.8268
	1,400	2.4438	1,535.1	1,738.6	1.8803		1,400	1.9963	1,533.9	1,737.0	1.8575
	1,500	2.5794	1,580.0	1,794.8	1.9097		1,500	2.1078	1,578.9	1,793.4	1.8870
	1,600	2.7145	1,625.7	1,851.8	1.9381		1,600	2.2188	1,624.7	1,850.5	1.9155
500 (T_{sat} = 467.0°F)						600 (T_{sat} = 486.2°F)					
	500	0.9919	1,139.4	1,231.2	1.4919		500	0.7944	1,127.5	1,215.7	1.4586
	600	1.1584	1,191.4	1,298.6	1.5588		550	0.8746	1,158.3	1,255.4	1.4990
	700	1.3038	1,236.3	1,356.9	1.6115		600	0.9456	1,184.8	1,289.8	1.5323
	800	1.4397	1,278.8	1,412.0	1.6571		650	1.0109	1,209.0	1,321.3	1.5613
	900	1.5709	1,320.6	1,465.9	1.6982		700	1.0726	1,231.9	1,351.0	1.5875
	1,000	1.6992	1,362.3	1,519.5	1.7363		750	1.1318	1,254.0	1,379.7	1.6117
	1,100	1.8257	1,404.4	1,573.3	1.7719		800	1.1893	1,275.6	1,407.6	1.6343
	1,200	1.9507	1,447.0	1,627.5	1.8056		850	1.2455	1,296.9	1,435.2	1.6558
	1,300	2.0746	1,490.4	1,682.3	1.8377		900	1.3008	1,318.0	1,462.5	1.6762
	1,400	2.1977	1,534.5	1,737.8	1.8683		950	1.3554	1,339.1	1,489.6	1.6958
	1,500	2.3200	1,579.4	1,794.1	1.8978		1,000	1.4094	1,360.2	1,516.7	1.7147
	1,600	2.4418	1,625.2	1,851.2	1.9262		1,050	1.4629	1,381.4	1,543.8	1,7330
							1,100	1.5160	1,402.6	1,570.9	1.7507
							1,150	1.5687	1,424.0	1,598.1	1.7678
							1,200	1.6212	1,445.5	1,625.5	1.7846
							1,250	1.6733	1,467.1	1,652.9	1.8009
							1,300	1.7253	1,489.0	1,680.5	1.8168
							1,350	1.7770	1,511.0	1,708.3	1.8323
							1,400	1.8285	1,533.2	1,736.3	1.8476
							1,450	1.8798	1,555.7	1,764.4	1.8625
							1,500	1.9310	1,578.3	1,792.7	1.8771
							1,550	1.9820	1,601.1	1,821.2	1.8915
							1,600	2.0329	1,624.2	1,849.9	1.9056

607

P, psia	T, °F	v, ft³/lbm	u, Btu/lbm	h, Btu/lbm	s, Btu/(lbm · °R)
800 (T_sat = 518.2°F)					
	550	0.6151	1,138.7	1,229.7	1.4467
	600	0.6774	1,170.4	1,270.6	1.4863
	650	0.7323	1,197.7	1,306.1	1.5190
	700	0.7828	1,222.7	1,338.6	1.5476
	750	0.8303	1,246.2	1,369.1	1.5734
	800	0.8759	1,268.9	1,398.5	1.5972
	850	0.9200	1,291.0	1,427.2	1.6195
	900	0.9631	1,312.8	1,455.4	1.6406
	950	1.0054	1,334.4	1,483.2	1.6608
	1,000	1.0470	1,355.9	1,510.9	1.6801
	1,050	1.0882	1,377.4	1,538.5	1.6987
	1,100	1.1289	1,399.0	1,566.1	1.7166
	1,150	1.1692	1,420.6	1,593.7	1.7341
	1,200	1.2093	1,442.3	1,621.3	1.7510
	1,250	1.2490	1,464.2	1,649.1	1.7674
	1,300	1.2886	1,486.2	1,677.0	1.7835
	1,350	1.3279	1,508.4	1,705.0	1.7992
	1,400	1.3670	1,530.7	1,733.1	1.8145
	1,450	1.4059	1,553.3	1,761.4	1.8296
	1,500	1.4446	1,576.0	1,789.9	1.8443
	1,550	1.4833	1,599.0	1,818.6	1.8587
	1,600	1.5217	1,622.1	1,847.4	1.8729
1,000 (T_sat = 544.6°F)					
	550	0.4535	1,114.9	1,198.8	1.3967
	600	0.5138	1,153.7	1,248.8	1.4450
	650	0.5636	1,185.2	1,289.5	1.4826
	700	0.6080	1,212.7	1,325.2	1.5141
	750	0.6489	1,238.0	1,358.1	1.5418
	800	0.6875	1,261.9	1,389.1	1.5670
	850	0.7245	1,284.9	1,419.0	1.5902
	900	0.7603	1,307.4	1,448.1	1.6121
	950	0.7953	1,329.6	1,476.7	1.6327
	1,000	0.8296	1,351.6	1,505.1	1.6525
	1,050	0.8633	1,373.4	1,533.2	1.6714
	1,100	0.8966	1,395.3	1,561.2	1.6897
	1,150	0.9296	1,417.2	1,589.2	1.7074
	1,200	0.9622	1,439.1	1,617.2	1.7245
	1,250	0.9945	1,461.2	1,645.2	1.7411
	1,300	1.0266	1,483.4	1,673.4	1.7573
	1,350	1.0584	1,505.7	1,701.6	1.7732
	1,400	1.0901	1,528.2	1,729.9	1.7886
	1,450	1.1216	1,550.9	1,758.4	1.8037
	1,500	1.1529	1,573.8	1,787.1	1.8186
	1,550	1.1840	1,596.8	1,815.9	1.8331
	1,600	1.2151	1,620.1	1,844.9	1.8473

P, psia	T, °F	v, ft³/lbm	u, Btu/lbm	h, Btu/lbm	s, Btu/(lbm · °R)
1,200 (T_sat = 567.2°F)					
	600	0.4016	1,134.4	1,223.6	1.4054
	650	0.4497	1,171.2	1,271.1	1.4492
	700	0.4906	1,202.0	1,310.9	1.4843
	750	0.5273	1,229.3	1,346.4	1.5143
	800	0.5615	1,254.6	1,379.3	1.5409
	850	0.5939	1,278.6	1,410.5	1.5652
	900	0.6250	1,301.9	1,440.7	1.5879
	950	0.6551	1,324.7	1,470.1	1.6091
	1,000	0.6845	1,347.1	1,499.1	1.6293
	1,050	0.7134	1,369.4	1,527.8	1.6487
	1,100	0.7418	1,391.6	1,556.3	1.6672
	1,150	0.7698	1,413.7	1,584.7	1.6851
	1,200	0.7974	1,435.9	1,613.0	1.7025
	1,250	0.8248	1,458.2	1,641.3	1.7193
	1,300	0.8519	1,480.6	1,669.7	1.7357
	1,350	0.8788	1,503.1	1,698.2	1.7516
	1,400	0.9055	1,525.7	1,726.8	1.7672
	1,450	0.9320	1,548.5	1,755.5	1.7824
	1,500	0.9584	1,571.5	1,784.3	1.7973
	1,550	0.9846	1,594.6	1,813.3	1.8119
	1,600	1.0107	1,618.0	1,842.4	1.8262
1,400 (T_sat = 587.1°F)					
	600	0.3176	1,111.9	1,194.2	1.3652
	650	0.3667	1,155.5	1,250.6	1.4172
	700	0.4059	1,190.3	1,295.4	1.4567
	750	0.4400	1,220.0	1,334.0	1.4893
	800	0.4712	1,246.9	1,369.0	1.5177
	850	0.5004	1,272.1	1,401.8	1.5432
	900	0.5282	1,296.2	1,433.1	1.5666
	950	0.5549	1,319.6	1,463.4	1.5886
	1,000	0.5809	1,342.6	1,493.1	1.6092
	1,050	0.6063	1,365.3	1,522.3	1.6290
	1,100	0.6312	1,387.8	1,551.3	1.6478
	1,150	0.6556	1,410.2	1,580.1	1.6660
	1,200	0.6798	1,432.7	1,608.8	1.6835
	1,250	0.7036	1,455.2	1,637.4	1.7006
	1,300	0.7272	1,477.7	1,666.1	1.7171
	1,350	0.7505	1,500.4	1,694.8	1.7332
	1,400	0.7737	1,523.2	1,723.6	1.7489
	1,450	0.7967	1,546.1	1,752.5	1.7642
	1,500	0.8195	1,569.2	1,781.5	1.7792
	1,550	0.8421	1,592.4	1,810.6	1.7939
	1,600	0.8647	1,615.9	1,839.9	1.8083

P, psia	T, °F	v, ft³/lbm	u, Btu/lbm	h, Btu/lbm	s, Btu/(lbm · °R)	P, psia	T, °F	v, ft³/lbm	u, Btu/lbm	h, Btu/lbm	s, Btu/(lbm · °R)
1,600 (T_{sat} = 605.0°F)						2,000 (T_{sat} = 636.2°F)					
	650	0.3026	1,137.9	1,227.5	1.3853		650	0.2056	1,095.9	1,172.0	1.3184
	700	0.3415	1,177.5	1,278.6	1.4303		700	0.2488	1,148.0	1,240.1	1.3785
	750	0.3741	1,210.2	1,320.9	1.4661		750	0.2805	1,188.3	1,292.1	1.4224
	800	0.4032	1,239.0	1,358.3	1.4964		800	0.3073	1,221.7	1,335.4	1.4576
	850	0.4301	1,265.4	1,392.8	1.5232		850	0.3312	1,251.2	1,373.8	1.4874
	900	0.4555	1,290.4	1,425.3	1.5476		900	0.3534	1,278.3	1,409.1	1.5139
	950	0.4797	1,314.5	1,456.5	1.5702		950	0.3742	1,303.9	1,442.4	1.5380
	1,000	0.5031	1,338.0	1,487.0	1.5914		1,000	0.3942	1,328.6	1,474.5	1.5603
	1,050	0.5259	1,361.1	1,516.8	1.6115		1,050	0.4134	1,352.6	1,505.6	1.5813
	1,100	0.5482	1,384.0	1,546.3	1.6307		1,100	0.4320	1,376.2	1,536.1	1.6011
	1,150	0.5700	1,406.7	1,575.5	1.6491		1,150	0.4502	1,399.6	1,566.2	1.6201
	1,200	0.5915	1,429.4	1,604.5	1.6669		1,200	0.4680	1,422.8	1,596.0	1.6384
	1,250	0.6127	1,452.1	1,633.5	1.6841		1,250	0.4855	1,445.9	1,625.6	1.6559
	1,300	0.6336	1,474.8	1,662.5	1.7008		1,300	0.5027	1,469.1	1,655.1	1.6730
	1,350	0.6544	1,497.7	1,691.4	1.7170		1,350	0.5197	1,492.2	1,684.6	1.6895
	1,400	0.6748	1,520.6	1,720.4	1.7328		1,400	0.5365	1,515.4	1,714.0	1.7055
	1,450	0.6952	1,543.7	1,749.5	1.7482		1,450	0.5531	1,538.8	1,743.5	1.7212
	1,500	0.7153	1,566.9	1,778.7	1.7633		1,500	0.5695	1,562.2	1,773.0	1.7364
	1,550	0.7353	1,590.3	1,808.0	1.7781		1,550	0.5858	1,585.8	1,802.6	1.7513
	1,600	0.7552	1,613.8	1,837.4	1.7926		1,600	0.6020	1,609.6	1,832.4	1.7660
1,800 (T_{sat} = 621.2°F)						2,500 (T_{sat} = 668.1°F)					
	650	0.2505	1,118.1	1,201.5	1.3526		700	0.1681	1,102.6	1,180.4	1.3105
	700	0.2906	1,163.4	1,260.2	1.4044		750	0.2032	1,156.0	1,250.0	1.3694
	750	0.3224	1,199.6	1,307.0	1.4439		800	0.2293	1,197.5	1,303.5	1.4127
	800	0.3500	1,230.6	1,347.2	1.4765		850	0.2514	1,231.9	1,348.2	1.4475
	850	0.3753	1,258.4	1,383.4	1.5047		900	0.2712	1,262.3	1,387.7	1.4771
	900	0.3988	1,284.5	1,417.3	1.5301		950	0.2896	1,290.1	1,424.1	1.5034
	950	0.4212	1,309.3	1,449.6	1.5534		1,000	0.3068	1,316.4	1,458.3	1.5273
	1,000	0.4426	1,333.3	1,480.8	1.5752		1,050	0.3232	1,341.7	1,491.2	1.5494
	1,050	0.4634	1,356.9	1,511.2	1.5957		1,100	0.3390	1,366.3	1,523.2	1.5702
	1,100	0.4836	1,380.1	1,541.2	1.6152		1,150	0.3543	1,390.5	1,554.4	1.5900
	1,150	0.5035	1,403.2	1,570.9	1.6340		1,200	0.3692	1,414.4	1,585.2	1.6088
	1,200	0.5229	1,426.1	1,600.3	1.6520		1,250	0.3837	1,438.1	1,615.6	1.6269
	1,250	0.5420	1,449.0	1,629.6	1.6693		1,300	0.3980	1,461.7	1,645.9	1.6443
	1,300	0.5609	1,472.0	1,658.8	1.6862		1,350	0.4120	1,485.3	1,676.0	1.6612
	1,350	0.5795	1,494.9	1,688.0	1.7025		1,400	0.4259	1,508.9	1,706.0	1.6775
	1,400	0.5980	1,518.0	1,717.2	1.7185		1,450	0.4395	1,532.6	1,735.9	1.6934
	1,450	0.6162	1,541.2	1,746.5	1.7340		1,500	0.4529	1,556.4	1,765.9	1.7089
	1,500	0.6343	1,564.6	1,775.8	1.7492		1,550	0.4663	1,580.3	1,796.0	1.7241
	1,550	0.6522	1,588.0	1,805.3	1.7640		1,600	0.4794	1,604.3	1,826.1	1.7389
	1,600	0.6701	1,611.7	1,834.9	1.7786						

P, psia	T, °F	v, ft³/lbm	u, Btu/lbm	h, Btu/lbm	s, Btu/(lbm · °R)	P, psia	T, °F	v, ft³/lbm	u, Btu/lbm	h, Btu/lbm	s, Btu/(lbm · °R)
3,000						4,000					
(T_{sat} = 695.3°F)							800	0.1053	1,098.2	1,176.1	1.2766
	700	0.1006	1,047.4	1,103.2	1.2329		850	0.1284	1,159.6	1,254.7	1.3378
	750	0.1483	1,117.0	1,199.3	1.3141		900	0.1463	1,206.0	1,314.3	1.3825
	800	0.1760	1,169.4	1,267.1	1.3691		950	0.1616	1,244.0	1,363.6	1.4181
	850	0.1975	1,210.5	1,320.1	1.4104		1,000	0.1752	1,277.0	1,406.7	1.4482
	900	0.2161	1,245.0	1,364.9	1.4439		1,050	0.1877	1,307.0	1,446.0	1.4746
	950	0.2329	1,275.6	1,404.8	1.4728		1,100	0.1994	1,335.2	1,482.8	1.4987
	1,000	0.2484	1,303.8	1,441.7	1.4984		1,150	0.2105	1,362.2	1,518.0	1.5208
	1,050	0.2630	1,330.5	1,476.5	1.5219		1,200	0.2211	1,388.4	1,552.0	1.5416
	1,100	0.2770	1,356.2	1,509.9	1.5437		1,250	0.2313	1,414.0	1,585.1	1.5613
	1,150	0.2904	1,381.2	1,542.4	1.5642		1,300	0.2411	1,439.2	1,617.7	1.5801
	1,200	0.3033	1,405.8	1,574.2	1.5836		1,350	0.2507	1,464.2	1,649.8	1.5981
	1,250	0.3159	1,430.2	1,605.6	1.6022		1,400	0.2601	1,489.0	1,681.6	1.6154
	1,300	0.3283	1,454.3	1,636.5	1.6201		1,450	0.2693	1,513.8	1,713.1	1.6322
	1,350	0.3403	1,478.4	1,667.3	1.6373		1,500	0.2783	1,538.5	1,744.5	1.6484
	1,400	0.3522	1,502.4	1,697.9	1.6540		1,550	0.2872	1,563.3	1,775.8	1.6642
	1,450	0.3638	1,526.4	1,728.4	1.6702		1,600	0.2959	1,588.1	1,807.1	1.6795
	1,500	0.3753	1,550.5	1,758.8	1.6859						
	1,550	0.3866	1,574.7	1,789.3	1.7013						
	1,600	0.3978	1,598.9	1,819.8	1.7163						
3,500						4,500					
	800	0.1364	1,136.7	1,225.1	1.3243		800	0.0801	1,052.1	1,118.8	1.2244
	850	0.1584	1,186.6	1,289.1	1.3742		850	0.1047	1,128.9	1,216.1	1.3002
	900	0.1764	1,226.3	1,340.5	1.4127		900	0.1227	1,183.9	1,286.1	1.3527
	950	0.1922	1,260.2	1,384.7	1.4446		950	0.1377	1,226.8	1,341.4	1.3926
	1,000	0.2066	1,290.6	1,424.4	1.4723		1,000	0.1507	1,262.8	1,388.3	1.4254
	1,050	0.2200	1,318.9	1,461.4	1.4972		1,050	0.1626	1,294.9	1,430.3	1.4536
	1,100	0.2326	1,345.8	1,496.5	1.5201		1,100	0.1736	1,324.5	1,469.0	1.4788
	1,150	0.2447	1,371.8	1,530.3	1.5414		1,150	0.1839	1,352.5	1,505.6	1.5019
	1,200	0.2563	1,397.2	1,563.2	1.5615		1,200	0.1937	1,379.5	1,540.7	1.5235
	1,250	0.2675	1,422.1	1,595.4	1.5807		1,250	0.2031	1,405.7	1,574.8	1.5437
	1,300	0.2785	1,446.8	1,627.1	1.5990		1,300	0.2121	1,431.5	1,608.2	1.5629
	1,350	0.2891	1,471.3	1,658.6	1.6166		1,350	0.2209	1,457.0	1,641.0	1.5813
	1,400	0.2995	1,495.7	1,689.7	1.6336		1,400	0.2295	1,482.3	1,673.4	1.5990
	1,450	0.3098	1,520.1	1,720.8	1.6500		1,450	0.2379	1,507.4	1,705.5	1.6160
	1,500	0.3199	1,544.5	1,751.7	1.6660		1,500	0.2460	1,532.5	1,737.3	1.6325
	1,550	0.3298	1,569.0	1,782.6	1.6816		1,550	0.2541	1,557.5	1,769.1	1.6485
	1,600	0.3395	1,593.5	1,813.4	1.6968		1,600	0.2620	1,582.6	1,800.7	1.6640

P, psia	T, °F	v, ft³/lbm	u, Btu/lbm	h, Btu/lbm	s, Btu/(lbm · °R)	P, psia	T, °F	v, ft³/lbm	u, Btu/lbm	h, Btu/lbm	s, Btu/(lbm · °R)
5,000											
	800	0.0598	996.0	1,051.3	1.1659		1,250	0.1806	1,397.4	1,564.5	1.5274
	850	0.0855	1,093.6	1,172.8	1.2605		1,300	0.1890	1,423.8	1,598.7	1.5471
	900	0.1038	1,159.5	1,255.6	1.3226		1,350	0.1971	1,449.8	1,632.2	1.5659
	950	0.1185	1,208.5	1,318.1	1.3678		1,400	0.2050	1,475.4	1,665.2	1.5839
	1,000	0.1312	1,248.1	1,369.5	1.4036		1,450	0.2127	1,500.9	1,697.8	1.6012
	1,050	0.1425	1,282.4	1,414.2	1.4338		1,500	0.2203	1,526.4	1,730.2	1.6179
	1,100	0.1529	1,313.5	1,455.0	1.4603		1,550	0.2276	1,551.7	1,762.3	1.6341
	1,150	0.1626	1,342.6	1,493.1	1.4844		1,600	0.2349	1,577.1	1,794.4	1.6499
	1,200	0.1718	1,370.5	1,529.4	1.5066						

Properties generated from the program *STEAMCALC*, Wiley Professional Software, Wiley, New York, 1984.

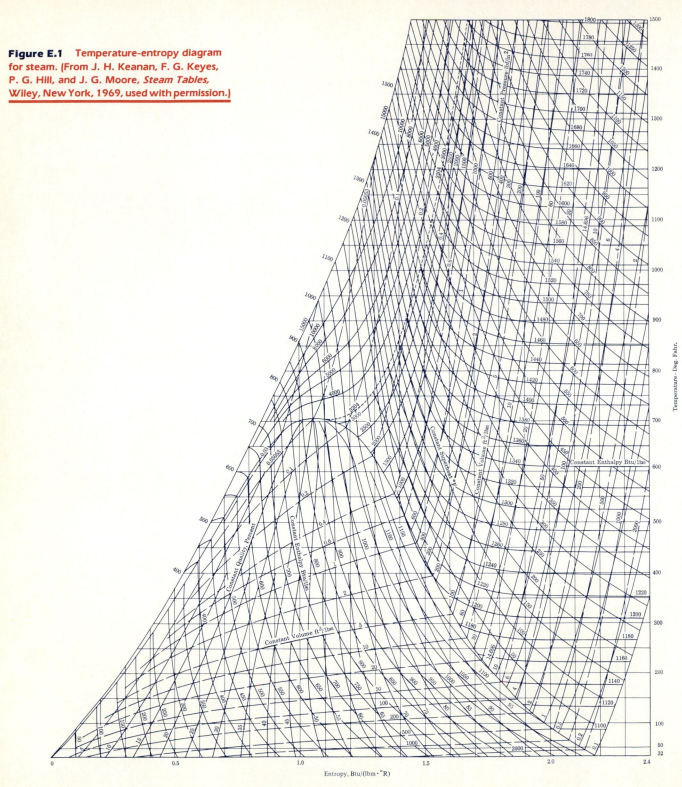

Figure E.1 Temperature-entropy diagram for steam. (From J. H. Keanan, F. G. Keyes, P. G. Hill, and J. G. Moore, *Steam Tables*, Wiley, New York, 1969, used with permission.)

Entropy, Btu/(lbm·°R)

Temperature - Deg. Fahr.

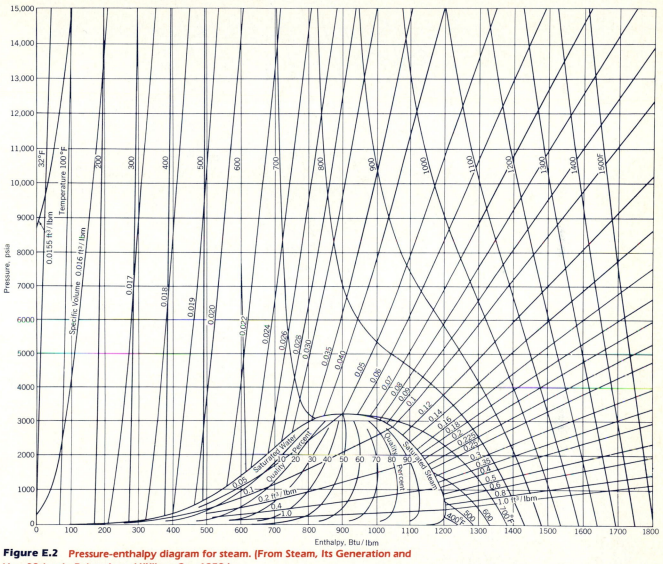

Figure E.2 Pressure-enthalpy diagram for steam. (From Steam, Its Generation and Use, 39th ed., Babcock and Wilcox Co., 1978.)

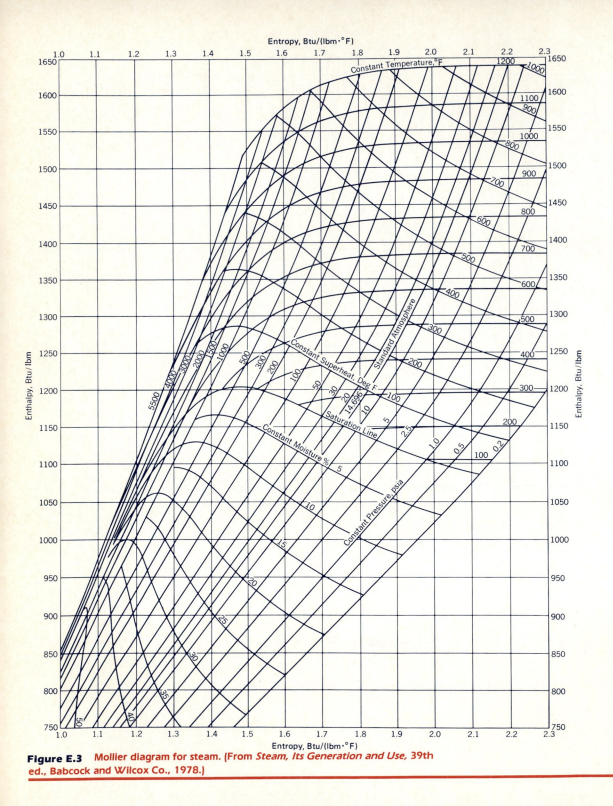

Figure E.3 Mollier diagram for steam. (From *Steam, Its Generation and Use,* 39th ed., Babcock and Wilcox Co., 1978.)

TABLE E.11 Properties of Saturated Refrigerant 12 — Temperature Table, USCS Units

Reference state: $h_l = s_l = 0$ at $-40°F$

T_{sat}, °F	P_{sat}, psia	Specific Volume, ft³/lbm			Internal Energy, Btu/lbm			Enthalpy, Btu/lbm			Entropy, Btu/(lbm · °R)		
		v_l	v_{lg}	v_g	u_l	u_{lg}	u_g	h_l	h_{lg}	h_g	s_l	s_{lg}	s_g
−150	0.2	0.009582	178.90	178.91	−22.703	78.461	55.758	−22.703	83.538	60.835	−0.06264	0.26976	0.20712
−140	0.3	0.009658	110.60	110.61	−20.659	77.315	56.656	−20.659	82.552	61.893	−0.05614	0.25824	0.20210
−130	0.4	0.009736	70.811	70.820	−18.617	76.186	57.570	−18.616	81.581	62.965	−0.04985	0.24746	0.19761
−120	0.6	0.009816	46.787	46.797	−16.574	75.071	58.497	−16.573	80.621	64.049	−0.04374	0.23735	0.19361
−110	1.0	0.009899	31.803	31.813	−14.529	73.966	59.437	−14.527	79.670	65.143	−0.03781	0.22784	0.19003
−100	1.4	0.009984	22.178	22.188	−12.481	72.869	60.388	−12.478	78.723	66.245	−0.03203	0.21887	0.18684
−90	2.0	0.010073	15.828	15.838	−10.428	71.777	61.348	−10.425	77.777	67.353	−0.02640	0.21040	0.18399
−80	2.9	0.010164	11.533	11.543	−8.357	70.673	62.316	−8.351	76.815	68.464	−0.02087	0.20232	0.18145
−70	4.0	0.010259	8.5656	8.5759	−6.286	69.577	63.290	−6.279	75.856	69.577	−0.01548	0.19467	0.17918
−60	5.4	0.010357	6.4722	6.4826	−4.208	68.477	64.269	−4.198	74.888	70.691	−0.01021	0.18738	0.17716
−50	7.1	0.010458	4.9675	4.9779	−2.121	67.372	65.252	−2.107	73.909	71.802	−0.00505	0.18041	0.17536
−40	9.3	0.010564	3.8672	3.8778	−0.024	66.260	66.236	0.000	72.910	72.910	0.00000	0.17376	0.17376
−30	12.0	0.010673	3.0500	3.0605	2.083	65.139	67.222	2.107	71.906	74.013	0.00498	0.16735	0.17233
−20	15.3	0.010787	2.4337	2.4445	4.200	64.007	68.207	4.231	70.878	75.109	0.00986	0.16121	0.17107
−10	19.2	0.010906	1.9630	1.9739	6.329	62.862	69.191	6.367	69.828	76.196	0.01466	0.15529	0.16994
0	23.8	0.011030	1.5988	1.6098	8.468	61.704	70.173	8.517	68.756	77.273	0.01937	0.14958	0.16895
10	29.3	0.011160	1.3137	1.3249	10.620	60.531	71.151	10.681	67.658	78.339	0.02402	0.14405	0.16807
20	35.7	0.011295	1.0882	1.0995	12.785	59.339	72.124	12.860	66.532	79.391	0.02859	0.13870	0.16730
30	43.1	0.011438	0.90797	0.91941	14.964	58.129	73.093	15.055	65.375	80.430	0.03311	0.13351	0.16661
40	51.6	0.011588	0.76235	0.77394	17.172	56.882	74.054	17.282	64.167	81.449	0.03759	0.12842	0.16601
50	61.4	0.011745	0.64395	0.65569	19.389	55.618	75.007	19.523	62.930	82.452	0.04201	0.12347	0.16548
60	72.4	0.011912	0.54673	0.55864	21.628	54.322	75.951	21.788	61.647	83.435	0.04638	0.11863	0.16501
70	84.9	0.012089	0.46631	0.47840	23.891	52.992	76.883	24.081	60.314	84.395	0.05072	0.11387	0.16459
80	98.8	0.012277	0.39926	0.41154	26.181	51.622	77.803	26.406	58.924	85.330	0.05503	0.10918	0.16422
90	114.4	0.012477	0.34299	0.35546	28.502	50.206	78.708	28.766	57.469	86.236	0.05933	0.10455	0.16388
100	131.8	0.012692	0.29541	0.30810	30.859	48.736	79.595	31.168	55.942	87.110	0.06362	0.09996	0.16357
110	151.1	0.012923	0.25494	0.26786	33.255	47.206	80.461	33.616	54.332	87.948	0.06790	0.09537	0.16328
120	172.3	0.013173	0.22019	0.23336	35.716	45.584	81.300	36.136	52.604	88.740	0.07223	0.09075	0.16298
130	195.6	0.013446	0.19028	0.20373	38.222	43.888	82.111	38.709	50.777	89.486	0.07657	0.08611	0.16268
140	221.2	0.013745	0.16433	0.17807	40.795	42.090	82.885	41.358	48.818	90.175	0.08095	0.08141	0.16236
150	249.2	0.014077	0.14166	0.15574	43.444	40.171	83.616	44.093	46.705	90.798	0.08539	0.07661	0.16200
160	279.7	0.014448	0.12165	0.13610	46.203	38.086	84.289	46.951	44.383	91.334	0.08995	0.07162	0.16157
170	312.9	0.014870	0.10393	0.11880	49.065	35.831	84.896	49.926	41.849	91.775	0.09461	0.06646	0.16107
180	348.9	0.015359	0.08800	0.10336	52.079	33.332	85.411	53.070	39.013	92.084	0.09945	0.06099	0.16044
190	387.9	0.015940	0.07352	0.08946	55.275	30.530	85.805	56.419	35.807	92.226	0.10451	0.05512	0.15963
200	430.0	0.016657	0.06012	0.07678	58.716	27.311	86.027	60.042	32.094	92.136	0.10989	0.04865	0.15855
210	475.4	0.017597	0.04731	0.06491	62.511	23.471	85.982	64.059	27.633	91.692	0.11577	0.04126	0.15703
220	524.3	0.018981	0.03423	0.05321	66.921	18.512	85.433	68.762	21.833	90.595	0.12253	0.03212	0.15466
232.7	582.0	0.02870	0	0.02870	75.69	0	75.69	78.86	0	78.86	0.1359	0	0.1359

Properties generated from the program *REFRIG*, Wiley Professional Software, Wiley, New York, 1985.

TABLE E.12 Properties of Saturated Refrigerant 12 — Pressure Table, USCS Units

Reference state: $h_l = s_l = 0$ at $-40°F$

P_{sat}, psia	T_{sat}, °F	Specific Volume, ft³/lbm			Internal Energy, Btu/lbm			Enthalpy, Btu/lbm			Entropy, Btu/(lbm · °R)		
		v_l	v_{lg}	v_g	u_l	u_{lg}	u_g	h_l	h_{lg}	h_g	s_l	s_{lg}	s_g
0.2	−144.9	0.009620	139.37	139.38	−21.662	77.875	56.213	−21.661	83.033	61.372	−0.05930	0.26380	0.20449
0.4	−130.6	0.009731	72.747	72.756	−18.745	76.257	57.512	−18.744	81.641	62.897	−0.05024	0.24812	0.19788
0.6	−121.5	0.009803	49.776	49.786	−16.889	75.242	58.353	−16.888	80.769	63.881	−0.04467	0.23887	0.19420
0.8	−114.7	0.009859	38.042	38.052	−15.496	74.487	58.991	−15.494	80.119	64.624	−0.04060	0.23227	0.19167
1.0	−109.2	0.009906	30.888	30.898	−14.368	73.879	59.511	−14.366	79.595	65.229	−0.03735	0.22712	0.18977
2.0	−90.7	0.010067	16.187	16.197	−10.569	71.851	61.282	−10.565	77.842	67.277	−0.02678	0.21096	0.18418
4.0	−69.7	0.010262	8.4890	8.4992	−6.222	69.542	63.321	−6.214	75.826	69.612	−0.01532	0.19443	0.17912
6.0	−56.0	0.010396	5.8196	5.8300	−3.385	68.042	64.657	−3.373	74.503	71.130	−0.00816	0.18459	0.17643
8.0	−45.7	0.010503	4.4508	4.4613	−1.216	66.893	65.677	−1.200	73.482	72.282	−0.00285	0.17750	0.17465
10.0	−37.2	0.010594	3.6140	3.6246	0.565	65.947	66.152	0.585	72.635	73.220	0.00141	0.17193	0.17334
20.0	−8.1	0.010929	1.8868	1.8977	6.733	62.644	69.377	6.774	69.627	76.401	0.01556	0.15419	0.16975
40.0	26.0	0.011379	0.97617	0.98755	14.081	58.621	72.702	14.165	65.847	80.011	0.03129	0.13559	0.16688
60.0	48.7	0.011724	0.65834	0.67006	19.094	55.787	74.881	19.224	63.097	82.321	0.04142	0.12412	0.16555
80.0	66.2	0.012021	0.49478	0.50680	23.038	53.496	76.534	23.216	60.821	84.037	0.04909	0.11565	0.16474
100.0	80.8	0.012292	0.39446	0.40675	26.364	51.512	77.876	26.591	58.812	85.403	0.05538	0.10882	0.16419
200.0	131.8	0.013497	0.18543	0.19892	38.673	43.578	82.251	39.172	50.441	89.613	0.07734	0.08528	0.16262
300.0	166.2	0.014703	0.11041	0.12512	47.964	36.711	84.675	48.781	42.840	91.621	0.09283	0.06845	0.16127
400.0	193.0	0.016135	0.06947	0.08560	56.263	29.629	85.893	57.458	34.771	92.229	0.10606	0.05328	0.15934
500.0	215.1	0.018225	0.04072	0.05895	64.670	21.120	85.789	66.356	24.888	91.243	0.11908	0.03688	0.15596
582.0	232.7	0.02870	0	0.02870	75.69	0	75.69	78.86	0	78.86	0.1359	0	0.1359

Properties generated from the program *REFRIG,* Wiley Professional Software, Wiley, New York, 1985.

TABLE E.13 Properties of Superheated Refrigerant 12 — USCS Units

Reference state: $h_l = s_l = 0$ at $-40°F$

P, psia	T, °C	v, ft³/lbm	u, Btu/lbm	h, Btu/lbm	s, Btu/(lbm · °R)	P, psia	T, °F	v, ft³/lbm	u, Btu/lbm	h, Btu/lbm	s, Btu/(lbm · °R)
0.4						0.6					
($T_{sat}=-130.6°F$)						($T_{sat}=-121.5°F$)					
	−100	79.601	60.485	66.377	0.20799		−100	53.005	60.466	66.351	0.20128
	−50	90.748	65.703	72.420	0.22371		−50	60.455	65.690	72.402	0.21702
	0	101.875	71.339	78.880	0.23857		0	67.885	71.330	78.867	0.23190
	50	112.989	77.356	85.720	0.25269		50	75.302	77.349	85.710	0.24602
	100	124.096	83.720	92.905	0.26614		100	82.712	83.715	92.898	0.25947
	150	135.199	90.397	100.404	0.27897		150	90.117	90.392	100.398	0.27230
	200	146.299	97.356	108.185	0.29123		200	97.520	97.353	108.180	0.28457
	250	157.397	104.570	116.221	0.30297		250	104.920	104.567	116.217	0.29631
	300	168.493	112.013	124.485	0.31422		300	112.319	112.011	124.482	0.30756
	350	179.620	119.662	132.957	0.32502		350	119.718	119.660	132.952	0.31836
	400	190.711	127.496	141.612	0.33539		400	127.116	127.494	141.608	0.32873

TABLE E.13 (Continued)

P, psia	T, °C	v, ft³/lbm	u, Btu/lbm	h, Btu/lbm	s, Btu/(lbm · °R)	P, psia	T, °F	v, ft³/lbm	u, Btu/lbm	h, Btu/lbm	s, Btu/(lbm · °R)
0.8						6					
($T_{sat} = -114.7°F$)						($T_{sat} = -56.0°F$)					
	−100	39.708	60.447	66.326	0.19650		−50	5.925	65.328	71.907	0.17833
	−50	45.309	65.677	72.384	0.21226		0	6.701	71.072	78.512	0.19354
	0	50.889	71.320	78.854	0.22715		50	7.464	77.160	85.447	0.20785
	50	56.458	77.342	85.701	0.24128		100	8.219	83.570	92.696	0.22141
	100	62.019	83.709	92.891	0.25474		150	8.969	90.279	100.238	0.23432
	150	67.576	90.388	100.392	0.26757		200	9.717	97.261	108.050	0.24663
	200	73.130	97.349	108.175	0.27984		250	10.462	104.491	116.107	0.25840
	250	78.682	104.564	116.213	0.29158		300	11.207	111.945	124.388	0.26967
	300	84.233	112.008	124.478	0.30283		350	11.950	119.603	132.871	0.28049
	350	89.782	119.658	132.949	0.31363		400	12.693	127.443	141.535	0.29087
	400	95.331	127.492	141.605	0.32400						
1						8					
($T_{sat} = -109.2°F$)						($T_{sat} = -45.7°F$)					
	−100	31.729	60.428	66.300	0.19278		0	5.0008	70.975	78.378	0.18861
	−50	36.221	65.663	72.366	0.20857		50	5.5789	77.089	85.348	0.20299
	0	40.692	71.311	78.841	0.22347		100	6.1500	83.517	92.621	0.21660
	50	45.152	77.336	85.691	0.23761		150	6.7154	90.237	100.179	0.22953
	100	49.604	83.704	92.883	0.25106		200	7.2779	97.227	108.001	0.24186
	150	54.051	90.384	100.386	0.26390		250	7.8386	104.463	116.067	0.25364
	200	58.496	97.346	108.171	0.27617		300	8.3980	111.921	124.354	0.26492
	250	62.939	104.562	116.208	0.28791		350	8.9564	119.581	132.841	0.27574
	300	67.380	112.006	124.475	0.29917		400	9.5142	127.424	141.509	0.28612
	350	71.821	119.656	132.946	0.30997						
	400	76.261	127.490	141.602	0.32034						
2						10					
($T_{sat} = -90.7°F$)						($T_{sat} = -37.2°F$)					
	−50	18.045	65.597	72.276	0.19702		0	3.9807	70.877	78.243	0.18473
	0	20.298	71.263	78.776	0.21199		50	4.4481	77.017	85.248	0.19919
	50	22.539	77.301	85.642	0.22616		100	4.9085	83.463	92.546	0.21284
	100	24.773	83.677	92.846	0.23964		150	5.3630	90.195	100.119	0.22580
	150	27.002	90.363	100.357	0.25249		200	5.8146	97.193	107.953	0.23815
	200	29.229	97.329	108.146	0.26476		250	6.2643	104.434	116.027	0.24994
	250	31.453	104.548	116.188	0.27651		300	6.7128	111.897	124.319	0.26123
	300	33.676	111.994	124.457	0.28777		350	7.1603	119.560	132.810	0.27205
	350	35.898	119.645	132.931	0.29857		400	7.6072	127.405	141.482	0.28244
	400	38.120	127.481	141.589	0.30895						
4						20					
($T_{sat} = -69.7°F$)						($T_{sat} = -8.1°F$)					
	−50	8.955	65.464	72.092	0.18532		0	1.9389	70.373	77.549	0.17226
	0	10.101	71.168	78.645	0.20040		50	2.1857	76.654	84.743	0.18711
	50	11.233	77.230	85.545	0.21464		100	2.4240	83.190	92.161	0.20099
	100	12.358	83.624	92.771	0.22816		150	2.6572	89.983	99.817	0.21408
	150	13.478	90.321	100.297	0.24104		200	2.8873	97.022	107.708	0.22652
	200	14.595	97.295	108.098	0.25333		250	3.1159	104.292	115.824	0.23837
	250	15.710	104.519	116.148	0.26509		300	3.3424	111.775	124.146	0.24970
	300	16.824	111.970	124.423	0.27636		350	3.5681	119.454	132.659	0.26055
	350	17.937	119.624	132.901	0.28717		400	3.7932	127.310	141.348	0.27096
	400	19.049	127.462	141.562	0.29755						

P, psia	T, °C	v, ft³/lbm	u, Btu/lbm	h, Btu/lbm	s, Btu/(lbm · °R)	P, psia	T, °F	v, ft³/lbm	u, Btu/lbm	h, Btu/lbm	s, Btu/(lbm · °R)
40						100					
$(T_{sat} = 26.0°F)$						$(T_{sat} = 80.8°F)$					
	50	1.0526	75.889	83.680	0.17425		100	0.43135	80.749	88.731	0.17023
	100	1.1811	82.625	91.368	0.18863		150	0.49005	88.152	97.220	0.18474
	150	1.3040	89.548	99.200	0.20202		200	0.54414	95.580	105.649	0.19801
	200	1.4235	96.674	107.210	0.21464		250	0.59548	103.110	114.129	0.21039
	250	1.5408	104.004	115.409	0.22661		300	0.64516	110.775	122.714	0.22207
	300	1.6567	111.530	123.793	0.23802		350	0.69374	118.584	131.421	0.23316
	350	1.7717	119.239	132.353	0.24893		400	0.74155	126.536	140.259	0.24374
	400	1.8859	127.118	141.078	0.25938						
60						200					
$(T_{sat} = 48.7°F)$						$(T_{sat} = 131.8°F)$					
	50	0.6727	75.066	82.534	0.16596		150	0.21368	85.370	93.278	0.16871
	100	0.7659	82.034	90.538	0.18093		200	0.24859	93.540	102.740	0.18359
	150	0.8524	89.099	98.563	0.19465		250	0.27909	101.504	111.833	0.19684
	200	0.9353	96.318	106.702	0.20747		300	0.30734	109.449	120.824	0.20906
	250	1.0158	103.711	114.990	0.21957		350	0.33409	117.449	129.814	0.22050
	300	1.0948	111.281	123.437	0.23107		400	0.35994	125.540	138.861	0.23132
	350	1.1729	119.023	132.045	0.24203						
	400	1.2502	126.925	140.806	0.25253						
80						400					
$(T_{sat} = 66.2°F)$						$(T_{sat} = 193.0°F)$					
	100	0.55730	81.410	89.661	0.17509		200	0.09100	87.605	94.341	0.16254
	150	0.62624	88.634	97.905	0.18919		250	0.11743	97.606	106.298	0.17994
	200	0.69093	95.953	106.182	0.20222		300	0.13680	106.457	116.583	0.19389
	250	0.75317	103.413	114.563	0.21446		350	0.15352	114.992	126.356	0.20629
	300	0.81383	111.030	123.078	0.22604		400	0.16884	123.435	135.932	0.21773
	350	0.87343	118.804	131.734	0.23707						
	400	0.93230	126.732	140.533	0.24760						

Properties generated from the program *REFRIG,* Wiley Prefessional Software, Wiley, New York, 1985.

Figure E.4 *Pressure-enthalpy diagram for refrigerant 12. (Freon' 12 is the Dupont trademark for refrigerant 12). (Copyright, E. I. Dupont de Nemours & Company, used with permission.)*

TABLE E.14 Properties of Saturated Ammonia — Temperature Table, USCS Units

Reference state: $h_l = s_l = 0$ at $-40°F$ and 1 atm

T_{sat}, °F	P_{sat}, psia	Specific Volume, ft³/lbm			Internal Energy, Btu/lbm			Enthalpy, Btu/lbm			Entropy, Btu/(lbm · °R)		
		v_l	v_{lg}	v_g	u_l	u_{lg}	u_g	h_l	h_{lg}	h_g	s_l	s_{lg}	s_g
−60	5.5	0.02277	44.933	44.955	−23.80	568.73	544.93	−23.78	614.66	590.88	−0.0580	1.5379	1.4799
−55	6.5	0.02288	38.509	38.532	−17.68	564.11	546.43	−17.66	610.53	592.88	−0.0428	1.5087	1.4659
−50	7.6	0.02299	33.152	33.175	−11.72	559.63	547.90	−11.69	606.53	594.84	−0.0281	1.4805	1.4524
−45	8.9	0.02310	28.663	28.686	−5.90	555.25	549.36	−5.86	602.63	596.77	−0.0140	1.4533	1.4393
−40	10.4	0.02321	24.877	24.901	−0.06	550.84	550.78	0.00	598.66	598.66	0.0000	1.4265	1.4265
−35	12.0	0.02333	21.679	21.702	5.58	546.60	552.18	5.64	594.89	600.53	0.0134	1.4008	1.4142
−30	13.9	0.02345	18.961	18.985	11.16	542.40	553.56	11.22	591.14	602.35	0.0264	1.3758	1.4022
−25	16.0	0.02356	16.642	16.666	16.67	538.24	554.91	16.74	587.40	604.14	0.0392	1.3514	1.3906
−20	18.3	0.02369	14.656	14.679	22.15	534.08	556.23	22.23	583.67	605.90	0.0517	1.3275	1.3792
−15	20.9	0.02381	12.947	12.971	27.60	529.93	557.53	27.69	579.92	607.61	0.0640	1.3042	1.3682
−10	23.7	0.02393	11.473	11.497	33.03	525.77	558.79	33.13	576.15	609.29	0.0762	1.2813	1.3575
−5	26.9	0.02406	10.196	10.220	38.44	521.59	560.03	38.56	572.36	610.92	0.0882	1.2588	1.3470
0	30.4	0.02419	9.0863	9.1105	43.86	517.38	561.24	44.00	568.51	612.51	0.1000	1.2368	1.3368
5	34.3	0.02432	8.1189	8.1433	49.28	513.13	562.42	49.44	564.62	614.06	0.1117	1.2151	1.3269
10	38.5	0.02446	7.2730	7.2975	54.71	508.85	563.56	54.89	560.68	615.57	0.1234	1.1938	1.3171
15	43.2	0.02459	6.5311	6.5557	60.16	504.52	564.68	60.35	556.67	617.03	0.1349	1.1728	1.3077
20	48.2	0.02473	5.8785	5.9032	65.62	500.15	565.77	65.84	552.60	618.44	0.1463	1.1521	1.2984
25	53.8	0.02488	5.3029	5.3277	71.10	495.72	566.82	71.34	548.47	619.81	0.1577	1.1316	1.2893
30	59.8	0.02502	4.7938	4.8188	76.59	491.24	567.84	76.87	544.26	621.13	0.1690	1.1115	1.2805
35	66.3	0.02517	4.3423	4.3675	82.11	486.71	568.82	82.42	539.97	622.40	0.1802	1.0916	1.2718
40	73.4	0.02533	3.9411	3.9664	87.65	482.12	569.77	88.00	535.62	623.61	0.1914	1.0719	1.2633
45	81.0	0.02548	3.5835	3.6090	93.21	477.47	570.68	93.59	531.18	624.77	0.2024	1.0525	1.2550
50	89.2	0.02564	3.2641	3.2898	98.79	472.76	571.55	99.21	526.67	625.88	0.2134	1.0334	1.2468
55	98.1	0.02580	2.9782	3.0040	104.39	468.00	572.39	104.86	522.07	626.93	0.2244	1.0144	1.2388
60	107.7	0.02597	2.7217	2.7476	110.01	463.18	573.18	110.52	517.40	627.92	0.2352	0.9956	1.2309
65	117.9	0.02614	2.4910	2.5172	115.64	458.29	573.94	116.21	512.64	628.85	0.2460	0.9771	1.2231
70	128.9	0.02632	2.2832	2.3095	121.30	453.35	574.65	121.92	507.79	629.72	0.2568	0.9587	1.2155
75	140.6	0.02650	2.0956	2.1221	126.97	448.34	575.31	127.66	502.86	630.52	0.2674	0.9405	1.2080
80	153.1	0.02668	1.9258	1.9525	132.66	443.27	575.92	133.41	497.84	631.25	0.2780	0.9225	1.2005
85	166.5	0.02687	1.7720	1.7989	138.36	438.13	576.49	139.19	492.72	631.91	0.2886	0.9046	1.1932
90	180.7	0.02706	1.6323	1.6594	144.08	432.92	577.00	144.99	487.51	632.50	0.2990	0.8869	1.1860
95	195.9	0.02726	1.5052	1.5325	149.82	427.64	577.46	150.81	482.20	633.01	0.3095	0.8693	1.1788
100	212.0	0.02747	1.3895	1.4169	155.58	422.28	577.86	156.66	476.78	633.44	0.3198	0.8519	1.1717
105	229.0	0.02768	1.2838	1.3114	161.36	416.84	578.21	162.54	471.25	633.78	0.3301	0.8346	1.1647
110	247.1	0.02790	1.1871	1.2150	167.17	411.32	578.49	168.44	465.60	634.04	0.3404	0.8173	1.1577
115	266.2	0.02813	1.0986	1.1267	172.99	405.71	578.70	174.38	459.83	634.21	0.3506	0.8002	1.1507
120	286.5	0.02836	1.0174	1.0458	178.85	399.99	578.84	180.35	453.93	634.28	0.3607	0.7831	1.1438

Properties generated from the program *REFRIG*, Wiley Professional Software, Wiley, New York, 1985.

TABLE E.15 Properties of Saturated Ammonia—Pressure Table, USCS Units

Reference state: $h_l = s_l = 0$ at $-40°F$ and 1 atm

P_{sat}, psia	T_{sat}, °F	Specific Volume, ft³/lbm			Internal Energy, Btu/lbm			Enthalpy, Btu/lbm			Entropy, Btu/(lbm · °R)		
		v_l	v_{lg}	v_g	u_l	u_{lg}	u_g	h_l	h_{lg}	h_g	s_l	s_{lg}	s_g
6.0	−57.5	0.02283	41.588	41.612	−20.74	566.41	545.68	−20.71	612.59	591.88	−0.0503	1.5233	1.4729
7.0	−52.8	0.02293	36.005	36.027	−15.01	562.10	547.09	−14.98	608.74	593.75	−0.0362	1.4960	1.4599
8.0	−48.6	0.02302	31.776	31.799	−10.03	558.36	548.33	−10.00	605.40	595.40	−0.0240	1.4726	1.4486
9.0	−44.8	0.02311	28.463	28.486	−5.61	555.04	549.43	−5.58	602.44	596.87	−0.0133	1.4520	1.4387
10.0	−41.3	0.02319	25.786	25.809	−1.52	551.94	550.42	−1.48	599.66	598.18	−0.0035	1.4333	1.4298
11.0	−38.1	0.02326	23.585	23.608	2.12	549.21	551.32	2.17	597.22	599.38	0.0052	1.4166	1.4218
12.0	−35.1	0.02333	21.741	21.765	5.47	546.69	552.16	5.52	594.97	600.49	0.0131	1.4014	1.4145
13.0	−32.3	0.02339	20.172	20.196	8.57	544.35	552.92	8.63	592.88	601.51	0.0204	1.3874	1.4077
14.0	−29.7	0.02345	18.820	18.843	11.47	542.17	553.64	11.53	590.92	602.46	0.0271	1.3744	1.4015
15.0	−27.3	0.02351	17.642	17.666	14.19	540.11	554.31	14.26	589.08	603.34	0.0335	1.3623	1.3958
16.0	−24.9	0.02357	16.607	16.631	16.76	538.17	554.93	16.83	587.34	604.17	0.0394	1.3510	1.3904
17.0	−22.7	0.02362	15.690	15.713	19.20	536.32	555.52	19.27	585.68	604.96	0.0450	1.3403	1.3853
18.0	−20.6	0.02367	14.872	14.895	21.51	534.57	556.08	21.59	584.10	605.69	0.0503	1.3303	1.3805
19.0	−18.6	0.02372	14.137	14.161	23.72	532.89	556.61	23.81	582.59	606.39	0.0553	1.3207	1.3760
20.0	−16.6	0.02377	13.472	13.496	25.84	531.27	557.11	25.93	581.14	607.06	0.0601	1.3117	1.3717
25.0	−7.9	0.02398	10.926	10.950	35.25	524.05	559.30	35.36	574.60	609.96	0.0811	1.2720	1.3531
30.0	−0.6	0.02418	9.2042	9.2284	43.25	517.85	561.10	43.38	568.95	612.33	0.0987	1.2393	1.3379
35.0	5.9	0.02435	7.9598	7.9842	50.25	512.37	562.62	50.41	563.93	614.33	0.1138	1.2113	1.3251
40.0	11.7	0.02450	7.0166	7.0411	56.52	507.42	563.94	56.70	559.36	616.06	0.1272	1.1868	1.3140
45.0	16.9	0.02465	6.2767	6.3013	62.20	502.89	565.09	62.41	555.16	617.56	0.1392	1.1650	1.3042
50.0	21.7	0.02478	5.6799	5.7047	67.43	498.69	566.12	67.66	551.24	618.90	0.1501	1.1453	1.2954
55.0	26.1	0.02491	5.1882	5.2131	72.28	494.76	567.04	72.53	547.57	620.10	0.1601	1.1273	1.2874
60.0	30.2	0.02503	4.7758	4.8009	76.80	491.07	567.87	77.08	544.10	621.18	0.1694	1.1107	1.2801
65.0	34.0	0.02514	4.4245	4.4496	81.06	487.58	568.63	81.36	540.80	622.16	0.1781	1.0954	1.2734
70.0	37.7	0.02525	4.1218	4.1471	85.07	484.26	569.33	85.40	537.65	623.05	0.1862	1.0810	1.2672
75.0	41.1	0.02536	3.8582	3.8836	88.88	481.09	569.97	89.23	534.64	623.87	0.1938	1.0676	1.2614
80.0	44.4	0.02546	3.6265	3.6520	92.51	478.06	570.57	92.88	531.75	624.63	0.2010	1.0550	1.2560
85.0	47.5	0.02556	3.4210	3.4466	95.97	475.15	571.12	96.37	528.96	625.33	0.2079	1.0430	1.2509
90.0	50.4	0.02565	3.2375	3.2632	99.28	472.35	571.63	99.71	526.26	625.98	0.2144	1.0317	1.2461
95.0	53.3	0.02575	3.0728	3.0985	102.47	469.64	572.11	102.92	523.66	626.58	0.2206	1.0209	1.2415
100.0	56.0	0.02584	2.9238	2.9497	105.53	467.03	572.55	106.01	521.13	627.14	0.2266	1.0106	1.2372
110.0	61.2	0.02601	2.6651	2.6912	111.33	462.03	573.37	111.86	516.29	628.15	0.2378	0.9912	1.2290
120.0	66.0	0.02618	2.4484	2.4746	116.75	457.33	574.08	117.33	511.69	629.03	0.2482	0.9734	1.2216
130.0	70.5	0.02633	2.2636	2.2900	121.86	452.85	574.71	122.50	507.31	629.80	0.2578	0.9569	1.2147
140.0	74.8	0.02649	2.1043	2.1308	126.69	448.59	575.28	127.38	503.11	630.48	0.2669	0.9414	1.2083
150.0	78.8	0.02664	1.9657	1.9923	131.27	444.51	575.78	132.01	499.07	631.08	0.2755	0.9269	1.2023
160.0	82.6	0.02678	1.8437	1.8705	135.63	440.60	576.22	136.42	495.18	631.61	0.2835	0.9132	1.1967
170.0	86.3	0.02692	1.7355	1.7624	139.80	436.82	576.62	140.65	491.42	632.07	0.2912	0.9001	1.1914
180.0	89.8	0.02705	1.6389	1.6660	143.80	433.18	576.98	144.70	487.77	632.47	0.2985	0.8878	1.1863
190.0	93.1	0.02719	1.5522	1.5793	147.64	429.66	577.29	148.59	484.23	632.82	0.3055	0.8760	1.1815
200.0	96.3	0.02732	1.4739	1.5012	151.33	426.24	577.57	152.34	480.79	633.13	0.3122	0.8648	1.1769
220.0	102.4	0.02757	1.3377	1.3653	158.34	419.69	578.03	159.47	474.15	633.61	0.3247	0.8436	1.1683
240.0	108.1	0.02782	1.2234	1.2512	164.93	413.46	578.39	166.16	467.79	633.95	0.3364	0.8240	1.1604
260.0	113.4	0.02805	1.1261	1.1541	171.13	407.51	578.64	172.48	461.69	634.17	0.3473	0.8056	1.1529
280.0	118.4	0.02829	1.0421	1.0704	177.01	401.80	578.81	178.47	455.79	634.27	0.3575	0.7884	1.1460

Properties generated from the program *REFRIG*, Wiley Professional Software, Wiley, New York, 1985.

TABLE E.16 Properties of Superheated Ammonia—USCS Units

		Reference state: $h_l = s_l = 0$ at $-40°F$								

P, psia	T, °F	v, ft³/lbm	u, Btu/lbm	h, Btu/lbm	s, Btu/(lbm · °R)	P, psia	T, °F	v, ft³/lbm	u, Btu/lbm	h, Btu/lbm	s, Btu/(lbm · °R)
6 ($T_{sat} = -57.5°F$)						12 ($T_{sat} = -35.1°F$)					
	−40	43.535	552.35	600.68	1.4943		−30	22.055	554.19	603.17	1.4207
	−20	45.716	559.98	610.73	1.5177		−10	23.184	562.14	613.63	1.4445
	0	47.881	567.62	620.78	1.5401		0	23.740	566.11	618.82	1.4558
	20	50.034	575.28	630.83	1.5615		10	24.293	570.06	624.00	1.4671
	40	52.177	582.97	640.90	1.5820		30	25.392	577.95	634.33	1.4886
	60	54.313	590.69	650.99	1.6019		50	26.481	585.83	644.63	1.5093
	80	56.442	598.46	661.13	1.6210		70	27.563	593.72	654.92	1.5291
	100	58.566	606.29	671.31	1.6395		90	28.640	601.63	665.23	1.5482
	120	60.700	614.18	681.57	1.6575		110	29.712	609.58	675.56	1.5666
	140	62.800	622.14	691.87	1.6750		130	30.785	617.58	685.94	1.5845
	160	64.920	630.18	702.26	1.6920		150	31.848	625.64	696.36	1.6019
	180	67.034	638.30	712.72	1.7087		170	32.912	633.77	706.86	1.6188
	200	69.146	646.50	723.28	1.7249		190	33.975	641.98	717.42	1.6354
	220	71.256	654.80	733.92	1.7408		210	35.036	650.26	728.06	1.6515
8 ($T_{sat} = -48.6°F$)						14 ($T_{sat} = -29.7°F$)					
	−40	32.514	551.64	599.77	1.4591		−20	19.320	557.56	607.61	1.4134
	−20	34.171	559.38	609.97	1.4828		0	20.291	565.59	618.16	1.4369
	0	35.811	567.12	620.13	1.5054		20	21.245	573.58	628.61	1.4591
	20	37.439	574.86	630.28	1.5271		40	22.188	581.52	639.01	1.4804
	40	39.057	582.61	640.43	1.5478		60	23.124	589.46	649.37	1.5007
	60	40.668	590.39	650.59	1.5677		80	24.053	597.40	659.72	1.5202
	80	42.272	598.20	660.78	1.5870		100	24.978	605.37	670.08	1.5391
	100	43.872	606.06	671.01	1.6056		120	25.899	613.37	680.47	1.5573
	120	45.467	613.98	681.29	1.6236		140	26.812	621.42	690.89	1.5750
	140	47.055	621.96	691.62	1.6411		160	27.732	629.54	701.38	1.5922
	160	48.650	630.02	702.04	1.6582		180	28.645	637.72	711.93	1.6090
	180	50.239	638.15	712.53	1.6749		200	29.557	645.98	722.55	1.6253
	200	51.826	646.37	723.09	1.6912		220	30.467	654.31	733.25	1.6413
	220	53.411	654.68	733.75	1.7071						
10 ($T_{sat} = -41.3°F$)						16 ($T_{sat} = -24.9°F$)					
	−40	25.896	550.92	598.84	1.4314		−20	16.844	556.94	606.82	1.3964
	−20	27.243	558.78	609.19	1.4554		0	17.704	565.08	617.50	1.4202
	0	28.569	566.61	619.48	1.4783		20	18.545	573.14	628.05	1.4426
	20	29.882	574.43	629.73	1.5002		40	19.376	581.16	638.53	1.4640
	40	31.185	582.25	639.96	1.5210		60	20.200	589.15	648.96	1.4845
	60	32.481	590.08	650.18	1.5411		80	21.017	597.14	659.36	1.5042
	80	33.770	597.93	660.43	1.5604		100	21.829	605.14	669.77	1.5231
	100	35.055	605.83	670.70	1.5791		120	22.637	613.17	680.19	1.5414
	120	36.335	613.78	681.01	1.5972		140	23.438	621.25	690.64	1.5591
	140	37.608	621.78	691.38	1.6148		160	24.245	629.38	701.16	1.5764
	160	38.888	629.86	701.82	1.6320		180	25.046	637.57	711.73	1.5932
	180	40.162	638.01	712.33	1.6486		200	25.845	645.84	722.37	1.6095
	200	41.433	646.24	722.91	1.6649		220	26.643	654.19	733.08	1.6255
	220	42.704	654.56	733.58	1.6809						

P, psia	T, °F	v, ft³/lbm	u, Btu/lbm	h, Btu/lbm	s, Btu/(lbm · °R)	P, psia	T, °F	v, ft³/lbm	u, Btu/lbm	h, Btu/lbm	s, Btu/(lbm · °R)
18 ($T_{sat} = -20.6$°F)						40 ($T_{sat} = 11.7$°F)					
	−20	14.918	556.32	606.01	1.3812		20	7.1963	567.73	620.99	1.3244
	0	15.688	564.56	616.81	1.4053		40	7.5595	576.62	632.58	1.3480
	20	16.445	572.71	627.49	1.4280		60	7.9143	585.31	643.90	1.3702
	40	17.189	580.79	638.05	1.4496		80	8.2600	593.86	655.00	1.3912
	60	17.925	588.84	648.55	1.4702		100	8.6005	602.30	665.97	1.4112
	80	18.655	596.87	659.01	1.4899		120	8.9369	610.69	676.85	1.4303
	100	19.379	604.91	669.46	1.5089		140	9.2701	619.06	687.68	1.4486
	120	20.100	612.97	679.92	1.5273		160	9.6008	627.42	698.49	1.4664
	140	20.818	621.07	690.41	1.5451		180	9.9294	635.81	709.31	1.4835
	160	21.533	629.22	700.94	1.5624		200	10.256	644.24	720.16	1.5002
	180	22.247	637.43	711.53	1.5792		220	10.582	652.72	731.05	1.5165
	200	22.958	645.71	722.18	1.5956						
	220	23.669	654.07	732.91	1.6116						
20 ($T_{sat} = -16.6$°F)						50 ($T_{sat} = 21.7$°F)					
	0	14.077	564.03	616.13	1.3918		40	5.9813	574.64	629.98	1.3180
	20	14.765	572.27	626.92	1.4148		60	6.2731	583.65	641.69	1.3410
	40	15.439	580.42	637.57	1.4365		80	6.5579	592.45	653.12	1.3625
	60	16.105	588.53	648.13	1.4573		100	6.8358	601.09	664.34	1.3829
	80	16.765	596.60	658.65	1.4771		120	7.1095	609.64	675.42	1.4024
	100	17.420	604.67	669.14	1.4962		140	7.3800	618.13	686.41	1.4210
	120	18.071	612.76	679.64	1.5147		160	7.6478	626.60	697.36	1.4390
	140	18.719	620.88	690.16	1.5325		180	7.9135	635.07	708.29	1.4564
	160	19.364	629.05	700.72	1.5498		200	8.1776	643.57	719.23	1.4732
	180	20.007	637.28	711.33	1.5667		220	8.4403	652.11	730.20	1.4896
	200	20.649	645.58	722.00	1.5831						
	220	21.289	653.95	732.74	1.5991						
30 ($T_{sat} = -0.6$°F)						75 ($T_{sat} = 41.1$°F)					
	0	9.2422	561.35	612.66	1.3386		60	4.0827	579.31	635.97	1.2852
	20	9.7206	570.04	624.00	1.3628		80	4.2850	588.79	648.26	1.3084
	40	10.189	578.55	635.11	1.3855		100	4.4812	597.97	660.16	1.3300
	60	10.646	586.94	646.04	1.4069		120	4.6717	606.93	671.77	1.3504
	80	11.096	595.24	656.84	1.4273		140	4.8589	615.76	683.19	1.3698
	100	11.541	603.50	667.57	1.4469		160	5.0431	624.49	694.48	1.3883
	120	11.982	611.74	678.25	1.4656		180	5.2252	633.18	705.70	1.4061
	140	12.420	619.98	688.93	1.4837		200	5.4056	641.86	716.88	1.4233
	160	12.855	628.24	699.61	1.5012		220	5.5845	650.54	728.05	1.4400
	180	13.289	636.55	710.32	1.5183						
	200	13.721	644.91	721.08	1.5348						
	220	14.151	653.34	731.90	1.5510						

TABLE E.16 (Continued)

P, psia	T, °F	v, ft³/lbm	u, Btu/lbm	h, Btu/lbm	s, Btu/(lbm · °R)
100					
(T_{sat} = 56.0°F)					
	60	2.9832	574.67	629.88	1.2424
	80	3.1459	584.93	643.14	1.2675
	100	3.3012	594.70	655.79	1.2905
	120	3.4514	604.13	667.99	1.3119
	140	3.5971	613.31	679.88	1.3321
	160	3.7400	622.33	691.54	1.3512
	180	3.8805	631.25	703.06	1.3695
	200	4.0192	640.12	714.49	1.3871
	220	4.1563	648.96	725.87	1.4041
125					
(T_{sat} = 68.3°F)					
	80	2.4597	580.84	637.74	1.2337
	100	2.5917	591.28	651.23	1.2582
	120	2.7180	601.21	664.08	1.2808
	140	2.8392	610.79	676.46	1.3018
	160	2.9574	620.12	688.52	1.3216
	180	3.0731	629.28	700.37	1.3404
	200	3.1869	638.34	712.06	1.3584
	220	3.2989	647.35	723.66	1.3757
150					
(T_{sat} = 78.8°F)					
	80	1.9997	576.49	632.00	1.2040
	100	2.1170	587.68	646.44	1.2303
	120	2.2275	598.17	660.00	1.2541
	140	2.3331	608.18	672.94	1.2761
	160	2.4351	617.84	685.43	1.2966
	180	2.5344	627.27	697.62	1.3159
	200	2.6318	636.54	709.59	1.3344
	220	2.7272	645.71	721.42	1.3520
175					
(T_{sat} = 88.02°F)					
	100	1.7762	583.89	641.41	1.2053
	125	1.9003	597.68	659.21	1.2364
	150	2.0165	610.53	675.84	1.2643
	175	2.1275	622.80	691.70	1.2898
	200	2.2348	634.70	707.07	1.3135
	225	2.3392	646.38	722.13	1.3360
	250	2.4413	657.94	737.00	1.3573
	275	2.5418	669.48	751.79	1.3778
	300	2.6411	681.04	766.57	1.3975

P, psia	T, °F	v, ft³/lbm	u, Btu/lbm	h, Btu/lbm	s, Btu/(lbm · °R)
200					
(T_{sat} = 96.3°F)					
	100	1.5191	579.87	636.09	1.1822
	125	1.6338	594.51	654.97	1.2153
	150	1.7400	607.94	672.34	1.2443
	175	1.8404	620.62	688.73	1.2707
	200	1.9369	632.82	704.50	1.2951
	225	2.0302	644.73	719.86	1.3179
	250	2.1213	656.48	734.98	1.3396
	275	2.2106	668.16	749.97	1.3604
	300	2.2987	679.83	764.91	1.3804
225					
(T_{sat} = 103.9°F)					
	125	1.4257	591.19	650.55	1.1956
	150	1.5243	605.26	668.73	1.2261
	175	1.6166	618.38	685.69	1.2533
	200	1.7049	630.90	701.89	1.2784
	225	1.7898	643.05	717.57	1.3017
	250	1.8723	654.99	732.95	1.3238
	275	1.9531	666.82	748.14	1.3448
	300	2.0324	678.62	763.24	1.3650
250					
(T_{sat} = 110.8°F)					
	125	1.2582	587.71	645.92	1.1771
	150	1.3513	602.48	664.99	1.2091
	175	1.4373	616.08	682.57	1.2373
	200	1.5190	628.94	699.22	1.2630
	225	1.5972	641.35	715.24	1.2869
	250	1.6729	653.48	730.88	1.3093
	275	1.7468	665.46	746.28	1.3307
	300	1.8194	677.39	761.56	1.3511
275					
(T_{sat} = 117.2°F)					
	125	1.1201	584.05	641.05	1.1594
	150	1.2092	599.60	661.13	1.1930
	175	1.2902	613.71	679.37	1.2223
	200	1.3667	626.94	696.49	1.2488
	225	1.4395	639.62	712.87	1.2732
	250	1.5098	651.95	728.79	1.2960
	275	1.5781	664.10	744.41	1.3177
	300	1.6450	676.15	759.86	1.3383

Properties generated from the program *REFRIG,* Wiley Professional Software, Wiley, New York, 1985.

Figure E.5 Psychrometric chart. (From American Society of Heating, Refrigerating, and Air-Conditioning Engineers, used with permission.)

APPENDIX F

Reynolds' Transport Theorem

The movement of mass through the system requires a transformation of fundamental statements presented in Sec. 4.2 to a form useful to control volume systems. The laws in that section do not change, but incorporation of the movement of fluid must be addressed. This appendix provides the transformation of the control mass expression to control volume form. This transformation should not be considered as a new physical law but rather as a manipulation of the conservation principle.

The expressions for the conservation of mass and the conservation of energy have been presented for a control mass. These equations are

$$\frac{d}{dt}\left(\int_V \rho \, dV \right) = 0 \tag{F.1}$$

$$\frac{d}{dt}\left(\int_V e\rho \, dV \right) = \dot{Q}_{CV} + \dot{W} \tag{F.2}$$

The left-hand sides of these equations represent the rate of change of mass and energy of the control mass. This rate of change of the control mass property is the quantity of interest in this transformation.

Figure F.1 presents a control mass (CM) which is passing through a particular component. The control volume (CV) is fixed to the component and coincides with the control mass at the initial time. The control volume also expands to raise the piston and performs work on the surroundings during the time interval. Thus, the boundary B of the control volume is fixed at the inlets and outlets of the system and expands to do work on the surroundings.

The control mass moves from its initial position at time t to its later position at $t + \Delta t$ as a result of the velocity of the fluid elements. The

Figure F.1 **The Reynolds' transport theorem transformation.**

Within the figure:
V

Control mass boundary at $t + \Delta t$

Control mass boundary at t and control volume boundary

Control mass boundary at $t + \Delta t$

Inlet

Inlet

Outlet

Control mass at t and control volume at t

Control mass at $t + \Delta t$

specific system diagramed indicates two inlets and a single outlet, although the specific number is arbitrary. During this motion, the fluid elements transport mass and energy into and out of the control volume. We consider the movement of the control mass through the control volume in a limiting process as the time interval Δt tends to zero.

The total derivatives in the conservation-of-mass and conservation-of-energy principles for a control mass are of interest. The property in these equations is represented in terms of the extensive property Λ and the corresponding specific intensive property λ as

$$\Lambda = \int d\Lambda = \int_V \rho\lambda\, dV \tag{F.3}$$

where $\lambda = 1$ and $\Lambda = m$ for mass and $\lambda = e$ and $\Lambda = E$ for total energy. The total derivatives are then given by

$$\frac{d\Lambda_{CM}}{dt} = \lim_{\Delta t \to 0} \frac{\Lambda_{CM,\,t+\Delta t} - \Lambda_{CM,\,t}}{\Delta t} \tag{F.4}$$

The control mass property is an integral over the volume occupied. The volume occupied by the control mass is expressed in terms of the control volume. The fluid has swept the control mass into a new volume at the outlets, and the control mass has left behind an old volume at the inlets. Therefore, since the control mass properties are volume integrals, the integral is reformulated in terms of the control volume. The terms in Eq. (F.4) are

$$\Lambda_{CM, t} = \Lambda_{CV, t}$$

(F.5)

$$\Lambda_{CM, t+\Delta t} = \Lambda_{CV, t+\Delta t} + \Lambda_{CM, out, t+\Delta t} - \Lambda_{CM, in, t+\Delta t}$$

Substituting these expressions into Eq. (F.4) yields

$$\frac{d\Lambda_{CM}}{dt} = \lim_{\Delta t \to 0} \frac{\Lambda_{CV, t+\Delta t} - \Lambda_{CV, t} + \Lambda_{CM, out, t+\Delta t} - \Lambda_{CM, in, t+\Delta t}}{\Delta t}$$

$$= \lim_{\Delta t \to 0} \frac{\Lambda_{CV, t+\Delta t} - \Lambda_{CV, t}}{\Delta t} + \lim_{\Delta t \to 0} \frac{\Lambda_{CM, out, t+\Delta t}}{\Delta t}$$

$$- \lim_{\Delta t \to 0} \frac{\Lambda_{CM, in, t+\Delta t}}{\Delta t} \quad \text{(F.6)}$$

The derivative of the control mass property is represented by three limits — the first is a change within the control volume while the second and third limits result from the flows in and out.

The first limit is the definition of the partial derivative of the control volume property, that is,

$$\lim_{\Delta t \to 0} \frac{\Lambda_{CV, t+\Delta t} - \Lambda_{CV, t}}{\Delta t} = \frac{\partial \Lambda_{CV}}{\partial t}$$

(F.7)

This is the change in the property within the control volume.

The second and third limits are similar and require a detailed consideration of the inlets and outlets. An expanded view of an outlet is shown in Fig. F.2. The velocity V is a vector denoted at an area element dA. The area orientation is represented by a normal vector **n** which is the

Control mass boundary
at t and control volume
boundary

Control mass boundary
at $t + \Delta t$

Figure F.2   Flow at the outlet.

outward normal to the area element. The angle between the velocity vector and normal vector is θ. This area sweeps out a volume of

$$dV = \Delta L \, dA \, \cos \theta \tag{F.8}$$

The volume integral of the control mass property within the second limit of Eq. (F.6) is

$$\frac{\Lambda_{CM, \text{ out}, \, t+\Delta t}}{\Delta t} = \frac{1}{\Delta t} \int_{V_\text{out}} \rho \lambda \, dV = \frac{1}{\Delta t} \int_{A_\text{out}} \rho \lambda \, \Delta L \, \cos \theta \, dA \tag{F.9}$$

The area of the integration is the boundary of the outlet. With the magnitude of the fluid velocity given as

$$\lim_{\Delta t \to 0} \frac{\Delta L \, \cos \theta}{\Delta t} = |\mathbf{V} \cdot \mathbf{n}| \tag{F.10}$$

The second limit is

$$\lim_{\Delta t \to 0} \frac{\Lambda_{CM, \text{ out}, \, t+\Delta t}}{\Delta t} = \int_{A_\text{out}} \rho \lambda |\mathbf{V} \cdot \mathbf{n}| \, dA \tag{F.11}$$

Note that

$$|\mathbf{V} \cdot \mathbf{n}| = \mathbf{V} \cos \theta \tag{F.12}$$

The absolute value sign is used since the volume swept out in Eq. (F.8) represents a positive quantity. The $\rho |\mathbf{V} \cdot \mathbf{n}| \, dA$ term in the integrand represents the mass flow rate for an area element,

$$d\dot{m} = \rho |\mathbf{V} \cdot \mathbf{n}| \, dA \tag{F.13}$$

Thus Eq. (F.11) could alternatively be represented as

$$\lim_{\Delta t \to 0} \frac{\Lambda_{CM, \text{ out}, \, t+\Delta t}}{\Delta t} = \int_{A_\text{out}} \lambda \, d\dot{m} \tag{F.14}$$

The third limit is considered in exactly the same manner as the second. The only difference is that the angle between the outward normal of the area elements and the velocity vector is larger than $90°$. Thus the absolute value sign ensures a positive value for the volume. Thus,

$$\lim_{\Delta t \to 0} \frac{\Lambda_{CM, \text{ in}, \, t+\Delta t}}{\Delta t} = \lim_{\Delta t \to 0} \frac{1}{\Delta t} \int_{A_\text{in}} \rho \lambda \, \Delta L \, \cos \theta \, dA$$

$$= \int_{A_\text{in}} \rho \lambda |\mathbf{V} \cdot \mathbf{n}| \, dA$$

$$= \int_{A_\text{in}} \lambda \, d\dot{m} \tag{F.15}$$

Combining the three limits in Eqs. (F.7), (F.11), (F.14), and (F.15) into the control mass derivatives in Eq. (F.6) yields

$$\frac{d\Lambda_{\text{CM}}}{dt} = \frac{\partial\Lambda_{\text{CV}}}{\partial t} + \int_{A_{\text{out}}} \rho\lambda|\mathbf{V}\cdot\mathbf{n}|\,dA - \int_{A_{\text{in}}} \rho\lambda|\mathbf{V}\cdot\mathbf{n}|\,dA$$

$$= \frac{\partial\Lambda_{\text{CV}}}{\partial t} + \int_{A_{\text{out}}} \lambda\,d\dot{m} - \int_{A_{\text{in}}} \lambda\,d\dot{m} \tag{F.16}$$

The last two integrals are area integrals over the crossing boundary. The velocity of the fluid elements crossing the inside boundary is zero (no fluid is permitted to cross — see Sec. 2.2.1), so these two integrals are represented by a single integral over the boundary B. With the definition of Λ given in Eq. (F.3), Eq. (F.16) becomes

$$\frac{d\Lambda_{\text{CM}}}{dt} = \frac{\partial}{\partial t}\left(\int_{\text{CV}} \rho\lambda\,dV\right) + \int_{B} \rho\lambda\mathbf{V}\cdot\mathbf{n}\,dA$$

$$= \frac{\partial}{\partial t}\left(\int_{\text{CV}} \rho\lambda\,dV\right) + \int_{B} \lambda\,d\dot{m} \tag{F.17}$$

The absolute value sign is removed in this expression, so the negative contribution at the inlet is incorporated. This desired relation between the derivative of a control mass property and the control volume property is termed the *Reynolds' transport theorem*. The change in the control mass property is equal to the change of the property within the control volume plus the net transport out of the control volume.

Conservation of mass and conservation of energy for a control volume are obtained from the original conservation statements with Eq. (F.17). These expressions are

$$\frac{\partial}{\partial t}\left(\int_{\text{CV}} \rho\,dV\right) + \int_{B} d\dot{m} = 0 \tag{F.18}$$

$$\frac{\partial}{\partial t}\left(\int_{\text{CV}} \rho e\,dV\right) + \int_{B} e\,d\dot{m} = \dot{Q}_{\text{CV}} + \dot{W} \tag{F.19}$$

If there are several separate inlet streams and several separate outlet streams, then these are expressed in terms of sums over the streams. Equations (F.18) and (F.19) become

$$\frac{\partial}{\partial t}\left(\int_{\text{CV}} \rho\,dV\right) + \sum_{\text{out}}\int_{A_{\text{out}}} d\dot{m} - \sum_{\text{in}}\int_{A_{\text{in}}} d\dot{m} = 0 \tag{F.20}$$

$$\frac{\partial}{\partial t}\left(\int_{\text{CV}} \rho e\,dV\right) + \sum_{\text{out}}\int_{A_{\text{out}}} e\,d\dot{m} - \sum_{\text{in}}\int_{A_{\text{in}}} e\,d\dot{m} = \dot{Q} + \dot{W} \tag{F.21}$$

These are recognized as Eqs. (4-34b) and (4-45), respectively. By noting that $\int_{A} d\dot{m} = \dot{m}$, the mass flow rate across the boundary at a stream, and denoting $\int_{A} e\,d\dot{m} = \dot{E}$, the energy transport across the boundary at a stream, Eqs. (F.20) and (F.21) become

$$\frac{\partial}{\partial t}\left(\int_{CV} \rho \, dV\right) + \sum_{out} \dot{m} - \sum_{in} \dot{m} = 0 \tag{F.22}$$

$$\frac{\partial}{\partial t}\left(\int_{CV} \rho e \, dV\right) + \sum_{out} \dot{E} - \sum_{in} \dot{E} = \dot{Q}_{CV} + \dot{W} \tag{F.23}$$

These are directly related to Eqs. (4-25) and (4-38).

APPENDIX G

Computerized Tables of Thermodynamic Properties

In the pocket inside the rear cover of this text is a floppy disk that contains computerized tables of the thermodynamic properties of steam, argon, nitrogen, air, and refrigerant 12. The program, called PROPERTIES, is in compiled code in a format that can be executed on the IBM PC, IBM/XT, and many compatible computers that operate on PC-DOS or MS-DOS, versions 1.1 and above. Before you try to use the programs, please read this appendix carefully. Then follow the instructions in Sec. G.3 for installing the disk-operating system (DOS) files on the disk and making the disk self-booting.

   The programs are completely menu-driven and are essentially self-explanatory. The tables in Apps. D and E for steam, gases, and refrigerant 12 were generated from expanded versions of these programs, so solutions generated from either the tables or the programs should be identical.

G.1   What the Programs Can Do

The programs are modified from more extensive commercial versions and are to be used as an aid in solving problems in this text. The programs can help avoid the single and double interpolations necessary if the tables of Apps. D and E are used for property data. Many of the problems are extremely tedious if the programs are not used as an aid in their solution. The computerized tables cover the range of temperatures and pressures necessary for solution of the problems in this text and allow consideration of more interesting and useful problems than is possible by using tabular data alone.

Each program provides results in the user's choice of either SI or USCS units.

Steam Tables*

The computerized tables provide the properties of steam in the subcooled, saturated, and superheated regions. Maximum temperature is limited to below 400°C (750°F) in the superheat regions and pressures are limited to below 6900 kPa (1000 psia) in all regions. Attempts to exceed these limits will cause a message to be displayed to that effect and a request that you reenter the input state data in an acceptable range. The program provides the functions $P_{sat}(T)$, $T_{sat}(P)$, and v, u, h, and s in the saturation region as a function of either T or P and quality. In the superheat region, v, u, h, and s are given as functions of T and P. In addition, the function $h(P, s)$ is provided in all regions, and $T(P, h)$ is provided in the superheated and subcooled regions. The properties are computed from the fundamental thermodynamic equations and thus do not contain any interpolation errors at any state.

Gas Tables*

Ideal gas properties for argon, nitrogen, and air are provided over the full range given in the tables in Apps. D and E. The complete statistical thermodynamic equations for ideal gases are used in the program. Rather than the function $s_0(T)$ as presented in the tables, the actual value of $s(T, P)$ is given by the program. At a given state, any two independent properties from the set of P, T, v, u, h, and s can be specified, and the rest will be computed and displayed. Note that T, h, and u are not independent for an ideal gas. The values of c_P, c_v, and $k = c_P/c_v$ are also displayed at the chosen state.

Refrigerant 12 Tables

The program for refrigerant 12 calculates properties in the saturated, superheated, and subcooled regions for states that do not have a pressure in excess of 1379 kPa (200 psia). At some temperature levels, states are outside the stated limits of accuracy of the program, and a message is displayed. This program also uses the fundamental thermodynamic relations to compute the properties. In the subcooled region, the program provides v, u, h, and s as a function of T and P. In the saturation region, an input of T or P and x, v, u, h, or s provides the remaining properties. In the superheated region, an input of any two properties from the set of P, v, and T provides all others, including c_P and c_v. If the region is unknown, this

---

* Some numerical values from the computerized table may vary slightly from those in the appendices.

particular program finds it from an input of P and T and displays the other properties at that state.

Each of the programs allows the state properties to be output to a printer if desired.

G.2   Hardware Requirements

The specific equipment required includes

1. IBM Personal Computer (PC, XT, and some compatibles, each with at least 128K of memory)
2. At least one $5\frac{1}{2}$-in double-sided disk drive
3. A color or monochrome monitor that is compatible with the computer used
4. An 80-column or wider printer (optional)

G.3   Creating a Self-Booting Version of the Program

A self-booting program is created by the following steps:

1. Place the DOS disk for your system in disk drive A (the left drive).
2. Turn on the computer; or if it is already on, press ⟨Ctrl⟩ and ⟨Alt⟩ simultaneously, and then press ⟨Del⟩.
3. You will see the command "Enter new date." Simply press RETURN twice.
4. You will now get an *A prompt,* which looks like this:

$$A\rangle$$

When this appears, remove the DOS diskette from the A drive. Put the PROPERTIES diskette into the A drive.
5. Type the following:

INSTALL

and then press RETURN.
6. Follow the instructions on the screen.

After this sequence is carried through one time, it need not be repeated for subsequent use of the program.

When the PROPERTIES program is to be used again, it can be inserted directly into drive A. When the computer is turned on, the program will automatically be loaded and started.

If the computer is already on, the PROPERTIES disk can be inserted in drive A, and the commands ⟨Ctrl⟩ and ⟨Alt⟩ pressed simultaneously, followed by ⟨Del⟩. The computer will be restarted, and the PROPERTIES program will be loaded and started automatically.

Index